Metropolitan Transportation Planning

Metropolitan Transportation Planning
Second Edition

JOHN W. DICKEY, Senior Author
Virginia Polytechnic Institute and State University

Walter J. Diewald
N. D. Lea & Associates, Inc.,
Washington, DC

Antoine G. Hobeika
Charles J. Hurst
N. Thomas Stephens
Robert C. Stuart
Richard D. Walker
All of Virginia Polytechnic
Institute and State University

Taylor & Francis
Publishers since 1798

Metropolitan Transportation Planning
Second Edition

4 5 6 7 8 9 0   E B E B   9 8 7 6 5 4

This book was set in Theme by Hemisphere Publishing Corporation.
The editors were Kiran Verma, Anne Shipman, and Elizabeth Dugger;
the designers were Victor Enfield and Sharon Martin DePass; the
production supervisor was Miriam Gonzalez; and the typesetter
was Peggy Rote.
Edwards Brothers, Inc. was printer and binder.

Library of Congress Cataloging in Publication Data

Main entry under title:

Metropolitan transportation planning.

   (McGraw-Hill series in transportation)
   Includes bibliographical references and index.
   1. Urban transportation policy. 2. Urban transportation—Mathematical models. I. Dickey, John W.,
date. II. Diewald, Walter J. III. Series.
HE305.M47   1983   388.4'068   82-23319
ISBN 0-89116-922-9

Contents

Preface

The authors involved in the creation of this book have all had experience in teaching all or part of introductory courses in transportation planning. These courses have been mostly in the civil engineering and urban and regional planning curricula at Virginia Tech. At the time we wrote the first edition, the lack of a text covering all the items we wanted to cover was a major problem. At that time, the available texts were somewhat outdated or were directed at one discipline such as engineering, planning, or economics.

Our dissatisfaction forced us to set several goals for the book. First, since no book can be kept completely up to date, especially in a fast-changing field such as transportation planning, we decided to provide a modular framework within which new information could easily be incorporated. Second, we felt the book should have an orientation toward solving urban transportation problems, regardless of the disciplinary backgrounds needed to do it. Third, and finally, since urban transportation problems pervade almost all aspects of life, there should be an interdisciplinary input to the book. We felt the last goal to be particularly important since we were convinced that no real-world metropolitan transportation problem could be solved without an interdisciplinary approach (this is verified in the discussion in Chap. 8).

In keeping with our first goal, we developed the book around the concept of the "transportation planning process." This is identified in Chap. 2 as having ten stages, starting with the identification of the problem(s) and ending with the implementation, operation, and maintenance of a plan, policy, or solution to alleviate that problem. We then developed each chapter or set of chapters to correspond with each stage in the process. This setup provided two advantages:

1. Each author could see where his contribution fit in the overall framework.
2. Each chapter or set of chapters corresponding to a stage in the process became modular. New information could be incorporated rather easily.

Concentration on a transportation planning process also allowed us to center on transportation problems, to meet our second goal. As noted, this concentration on process allowed for relative ease of input from the authors representing different disciplines.

The process-oriented approach is not without its drawbacks, however. Probably the greatest one is the emphasis on breadth over depth. This opens us to the criticism that we have not covered any single topic to the degree desired by a given discipline. Some civil engineering professionals, for example, will feel that we have slighted such important topics as drainage or structural design; some transportation planners will feel that we have not given enough attention to travel models; economists to economic analysis; sociologists and planners to social impacts; political scientists to politics and citizen participation; and so on. To all these disciplines we must confess our shortcomings and say that there are many other good books available that cover the individual topics in much more depth and with much greater skill. Our purpose was not to compete with these books, however, but to combine their thoughts into a broad, integrated spectrum of topics.

A more immediate problem is that of using this text in actual courses. It is a good introductory book in any of the disciplines listed above (and others). Yet there must be a commitment by the particular professor to emphasize more than the discipline in which he or she has the most experience and therefore feels most comfortable. For example, in an introductory course in civil engineering, primary emphasis may be given to the chapters on travel and capacity, benefit-cost evaluation, and general operation and maintenance. An introductory urban planning course, on the other hand, may emphasize the planning process, identification of problems and goals, data collection, and solution implementation. In either case, we feel that the students should be asked to read the other chapters so that they can see where the material given them fits in the larger picture. Guest lecturers from other disciplines could help in this context. Overall, we expect that there is enough material in the text to satisfy any one relevant discipline in a semester course. Thereafter, either emphasis would have to be given to other disciplines or a text going into more depth would have to be selected for the succeeding course or courses.

A final difficulty with the text is that because of its emphasis on breadth, many possibly important topics have been either left out completely or covered only briefly. Nothing is said, for instance, about goods movement in metropolitan areas, airports, water ports, or pipelines. Nor has anything been said about such important functional topics as cost allocation among modes or the impacts of international and national events (e.g., oil shortages). Restrictions on the length of the book precluded us from discussing these and many other topics we felt were not quite as important as the ones we included. Moreover, for the same reason, we could not go into any *detailed* analysis of national transportation policies and decisions that have played and probably will continue to play an all-important role in helping solve metropolitan transportation problems.

In this, the second edition, we have expanded several areas and deleted one. Much more attention has been paid to (1) public transportation characteristics (e.g., speed, capacity); (2) user costs; (3) air and noise pollution impacts; (4) post project evaluation; (5) actual decision making and community involvement; (6) transportation systems management; and (7) plans at the strategy, policy, program (or system), and project level. Perhaps the main expansion, however, was in the all-important area of finance, budgeting, and related legislation and organization. To make room for all this, however, we unfortunately had to drop the chapter on solution specification (detailed design).

Finally, we hope that the text will also prove suitable to professionals in the field who, with only a little basic background, can obtain a good picture of the present state of the art in metropolitan transportation planning.

One note to all readers: the bibliography at the end of each chapter is not limited solely to those references cited in the text of the chapter. Other references have been included so that the reader will be aware of further efforts related to the topic being discussed.

It is difficult, if not impossible, to remember, much less thank, many of the people who have helped create this book. Deserving special mention, however, are several people working under Kevin Heanue in the Federal Highway Administration's Office of Highway Planning. They responded with much valuable information. But then, almost everyone we contacted, in a variety of agencies, did the same. And there were the hundreds of students who gave us their comments as we subjected them to bits and pieces of the unfinished manuscript.

I guess there is no adequate way to thank the secretaries who have labored over our poor handwriting to develop a neatly typed manuscript. But particular thanks go to Janice Clements, who did most of the work. I continue to be amazed at the quiet, pleasant way in which she seems to get so much done. Others who contributed their time and talents include Katherine Golobic, Vanessa McCaffrey, and Joanne Smith. There were many others who also helped in the typing, and if we have forgotten them here, it is not from lack of appreciation.

Our final thanks go to B. J. Clark of McGraw-Hill and Fred Begell of Hemisphere for again allowing us the unique opportunity to publish a book that does not fall strictly into any one discipline, but is truly interdisciplinary.

John W. Dickey

Metropolitan Transportation Planning

1 The Changing Concerns of Metropolitan Transportation Planning

Those who have witnessed metropolitan transportation planning in a span of as little as 5 years know the large, and in some cases almost complete, transformation that can and has taken place. One day it seems as if the transportation system user is the only person of concern. The next day it seems as if the nonuser must be given equal if not greater concern. One day it seems as if travel time, cost, and safety are the only factors of importance. The next day the factors seem to be regional air pollution, the national economy, and the worldwide energy shortage. One day it seems as if highway engineers are perfectly capable of making complete decisions on highway planning and design. The next day, ten citizens' groups, five local planning bodies, the governor, several Federal executives and legislators, and the Supreme Court all seemed to be making a variety of relevant decisions. This, then, is the rapidly changing context in which the transportation planner is involved. It will be the objective of this chapter to illustrate the history of rapid change in metropolitan transportation planning and to show the need for a more fixed but modular *process* to deal with such change. The stages in this process, to be discussed in Chap. 2, form the framework for succeeding chapters in this book.

1.1 EARLY APPROACHES TO METROPOLITAN TRANSPORTATION PLANNING

Before the early 1950s, the overwhelming concern in planning for transportation in urban areas was costs for, and benefits to, the user. In fact, it could be said that there was relatively little in the way of "planning" for "transportation" in "urban"

1

areas. "Planning" consisted primarily of making straight-line projections of traffic counts and comparing the forecasted volumes with existing capacities. "Transportation," at least as far as government investment was concerned, consisted almost exclusively of highways; and "urban" areas were not treated too much differently from rural ones.

The first edition of one book on the topic, published in 1951 [1.1], demonstrates the state of the art and the thinking at the time. About 80 percent of that book was devoted to geometric design, soils, drainage, road materials, and so forth (as opposed to less than 5 percent of this text). About 15 pages of that book were devoted to urban traffic and parking surveys, with little or nothing on modeling endeavors, for example. The evaluation of alternatives was pictured almost entirely in economic terms and only with respect to the user. "Costs" were considered for roads and for vehicles; "benefits" were stated in terms of savings in distance, decreased costs of vehicle operation, decreased time to cover a given distance, and decreased accidents. (It is interesting to note that in those years many people were so happy to have a highway come to their area that they actually went out to meet highway officials to *give* part of their land for the right-of-way.)

1.2 METROPOLITAN TRANSPORTATION PLANNING IN THE 1950s

The early 1950s saw the advent of large-scale urban transportation studies. While there is some argument as to which was first, the study in Detroit or in San Juan, there is not much doubt that the former one [1.3], followed by a more elaborate effort in Chicago [1.4], was the most significant. Neither could have been done had it not been for the development of the digital computer, which at that time had reached the point where a very large mass of data could be manipulated and analyzed. An emphasis thus was placed on expanding and improving the *technical* side of the metropolitan transportation planning process. Data collection efforts were initiated that cost millions of dollars for a major metropolitan area. Surveys included extensive home interviews to determine both the travel patterns and demographic characteristics of respondents. Data on land use types, intensities, and location also were collected in large quantities.

The collected data then were employed to help develop methods, such as those discussed in the first part of Chap. 7, for forecasting travel on the highway and mass transit networks for a period of as much as 25 years into the future. The greatest emphasis, however, was on the *process* of doing transportation studies, and many people made extensive charts showing how such a study should be done. Yet the problems being attacked were assumed to be about the same. Creighton lists them as being [1.2, pp. 6-13]:

1. Accidents	5. Ugliness
2. Congestion	6. Strain and discomfort,
3. Inefficient investment	noise, and nuisance
4. Inaccessibility	7. Air pollution

Prime consideration was given to the first three factors, with the others becoming recognized as time progressed. The process did, nevertheless, focus specifically on *metropolitan* areas, gave some attention to transit, and resulted in fixed *plans* for investment over the succeeding 20 to 25 years. Thus we find a major breakthrough in metropolitan transportation planning, and such studies eventually were carried out throughout the world (see, for example, Ref. [1.5]). The *Highway Act of 1962* recognized the significance of the process that had been developed and thus required all metropolitan areas with central cities having more than 50,000 persons to have a comprehensive, coordinated, and continuous transportation plan.

1.3 CHANGES IN THE 1960s

The *Highway Act of 1962* required that a comprehensive, coordinated, and continuous transportation plan be implemented by 1965—otherwise no Federal funds, especially those for the Interstate system provided by Congress in 1956, would be forthcoming. Naturally, this requirement sent many city and state highway departments rushing to execute and complete their metropolitan transportation planning endeavors. At the same time, several new urban transportation and related problems became apparent, and these to some extent changed the direction of thinking away from the more straightforward planning process that had developed.

One of the first problems to become apparent was that of the state of mass transit in the cities. The first year in which expenditures for transit services nationwide exceeded revenues was 1962 [1.6]. And the gap continued to widen from that point in time. The *Urban Mass Transportation Acts of 1964 and 1966* recognized the financial plight of transit as well as other problems, such as outdated equipment, poor service, and so on. About $375 million thus was allocated for a program of grants, loans, and studies for mass transit.

Two results of these acts and the studies that followed were significant. First there came about a renewed interest in transit technologies (see Chap. 12 of this book). Through a series of investigations sponsored by the Urban Mass Transportation Administration and generally known as the "New Systems Studies," a variety of innovative technologies were identified and some tested. These included various "people-mover" systems for major activity centers, rail buses, dial-a-ride systems, and so on [1.7]. Also, many service and pricing ideas were proposed and demonstrated. These ranged from the simple, such as park-and-ride systems, to the more sublime, such as telephone-retrieved minibuses featuring stewardesses serving donuts, coffee, and other liquid refreshment. Many demonstrations were not successful from a financial standpoint, but they did at least add some technically feasible solution ideas to those involving only massive, capital-intensive highway construction.

The second result, perhaps more important from a humanistic standpoint, was the greatly increased recognition that a significant number of people in urban areas were not being served by an auto-highway-dominant form of transportation. Thus if transit were allowed to lapse, many people, primarily the poor, black, young, and old, would suffer disproportionately. These studies (most of which are abstracted in Ref. [1.8]) thereby provided a new dimension of considerations to metropolitan transportation planning—the disadvantaged.

The years that followed up to 1970 saw three acts of Congress occur that were of great significance to metropolitan transportation planning. The first was the *Urban Mass Transportation Act of 1970*. Of prime interest in that Act was the much-increased funding for mass transit—$10 billion over a 12-year period. This money helped close the financial gap for transit mentioned previously.

The other two acts were important because they helped to (1) expand the factors of concern in transportation planning, and (2) give the metropolitan areas or regions (as opposed to the localities) a review and consent power over spending on all Federal-aid projects related to urban transportation. The first resulted from the *National Environmental Policy Act of 1969*, which required that environmental impact statements be developed and approved for Federal-aid projects. It should be noted that the word "environmental" was interpreted in a very broad manner, including not only air, water, and noise pollution, but also social, economic, aesthetic, military, and other related impacts. In essence, planners and designers were asked to look at the *whole gamut* of impacts, not a small order by any means, but still an extremely important task.

The *Demonstration Cities and Metropolitan Development Act of 1966*, although coming before the *National Environmental Policy Act of 1969*, turned out to be a powerful, complementary tool in helping to foster metropolitan approaches to problems. It made it difficult for one locality to "go its own way" when dealing with problems such as air pollution that had obvious metropolitanwide implications.

1.4 CHANGES IN THE 1970s

The beauty and the bane of the *National Environmental Policy Act of 1969* was that it forced planners and designers to consider most of the significant impacts. The beauty was that persistent problems like air and noise pollution could no longer be ignored. Some plan, strategy, or design had to be developed to help alleviate the problems or else the project could easily be contested and halted in court. The bane was that Pandora's Box had been opened: many impacts are difficult to identify and define, much less reduce or enhance (whatever the case may require). Furthermore, many impacts are the antithesis of each other. Unless and until new technologies of consequence appear, there is no way to have a greater amount of comfortable and convenient travel with less air pollution or noise or energy consumption.

The 1970s thus brought more explicit identification of problems and goals, and a recognition of the multitude of external factors at state, national, and international levels. Certainly the energy "crisis" of 1973–1974 brought home the point that many of the factors affecting metropolitan transportation were outside the domain of metropolitan planners and decision-makers. This in turn brought recognition that state and national efforts had to take precedence. Federal laws then came into being that, for instance, required certain fuel efficiencies and reduced pollutant emissions on the part of newly manufactured automobiles.

Partly in response to this change of venue, the large data-collection and modeling efforts of the 1960s subsided. In response, there was an emphasis on more simplified travel analysis tools. This resulted on the one hand in the so-called "disaggregate" models and on the other in highly aggregated "policy" models. The former were advantageous in that only a small number of households needed to be included in the sample. These models also fitted more closely to microeconomic theory and thus were expected to be more accurate. The aggregate models, by contrast, required almost no new data and were particularly useful in assessing effects of, say, fuel rationing schemes, on overall metropolitan travel.

Evaluation in the 1970s generally attempted to take into account the wide variety of factors that had to be considered. Since some of these were difficult to quantify, a large burden was placed on the political process to resolve issues and make transportation decisions. This was made even more difficult because of the highly political interaction between the wide variety of stakeholders—the poor, the old, those with autos, conservationists, truckers, etc. One result was that almost any proposed transportation project of consequence landed in a court battle. This was exemplified by school-busing cases, which almost always were resolved in court and rarely were ever addressed directly by metropolitan transportation planners.

The transportation alternatives considered in the 1970s changed fairly drastically.

By the end of the decade construction funds were running low, especially because state and Federal highway trust funds were being squeezed by reduced revenues from gasoline taxes (owing to conservation) and rapidly rising construction costs. While several rapid rail systems were completed, it also was clear that there would be few more of these in the future.

Most of the attention thus turned to alternatives involving, in the case of transit, new buses and services, and, in the case of highways, road widenings, new traffic signal systems, and similar marginal changes in the network. Together both these kinds of improvements became incorporated into what was known as *transportation systems management* (TSM). This approach stressed better use of existing systems.

In another area, much more attention was given to human services transportation. It was found that literally hundreds of agencies provided transportation or requisite resources for transport for the poor, handicapped, elderly, unemployed, etc. The programs on which these efforts were based came from such diverse legislative backgrounds that the rules for funding and regulating them often were directly contradictory to each other. Some progress was made in overcoming these funding and administrative problems, especially through "brokerage" programs, where one agency attempted to coordinate travel and billing for several others.

By the end of the 1970s, implementation of plans and programs had become burdensome. Responsibilities for implementation were more dispersed than before. The approval process was long and complicated. There was stakeholder intervention at almost every stage, although some of this was highly useful. The courts were being employed to resolve many conflicts. And, as noted above, funding was becoming short and unstable. Most major projects were delayed or deleted, but there was, as a result, much more attention paid to smaller, unobtrusive changes that fit more closely with the existing urban environment.

From an organizational standpoint, much effort was made to concentrate decision making (or at least approval) for metropolitan transportation projects in the hands of a metropolitan planning organization (MPO). A main mechanism for this was annual approval of a transportation improvement program (TIP), which was required of every MPO and member jurisdictions. This specified the types and numbers of projects and programs to be carried out. Some of the MPOs were quite successful in planning and implementation of coordinated transportation projects and programs. Others were less auspicious, for many of the reasons suggested in the preceding paragraph.

1.5 THE 1980s AND ON

With the role of government being downplayed in the early 1980s, it is possible that the implementation process will become shorter and less complicated. It also appears that planners and elected officials are finding better ways to integrate various stakeholders into the decision-making process. Still, the courts have become one of the primary places for resolution of conflicts, and we do not anticipate that this will be diminished.

The other side of the picture has to do with funding. Unless some new ways are found to raise revenues in a fairly unobtrusive way, it is doubtful that people will approve more taxes for transportation-related purposes. This thus would seem to limit the type of alternatives to:

1. Minor construction
2. Transportation systems management
3. Maintenance

On the other hand, there may be rising use of the private sector to provide or substitute for transportation. With increased transit fares, for example, many taxi companies will become more competitive. Then there is the often discussed tradeoff between communication and transportation. In the past, both actually have grown together. Now, with the price of the former dropping because of new computer technology, and with the cost of the latter increasing, there may be a true substitution.

Lack of funding may also force governmental agencies to take a much closer look at the plethora of governmental programs that affect transportation. The starting point in the late 1970s for this effort was in human services transportation. It was obvious that many services could be combined beneficially if only the legal, regulatory, and funding conditions could be identified and integrated. Initial efforts have been made in this direction.

The realization also has begun to dawn that other governmental programs involving income transfers, tax expenditures, and loans have an enormous impact on urban transportation. Social Security, for example, probably puts more money in the pockets of the poor and elderly for spending on transportation (among other items) than any other governmental program. Similarly, the subsidy inherent in, for example, the investment tax credit, allows corporations and individual business people to purchase transportation vehicles and equipment at a reduced price. Some low-cost or guaranteed loans have the same effect (see Chap. 14). All impinge on the supply and demand for urban transportation, yet the relationships and consequences have barely been explored.

Because of the impact on urban transportation of state, national, and international programs and events, planning probably will turn more toward contingency assessments. These will attempt to show the effects of likely external events and governmental programs designed to ameliorate them. Associated with this "avoidance planning" will be the development of monitoring and early-warning systems to help indicate when the external events are imminent.

In sum, it appears that, because of continuing uncertainties, future transportation planning will be more defensive than offensive. In other words, the aim of planning will be more to preserve and maintain what exists than to create new and extensive construction programs. If this does occur as indicated, planning probably will become a more delicate art, with emphasis on powerful but simplistic tools, less costly secondary data, neighborhood conservation, smoothing of stakeholder participation, and coordination of diverse governmental efforts.

1.6 SUMMARY

The purpose of this chapter has been to provide an introduction to metropolitan transportation planning. The primary point is that such planning, despite the fairly constant charge to obtain the greatest "public good," has changed and probably will continue to change fairly drastically in context. In the early 1950s, when metropolitan transportation planning was being initiated in roughly the same mode as it is known now, the overwhelming concern was for the highway user: costs, travel times, and safety. Greater emphasis also was given to the technical side of studies: data collection, modeling, and the like.

The *Highway Act of 1962* was an important evolutionary step. It required continuous, comprehensive, and coordinated transportation planning for metropolitan areas. Shortly thereafter, mass transit became a focus of concern, and many new technologies were investigated. The late 1960s and early 1970s saw the advent of increased concern for social and environmental aspects of transportation and for citizen involvement in the transportation decision-making process.

The future of metropolitan transportation planning appears to be directed toward:

1. Single multimodal transportation systems
2. Increased alternatives and options
3. Increased recognition of the role of governmental programs outside the direct transportation sector
4. Increased identification of transportation effects and their ramifications for equity among groups of people
5. Better analysis tools and handling of uncertainty
6. Increased contingency or avoidance planning, with associated monitoring and early warning systems
7. Greater continuity in evaluation, with more stakeholder involvement and negotiation
8. Institutional changes emphasizing transportation as one component of overall development and neighborhood preservation

As mentioned, these changes most likely will alter the context but not the overriding purpose of metropolitan transportation planning. There thus appears to be a need to formulate some sort of modular framework or process for planning so that changes can be integrated without wholesale restructuring. The transportation planning process to be discussed in the next chapter is the framework we have adopted for this purpose.

BIBLIOGRAPHY

1.1   Ritter, L. J., Jr., and R. J. Paquette: *Highway Engineering*, Ronald Press, New York, 1951.
1.2   Creighton, R. L.: *Urban Transportation Planning*, University of Illinois Press, Urbana, 1970.
1.3   *Detroit Metropolitan Area Traffic Study:* Final Report (in 2 parts), Detroit, Mich., 1955.
1.4   *Chicago Area Transportation Study:* Final Report (in 3 parts), Chicago, Ill., 1959, 1960, and 1962.
1.5   The Metropolitan Transportation Committee: *Melbourne Transportation Study*, (2 vols.), Melbourne, Australia, 1969.
1.6   American Transit Association: *Transit Fact Book*, Washington, D.C., 1972.
1.7   U.S. Department of Housing and Urban Development, Office of Metropolitan Development: *Tomorrow's Transportation*, Government Printing Office, Washington, D.C., 1968.
1.8   U.S. Department of Transportation, Urban Mass Transportation Administration: *Urban Mass Transportation Abstracts*, National Technical Information Service, Springfield, Va., PB-213 212, October 1972.
1.9   Bosselman, F., and D. Callies: *The Quiet Revolution in Land Use Control*, Government Printing Office, Washington, D.C., 1971.
1.10   Manheim, M. L., and J. M. Suhrbier: *Incorporating Social and Environmental Factors in Highway Planning and Design*, Special Report 130, Highway Research Board, Washington, D.C., 1973.
1.11   Martino, J. P.: *Technological Forecasting for Decision-Making*, American Elsevier, New York, 1972.
1.12   Webber, M.: "On Strategies for Transport Planning," in *The Urban Transportation Planning Process*, Organization for Economic Cooperation and Development, Paris, 1971.
1.13   Owen, W.: *The Accessible City*, The Brookings Institution, Washington, D.C., 1972.
1.14   Schofer, J. L., and P. R. Stopher: "Specifications for a New Long-Range Urban Transportation Planning Process," *Transportation*, vol. 8, 1979.
1.15   Spielberg, F., E. Weiner, and U. Ernst: "The Shape of the 80's: Demographic, Economic and Travel Characteristics," *Transportation Research Record* (forthcoming).

1.16   Hamilton, N. W., and P. R. Hamilton: *Governance of Public Enterprise: A Case Study of Urban Mass Transit*, Lexington Books, Lexington, Mass., 1981.

1.17   Weiner, E.: "On the Future of Urban Areas and Urban Transportation Planning," Policy Analysis Division, U.S. Department of Transportation, Washington, D.C., Feb. 28, 1979.

1.18   Heightchew, R. E. Jr.: "On the Evolution of Urban Transportation Planning Theory and Techniques in the United States," Ph.D. dissertation, University of Maryland, College Park, Md., 1981.

EXERCISES

1.1   In your library locate the Federal-Aid Highway Act of 1962. What are its major elements? What developments since have made parts of it obsolete?

1.2   Add another "direction" to the list discussed in the summary. Describe that direction in about two paragraphs.

1.3   Write a short (750-word) position paper on "stakeholder participation" in the transportation planning process. In the paper defend one side of the argument as to whether such participation is valuable or not.

2 The Transportation Planning Process

One of the most interesting paradoxes in any profession is that which faces the engineer and planner who, although possessing a kit bag of mathematical and scientific tools, are called upon to perform a highly artistic task—that of problem solving. Many analysts can agree on and accept a given method of analysis as being logically consistent and reliable, yet few will concur on the time and place that the method should be used, or on the degree of judgment needed to derive a proper solution to a problem.

A great deal of the mystery that enshrouds these types of decisions is bound up in the problem-solving process, that nebulous procedure in which the planner or engineer starts with an original statement of the problem and finishes with a rather specific and fully implemented solution. It is because the problem-solving process usually is so vague that worthwhile solutions often are obtained only through some fortunate traversal between problem statement and implementation and operation. Therefore, it would seem logical that if the process could be mapped more precisely, better solutions might result since more persons then could follow the optimal path. It is toward this goal of better definition of the problem-solving process, especially in regard to transportation, that this chapter is directed.

2.1 MODELS

Before a discussion of the problem-solving process can proceed, it is important to delve first into the topic of models, since they are a natural component of that process. A "model," according to Krick [2.3, p. 85], is: "Something which in some respect resembles or describes the structure and/or behavior of a real life counterpart. There is some correlation between the model and its corresponding reality, although obviously a less than perfect correlation." Observe that this definition "so requires that we extend our interpretation of the word 'model' considerably beyond ordinary usage, which implies a three-dimensional replica of an object or person" [2.3, p. 84]. In fact, there are actually three types of models [2.4, p. 143]:

1. Iconic: Those that look like the reality, that is, are visual geometric equivalents (e.g., a model airplane)
2. Analog: Those in which there is a correspondence between elements and actions in the model and those in reality but no physical resemblance (e.g., a football play diagram)
3. Symbolic: Those that compactly and abstractly represent the principles of the reality (e.g., $F = ma$)

It is interesting to note at this point that words in themselves are symbolic models of some reality. For example, the word "tree" is a model of a variety of items ranging from dogwoods to sequoias.

In general, a model of any situation contains the following five sets of elements:

1. Variables over which the planner has complete control: X_i
2. Variables over which the planner has no control: Z_j
3. Variables over which the planner has indirect control: Y_k
4. General relationships between the above variables: R_m
5. Parameters (coefficients, constants, exponents, etc.) in the above relationships: P_n

Symbolically, a model M is represented by

$$M = \{X_i, Z_j, Y_k, R_m, P_n\} \quad \text{for some or all } i, j, k, m, n \tag{2.1}$$

where the brackets indicate a *set* of items.

An example may help to clarify Eq. (2.1). Suppose that the monthly revenue, r, from a given busline operation depends on the fare charged, f, and the monthly number of passengers, p, riding the buses. Revenue then would be the product of the fare ($/person) and the number of passengers (persons)

$$r = fp \tag{2.2}$$

It is also found that the number of passengers riding the bus in any month is a function of the number of inches of rain that month, i, and the bus fare, with the general relationship being

$$p = b/(i + 1)f^\phi \tag{2.3}$$

The "+1" after i is included so that "no rain" will not result in an infinite number of passengers (division by 0). The b and ϕ are parameters established from an analysis of past events. For example, in a hypothetical case it may have been found that the ridership is 10,000 passengers when there is no rain in a month and the fare is $1.00. This would give

$$10,000 = \frac{b}{(1 + 0)(1.00)^\phi} \quad \text{or} \quad b = 10,000$$

Similarly, let us assume that past data have shown that there are 40,000 passengers in a month when there is no rain and the fare is $.50. With the above value of b, these figures would lead to

$$40,000 = \frac{10,000}{(1+0)(0.50)^\phi} \quad \text{or} \quad (0.50)^\phi = \frac{10,000}{40,000} = 0.25$$

which gives $\phi = 2.0$.

Stopping here, we notice that the fare is a variable over which we as problem solvers hired by the bus company have control (an X_i), whereas we do not have control over the rain in a given month (a Z_j). Moreover, we have only indirect control over the number of passengers and revenue (Y_k's) since the weather influences their values. Continuing in our effort to derive a completed model, we find from Eqs. (2.2) and (2.3) and from the above discussion that

$$r = fp = f\frac{b}{(i+1)f^\phi} \tag{2.4}$$

which leads to:

$$r = f\frac{10,000}{(i+1)f^2} = \frac{10,000}{(i+1)f} \tag{2.5}$$

Searching further, we may find that (2.5) holds true only for high-income riders (a Z variable but stated as a category), while for lower income riders the proper relationship might be

$$r_l = \frac{12,000}{(i+1)f} \tag{2.6}$$

Equations (2.2) through (2.6) are all symbolic models and will provide the basis for the forthcoming discussions on the use of modeling in various stages of the problem-solving process. Equations (2.5) and (2.6) help to indicate that, in the most general situation, an indirectly controlled variable (like r) is always a function of a controllable variable (like f) and an uncontrollable variable (like i). Thus

$$Y_k = f(X_i, Z_j) \tag{2.7}$$

Equation (2.7) actually *defines* what is meant by the words "indirect control" in the sentences above Eq. (2.1).

2.2 THE PROBLEM-SOLVING PROCESS

As was remarked in the first section, the product of the problem-solving process usually is a fairly well detailed entity, yet the actual process itself often is a rather vague and undefined procedure, so that any representation of it would not necessarily be satisfactory to all concerned. Nevertheless, there is a need for explicitness, and many authors have attempted to blueprint in some way their feelings as to a realistic interpretation of the procedure by which a solution evolves after several transformations from a problem statement. One interpretation, which is a synthesis of those formerly proposed, is presented in Fig. 2.1. This diagram will provide the framework upon which the ensuing discussion will be built.

Except for two major *feedback* loops, the 10-stage process in Fig. 2.1 is a sequential process. Each link indicates both a *direction of movement* and a *flow of information* through the process, whereas each stage, I through X, indicates an action to be taken. Some actions probably would be done concurrently in a real world situation, as would be true of *collecting data* while at the same time *defining objectives and constraints*, but simplification at this point probably will not be detrimental to the accuracy of the overall description of the process. Briefly, the actions unfold in the following manner:

    I. An attempt is made to define as precisely as possible the nature of the problem and the domain within which it exists.

    II. Given the problem statement, the problem solver conceives of a set of goals and objectives that indicate the direction toward which any modification should be oriented in order to become a candidate for a satisfactory solution. He or she also conceives of a set of constraints which cannot be exceeded in any proposed solution.

    III. A model with the properties presented in Sec. 2.1 is made of the domain within which the problem has developed.

    IV. Data are collected in field studies or from information given in relevant publications. These data are used both to obtain a clear picture of the problem and problem domain and for establishing model parameters and present and past values of variables of interest.

    V. The model developed in (III) is calibrated (parameters are set) and used to determine the future magnitude of the problems (a) if no planned changes are made, or (b) if the solution proposed in stage VII were to be implemented.

    VI. An evaluation is made of the future situation with (a) no changes made and (b) proposed solutions. A decision is made among the alternatives.

    VII. Solutions (modifications) are generated and incorporated into the problem situation. These solutions are evaluated in (VI) and then (possibly) further modifications are made. This repetitive feedback procedure continues until some acceptable or perhaps optimal solution is reached.

    VIII. The details of the solution are evolved and specified by means of drawings, legal specifications, and so forth.

    IX. The solution proposed is implemented through such procedures as physical construction, financial and budgetary measures, policy statements, and so forth.

    X. The implemented solution is operated and maintained. Further data are collected to determine whether the solution has in fact led to the reduction of the problem as was predicted by the model. If this is not the case, the changes either in the model, in the modifications, or in both are needed. Continued problem solving is required as problems, goals, and so on change.

A simple example of the use of the problem-solving process can be created by referring back to the busline operation model in the previous section. Suppose that the bus company has the problem of having revenue that is too low (stage I) and realizes that the fare charged, weather, and income status of riders all have a bearing on the

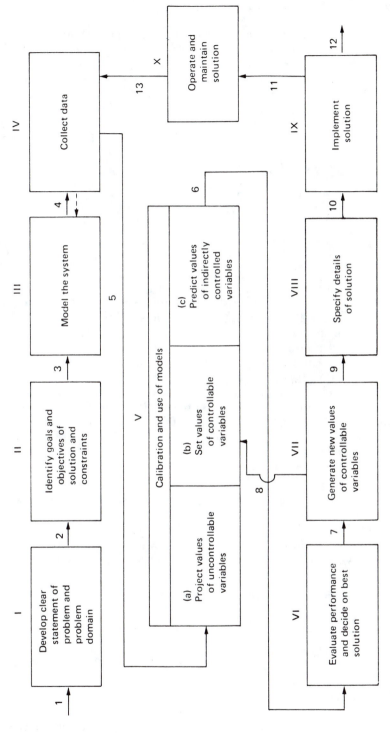

Fig. 2.1 The problem-solving process.

problem (the problem domain). The company would like to increase its net revenue as much as possible (objective) by modifying the fare structure while at the same time taking into account relevant governmental restrictions (constraints) on the magnitude of such modifications. The models the company uses (stage III) are those in Eqs. (2.5) and (2.6). It also is found at this stage that the local regulating agency requires that the fare, f, fall within the range between 10 cents and 40 cents. The net revenue for the *next* month is calculated based on the projection that the amount of rain for that month will be 1 inch and that the fare will remain as it is today—20 cents (stage V). Under these circumstances, the total revenue is the sum of that from both the higher and lower income riders, or, from Eqs. (2.5) and (2.6):

$$r_t = \frac{10,000}{(1+1)(0.20)} + \frac{12,000}{(1+1)(0.20)} = \frac{22,000}{(2)(0.20)} = \$55,000/\text{month}$$

where r_t is the sum of r_h (the revenue from high income people) and r_l. This revenue is not deemed adequate (stage VI), so that a modification of the fare to 30 cents is proposed (stage VII). This modification leads to a revenue of $22,000/(2)(0.30) = \$36,630$ (stage V and VI again), which still is not acceptable. Finally, after several modifications, a fare of 10 cents is suggested, which leads to the greatest possible net revenue $22,000/(2)(0.10) = \$110,000/\text{month}$ while *at the same time* keeping within the fare limitations of the regulatory agency.

To implement this fare change, the bus company must notify prospective passengers, get new change machines for the drivers, and, in general, specify many of the details (stage VIII) needed for the satisfactory fulfillment of the innovation. Finally, the fare change is brought into effect (stage IX) and operated and maintained (stage X). After a month of experience with the modification, it turns out that the revenue received is not $110,000/month but $100,000. To locate possible causes for this discrepancy, the bus company collects more data (feedback to stage IV) and finds that Eq. (2.5) really should be:

$$r_h = \frac{8,000}{(i+1)f} \tag{2.8}$$

By entering Eqs. (2.6) and (2.8) with the values for i and f, the company then is assured that $r_t = 8,000/(2)(0.10) + 12,000/(2)(0.10) = \$100,000$ per month. Similar procedures would be followed month after month to solve related problems as they arise.

2.2.1 Comments on the Problem-Solving Process

It is easy to be lulled into a sense of security when no security really exists. This is a distinct possibility in conjunction with the previous description of the problem-solving process: there are a multitude of difficulties concerning the process which appear only under further scrutiny. Several of these difficulties are listed and discussed below.

1. Various stages and links of possible importance in the problem-solving process may have been omitted from Fig. 2.1. It might happen, for example, that when an attempt is made to collect data pertinent to certain variables, it will be found that they cannot be obtained, thus requiring that an adjustment be made to the model formulated in stage III. A feedback link from stage IV to

III to account for this possibility should have been incorporated in Fig. 2.1, and, in general, a case probably could be made for the inclusion of a link between any two stages.

2. Statement of the problem (stage I) often is extremely difficult, first because the problem presented to the planner usually is not the actual problem and second because problem statements generally are too narrow in scope. A good example of these two observations is the "problem" of a lack of parking spaces in city centers. In many cities spaces are available, but the real problem is that they are too far away from desired final destinations. If enough spaces were provided at the proper locations, the city center probably would have to be transformed into a large parking lot, a situation which would be undesirable insofar as overall development is concerned. Thus, the parking problem is only one part of a broad development problem.

3. Explicit objectives and constraints (stage II) are not easy to establish. If a person claims he desires "efficient" transportation, do we really have any idea of what he wants?

4. Models (stages III and V) are models. They are not reality, but contain assumptions and simplifications which allow them to represent reality in a succinct manner. Yet, it is this succinctness which detracts from a model's accuracy. Is Newton's "model" of gravity a reliable predictor when friction exists and each mass is *not* concentrated at a point?

5. Relevant data (stage IV) often cannot be obtained, due either to the great expense involved or the unwillingness of certain parties to release the information. For example, the U.S. Bureau of the Census has certain well-specified disclosure rules.[1]

6. Future values of uncontrollable variables usually must be projected or extrapolated from trends or time series information (stage V). By a similar technique [2.24], it was found that the population of the United States would never exceed 148 million people.

7. Evaluation (stage VI) oftentimes is treacherous. Is a human life worth $34,000? How is appearance to be judged relative to cost?

8. Most modifications are the result of a *creative* process which many times cannot be replicated or emulated because of the variety of the human mind and the complexity of many problem situations. Resources limitations often prevent the testing of more than a few possible modifications.

These and many other comments are pertinent to the problem-solving process. They provide the major reasons for labeling the process as "nebulous" and "artistic."

2.2.2 The Distinction between Problem Solving and Planning

Because in the next section we will be concerned with the transportation planning process, it is necessary to discuss briefly the distinction between problem solving and planning. What usually distinguishes the two is (a) the time period involved and (b) the breadth of the situation being studied. For example, in planning we may be concerned with the location and capacity of a mass transit facility to be placed in a corridor of an urban area 15 years from now, whereas in problem solving our concern may be for the dimensions and performance of a vehicle to be placed in operation a year from now.

---

[1] See any *Census* reports available in most libraries.

This latter situation is narrower in scope, more detailed, and has a much shorter time dimension.

There are several natural consequences of the differences in time span and breadth found in planning and problem-solving endeavors. With respect to time span, for instance, it would be expected that forecasts of uncontrollable variables for planning purposes would have to be made further into the future, with the result being reduced reliability of the forecasted values. Obviously, a prediction of population for next year will have a much better chance of being correct than one for 20 years from now, especially because of the multitude of factors (such as wars, famines, and so forth) not considered in the predictive process but likely to exert an influence on growth at some time during a 20-year span.

Another consequence of the time period difference is that the problems and objectives relevant to a planning situation may themselves change during the period so that solutions based on *present* problems and objectives may be outdated before they are implemented. The interplay between communication and transportation might provide a good example for this situation. Traffic congestion in urban areas often is cited as a major problem, and solutions offered are those of more highways, better mass transit, and so forth. Yet, it may be that congestion will not be a problem in 20 years *even if nothing were done about it.* Television-type telephones may be developed to such an extent that many trips needed to achieve face-to-face communication would not be necessary, thereby freeing transportation facilities from this partial burden. Consequently, the objective of relieving congestion (the problem) would be antiquated. On the other hand, new problems (and corresponding objectives) of a type presently nonexistent (or unimportant) may arise within the planning period, and these should be anticipated in any planning endeavor.

A distinct advantage of the elongated time period is that presently nonexistent innovations which may lead to better solutions can be foreseen and used to complement or even replace the present solution before it is implemented. For example, if it is anticipated that public rental car fleets will be significantly developed several years from now, plans for later periods could include them. In fact, in some rare situations, even if such were not foreseen for the near future, enough time would be available for the planner to investigate the financial and technical feasibility of developing this system so that it *would be* available in several years.[2] The problem-solving process, as it is commonly understood, lacks the farsightedness to take advantage of forthcoming innovations.

A final advantage of the elongated time period is that anticipated solutions can be implemented *sequentially,* that is to say, on a *programmed* basis. Thus, instead of making improvements all at once, the planner will usually pick out a small subset to be accomplished in each subperiod of time (of, say, 5 years duration), and, if his choices are made properly, he or she can accrue substantial benefits over time from such a sequential procedure. Solutions arising from the problem solving process generally cannot be programmed for sequential implementation so as to achieve benefits comparable to those derived from planning.

The second major aspect that differentiates planning from problem solving is, as previously mentioned, the breadth of the problem addressed. For example, many cities

---

[2] Because of the time and expense involved, most planning organizations do not have an opportunity to make feasibility studies for possible innovations. On a national level, however, several studies have been done. See the program for the Northeast Corridor Project [2.9] and several reports from the Department of Housing and Urban Development [2.21, 2.22].

and towns are interested in providing more parking in their central areas, and parking garages and lot designs can be prescribed without too much difficulty. Nevertheless, parking is only one part of the overall transportation and urban growth situation and should be considered in this context. Future construction of mass transit facilities could reduce the need for extra parking. Planning, rather than problem solving, is the name usually given to the study of broader situations. Conversely, planning, because of its greater generality, can never be considered to have the depth of detail required for a final product, so that the problem-solving process can be considered to be an adjunct to the planning process in which specifications of each general concept of a solution generated from the planning process are detailed.

2.3 THE TRANSPORTATION PLANNING PROCESS

The transportation planning process, as brought out above, should bear a great similarity to the problem-solving process developed in the foregoing sections of this chapter. The two major differences are that the former process is directed at the production of a rather *broad* description of a *future* transportation system, whereas the problem-solving process is much more detailed and immediate in nature. Yet, as the time for implementation nears, transportation plans also must be specified to a much greater extent.

Generally speaking, the major thrust of any transportation planning study involves identification of travel-related problems. This is followed by determination of the demand for travel in the given situation. This in turn is followed by an attempt to find those presently available or anticipated future transport systems which would meet the established travel demands in an acceptable manner (that is, with as few adverse impacts as possible). Thereafter, as time progresses, plans become more precise—proposed budget allocations are made, exact locations are determined, work programs are fixed, and so forth—until the proposed system becomes a reality. This general transportation planning process, outlined only briefly at this point, will be discussed in more detail in succeeding sections of this chapter. This presentation subsequently will form an all-important framework to which the concepts of the ensuing chapters of this text can be related in a comprehensive and logical fashion. In fact, each succeeding chapter of this text will represent one or part of one stage of the transportation planning process as it will be presented below.

2.3.1 A General Description

Any planning process is initiated by an attempt to define as neatly and succinctly as possible the types of problems that do or will exist and the domain in which they are set. This first stage is shown along with the remaining ones in Fig. 2.2. Almost invariably these transportation service problems are both numerous and nebulous, and they range from such mundane matters as pot holes in streets to more complex situations of development of regions. In any case, most of these problems can be reduced to that of the high cost, in time, money, and comfort (or some other measure), of getting from origins to desired destinations. The factors affecting or affected by these problems (the problem domain) usually consist of a significant portion of both the manmade and natural habitat.

Given general problem statements and a description of the problem domain similar to that above, we can go on to identify the objectives and constraints (stage II, Fig. 2.2) that will guide our search for new solutions. Example objectives might be

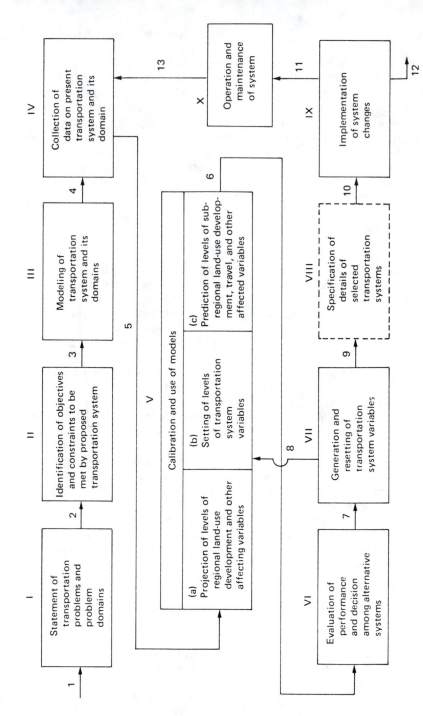

Fig. 2.2 The transportation planning process.

those of minimizing travel time throughout the region, maximizing the number of people who will use public transit facilities, or influencing certain kinds of growth in a region through transportation improvements. Constraints relevant to these objectives may be those of the yearly budget of the local or state highway departments, Federal regulations on removal of housing for construction purposes, and the quantity of certain subgrade materials available in the vicinity of the proposed improvement.

In stage III of the transportation planning process an attempt is made to model in an abstract manner the transportation system and its domain. Here the three most prominent types of models to be employed are: (1) those describing affected factors and their relation to transportation (land use development, air pollution, noise, etc.), (2) those describing travel behavior or the "demand" for transportation facilities, and (3) those describing the characteristics of various forms of transportation technologies (supply).

By having a general picture of these model types, we can proceed to a rational data collection effort (stage IV), the main purpose of which is to provide the information necessary to establish the parameters (constraints, coefficients, exponents, and so forth) for each model. The data collection stage has another important role, however— the development of more accurate statements of the problems, problem domains, objectives, and constraints. What previously may have been a general congestion problem at an intersection, after the gathering of more data, might be specified as 65.3 seconds of delay per vehicle between 4:30 and 5:30 p.m. on weekdays.

In stage V, the land use development, affected factors, travel demand, and transportation supply models are calibrated using the collected data and utilized in concert to predict the future state of environmental and transportation service factors *if no changes in the transportation system were made.* The consequences of such an alternative, when evaluated under stage VI of the planning process, usually are worse than those in the existing state: driving delays may be greater than at present, transit facilities may be expected to deteriorate, and the handicapped may be getting insufficient access to facilities. As a consequence of this anticipated poor performance, modifications to the present transport system must be generated (stage VII), and this generation must be done in such a manner that the objectives ascribed to in stage II are accomplished to the highest degree possible.

At this point in the process the modifications suggested should not be carried out directly, however, because of the extremely important feedback (arrow 8) that exists in most instances. The crux of this feedback is the *impact* of the modified transportation system on land use in various parts of the region, on the amount, direction, and mode of travel, and on other affected factors such as air pollution and appearance. This impact must be reevaluated in stage VI, at which point a need for further modification of the transportation system may arise. Finally, after several possible loops through these three stages (V to VII), a decision would be made among the alternative solutions that were generated. This chosen solution then would be set forth more specifically in stage VIII, although this step generally would not be considered part of planning but of detailed engineering or architectural design.[3] Implementation (stage IX) then would be next. Here the planner would be concerned with the financial, organizational, legal, and related aspects of implementation rather than directly with, say, construction procedures.

Continued operation and maintenance would follow (stage X). This stage would conclude the main effort of the planning process, yet there would remain a continual

---

[3] Thus the dotted lines around this stage. This step will not be discussed in the text.

need to collect information for the evaluation of the *actual*[4] performance of the transportation system. Moreover, planning should never be a one time endeavor, but instead should be considered a continual process, with solutions being altered as new aspects of problems arise, as goals and objectives change, and as improved technology becomes available.

2.3.2 Transportation Problems and Problem Domains

To define any type of problem and its accompanying domain often is one of the most difficult tasks to be faced by any planner. Transportation problems in particular exhibit this difficulty, first because most occur in space as well as in time and second because the problems are hierarchical in nature. The planner thus must be able to identify the areal region and various subparts of the region in which certain types of problems can be expected to arise at predicted times in the future. He or she must be capable of perceiving the whole spectrum of problems from the extremely localized ones to the regionally oriented ones. To accomplish these two tasks the planner usually would go through the following procedure:

1. Circumscribe a *region* wherein the activities that occur are as independent from those outside the region as possible. The resulting line of demarcation usually is referred to as a *cordon line*.
2. Divide the cordoned region into a set of *zones* or *subregions* whose areas are small enough that most problems can be pinpointed fairly accurately and yet large enough that the study does not become inundated with data. An example of a region and its zones for the Waco, Texas, area is shown in Fig. 2.3.
3. Finally, through contact with interested parties, help define various transportation service problem types which fall within the broad range from regional access problems to local congestion problems and, as part of this effort, attempt to identify those factors in both the man-made and natural environment which would either contribute to or be a derivative of transportation system deficiencies.

The output of this three-step process would be an identification of transportation service and related problems such as:

a. Much delay is incurred in traveling on routes through zone 2
b. Many people in zone 51 have no means by which to get to industrial jobs
c. Pedestrian travel is too restricted in downtown zones
d. Central-area stores are deteriorating
e. Growth in the region is not as rapid as that in other regions
f. Air pollution is prevalent in the region

The latter three problems generally would be considered to be contained in the transportation problem domain. All have significance here, however, in that they are related to the location, characteristics, and numbers of human beings and their activities. Growth in the region, for instance, depends on national economic activities, taxation, and local worker skills as well as transportation and other factors.

A particular subset of activities that have been found to influence and be influenced by transportation is *land use*. This term, interpreted in its broadest sense, refers not only to the activity classification—manufacturing, commercial, agricultural,

---

[4]This performance contrasts with the *model-predicted* performance determined in stage VI.

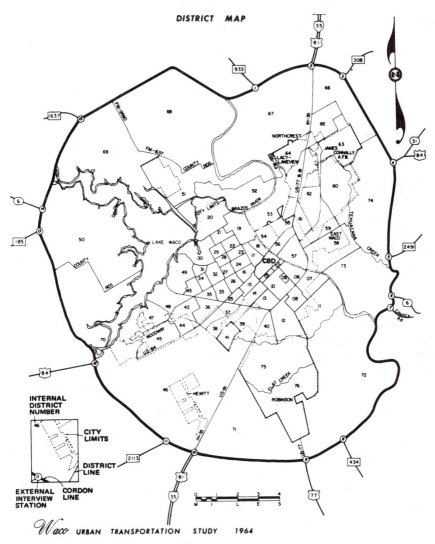

Fig. 2.3 Zonal map for cordoned area surrounding Waco, Texas. (*Waco Urban Transportation Study, 1964.*)

and so forth—but also to the intensity of use, measured by employment, population, acreage, etc., by the zonal location of uses, and by the suitability of land for various activities.

Since travel patterns generally are considered to be a function of the location and extent of human activities, we can infer that travel problems are related to land use. Moreover, an extended view of the picture would show transportation affecting human activities through the inducement of development in areas adjacent to transportation facilities. Taken together, these two concepts point to human activities (and particularly land use) as major elements in the domain associated with transportation problems.

2.3.3 Definition of Objectives and Constraints

If the problem were the first one listed, that of too much travel delay on routes in zone 2, the objective would be to minimize this delay as much as possible or to reduce it to some acceptable level. Yet this specification is not complete since we realize that there are many limitations or constraints on our endeavors to provide better solutions: time and money budgets must be adhered to, financing must be found, the political and social situation must be stable, and so forth. The output of stage II of the planning process therefore is a two-part product that, if exemplified by the above problem, would be: *minimize travel delay in zone 2 subject to given budgetary, financial, political, and social constraints.*

2.3.4 Modeling of the Transportation System and Its Domain

The development of models to be used to represent the interrelationships between the transportation system, external factors, local human activities, travel, and other affected or impact factors comprises stage III of the transportation planning process. Since the calibration and use of these models (stage V) is a relevant consideration here, both of these stages will be discussed together at this time.

Figure 2.4 shows six phases basic to the modeling stage of the transportation planning process. Each of these phases corresponds to a type of model (or set of models) needed to produce inputs for the testing and evaluation stage to follow. Perhaps the most important of the six from an instructive standpoint is that of the prediction of human activity (particularly land use development) levels, since this phase is linked directly with the other five.

Required as inputs to this phase are three types of information: (1) projections of future external variables such as population, employment, and land use totals; (2) determinations of present and future levels of other local affecting variables such as water, sanitary, and educational services; and (3) estimates of present and proposed transportation system performance, with the emphasis being on capacities, speeds, costs, and the like.

An activity model now can be employed to help estimate the effect of external forces on the location and extent of human activity throughout the region. These in turn can be used in helping to predict travel demand [phase (e)].

Travel demand forecasting usually is accomplished by means of a four-step procedure that, as noted above, uses predicted future zonal activity values as well as future transportation system performance characteristics (see Fig. 2.4). The first step, *trip generation*, involves the determination of the number of trips produced by or attracted to each zone in a given time period, usually one day. Total numbers of trips are accumulated for each zone regardless of destination. A typical relationship calibrated via regression or similar techniques using existing data might be

$$T_i = aP_i + bI_i + c \tag{2.9}$$

where T_i = number of daily trips produced in zone i
P_i = resident population of zone i
I_i = average income of people in zone i
a,b,c = empirically derived parameters

Notice that two activity variables are instrumental in the predictive relationship in Eq. (2.9). No transportation system characteristics are employed in Eq. (2.9), but

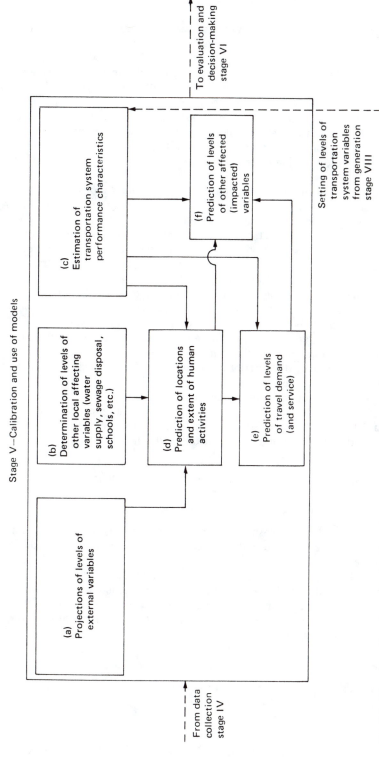

Stage V—Calibration and use of models

From data collection stage IV

(a) Projections of levels of external variables

(b) Determination of levels of other local affecting variables (water supply, sewage disposal, schools, etc.)

(c) Estimation of transportation system performance characteristics

(d) Prediction of locations and extent of human activities

(e) Prediction of levels of travel demand (and service)

(f) Prediction of levels of other affected (impacted) variables

To evaluation and decision-making stage VI

Setting of levels of transportation system variables from generation stage VIII

Fig. 2.4 The six major phases in the calibration and model use stage of the transportation planning process.

23

they will become necessary in the succeeding three steps of the travel prediction procedure.

In step two the trips "piled up" at each zone are "dealt out" to the other zones in the region in a manner similar to the way in which individual persons are known to allocate their trips to various destinations—usually on the basis of the relative attractiveness of each zone and its distance from the zone of origin. This allocation process, referred to as *trip distribution*, in most instances uses land use variables such as ground area and floor area as measures of "attractiveness" and transportation network travel times as indicators of "distance." Thus, the trip distribution phase presents an excellent example of the joint use of land use and transportation system variables in the prediction of travel.

The third phase of travel prediction is that of *mode choice*, a phase in which, as the name denotes, estimates are made of the proportion of the interzonal trips found from the trip distribution model that travelers will choose to make by each mode.[5] Again, it is easy to see that both activity and transportation system variables are important factors in the relationships underlying mode choice since the decisions people will make probably will depend both on the characteristics of the transportation system offered (service rate, cost, travel time) and the socioeconomic class of the traveler (e.g., higher income people are more likely to take their car instead of transit.)

The final phase of the travel prediction process, *trip assignment*, involves another choice, in this case between the various routes of each mode which may be available to the traveler. For repetitive trips (and most are) this choice generally is based on either relative travel time or cost. With this last type of prediction accomplished, we then would have completed a description of the future travel on all available routes of each mode of transport in service.

Phase (*f*) of the modeling process also is significant. It is well known that a transportation system has both beneficial and nonbeneficial impacts on factors other than human activities that generate travel (that is, on the nonuser). A complete list of factors would be extensive and varied, including such items as:

Air pollution	School district boundaries
Noise	Health care for the poor
Financial conditions of	Crime
private corporations	Property damage

Many of the relationships between transportation, travel, and the above items have not been established with finality. Yet more are becoming known as research progresses.

Still lacking in the overall description of the modeling process is a representation of the type and interrelationship of performance characteristics of the transportation system itself. An easy way to view the transportation system is to think of it as composed of four types of items: *vehicles, networks, terminals,* and *controls,* all mixed together in some quantity.

The product of the resultant mixture can be characterized by some of the following features:

Speeds	Dimensions
Capacities	Routes
Locations	Schedules, etc.

---

[5] Mode choice also can be the first or second step in the travel prediction process.

These features are not independent of each other, however, as is demonstrated by one relationship between the speed and volume of vehicles on an expressway [2.23, p. 334]:

$$V = 175S - 4.86S^2 \qquad (2.10)$$

where V = volume of vehicles (vehicles per hour) and
S = average speed of vehicles (mph)

Thus, to specify the performance of a particular transportation system, we must be able to identify both the relevant characteristics of the components in the system and the *relationships* between the characteristics. To accomplish this characterization we need models such as Eq. (2.10).

2.3.5 Collection of Data

While it may be true that in other fields of endeavor many models germane to the task can be employed without first investigating in detail the particular situation under study, this generally is not the case in most transportation planning efforts. The main reason for this situation is that parameters (constants, coefficients, exponents, and so forth) that remain stable from study area to study area have not been found. As a consequence, data must be collected in each transportation study to be undertaken in order to calibrate the necessary models, that is, to determine their parameters.

For example, to be able to use Eq. (2.9) for predicting future numbers of trips produced in each zone, we probably would have to compile information on present and past zonal trip productions, zonal population, and average zonal incomes and then use a regression technique (see App. A) to find a, b, and c. Knowing these numbers and also knowing the future values of P_i and I_i, we then would predict T_i.[6]

At this point we also should emphasize an additional purpose of data collection: to identify more precisely problems, problem domains, objectives, and constraints. In performing stages I and II of the planning process we probably would generate only vague ideas of existing problems and the directions toward which new solutions should be aimed. By gathering more information, we are able to take a problem such as "lack of transit service" and specify it more explicitly as "elderly people and school children have to wait too long during daylight hours for a bus." The corresponding objective might be to "maximize the nonauto service provided these people" and a likely constraint might pertain to the yearly budget which could be allocated to the proposed service.

To summarize briefly the major needs of the data collection stage, we might classify desired information into 10 categories:

1. Statements of problems (present and anticipated future)
2. Statements of objectives and constraints (present and anticipated future)
3. Types of models (present)
4. Types, locations, and intensities of human activities (past and present)
5. Transportation system features (present)
6. Travel (past and present)
7. Transportation technology (present and anticipated future)
8. Other affecting variables (present and anticipated future)
9. Evaluation procedures (present)
10. Implementation procedures (present)

---

[6] We also must assume that a, b, and c all will hold constant in the future, a rather daring assumption.

It should be noted that the assembling of this vast quantity of data is a strenuous task, and, as an added burden, the engineer and planner must keep abreast of innovations in areas such as information detection, handling, sorting, storing, and display techniques as well as procedures for coping with sampling problems, errors, mistakes, bias, and fabrication in data collection efforts. In fact, a summary of these and similar items would in itself constitute a supplemental set of information to be collected.

2.3.6 Evaluation of Performance and Decision Making

Suppose data have been collected for the models discussed in Section 2.3.4 and that these models have been calibrated and used in stage V of the planning process on the initial assumption that no change from the present (or committed) transportation system will be made in the future. The transformations that could come about under this "no change" situation would be those in human characteristics and activities, travel, and other affected factors, and these could be evaluated in stage VI (see Fig. 2.2). We would expect in this situation that the evaluated future performance of these factors generally would be worse than that which now exists; that is to say, we most likely would end up further from our stated objectives (stage II) because of probable increases in tripmaking (and consequent congestion), probable restrictions in access by the handicapped due to the lack of new transportation facilities, and probable increases in air pollution, noise, pedestrian accidents and so forth.

In any case, by making the initial supposition of no new facilities or services, we provide ourselves with a datum by which transportation modifications, to be generated in stage VII, can be judged. In other words, the performance of a modified transportation system in regard both to travel and its impact on the environment will lie somewhere between the previously noted datum and the other extreme at which there is a complete securement of the stated objectives. Given these upper and lower limits on performance, we then are able to judge each system modification when the return is made from the generation stage (stage VII) to the modeling stage (stage V) and then to the evaluation stage (stage VI).[7]

In going through the evaluation and decision-making process, the planner must face some of the most vexing difficulties in the planning process. For example, the first, and perhaps most paradoxical, question concerns the role of the planner in evaluating solutions and making decisions among them. It may be that such decisions should be left entirely to the political body in the region or to appointed citizen boards. The answer to this question is far from apparent, yet, as mentioned above, it is just one in a series of questions that may be imponderable. Several others are:

1. On what governmental level should evaluative decisions be made—local, state, or Federal?
2. What variables should be considered in the evaluation—lives saved, families to be relocated, etc.?
3. How should some of these variables be measured (e.g., appearance)?
4. How should possible future risk and uncertainty be handled?

Some techniques exist for providing a rejoinder to these queries, but on the whole they are resolved through an oftentimes imperceptible interaction that occurs between all concerned parties. Whatever the nature of the evaluation process involved, decisions

---

[7] See the corresponding feedback arrow (number 8) in Fig. 2.2.

among alternative solutions eventually are made, but rarely through a simple and concise approach.

2.3.7 Generation of the Transportation Alternatives

In the generation stage of the transportation planning process an attempt is made to bring together the four components of a transportation system (vehicles, networks, terminals, and controls) in such a way as to come as close as possible to fulfilling transportation and environmental objectives. In other words, it is the duty of the planner to help pick from among all of the available components those which, when evaluated, will prove to be most beneficial or effective.

It should be noticed immediately that involved in generation is the process of creativity. Few planners would deny that most of the better solutions for problems (especially those in transportation) have come to the surface through a series of intangible and untraceable steps that are an outgrowth of the creative force at work. However, it also happens that some of the worst ideas have come about through "creativity," so that in recent years many techniques have been developed as aids both in the development of better solutions and the prevention of poorer ones.

The most sophisticated of these techniques are those of mathematical programming—linear programming, nonlinear programming, dynamic programming, and so forth. The rudiments of several of these are presented in App. A. Mathematical programming is useful because it entails the optimization of some objective function when certain constraints are present. Clearly, if we had simple, nicely behaved, measurable objectives as well as similar natured constraints, we could arrive at a "best" solution. Unfortunately, most objectives and constraints usually are neither explicit enough nor simple enough for such an approach.

The most prevalent approach is that of trial and analysis, which is nothing more than a systematic search of a very limited set of possible alternatives.

After any one of these approaches has been used, what evolves is a broad identification of a modified transportation system composed of each of the four major components and their features. An example of such a system would be that of a hanging monorail, capable of handling 2,000 passengers per hour, running on elevated prestressed concrete beams, located in the Central Business District, with terminals costing $200,000 each, spaced at 400-yard intervals, and controlled by a railroad block-type subsystem.

Given a proposed system of this nature, we would have to return to the model use and evaluation stages of the planning process in order to determine, first, the impacts of the system on land use, travel and other affected factors and, second, the extent to which stated system objectives have been fulfilled. After going through these two stages, it may turn out that transportation system modifications again are needed. In the end, the accepted modifications would be added to (or in some cases substituted for) the committed but not already operational system.

2.3.8 Solution Specification

Up to this stage in the planning process the description of the present transportation system, the present domain, and future transportation system modifications has been a broad one, accenting only the most general features. As far as the transportation system is concerned, for example, we have been dealing only with rough estimates of costs, speeds, capacities, and so forth, these being obtained from average figures from past experiences or previous testing with similar systems. But it may be that there is little in the way of past data available on particular systems, as would be the case for downtown distribution and new aircraft systems discussed in Chap. 12.

In these situations, the only information we would have would be "guesstimates" of the costs and other characteristics involved. Thus, the task falls to the designer at the solution specification stage to specify the chosen solution much more exactly, that is, in as much detail as needed to construct and operate the actual system. For example, most highway designs require geometrics, house and property-line locations, highway curvature information, and fence-line and drainage-pipe specifications. As noted earlier, solution specification generally is not considered to be part of the planning process. It might more properly come under the heading of "design."

2.3.9 Implementation

The ninth stage of the transportation planning process is concentrated on bringing the plan into existence. The planner rarely has much to do with, say, actual construction, purchase of buses, or the like. But, in developing alternatives and evaluating them, the planner must take into account other major aspects relevant to implementation. These may include:

Overall management	Research and development
Finance	Coordination
Taxation	Capital budgeting
Pricing	Improvement programming
Cost allocation	Marketing
Among users	Laws
Between users and nonusers	

On a slightly different plane there is the whole subject of regulation, of which *land use control* is an example. Inherent in this item are such concepts as zoning, in which laws are made concerning the use of individual properties; subdivision regulations, which refer to the manner in which raw land under one ownership is converted into building lots; and eminent domain and condemnation, which are procedures for obtaining land and other property needed for a public use such as transportation. Taken together, the set of subjects presented in this and the previous paragraph constitute the major tools for implementing and ensuring the stability of any planned solution.

2.3.10 Operation and Maintenance

There is a need to keep constant surveillance over the performance of the system to ensure that the intended objectives are fulfilled to the degree originally thought possible. What is required at this stage, then, is a control process, indicated by arrows 11 and 13 in Fig. 2.2, in which there is a collection of new data, referenced to the operation and maintenance of the system, an evaluation of the newly recorded performance of the system, and specification of certain small modifications which will improve the performance to the desired level. Thus monitoring of operation and maintenance really involves a continual recycling through stages III to IX of the planning process, with the emphasis placed on refining the more detailed parts of previously obtained general solutions.

It also should be stressed that, with low-capital solutions becoming more prevalent, there is additional need for planning for maintenance and operation as well. One example is pavement management systems (PMS), in which plans are made and budgets set for the ongoing repair of streets and highways.

2.3.11   Comments on the Process

The continuous nature of the planning process should be reemphasized at this point. Plans should not be so hard and fast that they cannot be subject to some revision as the types of relevant problems, the specified goals and objectives, and the technology of transportation change and evolve through time.

Additionally, there usually is a considerable time span between proposal of new solutions and their implementation. New freeways, for example, would be at least 8 years in this stage. What is needed, then, are studies parallel to the long-term transportation planning study. These endeavors would be short-term and directed toward improvements that can be implemented within periods up to 3 years in length. Included here might be traffic engineering changes such as traffic signals, one-way streets, and additional turning lanes. Also included would be changes in bus routes, creation of special bus-only street lanes, and new ticketing arrangements (e.g., script fare systems). By working on both short- and long-term solutions, the transportation planning agency can improve both its flexibility and usefulness.

The planning process described above is not uniformly accepted by all in the profession. Many authors and practitioners feel, for example, that it is very difficult to get people and agencies to state goals (stage II) beforehand. Often they find it easier to say what they *do not* want, especially after having been given examples to which to respond. Under these conditions it might be more reasonable to start, as Harris [2.26], Calkins [2.25], and Boyce et al. [2.24] suggested many years ago, by developing projections (stage V) and then obtaining people's reactions (stages I, II, and VI).

In addition, it must be recognized that few planners enter the process from the beginning, nor do they personally stay long enough to follow through each and every step. This means that their participation actually may be in quite a different and incomplete order from that shown in Fig. 2.2. In truth, then, that process might (and could) be rearranged in a variety of ways, depending on the situation under study, the particular status of the planner, and a wide spectrum of other influencing factors. Figure 2.2 in fact might be most realistic (but least easy to explain) with each stage connected to all others in a spirograph-type arrangement.

2.4   DIFFERENT TYPES OF PLANNING

Transportation planning in metropolitan areas can take place at several different levels in what is known as the "policy chain." At the top is the policy level itself. Planning here often is called *policy* or *strategic planning*. The same general process discussed in this chapter holds, but the alternatives investigated usually are very broad. These might be exemplified by strategies to provide free transportation, where needed, to all school children or elderly; to "crack down" on drinking drivers; to reconstruct all streets every 10 years; or to opt for light rail systems rather than "heavy" (rapid) rail. Planning analyses would help to assess the broad benefits of these strategies versus others. They also would help narrow the field of options for more detailed studies.

The next level down is the *program* or *system level*. A program usually is defined as a set of projects, where the latter are the smallest entities which are treated separately for decision-making, design, and implementation purposes. A systems-level program thus may be illustrated by a group of projects including widening of a certain arterial, creation of an integrated traffic signal system in a CBD, and construction of a high-occupancy-vehicle lane in a freeway median. In the human services context, a program may involve provision of transportation by several agencies to several different clientele groups such as the mentally retarded, visually handicapped, and aged. In either case, program or system plans can be developed, analyzed, and evaluated by considering the benefits and costs of the interrelated projects in the set.

The *project level* itself is the next lowest. Project planning looks particularly at the merits of the individual endeavor. For highway and transit systems the projects usually are alternatives for a given corridor. A project could, for instance, be a rail line, a busway, or an improved arterial. In human services situations, a project might be the acquisition of vans with, say, special lifts for the handicapped in wheelchairs. Again, planning can be accomplished using the process suggested in this chapter.

Another variant of interest here is contingency planning. The same process can be employed, but the emphasis is on external events, such as an energy shortage or a major storm (e.g., a hurricane), that may create excessive hardships or damage. The focus is on alternatives that can be applied to ameliorate these excesses. In some situations a "worst case" alternative is suggested so that the most damaging situation can be assessed.[8] In other situations, continuing data collection is undertaken as a form of monitoring to predict, assess, and give early warning if the feared external event seems forthcoming.

2.5 SUMMARY

The objective of this chapter has been to present a systematic version of the transportation planning process useful in organizing the approach to solving both short- and long-term transportation and related problems. This process was envisioned as having nine stages:

1. Statement of problem and problem domain
2. Identification of goals, objectives, and constraints
3. Modeling of the system
4. Collection of data
5. Calibration and use of models
6. Evaluation of performance and decision among solutions
7. Generation of new solutions
8. Implementation of chosen solution
9. Operation and maintenance of solution

The final stage would be followed by continued checks to see if the solution were as effective as predicted. Long-term planning processes also should be accompanied by short-term, problem-solving efforts.

It is not suggested that the process discussed in this chapter is the way transportation planning should be or is being done. In different situations new stages could be added, existing stages mixed in order, and many other changes made that might lead to a process (and the solutions it generates) more effective than that portrayed here. There is a need, however, to focus on one type of process so that thinking about approaches to transportation problems can be systematized somewhat. With this idea in mind, we have organized future chapters of this book to correspond to the stages in the planning process presented here.

---

[8] Interestingly, it is very rare for a contingency plan to be made to take advantage of possible highly beneficial external events.

BIBLIOGRAPHY

2.1   Vuchic, V. R.: *Urban Public Transportation: Systems and Technology*, Prentice-Hall, Engle-wood Cliffs, N.J., 1981.

2.2   Gray, G. E., and L. A. Hoel (eds.): *Public Transportation: Planning, Operations and Management*, Prentice-Hall, Englewood Cliffs, N.J., 1979.

2.3   Krick, E. V.: *An Introduction to Engineering and Engineering Design*, Wiley, New York, 1965.

2.4   Woodson, T. T.: *Introduction to Engineering Design*, McGraw-Hill, New York, 1966.

2.5   Morlok, E. K.: *Introduction to Transportation Engineering and Planning*, McGraw-Hill, New York, 1978.

2.6   Institute of Traffic Engineers: *Transportation and Traffic Engineering Handbook*, Prentice-Hall, Englewood Cliffs, N.J., 1976.

2.7   U.S. Department of Transportation: *Evaluating Urban Transportation System Alternatives*, Washington, D.C., Nov. 1978.

2.8   American Association of State Highway and Transportation Officials: *A Manual on User Benefit Analysis of Highway and Bus Transit Improvements, 1977*, Washington, D.C., 1978.

2.9   U.S. Department of Commerce, Office of High Speed Ground Transportation, Transport Systems Planning Division: *Study Design*, Northeast Corridor Transportation Project Technical Paper No. 5, Washington, D.C., July 1966.

2.10   U.S. Department of Transportation: "Urban Transportation Planning," *Federal Register*, Thurs., Aug. 6, 1981.

2.11   U.S. Department of Transportation, Technology Sharing Division: *Planning and Coordination Manual (for Elderly and Handicapped Transportation Services in Region IV)*, Washington, D.C., Jan. 1979.

2.12   Hutchinson, B. G.: *Principles of Urban Transport Systems Planning*, McGraw-Hill, New York, 1974.

2.13   U.S. Department of Transportation: *Planning Handbook: Transportation Services for the Elderly*, Washington, D.C., June 1976.

2.14   U.S. Department of Transportation, Federal Highway Administration: *Design of Urban Streets*, Government Printing Office, Washington, D.C., Jan. 1980.

2.15   Murin, W. J.: *Mass Transit Policy Planning: An Incremental Approach*, Heath Lexington Books, Lexington, Mass., 1971.

2.16   Steiner, G. A.: *Strategic Planning*, The Free Press, New York, 1979.

2.17   U.S. Department of Health, Education and Welfare and U.S. Department of Transportation: *Planning Guidelines for Coordinated Agency Transportation Services*, Washington, D.C., April 1980.

2.18   Ministry of Housing and Physical Planning, National Physical Planning Agency: *Process Monitoring on Behalf of Physical Planning*, The Hague, Netherlands, circa 1979.

2.19   Hampshire County Council, County Planning Department: *Strategic Monitoring Report 1979*, Winchester, England, Nov. 1979.

2.20   U.S. Department of Transportation, Urban Mass Transportation Administration: *Transportation Energy Contingency Planning: Local Experiences*, Washington, D.C., June 1979.

2.21   U.S. Department of Housing and Urban Development: *Tomorrow's Transportation*, U.S. Government Printing Office, Washington, D.C., 1968.

2.22   Westinghouse Air Brake Company: *Study of Evolutionary Urban Transportation*, vol. 2, Federal Clearinghouse, Springfield, Va., Feb. 1968.

2.23   Wohl, M., and B. V. Martin: *Traffic System Analysis for Engineers and Planners*, McGraw-Hill, New York, 1967.

2.24   Boyce, D. R., N. D. Day, and C. McDonald: *Metropolitan Plan Making*, Monograph Series, no. 6, Regional Science Research Institute, Philadelphia, 1970.

2.25   Calkins, H. W.: "An Information System and Monitoring Framework for Plan Implementation and the Continuing Planning Process," Ph.D. dissertation, University of Washington, Seattle, 1972.

2.26   Harris, B.: "Introduction: New Tools for Planning," *Journal of the American Institute of Planners*, vol. 31, May 1965.

2.27   Schofer, J. L., and P. R. Stopher: "Specifications for a New Long-Range Planning Process," *Transportation*, vol. 8, 1979.

EXERCISES

2.1 Pick a transportation problem of interest to you and briefly describe how a plan could be developed to solve it, based on the process presented in this chapter.

2.2 Discuss in two to three paragraphs how you might rearrange the transportation planning process described in the chapter. For example, would you add more stages or put one before another? Justify your changes briefly.

2.3 What is the difference between the planning process in this chapter and the design process?

2.4 Find out about and analyze the planning process used for a particular transportation project in your community. Describe each step and show how the overall process was similar to or differed from that shown here. What are the major advantages/disadvantages of the two approaches?

2.5 Define "contingency plan" and give an example.

3 Transportation Problems

Almost all attempts to improve a given situation start with the recognition and statement of a problem. Where and when the problem first emerges is often difficult to tell; many are uncovered through the press, radio, television, and other mass media; still others are vocalized by citizen groups reporting to elected officials or in discussions between the elected officials themselves. Whatever the manner in which the problems are identified, it is clear that certain ones come to be held as more important or of broader scope than others and thus are given consideration at higher levels of government and industry. Lacking information about many problems specific to one community or state, we have chosen to highlight in this chapter those that are deemed to have broader import. Nevertheless, local difficulties are far from unimportant, and it is hoped that procedures similar to that evolved here can be utilized to identify local problems.

This chapter is divided into three parts. In the first, an attempt is made to define and clarify what is meant by a "problem" and its "domain." This is not an easy task by any means, with one of the primary difficulties being the complex relationship between a problem and each of its counterparts. As a consequence, in the second part of the chapter there will be a brief discussion about problem identification, problem hierarchy, and problem breadth. In the final part, a set of specific problems generally associated with transportation will be presented.

3.1 THE NATURE OF A PROBLEM

Perhaps the easiest way to avoid some of the complexity inherent in any discussion of problems is to begin with a definition. For purposes of this book, the following appears reasonable:

A problem for an individual or group of individuals is the difference between the desired state for a given situation and the actual state. This difference usually cannot be eliminated immediately (if ever).

This definition brings out certain fundamental features. First, and fairly obviously, a problem relates to a given individual or group and not necessarily to the whole population. This disparity in itself often is a source of misunderstanding since those people not directly affected may find it difficult to comprehend the nature and/ or magnitude of the problem.

Second, a problem arises out of an ambition to achieve the most desirable state for a given situation. However, the identification of what is "most desirable" is in itself a problem and seems to depend on the nature of goals, either conscious or unconscious, that people feel are significant. This connection of problems to goals thus implies that they are really two sides of the same coin. They are so completely related that it is difficult to say which is the more likely definition: the problem is to meet the particular goal or the goal is to solve the particular problem. Either statement might serve well under given circumstances.

The fact that the difference between the most desirable state (goal) and actual state cannot be eliminated immediately (if ever) leads the individuals or groups involved into the characteristic psychological conditions of anger, anxiety, yearning, or even agony usually associated with a problem. In fact, if one or more of these conditions do not occur, we, as problem solvers, probably would tend to doubt the existence of a "real" problem. If it were otherwise, the situation could be readily improved to the most desirable state and the anxiety associated with the problem would not have a chance to build up.

One other feature of problems that should be recognized is that when the actual state is improved to the point where it is equivalent to the most desirable state, aspirations may subsequently rise, again creating a gap between actual and desired states. This type of occurrence is common in areas of the world where there is a "wave of rising expectations" founded upon a steadily improving economy and a relatively stable political scene. As Thomas and Schofer [3.17, p. 65] remark:

The success or failure of a problem must not be judged solely upon the basis of the satisfaction or dissatisfaction of the individuals and groups affected by it. Very successful programs may result in the elevation of levels of aspiration so that dissatisfaction still remains.
. . . even if urban areas were instantly rebuilt today so that every element within them were evaluated as performing perfectly, then, after the passage of time, aspiration levels would change so that performance of certain elements would again be deemed unsatisfactory.

This dilemma is a discouraging one because it implies that most problems can never be solved "once and for all time." However, we are left with the consolation that a problem situation at least can be improved beyond its present state.

3.2 COMMENTS ON PROBLEMS

The first comment that can be made about problem statements is that they oftentimes are a reflection of cause and effect. As an illustration, consider the following quote taken from *Railway Track and Structures* [3.18, p. 26]:

Foremost among these problems is the lack of uniformity in the grade of ties, even when cut from stands of timber in the same localities, resulting in non-uniform shrinkage and uneven depth.

In reading this, one cannot help but think that the problem of "lack of uniformity in the grade of timber ties" really is not a very important one unless this factor *has an effect on* some other, more notable conditions such as ride comfort or safety. With this example, it is easy to recognize that many problems are stated in such a way that the actual difficulty, related to the real goal or desired state, is not mentioned.

The lack of a clear declaration of the desired state to be reached creates both advantageous and disadvantageous situations. It is advantageous in that it may lead to the discovery of a factor that previously had not been recognized as being relevant to the particular situation. A good illustration would be the previous statement on nonuniformity of timber railroad ties. Until this factor had been specified as a problem, at least tentatively, no attempt would have been made to study it, and comfort and safety might have remained below par. In fact, by using this type of problem statement as a stimulant to research, the analyst is likely to locate several causative factors which then can be used to help uncover possible solutions.

On the other side of the ledger, problem statements not related to ultimate goals can be disadvantageous because they tend to mislead the investigator. It may turn out, for example, that nonuniformity of timber railroad ties had no effect whatsoever on ride comfort and safety so that any efforts expended to create uniform shrinkage and even depths would be all for naught. Perhaps it is because past occurrences of this sort of misdirection have been numerous that the expression that "the proper identification of the problem leads one 90 percent of the way toward its solution" is so often quoted.

3.2.1 Problem Breadth

Another unfavorable aspect of problem identification made without reference to the ultimately desirable state is that the identification is apt to be concentrated on only a small set of factors and not the total set that adequately describes the actual problem situation. In other words, there is likely to be a lack of breadth in the problem statement. Krick, in his introductory book for engineers, warns that [3.19]:

> The more a total problem is subdivided into independent subproblems, the less effective the total solution is likely to be.

He then states that:

> Some of the most significant engineering contributions come about as a result of a much broader treatment of problems that were previously treated in a piecemeal fashion.

To illustrate such a situation he cites a transportation-related example in which City X, like many urban communities, is plagued by a severe parking problem in its downtown area. The proportion of its commercial district allotted to parking has reached 40 percent, which has prompted city officials to hire a consulting engineer to design a 600-car, multistory parking facility, which is to be one of a series of similar facilities to be constructed in the congested area.

At this point the engineer notes that not only has a limited problem (parking) been given but the solution (car park) as well. The "real" problem is that of transferring a large segment of the population between its place of residence and business. This broad formulation opens up a whole realm of promising solutions, one of which might be a large-scale, high-speed transit system. Of course, nothing in the formulation precludes the possibility of a radically different type of urban community that reduces the need for transportation.

From this example it becomes clear that many problem solvers have not attacked the most significant problem but instead have narrowed their analysis to a smaller portion of the situation. The result, all too often, is that the solution to the more parochial problem is in direct opposition to the solution of the large one and actually hinders further attempts at solving it.

This is not to deny the importance and difficulty of subdividing large problems into smaller, more manageable ones. This task, to be discussed later, is necessary since many real world situations are far too complex to be handled *in toto*. What is of concern here is, however, that the problem not be stated so narrowly at the beginning that its full scope and impact remains hidden.

Many transportation problems have extremely broad implications, perhaps broader than many have imagined. This is the primary reason why we have attempted to establish a wide base for modeling and estimation in the chapter on the transportation planning process, and it is also the reason why the *domain* of a problem must be considered. The previous example of City X's parking situation provides a good illustration of this point. Even if the "correct" problem of "transferring a large segment of the population between its place of residence and business" is recognized, a solution may bring with it certain problems that presently do not exist. In City X's case, for instance, if the city council does decide on mass transit instead of the car park, they may eventually face a financial problem since most mass transit systems have had fiscal difficulties in the past.

Yet there is also a plus side: air pollution, noise, and automobile congestion in the CBD (central business district) may be relieved to some extent. Whatever the final result, it is important for the council to acknowledge some consequences that were not of particular interest when the "problem" was initially stated. Those factors of indirect concern that either affect or are affected by the transportation system comprise the transportation *problem domain*. The second type are also known as "concomitant outputs" and the first as "externalities" in various academic disciplines. These factors in the problem domain generally have to be given as much or more attention as any of those of direct concern. Oftentimes it is even possible to think of *using* transportation to help achieve a desirable state for the problem domain.[1] In fact, if travel is viewed, as it is by many economists, as a derived demand,[2] that is, a demand that is brought about by attempts to achieve something else (e.g., a job or an education or better health through a visit to the doctor), then it follows that *all* consequences of a transportation system are concomitant and thus part of the transportation problem domain. Even if this concept of travel were not felt relevant, we still should be aware that factors (and associated problems) in the domain are of extreme importance in a transportation analysis and should, in the interests of breadth of approach, be given full consideration.

3.2.2   Identification and Classification of Problems

Even if the investigator were desirous of giving problems in the problem domain full consideration, the difficulty to be faced would be that of identifying and classifying all of the problems of concern. Without any stretch of the imagination, it is possible to make a list of at least 50 problems of importance: air pollution, noise, odor, safety, travel time, capacity, and so forth. With greater thought we may even be able to expand the list into the hundreds. The difficulties here are (1) we never know

---

[1] A good example of such a situation would be the busing of school children to achieve racial balance in schools.

[2] For a discussion of derived demand for travel, see W. Oi and P. W. Shuldiner, *An Analysis of Urban Travel Demand*, Northwestern University Press, Evanston, Ill., 1962.

whether we have a complete list and (2) we never know whether we have listed a problem more than once (through an inadvertent rephrasing). In the words of the mathematician, we do not know whether or not we have achieved an exhaustive and mutually exclusive set of problems.

How to obtain such a set is not at all clear at this time. The approach used in this book has been to divide all related problems into three classes: (1) those that are direct transportation service problems—that is, that affect mainly the user, (2) those in the problem domain *affected* by transportation, and (3) those in the problem domain *affecting* transportation. The hope is that these three together will form an exhaustive set and separately will form mutually exclusive sets. However, we are sure that many overlaps and additions eventually will be found, thereby requiring the creation of a new and better classification system.[3] More will be said about this awkward situation while discussing goals in Chap. 4, but for now the pinpointing of identification difficulties is sufficient.

3.2.3 Problem Hierarchy

Required now is a subdivision of the above three classes into more tangible and workable subcategories. We could, for example, trace down a "tree" as in Fig. 3.1 as follows:

The sum total of all transportation problems is divided into those that are direct service problems, and those that affect, or, are affected by transportation. The service problems are further subdivided into the more specific categories of "congestion," "inadequate capacity," "lack of safety for the user," and so forth. The third of these categories is broken down into "too many accidents involving other motor vehicles," "too many accidents involving fixed objects," and so forth. And so the process continues.

Near the last stage, most people would feel they had reduced the problem to a level at which they could really cope with it. They could, for instance, try to predict the effect a new expressway would have on the number and location of accidents involving fixed objects, and this type of analysis would be sufficient for most investigations. Any studies made on higher level problems (e.g., lack of safety for user) may not be detailed enough, whereas any on lower levels (e.g., accidents involving telephone poles) may be too time- and effort-consuming to be of any value.

The points to be made here are that, first, problems come in a hierarchy, in other words, at different levels of generality. They range all the way from "inadequate transportation" to "a recurrent pothole at such and such a street at such and such a time, etc." Second, various problems are "solved" at different levels in the hierarchy. If work is being done on an overall regional transportation problem, it may not be possible to be concerned with anything more specific than "poor accessibility."

On the other hand, if a study involves a particular route location problem, it may well be concerned with the types of accidents that occur (e.g., involving pedestrians, fixed objects, or motor vehicles) as well as ugliness and unhealthy conditions that may be fostered. The analysis of the problem, and the corresponding solution, may turn out to be very specific in this situation. Whatever the situation, the concept should be clear: each and every problem falls somewhere in a hierarchy and usually is solved at that level.

---

[3] A classification according to (1) man and groups (of men), (2) the man-built environment, (3) the natural environment, and (4) activities of man can be found in Sec. 4.3.

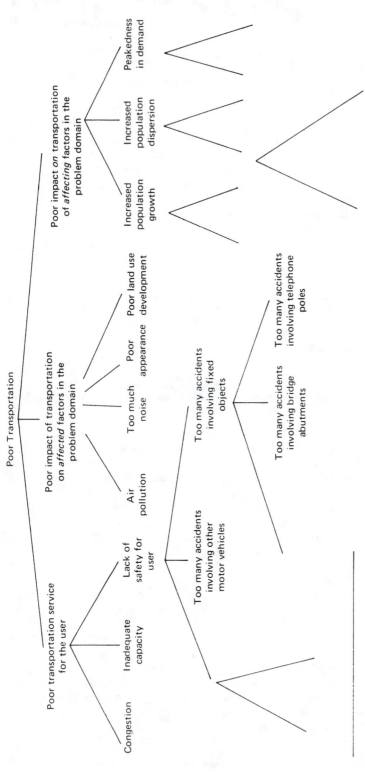

Fig. 3.1 A hierarchical "tree" of problems.

3.2.4 Future Problems

The final remark seems a bit obvious, but actually has deceptively complex connotations: the problems we should be solving are not necessarily those that exist right now but those that will exist at the time the solution will take effect (and thereafter). The lag time between the formulation of a solution to many transportation problems and the implementation of that solution in the field may (and often does) exceed 10 years (especially when large-scale construction is involved). The planner is forced to solve the problem that will exist *at that time*, not necessarily now.

As a practical illustration of this idea, consider the construction of a freeway. If planners were to plan the freeway for today's traffic (and the problems that it creates), they might risk the possibility of the facility being overloaded (congested) on the day of its opening. The main cause of this dilemma would be the buildup of traffic over the time period of construction, to say nothing of the additional problem created by the traffic "induced" by the new facility itself.

Working with future problems, although theoretically correct, does present at least one difficulty with which few if any researchers or practitioners have been able to come to grips: the importance attached to a given problem may change in the future. If we were to look 20 years in the past, for example, we probably would find that very little attention was directed toward such problems as air pollution, noise, lack of public transport needed for the poor to get to employment opportunities, and so forth. Even if a farsighted plan from the past did happen to take these problems into account, they probably would not have been given a very high priority. Reinterpreted, these statements indicate that in the past the difference between the desired and actual states of certain factors was very small. Now, because of probable adverse changes in the levels of these factors and also to increased public awareness of their existence and importance, the "problems" have become more severe, that is to say that difference between the desired and actual states has increased.

The question of how to predict changes over time in the perceived severity of problems actually has received little attention to date. As a consequence, this all-important consideration is left almost entirely to the subjective judgment of both the person or persons attempting to solve the problem (whatever it is at a given time) and those attempting to make decisions among alternatives proposed as solutions to the problem. The dilemma of the constantly varying importance of problems is, to our way of thinking, a frustrating one and adds even more significance to the statement that "the proper identification of the problem takes one 90 percent of the way toward its solution."

3.3 TRANSPORTATION PROBLEMS

As has been suggested several times previously in this chapter, the scope of transportation problems is extremely broad, so broad in fact that it often is difficult to identify all of the problems directly and indirectly influenced by transportation. In this section we will attempt to list and document a few of these problems and hope that the readers will be able to fill in the remainder from their own experience or particular situation.

The types of problems to be discussed are divided into three general classes:

1. Those that are direct transportation service problems
2. Those in the problem domain *affected by* (or impacted by) transportation
3. Those in the problem domain that *affect* transportation

Quotes and statistics taken from a variety of sources will help to illustrate each type of problem. The reader should be aware that many of the quotes are somewhat exaggerated for purposes of emphasis and may show only one side of an issue or problem. Some may even contain incorrect or misleading information. Still others may be stated in a highly emotional way to appeal to the consciences of the problem-solver and decision-maker. What is of importance here, however, is that the planner recognize that these are the conditions in which real problems often show themselves, so that it becomes necessary to cut away the unneeded verbiage and misleading information to get to the heart of the situation. A similar approach should be used in reading through the sections to follow.

3.3.1 Transportation Service Problems

Those problems that have received the most attention are those that affect the *user* of the transportation system. Congestion, delay, the high cost of travel, safety, and lack of privacy (for mass transit) are but a few of the many difficulties that plague the traveler. These and similar problems will be discussed in this section.

3.3.1.1 Congestion   Perhaps the first problem that comes to mind in regard to transportation is congestion, which generally can be equated to long travel times and to delays in movement.

Buchanan effectively summarizes the problem of congestion in his much publicized report on *Traffic in Towns* [3.4, p. 22]. In it, he states that:

> The problems of movement are equally familiar so that there is little need to emphasize the frustrations and irritations of traffic jams, the waste of fuel, the waste of time, and the vast and essentially unproductive effort by police, wardens, and others engaged in many capacities in regulating traffic. A motor vehicle, even in its heaviest and clumsiest form, is capable of moving at a mile a minute, yet the average speed of traffic in large cities is about 11 mph.

Just how bad is the congestion problem in measurable terms? Data collected by Vuchic for various transit systems in the 1970s, presented in Table 3.1, show that speeds in several large cities were extremely low, particularly for regular buses on CBD streets. Even for those buses on reserved lanes, average speeds rarely exceeded 7 mph. Express bus operations resulted in faster speeds, but in the case of Cleveland's Clifton Boulevard only to 11.2 mph. Rail "rapid" transit, on its own right-of-way, had a system-wide average speed of 26.7 mph in Chicago, but less than 19 mph in Philadelphia and Toronto.

Table 3.2 shows motor vehicle speeds in four different years in the Phoenix metropolitan area. These have decreased somewhat over the years in almost all cases, and so "congestion" probably appears to be an increasing problem to motorists in that area.

3.3.1.2 Inadequate capacity   A second problem, and one that is of great significance in regard to congestion, is that of capacity: there must be sufficient facilities available to accommodate the demand for travel when and where it occurs. Examples of problems evolving when capacity becomes a scarce commodity are prevalent in most urban areas. In fact, lack of capacity generally means delays in travel and thus is one aspect of congestion as the traveler sees it.

The capacity problem is one that seems to be getting worse in many places as time progresses. Table 3.3 shows the ratio of peak hour volumes to design capacities on several of Chicago's main north-south arteries. Over a 20-year period the ratios associated with five of the eight arteries increased. The situation on Cicero Avenue is especially difficult since the volume grew so rapidly in only 2 years that it eventually

Table 3.1 Operating Speeds on Selected Transit Systems

Mode and city	Facility	Operating speed (mph)
Regular bus		
New York	Hillside Ave.	5.6
San Francisco	Market St.	5.0
Cleveland	Euclid Ave.	6.2
Chicago	Washington Blvd.	6.2
Atlanta	Peachtree St.	5.6
Express bus		
St. Louis	Gravois St.	19.9
Cleveland	Clifton Blvd.	11.2
Streetcar		
San Francisco	Market St.	8.1
Streetcar		
Belgrade	Bul. Revolucije	9.3
Private row	Stadtbahn Linie A	14.3
Light rail		
Koln	At Neumarkt	15.5
Rapid rail		
Chicago	Whole system	26.7
Philadelphia	Whole system	18.6
Toronto	Whole system	18.4

Source: [3.5, pp. 449, 572, 573, adapted by permission of Prentice-Hall, Inc., Englewood Cliffs, N.J.].

Table 3.2 Average Peak-Period (Peak Direction) Speed by Jurisdiction in Maricopa County, Arizona

Jurisdiction	Speed (mph)				Change in speed (1966–1979)
	1966	1970	1976	1979	
Glendale					
Major streets	31.0	33.6	33.*	32.0	1.0
Mesa					
Major streets	38.9	39.7	35.*	33.3	−5.6
Freeway	NA†	NA	NA	49.3	NA
Phoenix					
Major streets	28.4	29.0	29.*	29.4	1.0
Freeway	58.4	58.0	49.2	48.8	−9.6
Scottsdale					
Major streets	32.1	32.1	35.*	32.2	0.1
Tempe					
Major streets	33.8	33.1	26.*	27.4	−6.7
Freeway		66.9	54.3	53.9	NA
Maricopa County					
Major streets	43.5	44.5	43.*	41.2	−2.3
Total					
Major streets	31.5	32.2	32.0	31.2	−0.3
Freeway	58.4	58.5	49.9	49.8	−8.6

*Estimated for all streets surveyed in 1979 from a more limited number of streets surveyed in 1976.
†NA: Not applicable.
Source: [3.2, p. 43].

**Table 3.3 Relationship of Peak-Hour Traffic Volume
to Design Capacity on North–South Arterials at Congress
Street Screenline: Chicago, 1959, 1961, and 1978/1979**

Arterial	Peak-hour design capacity		Peak-hour volume (5:00 to 6:00 P.M.)			Relationship of peak-hour volume to design capacity		
	1959–1961	1978–1979	1959	1961	1978/1979	1959	1961	1978/1979
Cicero	2,130	2,130	1,730	2,290	2,133	0.81	1.08	1.00
Laramie	1,640	1,640	1,790	1,540	1,744	1.09	0.94	1.08
Central	1,640	1,640	900	1,260	1,596	0.55	0.77	0.97
Austin	1,400	2,130	1,100	1,650	1,701	0.78	1.18	0.80
Ridgeland	1,550	1,550	1,200	1,200	1,246	0.77	0.77	0.80
Oak Park	1,110	1,550	1,080	930	1,139	0.97	0.84	0.73
Harlem	1,550	2,300	1,500	1,930	2,418	0.97	1.24	1.05
Des Plaines	1,000	1,800	800	900	1,201	0.80	0.90	0.67
Total	12,020	14,740	10,100	11,700	8,208	0.84	0.97	0.90

Source: [3.6, p. 79] for 1959 and 1961. The volume data for 1978/1979 as well as information to compute capacities were supplied by the Chicago Area Transportation Study.

exceeded the *predicted* capacity of the street. One can easily imagine the tieups and resulting frustrations that would result from such a situation.

The relative capacity of various modes has been a subject of much interest, and many critics of planning procedures in urban areas have jumped on the highway building program because highways just do not seem to be able to carry the great loads forced upon them. Higbee, for example, says [3.10, p. 203]:

> These are some of the rubber-tired facts of modern life: A single lane of surface street, subject to cross traffic in front and marginal friction along the sides, will allow about 1,600 persons in private automobiles to pass a given point in one hour. Trains running on a single track, whether above or below ground, can carry from 40,000 to 60,000 persons per hour depending on whether they run as locals or expresses.
>   Translated into space requirements these figures are staggering. A single railroad track, used for trains that stop briefly every half to three quarters of a mile, is worth 25 lanes of ordinary street. A single railroad track, used for trains that stop briefly every one to three miles, is worth 23 lanes of turnpike or elevated express highway.

Mumford comments on the relationship between pedestrian and vehicular capacity [3.9, p. 253]:

> As for the pedestrian, one could move a hundred thousand people, by the existing streets, from, say, downtown Boston to the Common (a nearby park and historical place), in something like half an hour, and find plenty of room for them to stand. But how many weary hours would it take to move them in cars over these same streets.

3.3.1.3 High user cost Another problem, and one for which little relief is in sight, is that of the high cost of transportation to the user. Many measurements of operating costs for automobiles have been made. The Federal Highway Administration reports these figures periodically, as can be seen in Table 3.4. All expenses have gone up, except for motor oil. All of us are aware that gasoline prices have risen rapidly— about 255% from 1970 to 1980. But we may not be aware that insurance has risen by 95%. It is interesting to note at this point that as speed increases, the total cost per

Table 3.4 10-Year Cost of Operating an Automobile, by Size of Car: 1979

Item	Total	Costs, excluding taxes	Depreciation	Repairs, maintenance	Tires	Gasoline	Insurance	Parking	Taxes and fees
				Total cost ($1,000)					
Standard	24.6	23.0	6.3	4.8	0.6	5.4	2.5	3.2	1.6
Compact	21.7	20.4	5.2	4.2	0.5	4.8	2.3	3.2	1.3
Subcompact	18.5	17.3	3.8	3.4	0.5	4.0	2.2	3.2	1.1
Passenger van	36.2	34.1	10.2	5.3	0.6	7.2	7.2	3.2	2.1
				Cents per mile cost					
Standard	24.58	22.98	6.26	4.80	0.58	5.44	2.54	3.20	1.60
Compact	21.72	20.39	5.18	4.16	0.52	4.83	2.28	3.17	1.33
Subcompact	18.45	17.93	3.81	3.43	0.51	3.96	2.21	3.17	1.12
Passenger van	36.19	34.09	10.21	5.28	0.60	7.25	7.17	3.20	2.10

Source: [3.35].

vehicle mile decreases, at least up to about 30 mph (3.14). After that, costs increase. This change at 30 mph brings to light an interesting tradeoff between travel time and operating cost. Apparently people are willing to pay the extra cost over and above that at 30 mph to travel on freeways at speeds up to 70 mph.

3.3.1.4 High facility cost and low rate of return Along with user operation cost, there must also be a concern for the cost of the service provided to the user. We must not forget that most transportation services, whether they come from the public or private sector, are businesses and must be operated as such. If the service is a public one, there must be some assurance that it is not too great a burden on already over-taxed sources of revenue; if a private one, it should return a profit to its owners, stock-holders, and other investors competitive with other investment potentials.

The transit industry is one of the most provocative examples of why there should be a concern for returns on investment. Mumford puts his finger on the problem when he discusses transit in his book on *The Highway and the City* [3.9, p. 255]:

> In order to maintain profits, or in many cases to reduce deficits, rates have been raised, ser-vices have decreased, and equipment has become obsolete, without being replaced and im-proved. Yet mass transportation, with far less acreage in roadbeds and rights of way, can deliver at least ten times more people per hour than the private motorcar. This means that if such means were allowed to lapse in our metropolitan centers—as the inter-urban electric trolley system, that complete and efficient network, was allowed to disappear in the nineteen-twenties—we should require probably five to ten times the existing number of arterial high-ways to bring the present number of commuters into the city, and at least ten times the existing parking space to accommodate them. In that tangled mass of highways, interchanges, and parking lots, the city would be nowhere: a mechanized nonentity ground under an endless procession of wheels.

Until about 1970, transit was suffering serious declines, and present financial conditions are much worse than before. Table 3.5 shows several sets of nationwide data on transit for the years 1940 and 1978. One fact that stands out strongly is that total operating revenue (for all modes) increased 223 percent in the 38-year period, but the total payroll went up 661 percent in the same time. Thus, while labor is not the only cost associated with transit, it is a large part ($2.7 billion out of $4.8 billion in 1978), and it is obvious that no business can continue to operate when its costs continue to outstrip its revenue. The future existence of transit now seems to hinge on either an increase in revenues or decrease in costs to bring about a better return on the expenditures.

While the return on investment for transit seems to be endangered at this time, it is not necessarily true that all highway construction endeavors are worthwhile from an investment standpoint. The cost side of the ledger has been particularly disturbing in many instances. The cases presented in Table 3.6 are representative examples of large-scale highway construction expenditures. The Westway Highway stands out at the most expensive facility, and it is difficult to imagine how any highway could cost $229 million a mile. Similarly, it is difficult to imagine how sufficient traffic can ever be generated, even in the middle of New York, to develop "revenues" in the form of reduced user costs to justify the facility economically. Whatever the economic feasi-bility of this route, the important point is that this factor must be given strong con-sideration; otherwise, the stability of the whole system might be undermined.

3.3.1.5 Lack of safety for user One problem that often has a great deal of per-sonal agony associated with it is that of safety. In fact, accidents, especially those on the highway, are so common that it is a rare person who has not had a relative or friend affected by an unfortunate occurrence of this type. Also unfortunate is the fact that few of us become concerned about safety until a situation strikes close to home.

Table 3.5 Trends in Urban Public Transit Characteristics

Public transit characteristics (excluding commuter rails)	Subway and elevated rail	Light rail	Trolley bus	Motor bus	Total
Revenue passengers (in millions)*					
1940	2,282	4,182	419	3,620	10,504
1978	1,415	80	51	4,406	5,963
Change	−867	−4,102	−368	786	−4,541
Percent change	−38	−98	−88	22	−43
Operating revenue (in millions of dollars)					
1940	128.3	327.8	25.0	255.9	737.0
1978	664.9	27.4	−14.6	1,617.4	2,381.1
Change	536.6	−300.4	−10.4	1,415.5	1,644.1
Percent change	518	−92	−42	553	223
Vehicle · miles operated (in millions)					
1940	470.8	844.7	86.0	1,194.5	2,596.0
1978	363.5	19.5	13.3	1,630.5	2,028.3
Change	−107.3	−825.2	−72.7	436	−567.7
Percent change	−23	−98	−85	37	−22
Number of employees					
1940	—	—	—	35,000	203,000
1978	—	—	—	52,866	165,400
Change	—	—	—	17,866	−37,600
Percent change	—	—	—	51	−19
Payroll (in millions of dollars)					
1940	—	—	—	—	360.0
1978	—	—	—	—	2,740.6
Change	—	—	—	—	2,380.6
Percent change	—	—	—	—	661
Vehicles owned					
1940	11,032	26,630	2,802	—	75,464
1978	9,567	944	593	—	64,013
Change	−1,465	−25,686	−2,209	—	−11,451
Percent change	−13	−96	−79	—	−15

*Data in 1940 is revenue passenger rides. That for 1978 is linked transit passenger trips.
Source: [3.1, various pages].

Table 3.6 Costs of a Sample of Urban Interstate Highways

Facility	Total cost (in $millions)	Miles	Cost per mile
Complete or partially complete (costs in various years)			
I-66 Arlington, Va.	233	10.1	23
I-635 Kansas City, Kans.	25	16.5	2
I-270 St. Louis	153	35.6	4
I-695 Baltimore	154	30.8	5
In design stage (costs in 1979 $millions)			
I-478 Westway, New York City	1400	6.1	229
I-90 Seattle	904	7.5	120
I-105 Century, Los Angeles	1600	17.2	93
I-696 Vine St., Philadelphia	168	2.5	67
I-670 Columbus, Ohio	199	10.5	19
I-595 Ft. Lauderdale, Fla.	457	13.4	34

Source: Mr. Donald Martilla, Federal Highway Administration.

The statistics show that automobile injuries accounted for more than 4 million injuries in 1979 and, more important, for more than 51,000 fatalities. The number of fatalities has increased from more than 38,100 in 1960. Some interesting facts presented by Joan Claybrook, former Administrator of the National Highway Traffic Safety Administration, add some depth to these figures [3.15]:

> ... Auto crashes are the largest killer of people under age 35 in the United States, the largest cause of paraplegia and a major cause of epilepsy. On the average, a highway fatality occurs every 10 minutes and an injury every eight seconds. As many Americans are killed on the highway each year as were killed in the entire Vietnam War (about 52,000) .... In dollars, the cost to the nation has reached $50 billion annually in medical and rehabilitation costs, lost wages, welfare and property damage, not to speak of incalculable personal loss. In terms of public health costs, auto crashes are second only to cancer.

Table 3.7 gives a breakdown of yearly accident figures by various categories. A particularly discouraging aspect of these figures is that the number of accidents in several of the classes had been decreasing for a period of time, but now seems to be holding steady. This change may be attributed to a growing population and a related increase in automobile usage, but the problem is that, despite a general reduction in the number of deaths per 100 million vehicle miles of travel (see [3.23]), the number of deaths per 100,000 population had until recently been increasing. The implication of this differentiation is that each person is becoming more exposed to death and injury on the highway and thus views the problem as a worsening condition.

An intermodal comparison of accident data also brings out some interesting features. The accident rate per 1000 million passenger · miles for auto and taxi travel exceeds by far that for any of the other modes. The rate for motor vehicle travel is roughly eight times that for bus travel, yet both types of vehicles use the same facilities! This considerable difference under similar circumstances opens the way for some interesting speculation as to the causes of highway accidents.

3.3.1.6 Requirement of user operation The problems of congestion, cost, and safety are explicit, confronting us almost every time we travel. Many other problems are not as obvious, but still may have an equivalent or greater amount of importance.

Table 3.7 Motor Vehicle Deaths by Type of Accident and Number of Nonfatal Injuries, 1938–1979

Year	Total deaths*	Pedestrians	Other motor vehicles	Railroad trains	Bicycles	Fixed objects	Other collisions	Deaths from noncollision accidents	Approximate totals of nonfatal injuries†
1938–1942 (ave.)	33,550	12,440	9,500	1,620	750	1,050	350	7,840	1,180,000
1945	28,076	11,000	7,150	1,703	500	800	293	6,600	1,000,000
1950	34,763	9,100	11,650	1,541	440	1,300	209	10,600	1,200,000
1955	38,426	8,200	14,500	1,490	410	1,600	105	12,100	1,350,000
1960	38,137	7,850	14,800	1,368	460	1,700	85	11,900	1,400,000
1965	49,000	8,800	20,700	1,600	680	2,200	120	14,900	1,800,000
1970	54,800	10,400	23,300	1,530	820	4,450	100	14,200	2,000,000
1975	46,800	8,600	20,500	1,000	1,000	3,100	100	11,900	1,800,000
1979	51,900	9,400	22,700	900	1,000	3,500	100	14,300	2,000,000

*Yearly totals do not quite equal sums of the various types because totals for most types are estimated, and these have been made only to the nearest 10 deaths for some types and to the nearest 50 deaths for others.

†Estimates of injuries that were disabling beyond the day of accident.

Source: [3.23, various years]. Note: Deaths are based on data from the national Vital Statistics Division, state traffic authorities, and Interstate Commerce Commission.

Primary among these latter types is that of requirements for operation that several present-day transportation systems (in particular, the auto-highway system) put on the user. If a transportation system is created in which the user has to play an active, physical role in his own travel, then it stands to reason that many people physically disabled for one reason or another will not be capable of using such a system. Moreover, the unfortunate ones generally are those already disadvantaged in some way—the young, handicapped, and so forth. As a Department of Housing and Urban Development (HUD) report brings out [3.12, p. 17]:

> The handicapped, the elderly, and the young . . . suffer from a transportation system that makes the individually owned automobile almost a necessity unless they are able to pay for someone to drive them. . . . By 1980, over 100 million persons will be under 18 or over 65 years old. The problem of the unserved is not limited to central cities. In suburban areas there is frequently no public transit. For all but two- and three-car families, intrasuburban transportation to shopping and to recreation is almost impossible. Even where there are two cars, someone must always assume the burden of driving for the rest of the family.

3.3.1.7 Lack of privacy Another less obvious problem with many transportation systems (in particular, mass transit systems) is that they do not allow for personal privacy. One could hardly feel happy about being squeezed together with the rest of humankind in a crowded subway, elevated train, or bus. Certainly such an occurrence is degrading to the mass transit traveler and has the perhaps undesired effect of making automobile travel all the more appealing.

Even in a place such as New York City where travel by auto is relatively expensive and time-consuming, a person who drives at least has a chance to think in peace, and, perhaps more importantly, to choose the people with whom to come in close contact. The statement to follow is all but priceless in describing some of the benefits associated with privacy in the automobile [3.24, p. 11]:

> I ask myself why, as an idealogical urbanite, I am in favor of mass transit and yet I rarely use it. That is, when I can avoid it, I do. There are little things that the auto gives you. It delivers a lot of service and chronologically it's important. For one thing, it's a portable closet. You pick up stuff and you throw it in the car and then you can sort of go on to your next errand. It's a marvelous thing that no real mass transit has attempted. The only consolation is the rental locker which is developed in the most densely used parts of the mass transit. One of the nice things about the auto is that it's not transparent—it gives you a lot of shield to pick your nose or scratch your crotch (even more easily than you can pick your nose, unless you have a little window shade in your car). The point is, you quickly move away, you don't stay with the same traffic—it gives you independence or just shielding in a general way. Let's say, you have to pick up something downtown. If you're going on mass transit, maybe you'll have to change your clothes whereas you can just hop in your car and really not be in any kind of social situation while you're there but run in, run out.

The problem of lack of privacy in transportation vehicles, especially in those designated by that impersonal word "mass" transit, does not seem to be improving. Instead, as Table 3.8 seems to indicate for five recent rail rapid transit installations, hourly loads can run to 65,000. One can hardly have much feeling of privacy when crushed in this bulk of humanity.

3.3.1.8 Discomfort A set of factors somewhat complimentary to privacy in creating a general feeling of comfort and physical satisfaction comprises those elements affecting the physical senses of the user: noise, appearance, temperature, humidity, precipitation, air flow, smell, dirt, sway, jerk, vibration, and so forth. The list of these features could be quite extensive, but the main idea is that these are the ones that relate to the user's basic "creature comfort." The automobile ranks high on creature comfort just as it did on privacy. With good heating, air conditioning, and ventilation systems; with stereo tape cartridges and FM radios; and with bucket seats,

Table 3.8 Peak-Hour Passenger Volumes: Selected Rail Rapid-Transit Lines

City	System–line	Trains/hr	Veh/train	Spaces/veh	Utilization*
New York	Queens–53rd St.	32	10	227	65,340
Toronto	Yonge St.	26	8	269	50,400
Chicago	North Side	30	8	157	33,930
Tokyo	Teito	30	10	222	59,940
Munich	S-Bahn	18	7	171	18,450

*In people per one-track line per hour.
Source: [3.5, pp. 574-575, adapted by permission of Prentice-Hall, Inc., Englewood Cliffs, N.J.].

arm rests, and leather interiors, it is easy to see why the automobile is such an attractive means of transportation.

In contrast, many mass-transit vehicles can only be called "shabby." One of the main reasons for this is their age. Taking 1978 as an example year, we find that 3805 new buses were delivered to U.S. operators to integrate with an existing fleet of 52,866. This means that no more than 7% of old buses were replaced that year. At that rate it would take at least 14 years to renew the whole fleet. The figures for other modes are even more distressing [3.1, pp. 35, 37]:

Heavy rail—172 new vehicles for a fleet of 9,567
Light rail—35 new vehicles for a fleet of 944
Trolley coach—0 new vehicles for a fleet of 593

There is thus a definite problem with urban mass-transit systems from the standpoint of comfort, and much needs to be changed in this area of concern.

3.3.2 Problems in the Problem Domain Affected by Transportation

The foregoing discussion of transportation service problems has been far from complete. A more exhaustive presentation might identify many additional problems of immediate and obvious importance both to the user of a system and, subsequently, to those who must plan for the user. Yet, as has been emphasized before, the discussion in this chapter must not be limited to the service problems themselves, but also to the less direct problems that may influence or be influenced by a transportation system. The latter of these two types of problems is explored in this section.

3.3.2.1 Energy consumption Petroleum shortages in 1973-1974 and 1979 brought about an overwhelming recognition that energy is basic to life support and that its future is not at all certain. Some forecasts were made at the time that petroleum supplies might run out by the year 2000 and that, since many other energy sources (nuclear, solar, coal, etc.) were not being tapped for a variety of economic, political, and environmental reasons, the overall energy future looked bleak.

Transportation, of course, is one of the major culprits in energy consumption. As a proportion of total energy consumption, it has remained around 25% for the last 30 years. But, as can be seen in Table 3.9, it accounts for over 50% of petroleum consumption, a comparatively large proportion of which has to be imported. In 1979, for instance, net imports represented 47% of petroleum consumption in the United States. The value of this, as portrayed in Fig. 3.2, is quite staggering, although there is

Table 3.9   U.S. Energy Consumption by Product (Quadrillion Btu)

Item	1950	1960	1970	1975	1979	1980
Total energy*	33.9	44.1	66.8	70.7	78.0	76.2
Transportation	8.7	10.9	16.3	18.2	19.8	18.6
Petroleum	13.5	19.9	29.5	32.7	37.0	34.3
Transportation				17.6	19.2	18.0
Natural gas	6.0	12.4	21.8	20.0	19.9	20.4
Transportation				0.6	0.5	0.6
Per capita (million Btu)	223	245	328	332	354	343
Net imports of petroleum as % of consumption†	13	19	24	38	47	41

*Energy "consumed" by electric utilities is not counted in the total but is in those for transportation, petroleum, and natural gas.
†Calculated from data in [3.16, p. 49].
Source: [3.16] and the Statistical Abstract of the United States (annual).

some indication in Table 3.9 that, mainly through conservation efforts, the rapid rise in net imports finally is reversing. This would mean less dependence on countries whose futures might be unstable, as well as less overall depletion of resources.

3.3.2.2 Air pollution   Since the early 1960s the motor vehicle has been recognized as a significant contributor to the nation's air-pollution burden. In many urban areas transportation facilities (predominantly those associated with the passenger car)

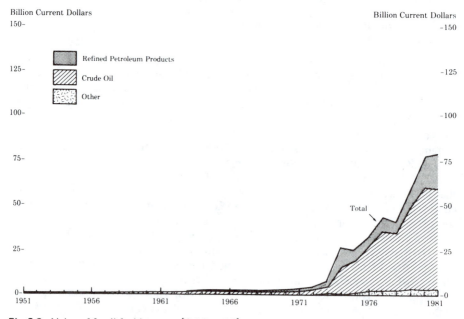

Fig. 3.2   Value of fossil fuel imports. [3.16, p. 24].

are major sources of carbon monoxide, hydrocarbons, and oxides of nitrogen. Also of increasing concern are particulate emissions such as lead and sulfur compounds from these sources.

A 1971 survey of mobile emission sources in urban areas showed that transportation facilities (excluding aircraft) accounted for 55 to 75% of the nitrogen in those areas [3.27]. Further chemical reactions in the atmosphere result in the generation of substantial concentrations of the secondary derivatives of those emissions—photochemical oxidants and "smog." Table 3.10 illustrates the impact of the automobile on the total emissions of four major pollutants.

As can be seen in Table 3.11, various air pollutants such as sulfur oxides and carbon monoxide can cause acute and chronic leaf injury, irritate the eyes and upper respiratory tract, may be cancer-producing, and in some cases impair mental processes. These adverse effects, and others on visibility, corrosion of metals, and destruction of building material, obviously are problems of a serious nature.

3.3.2.3 Crime[4]   An increasingly serious problem related to transportation is that of crime. This can occur as a result of theft or misuse (e.g., vandalism) of transportation facilities or through utilization of transportation in committing a criminal act. To illustrate the former, known motor vehicle thefts in the United States increased from 660,000 in 1967 to 1,097,000 in 1979, or from 334 per 100,000 population to 498. The rate in the latter year for cities over 250,000 people was 1054, compared to 594 for cities from 50,000 to 99,999 in population. These figures include neither one million thefts *from* motor vehicles nor more than 675,000 bicycle larcenies. About 143,000 persons were arrested for motor vehicle thefts in 1979.

It is difficult to determine how many crimes involved use of a motor vehicle (including transit), but of 453,000 robberies reported in 1979, 224,000 were listed as "highway" and 17,000 as "gas station." These, of course, are highly likely to depend on motor transportation as a means of get-away. An interesting and unique side problem is "road rustling." In 1980, Bedford County in Virginia "lost" 75 yards of gravel

[4] Data taken from U.S. Bureau of the Census: *Statistical Abstracts of the United States: 1980*, Washington, D.C., 1980, pp. 179–204.

Table 3.10 Mobile Source Contributions to Nationwide Pollutant Emission Levels in 1977 (in % of Total Emissions)

	Total suspended particulates	Carbon monoxide	Hydrocarbons	Nitrogen oxides
Automobiles	3.83	50.73	23.65	16.03
Light-duty trucks	.74	10.75	5.64	3.85
Heavy-duty trucks	1.62	13.93	4.97	9.07
Motorcycles	0	.64	.74	0
Off-highway vehicles	.79	5.88	1.99	5.57
Rail	.47	.26	.64	3.06
Aircraft	.54	.73	.78	.48
Vessel	.24	1.48	1.59	.63
Total	8.25	84.40	40.00	38.69

Source: National Commission on Air Quality, *To Breathe Clean Air*, Washington, D.C., March 1981, p. 192.

Table 3.11 Effects Attributed to Specific Pollutants

Air pollutant	Effects
Particulates	Speed chemical reactions; obscure vision; corrode metals; cause grime on belongings; aggravate lung illness
Sulfur oxides	Causes acute and chronic leaf injury; attack wide variety of trees; irritate upper respiratory tract; destroy paint pigments; erode statuary; corrode metals; ruin hosiery; harm textiles; disintegrate book pages and leather
Hydrocarbons (in solid and gaseous states)	May be cancer-producing (carcinogenic); retard plant growth; cause abnormal leaf and bud development
Carbon monoxides	Cause headaches, dizziness, nausea; absorbed into blood, reduce oxygen content; impair mental processes
Nitrogen oxides	Cause visible leaf damage; irritate eyes and nose; stunt plant growth even when not causing visible damage; create brown haze; corrode metals
Oxidants: ozone	Discolor upper surface of leaves of many crops, trees, shrubs; damage and fade textiles; reduce athletic performance; hasten cracking of rubber; disturb lung function; irritate eyes, nose, throat; induce coughing
PAN (peroxyacetyl nitrate)	Discolors upper surface; irritates eyes; disturbs lung function

Sources: (1) HEW, National Air Pollution Control Administration, *The Effects of Air Pollution,* no. 1556, revised 1967. (2) NAPCA, *Air Pollution Injury to Vegetation,* No. AP-71, 1970. (3) American Association for the Advancement of Science, *Air Conservation,* pub. no. 80, 1965. (4) National Tuberculosis and Respiratory Disease Association, *Air Pollution Primer,* 1969.

access road. Highway department officials believe it was stolen, since big chunks apparently were dug out with a front-end loader. All that was left was a rough dirt road.[5]

3.3.2.4 Noise A problem that bears similar characteristics to that of air pollution is noise. It has that same ability to pervade the environment close to the transportation system, it is irritating, and it may, in the long run, be harmful to a person's health. Research is still progressing on the noise problem and its impact, but Buchanan has placed a great deal of emphasis on it in his study [3.4, p. 25]:

> In addition to danger and anxiety, the motor vehicle is responsible for a great deal of noise. This has recently been under consideration, along with other aspects of noise, by an official committee set up by the Minister for Science. In their report, the committee concluded that "in London (and no doubt this applies to other large towns as well) traffic is, at the present time, the predominant source of annoyance, and no other single noise is of comparable importance." The committee distinguished five main kinds of noise from vehicles: propulsion noises (from engines, gears, transmissions, and exhausts), horns, brake squeal, door slamming, and loose loads or bodies.
>
> Our own conclusion, based on observation and many discussions, is that traffic noise is steadily developing into a major nuisance, seriously prejudicial to the general enjoyment to towns, destructive of the amenities of dwellings on a wide scale, and interfering in no small degree with efficiency in offices and other business premises. But again, this is something we have mostly grown up with, and we tend to take it very much for granted.

More detailed studies and measurements seem to bear out some of these conclusions. Judged in comparison to some common noise levels, such as conversation at 55–60 decibels or a vacuum cleaner at 85, transportation noises can be significant.

[5] "Road Rustlers? Highway Robbery Suspected in Bedford," *The Roanoke Times and World News,* Aug. 21, 1980.

(Table 3.12). Airplanes, of course, contribute heavily. A four-engine jet plane at take-off produces about 115 decibels even at a distance of 500 ft. This sound pressure level is almost equivalent to that from rock music with amplifiers in a closed room. Motor buses, subway and railroad trains, and heavy trucks all contribute average noise levels of 85 decibels or above at a distance of 20 ft. This means that the passers-by and those living in closely abutting buildings are continually subject to noises equivalent to that of a vacuum cleaner. Obviously, when such noise continues, rest and sleep is difficult, radios and televisions are not heard, and in some instances physical damage to the ear may occur. In any case, annoyance and frustration is built to produce an undesirable situation.

3.3.2.5 Visual intrusion and poor appearance One aspect of the impact of transportation on its environment is the visual aspect. Naturally, nothing creates more arguments than a discussion of what is "good looking" and what is not. Nevertheless, it is important to try to design transportation facilities that have a nice appearance, both as they stand by themselves and in the context of the environment in which they are placed. As Mumford says about highways [3.9, p. 247]:

> In many ways, our highways are not merely masterpieces of engineering, but consummate works of art: a few of them, like the Taconic State Parkway in New York, stand on a par with our highest creations in other fields. Not every highway, it is true, runs through country that offers such superb opportunities to an imaginative highway builder as this does: but then not every engineer rises to his opportunities as the planners of this highway did, routing the well-separated roads along the ridgeways, following the contours, and thus, by this single strategem, both avoiding towns and villages and opening up great views across country, enhanced by a lavish planting of flowering bushes along the borders. If this standard of comeliness and beauty were kept generally in view, highway engineers would not so often lapse into the brutal assaults against the landscape and against urban order that they actually give way to when they aim solely at speed and volume of traffic, and bulldoze and blast their way across country to shorten their route by a few miles without making the total journey any less depressing.

The problem of visual intrusion can take many forms other than just the simple "brutal assaults against the landscape" and the buildings that comprise a community. There is the visual intrusion by the multitude of signs and signals—directional, one

Table 3.12 Average Noise (Sound Pressure) Levels*
of Some Transportation Sources

Source	Decibel level	Source	Decibel level
Heavy trucks	86	Subway trains	90
Motor buses (starting)	85	Old street cars	88
Trolley buses	75	Railroad trains	
		(diesel, steam)	85
Light trucks	74	New PCC cars	75
Automobiles	71	Electric railroad trains	75
20,000-lb thrust 4-		10,000 MP 4-engine	
engine jet airliner		propeller aircraft	
at takeoff (500 ft		at takeoff (500 ft	
away)	115	away)	99

*At 300 cps re 0.0002 microbar. Measurements made 20 ft from source except in the case of steam and diesel trains. Adapted from C. M. Harris (ed.), *Handbook of Noise Control*, McGraw-Hill, New York, 1957, pp. 35-2, 35-3.

way, no parking, and so forth—that often crop up on city streets and give that "scattered debris" look. Then there is the all too familiar intrusion of parked vehicles in the spaces between city buildings. It is unfortunate but true that the recent great increase in the number of automobiles in the United States and elsewhere has brought a corresponding need to store them, and usually the only place where this can be done is in the already crowded confines of city streets and open spaces.

3.3.2.6 Excessive right-of-way and relocation requirements In addition to the problems of poor appearance and visual intrusion brought about by urban transportation facilities, there are also the difficulties associated with the excessive right-of-way and relocation requirements of most facilities. These two problems arise because of the need for a commodity especially valuable in urban areas: land. First, rights-of-way are required for the facility itself, and acquisition is expensive, both in initial cost and from the standpoint of productive land taken from the tax rolls. The second part of the problem is that people usually are living on the desired land, and relocation often is a bitter and trying experience, both to those who must move and leave what may have been fairly desirable and inexpensive quarters and also to those who must face the thankless task of reestablishing those displaced. These difficulties are demonstrated all too clearly by the often heard complaint, particularly by some area residents, that "highway building is Negro removal." Mumford is especially critical of the work of the highway engineer in urban areas [3.9, p. 246]:

> Unfortunately, highway engineers, if one is to judge by their usual performance, lack both historic insight and social memory: accordingly, they have been repeating, with the audacity of confident ignorance, all the mistakes in urban planning commited by their predecessors who designed our railroads. The wide swaths of land devoted to cloverleaves, and even more complicated multi-level interchanges, to expressways, parking lots, and parking garages, in the very heart of the city, butcher up precious urban space in exactly the same way that freight yards and marshalling yards did when the railroads dumped their passengers and freight inside the city.

He continues by giving some pertinent examples:

> Like the railroad, again, the motorway has repeatedly taken possession of the most valuable recreation space the city possesses, not merely by thieving land once dedicated to park uses, but by cutting off easy access to the waterfront parks, and lowering their value for refreshment and repose by introducing the roar of traffic and the bad odor of exhausts, though both noise and carbon monoxide are inimical to health. Witness the shocking spoilage of the Charles River basin parks in Boston, the arterial blocking off of the Lake Front in Chicago (after the removal of the original usurpers, the railroads), the barbarous sacrifice of large areas of Fairmount Park in Philadelphia, the partial defacement of the San Francisco waterfront, even in Paris the ruin of the Left Bank of the Seine.

Quite obviously, it is difficult to take almost any land in urban areas since some people are bound to be affected adversely, either directly because they must move, or indirectly because a favorite or closest social, recreational, or commercial spot has been eliminated. The hardship and disturbance associated with relocation may never be completely eradicated. Added to all this is the fact that streets, railroads, and parking facilities generally consume more urban land than any other category of land use and, leaving aside single-family residences, consume more land than all other uses *combined* [3.30]. One cannot help but wonder how transportation manages to take up more urban space than almost all other "living" needs added together.

3.3.2.7 Inappropriate or undesirable land development Another potential problem is in the relationship between land development and transportation (or, more specifically, the potential access to land that transportation provides). Increases in land value, and hence development, go hand-in-hand with access. The question of whether these changes are beneficial or not is difficult to answer. If the situation falls under the latter category, then of course we are faced with a problem.

The increase in land use intensity around route 128, the circumferential skirting Boston, is an interesting case. Bone and Wohl indicate fairly clearly [3.32] that there has been a substantial out-migration of industry from the central Boston area to the belt of land straddling the circumferential. In fact, a Wilbur Smith report summarizing the earlier Bone and Wohl study states that [3.33]:

> As of September, 1957, there were 99 new industrial and commercial plants located along the highway, costing over $100 million and employing 17,000 persons; more than 70 plants were previously located within a four-mile radius of Boston.

The effects of these moves are both advantageous and disadvantageous. On the positive side, industries probably were able to acquire relatively inexpensive land and ship their goods at a lowered cost. Surrounding counties most likely also benefited since they obtained new sources of public income from their taxes. On the negative side, however, travel of employees to their jobs became much more difficult and the City of Boston lost many of its highly valued revenue sources. These two latter aspects definitely must be regarded as problems that arise from changes in transportation.

Mumford sees development problems and lays the blame partially on the American people as well as on the engineer [3.9, p. 245]:

> As long as motorcars were few in number, he who had one was a king; he could go where he pleased and halt where he pleased; and this machine itself appeared as a compensatory device for enlarging an ego which had been shrunken by our very success in mechanization. That sense of freedom and power remains a fact today only in low-density areas, in the open country; the popularity of this method of escape has ruined the promise it once held forth. In using the car to flee from the metropolis the motorist finds that he has merely transferred congestion to the highway and thereby doubled it. When he reaches his destination, in a distant suburb, he finds that the countryside he sought has disappeared: beyond him, thanks to the motorcar, lies only another suburb . . . In short, the American has sacrificed his life as a whole to the motorcar, like someone who, demented with passion, wrecks his home in order to lavish his income on a capricious mistress who promises delights he can only occasionally enjoy.

From these statements and examples we would have a difficult time not concluding that there are development problems that are a direct function of transportation and thus should be given considerable attention in the transportation planning process.

3.3.2.8 Moral, religious, biological, and other related problems
With transportation (or the lack of it) already being held responsible for several of the major physical, social, and economic problems in our cities and rural areas, it would seem that all of its negative aspects had been uncovered. Nonetheless, some additional evidence, albeit not too prevalent or easy to trace, indicates that still more problems may be attributed to transportation. For example, Abigail Van Buren has said that the automobile has been one of the most influential forces in shaping the morals of American youth. She says:[6]

> . . . today almost every boy, upon reaching the legal age to drive, has a car of his own, or can borrow one at a moment's notice. And if he can't get a car, the girl can. The automobile has become the modern tribal symbol of manhood. With five gallons of gas, in 20 minutes our young people can be transported into a private world of their own to enjoy hours of uninterrupted privacy. If that doesn't spell trouble for healthy normal adolescents, I don't know what does.

---

[6] *Family Circle,* vol. 76, no. 1, January 1970, p. 37.

Another interesting effect comes from the religious side of life. A minister[7] has said that after a highway bisected his parish area, people from the sector on the far side of the facility from the church attended services less often than they did before the highway was built. So it seems (again on very superficial evidence) that transportation facilities may even affect our choices and strengths of religious activity.

A final example comes from a third, completely unrelated area of concern—the natural environment. Rachael Carson, whose book, *Silent Spring,* attracted the attention of both conservationists and nonconservationists alike, has written that [3.25, p. 69]:

> There is a steadily growing chorus of outraged protest about the disfigurement of once beautiful roadsides by chemical sprays, which substitute a sore expanse of brown, withered vegetation for the beauty of fern and wildflower, of native shrubs adorned with blossom or berry.

She continues by giving an example from her own personal experience:

> I know well a stretch of road where nature's own landscaping has provided a border of alder, viburnum, sweet fern, and juniper with seasonally changing accents of bright flowers, or of fruits hanging in jewelled clusters in the fall. The road had no heavy load of traffic to support; there were few sharp curves or intersections where brush could obstruct the driver's vision. But the sprayers took over and the miles along the road became something to be traversed quickly, a sight to be endured with one's mind closed to thoughts of the sterile and hideous world we are letting our technicians make.

It is obvious that Miss Carson feels strongly that someone is causing environmental problems that, with some intelligent effort, could be alleviated. As is usually the case, the responsibility for making this effort falls mostly on the transportation engineer or planner. Thus we see once again that transportation (or some small aspect of it) can affect parts of our environment in many ways we do not immediately realize.

3.3.2.9 Unequal impact upon certain population groups To talk about the *overall* significance of various problems related to transportation in an urban area is not necessarily the same as talking about their significance to any one individual or group of individuals. Unfortunate as it is, the impact (both good and bad) of transportation does not appear to fall evenly across the whole urban area:

1. The nonuser may be subject to the noise and air pollution caused by the automobile user.
2. The trucker and public may gain competitively at the expense of the railroad when new highway facilities are built.
3. The poor person may receive much worse transit service than the rich.

This list could be extended considerably, but the important point is that not all people stand to gain from transportation, and this differential can be a problematic situation.

A major irony is that metropolitan transportation systems tend to leave unserved those who need the service most—the poor, elderly, young, and secondary worker. In 1977, as an example, more than 51% of American households with an annual income of less than $3000 had no automobile available to them. This compares to less than 2% for those households with incomes greater than $25,000. The figure for Blacks was 35%, compared to the metropolitan average of only 16% and the center-city average of 26% [3.38].

Tomorrow's Transportation focuses on the job implications of some of these figures [3.12, p. 15].

---

[7] Personal communication.

The beeline distance between South Central Los Angeles and Santa Monica, a center of employment, is 16 miles; to make the trip by public transportation takes an hour and 50 minutes, requires 3 transfers and costs 83 cents one way. The Department of Housing and Urban Development (HUD) demonstration project in Watts has shown that when direct transportation service was provided for residents of that district to jobs and other opportunities in other parts of the city, ridership increased from 800 to 2,800 daily in 3 months. Many of the new riders were bound for work.

As more central business district jobs become white-collar, and an ever larger proportion of unskilled and semiskilled jobs move to outlying sections, poor people are more disadvantaged than ever by public transportation systems which focus on central business districts and also stop at city limits. A New York study reports, "The employment of suburban areas of both poverty and nonpoverty workers residing in the areas studied in New York City (poverty areas) appears to be almost insignificant." One reason is an often cited figure: It would cost a resident of central Harlem in New York some $40 a month to commute by public transportation to an aircraft factory in Farmingdale, Long Island.

The poor are not only isolated from jobs, but also from social and health services, recreation areas, and social contacts outside the immediate neighborhood. A HUD demonstration project in Nashville, Tenn., has provided bus service for outpatients and employees linking nine major medical centers with downtown Nashville and a hospital connecting service. In the first 2 months of actual operation, the medical center express service line showed a 61-percent increase in passengers, while the hospital connecting service line showed a 73-percent increase in ridership.

Another large group of people who may find extreme problems with transportation is the disabled. Many of these also are poor, which makes it doubly difficult for them to own and operate a private motor vehicle. This places them at the mercy of transit systems that traditionally have not catered to their needs. As Helen Meier states [3.22, p. 5]:

To disabled people, ... transit systems represent barriers to mobility, loss of civil rights and one reason why jobs are hard to find and hold. Transit systems are designed so that many disabled persons cannot use them, or use them with difficulty, discomfort and even physical danger ....

...As recently as ten years ago, there was no public provision of transit for the disabled .... Some social service agencies provided limited transit services for agency clients. This left large numbers without any form of transportation.

Quite obviously any study having as its purpose the betterment of transportation must identify special groups of interest and not treat the population of the urban area homogeneously. Otherwise problems such as those quoted above may unsuspectingly arise for one or more sets of people and cause them much psychological distress and perhaps even physical harm.

3.3.3 Problems in the Problem Domain Affecting Transportation

Up to this point we have concentrated on the impacts of transportation on various aspects of the physical environment, human characteristics, and human activities. Yet it is not difficult to reverse the roles of these entities, with transportation being the affected entity rather than the affecting one. What, then, are some of the factors that create changes (or a need for change) in transportation? Those identified most frequently are overall population growth, the constrasting forces of urbanization and suburbanization, increasing real income, with corresponding increases in automotive user prices and ownership (and corresponding leveling in transit ridership), and, finally, changes in lengths, numbers, and times of trips. Each of these will be briefly discussed in turn in the paragraphs to follow.

3.3.3.1 Increased population growth and dispersion Perhaps one of the foremost problems transportation planners must deal with is the ever increasing number of persons who must be served. The two charts in Fig. 3.3 show both the growth and rural–urban distribution and population in the United States since 1790. The trends definitely are upward, although the point has almost come at which the urban–rural split has stabilized. If we are experiencing difficulty now providing services for the present population, think of the problems when, by the year 2000, there will be 50 million additional people in the country, 75% of whom will be in urban areas.

The distribution as well as the amount of population is also quite important in relation to transportation needs. Obviously, if people are massed in several small areas of the country, the use of transit becomes feasible. But transportation also becomes

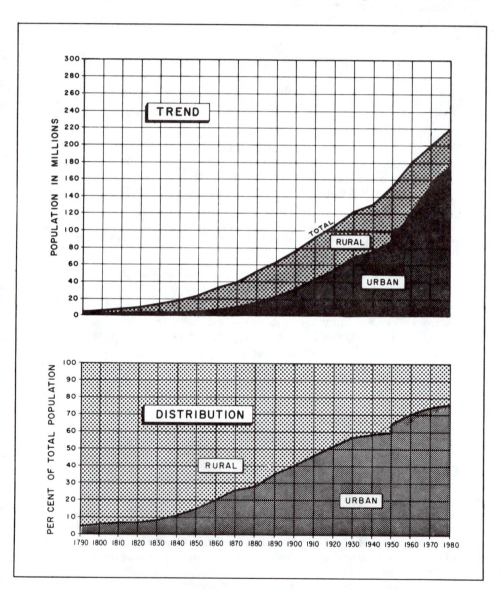

Fig. 3.3 Population trends: United States, 1790–1980 [3.8, various years].

more expensive, as well demonstrated by the case of the Westway in New York City. Thus, a switch from a predominately rural to predominately urban population has varied but significant impacts on transportation.

Similarly, the trend toward suburban and small-town as opposed to central-city living has significant import for urban transportation. The greater spread of activities implied by the suburbanization process automatically means longer distances to be traveled to get to any one activity and, correspondingly, fewer people taking the same paths of travel. Mass transit, in its present form, does not appear to be as useful in this type of situation, and highways, while relatively inexpensive on a per-mile basis, are needed in greater lengths, creating an additional cost that tends to cancel out the comparatively low suburban land costs. Thus, suburbanization, increasing as it is in the United States [3.8], can create real problems for the transportation planner in attempting to keep within ever-present budgets.

3.3.3.2 Rising incomes and prices Incomes have risen rapidly in the United States over the period from 1960 to 1980. In the earlier years, disposable income—that available, after taxes, for people to spend—averaged $1934 per capita. In 1979 the figure was more than four times as large, at $7367 per capita [3.34]. In the same period, however, prices also rose. The consumer price index for all items increased from 88.7 in 1960 (with the index set at 100.0 in 1967) to 217.4 in 1979. Still, this is only a 245% rise compared to the 391% lift in disposable personal income [3.7]. People thus are much better off, in terms of income, than they were in earlier days.

Of course, as noted in the earlier section on user costs, the prices of individual items, such as gasoline, insurance, and the like have gone up faster than the overall consumer price index, at least from 1973 to 1980. This means that consumer expenditures for auto purchase and travel are particularly affected.

3.3.3.3 Increased automobile ownership There cannot be much doubt that the automobile is both a popular and a useful means of transport. Its innate popularity is of importance here, especially because there seems to be no end to the desire of Americans to purchase more and more cars. Figure 3.4 illustrates the rapid growth of the motor vehicle in the United States. When a comparison is made of this growth to that of population, one immediately recognizes that cars are increasing more rapidly than people: in fact, at almost twice the rate! Consequently, if one has trouble comprehending an increase in population of 50,000,000 by the year 2000, he or she might have even more difficulty comprehending an additional 100,000,000 motor vehicles. Of course, present trends are not expected to continue, but the numbers of vehicles still will increase markedly, creating a deluge in the city that will be difficult to divert or decrease in scale.

The other half of the vehicle growth picture is the decline of transit ridership. These two factors go hand in hand as can be demonstrated by comparing numbers in Fig. 3.4 with those in the top part of Table 3.5. In the period since 1940, motor vehicle registrations have risen sharply, whereas the number of passengers per year on all modes of transit except bus has dropped off correspondingly. For the transportation planner to be successful in developing a suitable transportation system, an effort will have to be made to overcome or circumvent the "natural" tendencies described above and make transit a viable mode of travel in the face of enlarged car ownership. To accomplish such a task will not be easy.

3.3.3.4 Peakedness in the amount and timing of travel Two other "natural" problems that have a direct effect on the need for transportation are the quantity and temporal pattern of travel. Perhaps because of the rise in per capita income or because of the transportation system itself, the amount of travel in the United States has risen

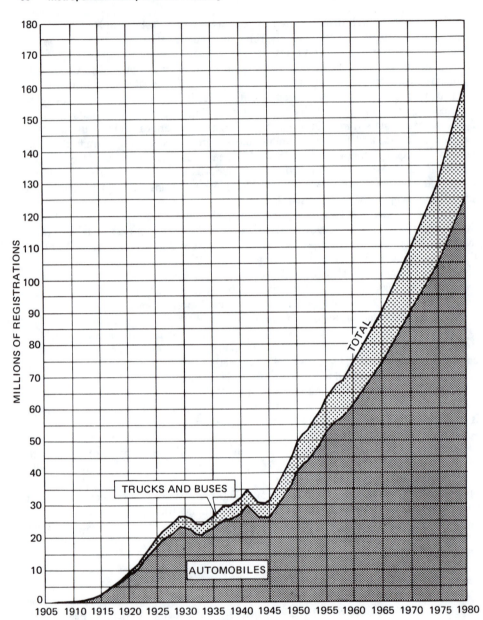

Fig. 3.4 Motor vehicle registration trends, United States [3.8, p. 29].

sharply. With 1940 considered as a base of 100, the vehicle miles of travel in 1980 stood at the 498 level [3.11, p. 33]. By way of comparison, the gross national product of the country stood at 420 while total population was at 167. Thus, we see a greatly increased desire to travel that must be met by the transportation system, and, with travel outstripping all other relevant quantities, it appears that a greater amount of resources must be expended to keep up with the rapidly expanding demands for transportation.

To make matters worse, we must also allow for the significant hourly peaking in travel that takes place in urban areas. As can be seen in Fig. 3.5, there usually are two general peaks—the rush hours—in which travel is much heavier than at other times. The evening peak in Chicago, for example, may amount to approximately 11 percent of the total trips in a day. Yet, if there were no peaks, we would only have to provide for 1/24 or about 4 percent of all daily trips in 1 hour. Temporal variations in trip-making thus require that we provide for roughly three times as many trips as would occur otherwise.

The situation for transit is even more difficult. Referring again to Fig. 3.5, we see that almost all transit trips occur during the peak periods, with the result being that most of the transit stock will lay idle during the rest of the day. Moreover, the cost of labor for transit is affected greatly since it usually takes more than one shift of drivers and other personnel to cover the two peaks.

While the data employed to develop Fig. 3.5 are old (about 1955), it is interesting to note that the peaks in many places have not declined over time. This is particularly true of auto/taxi *vehicular* trips between 1970 and 1980 (Fig. 3.5 shows *passenger* trips). As an example, on four bridges leading into Washington, D.C., from the Virginia side, inbound *passenger* morning peaks declined as a percentage of daily traffic on all but one. The auto *vehicular* traffic percentage, on the other hand, increased on three bridges in that same period of years [3.37, 3.38]. This might be contrary to expectations, given the emphasis in Washington on staggered work hours, car pooling, and the like.

3.4 SUMMARY AND CONCLUDING COMMENTS

Table 3.13 summarizes much of the material concerning the effect of many factors in the problem domain on the need for transportation. Most apparent are the extremely rapid growths of travel, vehicle registrations, and gross national product (GNP). It appears that there is a strong correlation between the first two entities and

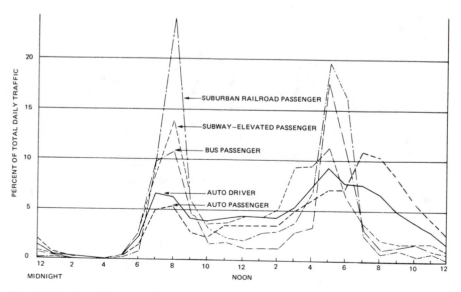

Fig. 3.5 The "peak" problem [3.35, p. 165].

Table 3.13 Summary Transportation Indices in Relation to Economic Growth, 1940-1959 (1950 = 100)

Year	Total population*	Urban population	Gross national product†	Manufacturing workers	Total vehicle registrations	Vehicle miles of travel	Total transit passenger rides
1940	100	100	100	100	100	100	100
1945	106	100	153	141	96	83	178
1950	115	119	154	139	151	152	132
1955	125	125	188	154	193	200	88
1960	136	161	212	153	228	238	72
1965	146	175	266	164	279	294	63
1970	154	192	309	176	335	371	56
1975	161	213	346	167	410	440	43
1980	167	220‡	430‡	185‡	498‡	564‡	60‡

*Includes Armed Forces.
†In constant dollars.
‡Preliminary estimate.
Source: [3.11, various pages, and estimates by the author].

the latter one so that we can expect that if economic conditions continue on the rise, there will be an ever increasing number of cars in use for an ever increasing amount of travel. In addition, with population gaining as it is, especially in urban areas, we might anticipate great pressures on urban transport systems as they attempt to handle the increased burdens placed upon them. These pressures represent the essence of all the factors in the problem domain affecting transportation.

Factors in the problem domain were only one of three types discussed in this chapter, however. The other two were: (1) those relating directly to transportation service, and (2) those in the problem domain affected or impacted by transportation. The distinction between the three types probably is not entirely clear, but this is to be expected when one considers the subject matter involved. What is important at this stage, however, is that transportation planners try to make themselves cognizant of the great range of consequences that result from the many decisions made concerning transportation systems of various types. Perhaps for a given situation the kinds of problems discussed in this chapter are not significant, or perhaps the real problems have not been covered here, or perhaps certain problems have not been given enough attention. Whatever the case, we must recognize (1) that the problem identification stage of the planning process is critical, (2) that too few transportation studies have expended enough effort in trying to pinpoint actual problems (going on the assumption that problems are more or less universal), and (3) that problems will change, acquiring different degrees of importance, as time progresses. The upshot of these conditions is that planners must be prepared to devote resources to problem identification; otherwise they open themselves to the mistrust of citizens who have made comments such as those that follow [3.24, p. 8]:

> I don't think there has been any planning, meaningful planning, anywhere in the United States.
> A number of people think they're engaged in meaningful planning. I think that's what annoys us. When I draw up a ——·—— plan and produce it on a slick brochure then that's profundity—you better believe it because it has my name at the bottom of it. How many of *those* have we seen?

I think most of the planning is very, very opportunistic for the guys who are doing the planning, trying to get some money and really no thought given to how it's going to fit into a total scheme of things—no thought given to how these people are going to live there, and it goes on and on.

BIBLIOGRAPHY

3.1   American Public Transit Association: *Transit Fact Book 78–79*, Washington, D.C., 1979.
3.2   Maricopa Association of Governments, Transportation and Planning Office: *Travel Speed Study, Phoenix Metropolitan Area, 1957 Thru 1979*, Phoenix, Dec. 1980.
3.3   U.S. Bureau of Labor Statistics: *Monthly Labor Review*, Washington, D.C. (monthly).
3.4   Traffic in Towns. The specially shortened edition of the Buchanan Report, Penguin Books Ltd., Hammondsworth, England, 1963.
3.5   Vuchic, V.: *Urban Public Transportation: Systems and Technology*, Prentice-Hall, Englewood Cliffs, N.J., 1981.
3.6   Meyer, J. R., J. F. Kain, and M. Wohl: *The Urban Transportation Problems*, Harvard Press, Cambridge, 1965.
3.7   U.S. Bureau of Labor Statistics: *Handbook of Labor Statistics*, U.S. Government Printing Office, Washington, D.C. (annually).
3.8   U.S. Bureau of the Census: *Census of Population*, vol. 1 (every 10 years).
3.9   Mumford, L.: *The Highway and the City*, Mentor, New York, 1953.
3.10 Higbee, E.: *The Squeeze—Cities Without Space*, Apollo Editions A-43, William Morrow & Co., New York, 1960.
3.11 U.S. Bureau of the Census: *Statistical Abstract of the United States: 1980* (101st edition), Washington, D.C., 1980.
3.12 U.S. Department of Housing and Urban Development, Office of Metropolitan Development: *Tomorrow's Transportation*, U.S. Government Printing Office, Washington, D.C., 1968.
3.13 U.S. National Highway Traffic Safety Administration: *Fatal Accident Reporting System Annual Report*, Washington, D.C. (annually).
3.14 Claffey, P. J.: *Running Costs of Motor Vehicles as Affected by Road Design and Traffic*, NCHRP Report 111, Washington, D.C., 1971.
3.15 Claybrook, J.: "Calm in the Sky; Mayhem on the Road," *The Washington Post*, Sept. 14, 1981.
3.16 Energy Information Administration, U.S. Department of Energy: *1981 Annual Report to Congress*, vol. 2, *Energy Statistics*, U.S. Government Printing Office, Washington, D.C., May 1982.
3.17 Thomas, E. N., and J. L. Schofer: *Strategies for Evaluation of Transportation Plans*, NCHRP Report 96, Highway Research Board, Washington, D.C., 1970.
3.18 Devalle, J. W.: "Concrete Ties for Timber Trestles," *Railway Track and Structures*, vol. 64, no. 9, Sept. 1968.
3.19 Krick, E. V.: *An Introduction to Engineering and Engineering Design*, Wiley, New York, 1965.
3.20 U.S. Department of Energy, Energy Information Administration: *Monthly Energy Review* (monthly).
3.21 Berry, D. S. et al.: *The Technology of Urban Transportation*, Northwestern University Press, Evanston, Ill., 1963.
3.22 Meier, H.: *Accessible Public Transit*, United Cerebral Palsy Association of San Francisco, San Francisco, 1981.
3.23 National Safety Council: *Accident Facts*, Chicago (annually).
3.24 Barton-Aschman Associates, Inc.: *Guidelines for New Systems of Urban Transportation*, vol. II: *A Collection of Papers*, National Technical Information Service, Springfield, Va., April 1968.
3.25 Carson, R.: *Silent Spring*, Fawcett, Greenwich, Conn., 1962.
3.26 Highway Research Board: *Highway Capacity Manual, 1965*, Special Report 87, Washington, D.C., 1965.
3.27 League of Women Voters of the United States: "A Congregation of Vapors," *Facts and Issues*, Washington, D.C., Sept. 1970.
3.28 U.S. Department of Labor, Bureau of Labor Statistics: *Monthly Consumer Price Indices Report*, Washington, D.C. (monthly).
3.29 U.S. Environmental Protection Agency: *1972 National Emissions Report*, Washington, D.C., June 1974.

3.30 Owen, W.: *The Metropolitan Transportation Problem*, rev. ed., The Brookings Institution, Washington, D.C., 1966.

3.31 Ritter, L. J., and R. J. Paquette: *Highway Engineering*, 3rd ed., The Ronald Press, New York, 1967.

3.32 Bone, A. J., and M. Wohl: "Massachusetts Route 128 Impact Study," Highway Research Board Bulletin 227, 1959.

3.33 Wilbur Smith and Associates: *Future Highways and Urban Growth*, New Haven, Conn., 1961.

3.34 U.S. Bureau of Economic Analysis, *Survey of Current Business*, Washington, D.C. (monthly).

3.35 Chicago Area Transportation Study: *Final Report, Part I*, Chicago, 1959.

3.36 U.S. Department of Transportation, Federal Highway Administration: *Cost of Operating an Automobile*, Washington, D.C. (periodic).

3.37 District of Columbia, Department of Highways and Traffic: *D.C. Cordon Traffic Survey 1970*, Washington, D.C., 1970.

3.38 Metropolitan Washington Council of Governments, National Capital Region Transportation Planning Board: *1980 Metro Core Cordon Count of Vehicular and Passenger Volumes*, Washington, D.C., 1980.

3.39 U.S. Bureau of the Census: *Annual Housing Survey*, part A: *General Housing Characteristics of the United States and Regions* and part C, *Financial Characteristics of the Housing Inventory*, Washington, D.C. (annually).

3.40 U.S. Environmental Protection Agency: *Compilation of Air Pollutant Emission Factors*, AP-42, 2nd ed., supplement 5, Washington, D.C., Dec. 1975.

EXERCISES

3.1 Add another problem statement to the ones described in this chapter. In doing so, obtain (1) some relevant data for at least two time periods and (2) quotes from "eminent" scholars and/or politicians indicating that there is such a problem.

3.2 Find a quote that you feel identifies the "wrong" problem. This might be because of lack of breadth of the statement or giving the suggested answer as part of the problem. Describe in two paragraphs how you think the quote could be improved.

3.3 Give an example of a problem that might exist in the future but does not now. Describe in about 250 words.

3.4 Take a particular problem and set up a diagram showing (1) those variables affecting the problem, (2) those affected by it, and (3) the interrelationships between them. The latter could be done simply by drawing an arrow to indicate that one variable affects another.

3.5 In Sec. 3.3.3.4 it was stated that some peaking of traffic volume actually has increased in recent years. In about 250 words, explain why such increased peaking could occur in the example given about the four bridges leading into Washington, D.C.

3.6 Nothing has been said in this chapter about (external) problems inherent in organizing and financing transportation improvements. List six of these and describe in one paragraph each.

3.7 What external technological changes (e.g., in auto engines) might increase future problems in urban transportation? Discuss two of these, including any evidence (data, quotes, etc.) that such will occur.

4 Transportation Goals and Objectives

The starting point of all planning and programming is some definition of the result, toward the achievement of which efforts are to be directed. "Goals" are desired ends expressed in the broadest sense, derived from a consideration of "values" and conducive to a further delineation of program objectives, alternative approaches, and definitive plans and schedules of action.

Transportation programs have generally been characterized by a lack of clear statements of broad overall goals, probably because the complexity of the transportation condition makes such statements difficult to compose, and because programs have generally been devised to meet a given "problem," such as low revenues, or congestion, or antiquated equipment.

Nonetheless, there is an existing structure of transportation goals at both national and local levels. However loosely they may be defined and however lacking in cohesive, assembled form, overall goals have been expressed in the development of legislation, in party platforms (see Fig. 4.1) and the statements of elected leaders, in comprehensive plans, in judicial decisions, in the stated aims of interest groups and in other reflections of consensus. Although much needs to be done to expand and refine (or perhaps reconsider) existing transportation goals, an immediate problem is how to understand those that do exist and apply them in the guidance of efforts to improve urban and rural conditions.

Understanding and pursuing broad goals is especially critical in the planning of transportation, for it is all too easy (as the record bears out) to view transportation simply as a supporting "service" to be projected in response to demands emanating from other actions or decisions. To the contrary, as was shown in the previous chapter, transportation is an element of the urban structure so pervasive in its influence that it

65

The sensible approach to problems facing us in the seventies.

High-quality education is a necessity Virginia needs more and better schools. highly qualified teachers. additional emphasis on early childhood education. a complete special education program. a full system of community colleges. vocational education. and increased opportunities for our children in higher education.

Efficient and responsive government at lowest cost must be insured by a series of studies aimed at effective tax collection procedures and a reorganization of mushrooming State agencies Areas like Southwest and Northwest Virginia need to be brought into closer contact with State government

Safe streets and colleges require firm action if disturbances should erupt in Virginia. well-trained law enforcement officials. concern for the rights of society as well as those of the accused. immediate steps to combat drug abuse. and a concerted attack on the growing death rate on our highways

A comprehensive transportation policy which treats all Virginians fairly and gives proper emphasis to secondary roads has become essential

Economic development can lighten local tax burdens A first-rate effort to attract industry to provide new jobs and to promote Virginia's ports railroads. and airfields is imperative

The welfare system must be reformed to provide training and rehabilitation for recipients so they can be removed from the rolls and lead useful and productive lives

The quality of rural life must be enhanced. Affiliation of hospital complexes in Roanoke and Winchester with the U. Va. medical school to provide more doctors and health care. a workable system of tax deferral on farm and open space land. and increased agricultural research and education can help stem the urban crisis

Battle for Governor The sensible choice

Fig. 4.1   An example of goal statements in political situations.

must be considered an area for key decision-making in the shaping of both urban form and environment. Transportation decisions are, furthermore, only infrequently related to the immediate future and, hence, seldom can be readily reversed if they turn out to be "wrong," in the sense that they fail to contribute to or even hamper the achievement of high-level goals in urban improvement.

In this chapter a loosely structured hierarchy of urban values and corresponding transportation goals and objectives is described. The intent is to summarize the principal goals that appear to be operative in recent thinking about transportation and development activities and then indicate briefly (and as a prelude to Chap. 10) how these goals might vary from person to person.

It should be stated again, as in Sec. 2.3.12, that many planners do not believe it is possible to obtain goals directly from people, groups, and/or governmental agencies. This belief is based on disappointing experiences in which attempts were made to elicit goals. Often it was found that people could not verbalize their broad desires (if in fact they had any). Mostly they were able to respond (usually negatively) to concrete project proposals set before them. Goals, then, appeared easier to define *after the fact* rather than beforehand for plan development purposes. We will not argue either way here, but present available techniques for identifying goals, some of which (Sec. 4.6) are based on induction (that is, on generalization from responses to particular proposals).

The chapter is divided into seven sections. First, a brief résumé of basic urban values and goals is presented. Second, two basic transportation goals, one for those factors of direct concern to transportation service and the other for those factors impacted by transportation, are discussed in general terms. These are then used in the third section as a guide for the identification of a select set of more detailed goals. In the fourth section, these detailed goals are transformed into specific objectives containing relevant criteria. Thereafter follows a discussion of the weightings of importance of different objectives and variations in these weightings for different groups of people. The final section contains summary comments on the difficulties associated with identifying transportation problems and goals.[1]

4.1 BASIC URBAN VALUES AND GOALS

All goals and objectives stem from basic *values* that are important to people. There are many ways of describing these values. A "good" city is popularly referred to in everyday terms such as "vital," "warm and friendly," "dynamic," "safe," "exciting," "full of opportunity," and "beautiful." A "bad" city is described as "dirty," "ugly," "dangerous," "hostile," "impersonal," "confusing," "overwhelming," "time-consuming," "congested," or "wasteful." A more formal listing of abstract values would include terms such as "freedom," "liberty," "dignity," "health and safety," "amenity," "diversity," "economy," "ownership," "mobility," and "affluence."

In most general terms, then, a value can be defined as:

Value: An element of a shared symbolic system (referred to as a value system), acquired through social learning, which serves as a guide for the selection from among perceived alternatives of orientation.

Indications of the types of values held by individuals in a given society should be of great concern to planners, particularly since these values form the basis for the development of goals for the activities and characteristics of the population and the use of resources in a region. For instance, at the highest and most general level, there should be little disagreement that governmental interest in urban development relates to two major goals:

First, that the quality of life be improved in the whole variety of ways that reflect the common values so often expressed in personal reactions to "the city."

Second, that the metropolitan area itself be strengthened in terms of its productive capacity, its democratic institutions, and its ability to allocate and use its natural and human resources to best advantage in each community.

As can be seen, these goals rest heavily on inherent values, and this fact serves to emphasize the definition of a goal as follows:

Goal: An articulation of values, formulated in light of identified issues and problems, toward the attainment of which policies and decisions are directed.

Two points regarding this definition need further elaboration. For one, goals are an outgrowth of identified *issues and problems* and usually do not stand on their own. We would not, for example, be concerned about "strengthening democratic institutions" in metropolitan areas, as stated in the second value above, if we happened to

---

[1] Much of the wording in the beginning sections of this chapter is taken from the report in [4.27], which we felt described even better than we could ourselves the concept of the nature and structure of goals.

be working for a totalitarian government. But, as a second point, we should not *limit* our set of goals to correspond only to problems that presently exist because in many cases solutions which are proposed bring about unanticipated new types of problems. To repeat an example given in the previous chapter, we cannot afford to narrow our consideration of parking in the central business district only to that problem. If we did, we might overlook additional problems such as that of the transfer of large segments of the population between their places of residence and places of business, or that of noise, or air pollution, or congestion near parking facilities, and so forth. The set of goals associated with transportation must therefore relate to more than existing problems; otherwise, proposed solutions will bring about new and unexpected difficulties.

Creating an amalgam of values and goals is a complicated process that has by no means been fully developed in the field of transportation planning and design. Among the complications is the fact that values or goals are not necessarily mutually supporting. Indeed, they are often in polar-like opposition, so that the problem in goal formulation is one of emphasis and balance among values. The process is also complex because points of balance are constantly shifting in time and from place to place. Technological, economic, social, and institutional changes are continually altering the extent to which individual values may be capable of being achieved. This means that the establishment of goals and the determination of relative emphasis to be placed upon various goals is a continuing process that must be applied at many levels and in many places.

Another apparent difficulty in developing goals is that they, like problems, seem to be hierarchical.[2] Statements of goals usually begin with broad generalizations and then are developed at more detailed levels to guide the various types and stages of planning and design activities. However, we should realize that goals are intended to be broad, extensive statements that form the basis for comprehensive concepts for our undertakings. Thus, we should not be quick to give in to the tendency to try to avoid generalizations inherent in goal statements and thus avoid the accusation of a lack of specificity and "practicality."

4.2 BROAD GOALS FOR TRANSPORTATION

The two broad goals for urban development presented in the previous section probably would be at the top of any goal hierarchy. Proceeding down from these would be a wide variety of goals pertaining to various aspects of human characteristics, human behavior, and the physical environment. We, of course, are particularly interested in transportation in this context, and in the previous chapter we subdivided transportation problems into three classes: (1) those that are "direct" or transportation service problems; (2) those in the problem domain *affected by* transportation; and (3) those in the problem domain *affecting* transportation.

A similar classification can be employed for transportation goals and will be developed in this section. However, it should be realized at this point that the setting of goals for the last class of transportation problems probably is not a worthwhile exercise. The reason is that goals can logically be created only for factors over which some control can be exerted. Since we as local transportation planners or engineers can do relatively little to influence factors such as overall population growth, income

---

[2] For a hierarchical structure of problems for which a corresponding pattern of goals could be constructed, see Chap. 3, Fig. 3.1; also see Fig. 4.2 in this chapter.

levels, and so forth, the discussion to follow on goals for transportation will not relate to these types of factors.

4.2.1 Goals for Direct or Transportation Service Factors

The primary, although certainly not exclusive, purpose of transportation is to serve the user or potential user of the system, that is, to provide accessibility to land and mobility between desired trip ends.

Beneficial transportation for the user would be most obvious in the form of increased access to opportunities for achievement, most notably in employment, purchase of goods, health care, and education. In the short run transportation development should aim to compensate for disparities inherent in such current phenomena as shifts in employment and living patterns which tend to reduce ready access to a variety of opportunities. To do this requires the provision of transportation to both centrally located employment, commercial, health, and educational centers and to new centers in outlying areas. It also means designing transportation improvements to enable a range of choice and diversity of urban experience.

Efficiency and economy in the use of public and private corporation funds also is of direct importance in transportation service, especially since these are virtually standard requirements in all forms of investment. It is apparent, for example, that many public and private transportation operations have had to maintain a continuing vigil on expenses so that they could achieve a respectable rate of return on their investments. Similarly, it is clear that funds from local, state, and Federal sources are limited and must be allocated to their best use.

4.2.2 Goals for Factors in the Problem Domain Affected by Transportation

An environment that responds to human needs and sensibilities and to the requirements of enterprise should be sought, thus helping to enlarge urban opportunity. This goal is concerned with rational arrangement, amenity, and fostering of the forms of development that are characterized by the variety, diversity, and ease of contact that they offer.

Rational arrangement makes it possible for people to understand the city's layout and be able to move about more readily. Thus, it is easier to plan and locate enterprises and institutions in desirable relationship to the rest of the city's activities.

Amenity has to do with positive qualities of convenience, safety, healthfulness, and beauty as opposed to the negative qualities of inconvenience, hazard, pollution, and ugliness.

Variety and ease of contact are two of the most dominant motivations in the growth of urban areas. They are concerned not only with transportation efficiency, but also with the design of large-scale development. It is the latter with which this goal is chiefly involved. For example, there is a strong trend (apparently reinforced by both internal and external economies) toward the development of new multipurpose centers or subcenters (business, educational, research, housing) throughout metropolitan areas. Despite the improvement in variety and ease of contact that such centers offer, their development is frequently inhibited by lack of transportation facilities.

Conservation of natural resources is an associated goal reflecting society's growing awareness of the limitations of air, water, land, and other natural resources needed in urban development and maintenance of all forms of life. Transportation planning can contribute to this goal in its support of concepts fostering the multiple use of land (including both multiple-use projects and conservation of land used for transportation facilities). The provision of adequate transportation services supporting high intensity of land use in and around special resource areas (water frontages, areas of unusual aesthetic or historical value) will permit greater use of these resources. Different types of transportation facilities and different designs (for example, elevation or depression according to microclimatological requirements) may be devised or selected in the interest of better natural resource management.

Transportation, as a facet of spatial organization, can play a major, although not an exclusive, role in the accomplishment of all of these goals. It may do this in terms of pricing (making available services within the ability of people to pay), in terms of speed (placing distant locations in improved time relationships), and in terms of other operating characteristics. It may also do this in terms of availability of service, a function of system design. It may contribute to equitable access through its influence on the arrangement of land use. Quite obviously, all of these goals are important, and methods needed to achieve them are of general concern to the public.

4.3 IDENTIFICATION OF MORE DETAILED GOALS

The two sets of goals brought out above, one for transportation service and the other for transportation impact on the environment, provide very general directions toward which transportation system modifications and control should evolve. More detailed and tangible goals and objectives are needed, however, so that solutions to certain specific problems can be evaluated. To develop such a detailed set is not easy since many important considerations do not present themselves until solutions with characteristics different from the present situation are suggested. Moreover, if adequate thought is given to goal identification procedures, the resulting list of goals can become quite long, and additional efforts must be devoted to ensure that the items on the list are fairly exclusive and exhaustive and that they are representative of the group of people they are intended to serve.

Dickey and Broderick [4.31] have developed one technique useful for being more exhaustive in identifying transportation service, impact, and affecting factors. Their technique involves a classification with four major components: (1) man (and groups), (2) the natural environment, (3) the manmade environment, and (4) activities.

Each of these components are further subdivided, as can be seen in Tables 4.1 to 4.4. These tables also serve as a checklist. By way of illustration, an often forgotten group in many urban planning and design situations consists of those people who do not at present live in the urban area being planned or designed. This group would have to be considered, for example, in planning for new towns or new developments involving vacant land within the boundaries of an existing urban area. This group is element I-3 in Table 4.1. Two other groups of possible interest consist of those individuals not born yet or those living now who may not be when actual developments take place in the distant future. These groups are included in the category of Individuals By Age (including unborn, those that will die, etc.), item I-1 in Table 4.1.[3]

As mentioned, the elements presented in Tables 4.1 to 4.4 serve as a good checklist. Yet, it is also important to determine the *characteristics* of interest for each element in these tables. Dickey and Broderick searched the *Thesaurus* [4.35] and produced a list of 49 characteristics that could possibly describe each element (see Table 4.5). These characteristics then were cross-referenced with each element to

---

[3] With the recent emphasis on advocacy planning, it is surprising that the planner often forgets he or she is one of the few advocates for the future generation.

Table 4.1   Man (and Groups)

Component I
Individuals and/or households

I-1	By age (including unborn, those that will die, etc.)
I-2	By race, religion, color, ethnic background
I-3	By locality (and future locality)
I-4	By sex
I-5	By employment category
I-6	By political leaning
I-7	By income
I-8	By educational background
I-9	By personality types (including deviants)
I-10	By occupation
I-11	By social status
I-12	By leisure pursuits
I-13	By power/control

Firms and institutions

I-14	Firms
I-15	Institutional groups
I-16	Governmental agencies, legislatures, and judiciaries
I-17	Social groups and clubs
I-18	Political groups
I-19	By locality (and future locality)
I-20	Military organizations
I-21	Unions
I-22	Peer groups

produce a matrix of items (possible goals or problems) like that suggested in Table 4.6. As an example of the use of this kind of table, consider element III-6 of Table 4.3, the "transportation system," and characteristic C-3 of Table 4.5, "technical." If we combine these two entities we might be led to think of those people under 16 years of age, those handicapped, and those too old to be capable (technically) of driving an automobile. One detailed goal, then, for a transportation system may be to reduce the

Table 4.2   Elements of the Natural Environment

Component II

II-1	Earth materials
II-2	Physiographic system (including land surface, etc.)
II-3	Hydrologic system (land-related surface and subsurface waters, etc.)
II-4	Climate (micro and macro)
II-5	Vegetation (forests, flowers, grass, etc.)
II-6	Wildlife (aquatic animals, land animals, insects, etc.)
II-7	Marine and estuarine systems
II-8	Time
II-9	Atmosphere

Table 4.3   Elements of the Manmade Environment

Component III	
III-1	Food, drink, tobacco, drugs
III-2	Clothing
III-3	Raw materials; intermediate and final goods (including crops, domestic animals, etc.)
III-4	Housing (including institutional)
III-5	Communication facilities (including mail, television, telephone, radio, etc.)
III-6	Transportation facilities (including vehicles, guideways, terminals, and controls)
III-7	Educational and cultural facilities (including schools, museums, libraries)
III-8	Water supply, sewage disposal, solid waste disposal, and drainage facilities
III-9	Health facilities (including hospitals, mental institutions, nursing homes)
III-10	Energy creation and supply facilities (including electric, coal, oil, natural gas, etc.)
III-11	Production facilities (including office buildings, machinery, storage areas, warehouses)
III-12	Sales, administrative, and service facilities (including wholesale and retail)
III-13	Military facilities (including bases, training camps, storage areas, etc.)
III-14	Governmental, police, fire, judicial, and welfare facilities
III-15	Leisure and recreational facilities (including parks, clubs, fraternal organizations, etc.)
III-16	Information
III-17	Monetary capital (stocks, bonds, cash, etc.)
III-18	Laws (including police power, eminent domain, zoning, etc.)
III-19	Energy

technical requirement for operation (that is, user operation). In a similar way, a variety of other goals can be identified.

This identification technique has several drawbacks:

1. Not all characteristics in Table 4.5 are descriptive of the elements in Tables 4.1 to 4.4 [e.g., "nicely shaped" (C-34) and "unions" (I-21) do not go together].
2. The elements (and characteristics) are not mutually exclusive [e.g., individuals of certain age classes (I-1) are also of certain races and religions (I-2)].
3. The number of items in Table 4.6 can become quite large.

These drawbacks can be reduced somewhat through various means, but what is important is that the technique can be employed to identify *ahead of time* a large percentage of possible goals of relevance to transportation planning and design.

4.3.1 Example Sets of Goals

Many different sets of goals have been proposed in transportation and related studies. As expected, the goals are specified at different levels of the hierarchy and in many cases are somewhat vague in nature. Further, there is no general agreement on

Table 4.4 Activity Elements and Agents

Component IV		
Agent		**Activity**
IV-1	Individuals and households	Income producing
IV-2		Child raising and family
IV-3		Educational and intellectual
IV-4		Spiritual development
IV-5		Social
IV-6		Recreation and relaxation
IV-7		Clubs
IV-8		Community service and political
IV-9		Associated with food, shopping, health, etc.
IV-10		Travel
IV-11	Firms	Goods producing
IV-12		Service
IV-13	Institutions	Human development
IV-14		Basic community service
IV-15		For welfare and special groups
IV-16	All (long-term)	Migration
IV-17		Investment
IV-18		Crime, war

Table 4.5 Characteristics to be Used in Conjunction with Elements in Tables 4.1 to 4.4

C-1	Religious-moral-ethical		C-26	Beautiful
C-2	Free		C-27	Quiet
C-3	Technical		C-28	Healthy
C-4	Stable		C-29	Safe
C-5	Private		C-30	Informed
C-6	Cheap		C-31	Liberal
C-7	Accessible		C-32	Upper class
C-8	Active		C-33	Polluted
C-9	Defended (militarily)		C-34	Nicely shaped
C-10	Large		C-35	Dark
C-11	Comfortable		C-36	New
C-12	Wealthy		C-37	Fragrant
C-13	Just		C-38	Hot
C-14	Happy		C-39	Windy
C-15	Parochial		C-40	Wet
C-16	Natural		C-41	Flexible
C-17	Numerous		C-42	Open
C-18	Organized		C-43	Biased
C-19	Time consuming		C-44	Hungry
C-20	Law abiding		C-45	Thirsty
C-21	Tasty		C-46	Angry
C-22	Exciting		C-47	Powerful
C-23	Affiliative		C-48	Fearful
C-24	Symbolic		C-49	Productive
C-25	Inducive to communication			

Table 4.6 Urban Element–Performance Characteristic Items

	Performance characteristics				
Urban elements	C-1 Religious-Moral-Ethical	C-2 Free	C-3 Technical	...	C-49 Productive
Man (and groups)					
I-1 Individuals and/or households by age					
⋮					
I-22 Peer groups					
Natural environment					
II-1 Earth materials					
⋮					
II-9 Atmosphere					
Manmade environment		Items			
III-1 Food, drink tobacco, drugs					
⋮					
III-19 Energy					
Activities					
IV-1 Income producing					
⋮					
IV-18 Crime, war					

major categories of goals (as, for example, the transportation service and impact categories employed in this book). These problems, while somewhat significant, can be tolerated if one does not become overly concerned with precision in an endeavor that is not relatively well understood or agreed upon. Greater precision is needed, however, when objectives and criteria (measures) are developed for each goal (Sec. 4.4).

An example on a broad level of goal specification (actually potential benefit specification) can be found in the (San Francisco) Bay Area Rapid Transit District's *Composite Report,* which was presented to the voters in the region just before a bond issue referendum:

1. It (the BART system) would aid future growth by (a) maintaining and encouraging concentration of business and industry and lessening sprawl, (b) improving living and working conditions, (c) preserving and increasing property values, and (d) permitting more economic use of land.
2. It would benefit state and local governments by (a) reducing the need for highway funds in the central cities and releasing them for suburban areas, (b) containing urban sprawl, thereby lessening costs of public services, (c) protecting and increasing public revenues by inducing greater economic growth, and (d) reducing usurpation of tax and job-producing lands by highway facilities.
3. It would benefit families and individuals in the three counties by (a) increasing mobility and job potentials of users, (b) providing transportation for those without automobiles, and (c) expanding social, educational, and recreational opportunities.

As noted, these goals are somewhat broad but do seem to cover most important aspects.

In Dallas, 87 people from diverse backgrounds were chosen to participate in a 3-day conference that produced a set of goals to be considered by citizens in town meetings. The resulting goals for transportation are reproduced in Table 4.7. A general goal statement is made first, followed by a set of specific recommendations. Many of the latter are perhaps overly specific and presuppose a solution. The fifth goal, for example, stresses more adequate taxi service as a need. Yet it may be that the kind of service desired could be fulfilled by other means (e.g., mini-bus). The main point is nevertheless that some general directions for improvement in Dallas' transportation system have been determined.

Table 4.8 shows impacts (goals) that state agencies were required by the Federal Highway Administration to analyze for highway projects. Most of these could be deduced from the Dickey-Broderick methodology.

Table 4.7 Transportation Goals for Dallas, Texas

General goal

Dallas must recognize and improve its position as a major transportation and communication center. In order that we may continue to grow and compete successfully with other metropolitan regions, we should work constantly to improve transportation and communications facilities. Within the city and the region, people must be able to move rapidly, pleasantly, safely, and economically from their homes to work, to schools, to shopping areas, and to recreational and cultural facilities. Transportation of goods within the city and region should be efficient without interfering with the citizen's enjoyment of his city.

Specific goals

1. Continue to expand and improve transportation service to the metropolitan areas and nations at reasonable rates.
2. Secure with the support of other governmental units in the region, enabling legislation for a Transit Authority or Authorities which would:
   a. Serve as large an area initially as is practical and be designed ultimately for the entire metropolitan region. Membership would include representatives of the areas served.
   b. Study the technology of rapid transit to select the system or systems which can best satisfy our needs.
   c. Assume ownership and operation of the Dallas Transit System and extend its services to satisfy as many needs as the Authority can justify economically.
   d. See to it that Dallas and other municipalities protect and, if possible, preserve at today's price, right-of-way for future rapid transit.
   e. Consider subsidy of public transit by the metropolitan region.
3. Bring the Dallas-Fort Worth airport to its fullest potential as a regional and world air center. Develop more private aircraft and short-hop commercial facilities.
4. Maintain a perpetual list of the community's needs in communications and take effective action to assure postal, telephone, and telegraph services which meet the needs.
5. Make available adequate taxi service in all parts of the city and at all hours.
6. Design transportation facilities and services to satisfy the needs of users without dissatisfying other people. For example, when transportation changes are needed, sufficient right-of-way should be acquired to protect adjacent land.

Source: [4.36, pp. 12-13].

Table 4.8 Impacts (Goals) to be Considered in Highway Project Analyses

1. Natural, ecological, or scenic resources (including energy)
2. Relocation of individuals and families
3. Social impacts, with respect to access to jobs, schools, churches, parks, hospitals, shopping, and community services for:
   (a) Elderly
   (b) School-age children
   (c) Those dependent on public transportation
   (d) Handicapped
   (e) Illiterate
   (f) Nondrivers
   (g) Pedestrians
   (h) Bicyclists
   (i) Low-income
   (j) Racial, ethnic, or religious groups
4. Air quality
5. Noise
6. Wetlands and coastal zones
7. Stream modifications or impoundments
8. Flood hazards
9. Construction effects
10. Sites of historic and cultural significance
11. Parks

Source: U.S. Dept. of Transportation, Federal Highway Administration, *Federal-Aid Highway Program Manual,* vol. 7, Chap. 7, Sec. 2, "Environmental Impact and Related Statements," Washington, D.C., Dec. 30, 1974.

4.4 GOAL-RELATED OBJECTIVES AND CRITERIA (MEASURES)

Up to this point little has been said about measurable entities that can be used to give explicit assessment of suggested solutions for metropolitan transportation. These more tangible entities are known as *objectives* and contain measurable attributes or quantities known as *criteria:*

Objective: A specific statement denoting a measurable end to be reached or achieved for a particular group of people, sometimes in a particular span of time.

Criterion (measure): An explicit attribute or characteristic used for the purpose of comparative evaluation.

Associated with each goal should be at least one strongly defined objective and a corresponding criterion (or criteria). In a study done for the Baltimore region [4.37], Ockert and Pixton identified a set of goals and associated measures. One goal, illustrated in Table 4.9, was that of provision of access to activities. Several possible *criteria* were listed and one was picked as a single measure of success in meeting the goal. As can be seen, the chosen *criterion* is very specific in terms of (*a*) income levels associated with the jobs and (*b*) travel times and headways associated with the transportation system. The *objective* would be to minimize the number of employment opportunities without "adequate" (as defined explicitly in the table) transportation. No time period to reach a certain level of the objective was given.

Some additional examples of criteria are presented in Table 4.10. These are

Phase I specifications part of the Federal Clean Car Incentive Program sponsored by the Environmental Protection Agency. The criteria are divided into six categories— emissions, safety, performance, serviceability, fuel availability, and noise level. Within each of these categories are sets of desired measures to be employed in testing the success of automobile designs.

An important point to be made here is that it is not necessary to have *quantifiable* criteria, just measurable ones. Guilford [4.25] defines these terms as:

Measurement: The *assignment of numbers* to objects according to logically accepted rules.

Quantification: The *ordering* of something according to quantity or amount.

Since it is possible to assign a number to something without ordering it (e.g., male = 1, female = 2), we can employ such criteria as "presence (or lack) of food services in a terminal," or "operation of vehicle by user required or not required." These criteria are nominal (name or categorical) in scale, not ordinal (e.g., army ranks), interval (e.g., °F), or ratio (e.g., yards) in scale. While nominal measures are not as precise in nature as the latter three, they are useful in situations where well-defined and accepted measures are not available. Moreover, because quantification is not a strict necessity in developing criteria, we can feel free to use completely subjective measures derived, for example, from responses of citizens concerned with the appearance of a proposed elevated highway. These responses could be measured on a semantic differential scale [4.26], which has a range from −3 for "very ugly" to +3 for "very beautiful." A 0 rating would indicate indifference—"neither ugly nor beautiful."

Table 4.9 Example of a Goal and Related Criteria

Goal: Increase activity access

Possible criteria (measures):	Units
Number of employment opportunities without adequate transportation	Jobs
Accumulated shortage of parking space	Parking spaces
Number of employment opportunities directly served by expressways	Jobs
Number of employment opportunities directly served by rapid transit	Jobs
Number of shopping trips without direct arterial highway access	Shopping trips

Suggested single criterion:

Number of employment opportunities without adequate transportation

Formulation:

$$J = J_{wc} + J_{bc}$$

where J = number of employment opportunities without adequate transportation (jobs)

J_{wc} = number of high income (> $7,500/yr) jobs > 5 min.
driving time from an arterial street and > 5 min.
walking time from transit stop having < 30 min. headways (jobs)

J_{bc} = number of low income (< $3,500/yr) jobs > 5 min.
walking time from transit stop having ⩽ 30 min. headways (jobs)

Adopted from [4.37, p. A-4].

Table 4.10 Example Criteria and Standards

Criteria	Standard
I. *Emissions* (for 4,000 miles)	
1. Hydrocarbons: grams per vehicle mile measured by the 1975 Federal Test Procedure	0.41
2. Carbon monoxide: grams per mile	3.4
3. Oxides of nitrogen: grams per mile	1.0
4. Evaporative hydrocarbons: grams per test	2.0
5. Smoke: percent capacity during (*a*) acceleration and (*b*) lugging	(*a*) 30 (*b*) 15
II. *Safety*	
1. Compliance with Federal Motor Vehicle Safety Standards and inherent safety of the vehicle and power plant	
III. *Performance*	
1. Startup time: seconds at 60° F	30
2. Acceleration: seconds on level surface from 0–60 mph	16
3. Top speed: mph for 1 mile on level surface	75
4. Range: miles at (*a*) 50 mph and (*b*) 65 mph	(*a*) 200 (*b*) 150
IV. *Serviceability*	
1. Equivalent to 1972 model year vehicles	
V. *Fuel availability*	
1. Million vehicle-miles per year fuel quantity available (also must be capable of being stored and dispensed by existing methods)	2.5
VI. *Noise level*	
1. Maximum dBA at 50 ft	80

Adopted from *Commerce Business Daily*, Issue No. PSA-5529, March 16, 1972.

The creation of measurable objectives and corresponding criteria completes the makeup of a second type of hierarchy which starts with values and goals. Figure 4.2 gives a general schematic representation of this other kind of hierarchy and emphasizes the idea that there is not necessarily a one-to-one relationship between criteria, objectives, goals, and values. Several criteria may relate to one objective, one objective to several goals, and so forth.

4.5 CONSTRAINTS AND STANDARDS

A complete discussion of values, goals, objectives, and criteria for transportation must also take into account constraints and standards (desired levels). These two elements are relevant since there often are certain criteria that people would like to minimize or maximize, such as accidents, while there are others whose values only have to be kept above or below a certain level. A government transportation budget provides a good example. Often a set amount of funds is allocated for, say, highway improvements. The aim then is to go as far as possible toward relevant goals while keeping within the limit of the budget. The budget in this example would be a *constraint*, not a goal or objective.

A *standard* is a *particular desired level* of the criterion which should not be exceeded or undercut (or both), depending on the particular situation. An example of a standard relates to temperature control. Usually, it is desirable to have the temperature in a vehicle kept at about 70°F, neither above nor below. Hence, an attempt would be made to design the heating and/or air conditioning systems to achieve this level. Examples of other standards are those associated with the previously discussed criteria for the Federal Clean Car Incentive Program. These can be seen in Table 4.10. Whether in a certain situation one would be dealing with objectives and criteria or constraints and standards is difficult to tell but seems to depend on how much importance people attach to levels of a criterion beyond its commonly accepted value. If, for example, it were not important as to how much a budget is underspent, then no one is expected to *minimize* expenses and the budget should be taken as a constraint to spending, not an objective for it.

4.6 WEIGHING THE IMPORTANCE OF OBJECTIVES

Not only must transportation planners be concerned with identifying and listing the various goals, objectives, constraints, and standards relating to a given problematic situation, but in some cases they must also determine the relative weights of importance carried by the stated goals and objectives. For instance, the goal of creating a transportation system that "offers novelty" has evoked little attention to date, whereas, offering low door-to-door travel time has been of such interest that it probably can be considered the prime mover in stirring people to act in solving their

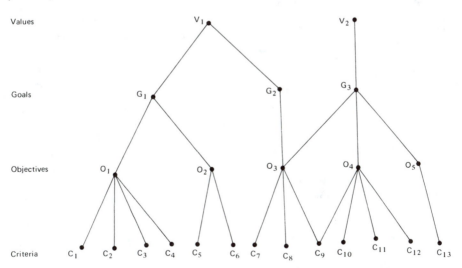

Fig. 4.2 Hierarchical interrelationships among values, goals, objectives, and criteria [4.20, p. 2-15].

transportation problems.[4] No attempt will be made in this section to make specific judgments about the relative weights of importance of various objectives, but several illustrations will be given to demonstrate the contributions which can come from efforts to define weightings of importance.

Actually, except for some preliminary studies, relatively little is known from a research standpoint about the weights of importance of different transportation goals. Perhaps the most interesting procedure developed so far for determining weights is that by R. L. Wilson (in [4.16, Chap. 11]) in his study of the desires of people in Greensboro and Durham, North Carolina.

Wilson presented to a sample of people in each city a chart or "game" board like that shown in Fig. 4.3. He also gave each respondent a number of "markers" with which to "pay" for lower density development and for "the closeness of neighborhood things to the house." For example, a 60 ft by 100 ft lot cost 18 markers while a 3-min *walk* to a grocery store cost 5 markers and a 10-min *drive* cost only 2 markers. The number of markers associated with each item on the game board was supposed to be in rough proportion to the expenditures needed to provide that item, and the total number of markers given to the respondents was supposed to be in rough proportion to the average income of residents in the particular city—40 markers in Greensboro and 36 in Durham.

The results of one survey are shown in Table 4.11. It seems that most people in these two cities were desirous of obtaining low-density living (categories 4 and 5), even to the point where they would be willing to "spend" over half their markers for this privilege. This result in itself is of extreme importance to transportation planners since it implies that people generally will spend the travel time gains they achieve from transportation improvements by purchasing homes in lower density (usually suburban) developments. For the most part, this move to lower density has been what has happened in American cities—a direct outgrowth of the importance attached to the goal of lower living density.[5]

The rank ordering or choices as to what neighborhood facilities people would like to have nearby also is interesting. In both Greensboro and Durham the religious building is placed at or near the top in importance. This outcome is almost completely unanticipated and should prove to have some interesting ramifications for neighborhood layout and circulation design. Following the religious building in importance are those kinds of facilities visited most frequently—elementary schools, grocery stores, and so forth. Those places not visited often, such as the movie theater, the shoe store, and the library, generally are found at the bottom of the list. These rank orders give the planner a rough idea of how much importance should be attached to the objectives of connecting places of various types together by means of transportation.

Another interesting conclusion from the Wilson study comes from an inspection of the third column in Table 4.11. The group of respondents in Durham were given a second turn at allocating their markers but with no constraint on how many they could spend. The results indicate that, *with unlimited resources,* people in Durham seemed to have an increased desire for access to shopping centers, neighborhood parks, playgrounds, and community centers. When resources are binding, however, these activities are somewhat subjugated to the more mundane matters of getting to the grocery and drug stores. We can conclude, therefore, that (1) desires (the importance

---

[4] See, for example, the introductory discussion in the "Red Book" [4.11] of the American Association of State Highway and Transportation Officials.

[5] See Chap. 3, Sec. 3.3.3.1, for further verification of this point.

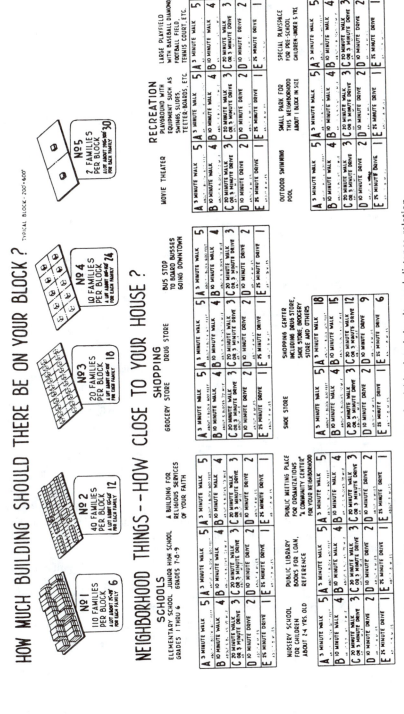

Fig. 4.3 "Game board" used to evaluate aspects of neighborhood density and distance relationships. Original, 28 × 45 in., included five photographs of building types typical of the five densities at the top of board to assist in conveying concept of relative densities [4.16, p. 389].

**Table 4.11   Rank Order, Percentage of Respondents Who Chose
Neighborhood Facilities in Figure 4.3**

Proportion of responses in each density category (percent)

	1	2	3	4	5	Total %	Total no.
Greensboro	1.1	2.7	23.2	55.7	17.3	100.0	185
Durham	0	1.9	24.7	35.8	37.7	100.0	162

Greensboro 40 markers	Durham "A" 36 markers	Durham "B" no limit
1. Religious building	1. Bus stop	1. Religious building
2. Elementary school	2. Grocery store	2. Shopping center
3. Grocery store	3. Religious building	3. Bus stop
4. Junior High school	4. Drug store	4. Neighborhood park
5. Bus stop	5. Elementary school	5. Elementary school
6. Shopping center	6. Shopping center	6. Junior high school
7. Drug store	7. Junior high school	7. Playground
8. Library	8. Movie theater	8. Movie theater
9. Community center	9. Neighborhood park	9. Community center
10. Swimming pool	10. Library	10. Swimming pool
11. Neighborhood park	11. Playground	11. Library
12. Playground	12. Community center	12. Playfield
13. Movie theater	13. Swimming pool	13. Grocery store
14. Playfield	14. Playfield	14. Drug store
15. Nursery	15. Nursery	15. Preschool play space
16. Preschool play space	16. Preschool play space	16. Nursery
17. Shoe store	17. Shoe store	17. Shoe store

Source: [4.16, p. 391].

of goals or objectives) will change as incomes and total resources increase, and that
(2) what people may desire ultimately is somewhat different from what they may
desire when forced to stay within the constraints (especially budgetary) of the real
world. Both these considerations should play an important role in the planner's thinking when proposing future transportation systems.

Despite the fact that many of Wilson's findings are of interest, they may not apply
elsewhere and we must conclude that it is the technique which is important at this
point. More mathematically sophisticated techniques are available, but this one has the
advantages that it is simple to understand and apply and takes into account constraints
as well as objectives. This technique also exemplifies a procedure in which people are
given choices and asked to respond, as opposed to being asked directly about their
goals (and their importance). This approach fits more closely with one philosophy suggested in the introduction to this chapter, which holds that people, groups, and
agencies find it difficult to express their goals directly.

4.7 VARIATIONS IN WEIGHTINGS BETWEEN
GROUPS OF PEOPLE

Another study dealing with weightings of importance was that by Golob et al.
[4.33]. This study is of double interest here in that it was concerned with potential

user preferences for a public transportation system and with variations in these preferences among different user groups. Using semantic differential scales (see Sec. 4.4 and [4.26]) and paired-comparison techniques [4.38], the investigators determined relative weights of importance for a large number of system characteristics. This determination was accomplished through a sample survey of families in a residential suburb of Detroit. The results of the survey of 786 persons are summarized in Fig. 4.4.

It is interesting to note that "lower fares" is not given the highest preference among all characteristics. In fact, it ranks eighth on the list behind such factors as "less

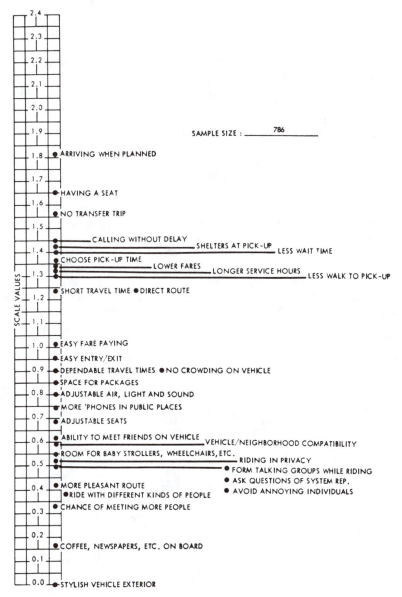

Fig. 4.4 Weightings of importance given system characteristics by the total population. [4.33].

wait time" and "arriving when planned." Also, "short travel time" ranks lower in importance than comfort factors such as "having a seat" and "shelters at pick-up." It would seem, then, that on an overall basis, transit users would like a type of system in which dependability ("arriving when planned," "calling without delay," etc.), basic creature comforts ("having a seat," "shelters at pick-up," etc.), and reduced excess time ("no transfer time," "less walk to pick-up," etc.) are stressed, even over travel cost and time—the two items of greatest concern in most traditional transportation studies. Of course, the survey by Golob et al. posed questions about many characteristics outside the experience of the average transit patron (e.g., "calling without delay"), so that it may be that *actual* responses may differ from those on the survey. Yet the relative weightings of factors do suggest the need for a broader outlook in most transportation studies.

In looking at variations in weightings between potential user groups, Golob et al. determined that while there were some minor differences between certain groups— nondrivers, housewives, and so on—only three groups were really distinct. These were elderly, low-income, and young people. The weighting scales for the first two groups are displayed in Figs. 4.5 and 4.6.

In the first figure it can be seen that the elderly focus more of their attention on the special physical problem they face: getting on and off the vehicle. They appear to want to be able to sit down and not to transfer. Also, since time generally is less important to them and cost is more important, they put a lesser value on "short travel time" and a greater on "lower fares." It should be noticed, too, that the preference scale is more dispersed for the elderly than for the total population. This, the authors indicate, shows that a greater proportion of respondents have similar preferences.

Low-income people are heavy users of present transit systems and seem to be most concerned about waiting on a corner for pick-up. They thus focus on waiting times, shelters at pick-up, and longer service hours (see Fig. 4.6). Interestingly, "lower fares" does not rank any higher in order for low-income people than for the population as a whole (although it does have a higher scale value). The conception that poorer people are concerned almost entirely with cutting travel expenses thus could be misleading to those concerned with making policy regarding transit regulations for a city.

The above results show that in transportation planning efforts, attention ought to be given to the variety of groups of people that may be affected. Otherwise, the system will not serve its intended purpose and may be a source of dissatisfaction.

4.8 COMMENTS ON TRANSPORTATION PROBLEMS AND GOALS

The preceding material in this chapter has been presented in an effort to uncover and establish thinking about directions and desires for transportation and for the factors it affects. The discussion ranged from the most general aspirations for urban development—basic urban values and goals—through descriptions of broad goals for transportation, then to more detailed goals, objectives, and criteria, and finally to the weighting of importance of objectives. This procession from the most general levels to the more specific and applicable criteria follows to some extent the thought process that should, and usually does, take place in the minds of the planner and finally the decision-maker as they attempt to come to grips with the perplexing, multifaceted problems so common to transportation and related endeavors.

Perhaps the most important points that emanate from the discussion in this

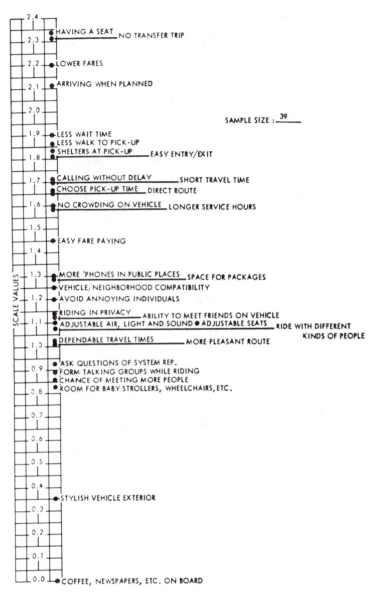

SAMPLE SIZE : 39

HAVING A SEAT — NO TRANSFER TRIP

LOWER FARES

ARRIVING WHEN PLANNED

LESS WAIT TIME
LESS WALK TO PICK-UP
SHELTERS AT PICK-UP — EASY ENTRY/EXIT

CALLING WITHOUT DELAY — SHORT TRAVEL TIME
CHOOSE PICK-UP TIME — DIRECT ROUTE
NO CROWDING ON VEHICLE — LONGER SERVICE HOURS

EASY FARE PAYING

MORE 'PHONES IN PUBLIC PLACES — SPACE FOR PACKAGES
VEHICLE/NEIGHBORHOOD COMPATIBILITY
AVOID ANNOYING INDIVIDUALS

RIDING IN PRIVACY — ABILITY TO MEET FRIENDS ON VEHICLE
ADJUSTABLE AIR, LIGHT AND SOUND ● ADJUSTABLE SEATS — RIDE WITH DIFFERENT KINDS OF PEOPLE
DEPENDABLE TRAVEL TIMES — MORE PLEASANT ROUTE

ASK QUESTIONS OF SYSTEM REP.
FORM TALKING GROUPS WHILE RIDING
CHANCE OF MEETING MORE PEOPLE
ROOM FOR BABY STROLLERS, WHEELCHAIRS, ETC.

STYLISH VEHICLE EXTERIOR

COFFEE, NEWSPAPERS, ETC. ON BOARD

Fig. 4.5 Weightings of importance given system characteristics by the elderly. [4.33].

chapter are that (1) the need for goals and objectives is an ever present one, and (2) existing efforts aimed at formulating and utilizing goals and objectives are not very systematic endeavors and, in fact, are not even very commonplace. These two points, taken together, are disturbing. It is not difficult to get most people to agree that it is an eminently reasonable procedure to set goals and objectives if only for the purpose of preventing us from "not being able to see the forest because of the trees." Some significant considerations somehow always escape the scrutiny of even the most

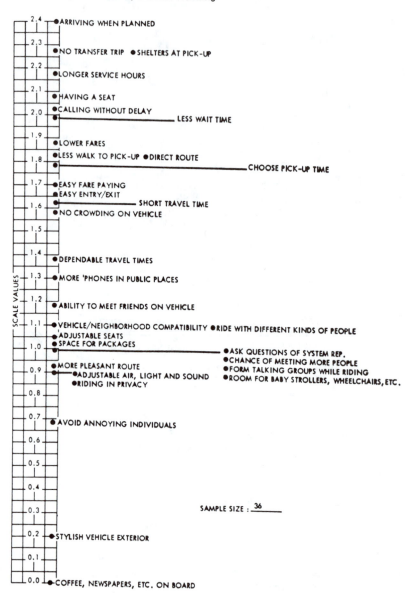

Fig. 4.6 Weightings of importance given system characteristics by those with low incomes. [4.33].

well-intentioned and intelligent planners and engineers. A goal checklist often prevents such occurrences. A more important reason for goals and objectives is, however, that the real needs of the citizenry be served. If we simply *assume* what these needs and desires are, we run the risk of planning, designing, and constructing a transportation system for the wrong purpose. To have to rebuild the system or operate it in an unintended fashion is a luxury that can be ill-afforded.

There are many reasons, nonetheless, why explicit goals and objectives do not exist, or when they do exist, why they are not readily applicable for planning. Many of these reasons are a direct outgrowth of the difficulties alluded to in this and the previous chapter on transportation problems.

First, there is the general confusion over problem or goal identification. A primary difficulty here is the inadvertent statement of problems or goals related to means, not ends. A good example of such a situation was that in the previous chapter where the problems of safety and comfort on the railroad were partially transformed into the "problem" of lack of uniformity in timber railroad ties. An illustration more relevant to broader goal statements might evolve about a decision to build more parking garages in a downtown area. It would then be all too easy to say that "our goal is to provide 3,000 more garage parking spaces downtown by the year ———." This type of goal is a *means* to an end, not an end itself (which may be "more employment downtown," or "revitalization of center city shopping," or something else). One way to help avoid substituting means for ends as goals is to continually ask the question, "Why do we want to achieve goal X"? If an answer such as ". . . because we want to do Y" is readily forthcoming, this would imply that Y is the actual desired end, not X. If the answer is ". . . just because," then it is likely that X is an actual end in itself.

Difficulties involved in exposing *all* problems or goals also hinder efforts to be more systematic. Table 4.8 is a result of an endeavor to be as comprehensive as possible about goal identification. Yet without a doubt, anyone could add more goals or restate those already in the table in a more definitive manner. One goal not given adequate treatment in that table, for instance, is that of "desirable impact on religious activities." Noting that people in Durham and Greensboro, North Carolina, placed a high value on nearness to a religious building and that church attendance is affected by highways, we might conclude that such a goal would be a desirable addition. The fact that some important factors often go unnoticed indicates rather well why a broad attack on problems (Sec. 3.2.1) is difficult to achieve.

Still another difficulty in being systematic about problems and/or goals and objectives is that the weightings of importance given to various factors are subject to change. Everyday experience tells us that different people attach different amounts of importance to various objectives and that, even for one person, the extent of importance varies over time (a condition which at the extreme results in fads). Naturally, these variations are perplexing to those who must plan and design for the benefit and welfare of the public, but they do lead to two considerations that should be taken into account;

1. Because there are variations in weightings between people, any solution for transportation problems should incorporate a flexibility and diversity somewhat commensurate with these variations.[6] A single plan or design, intended for universal application, probably is not realistic.
2. Because there are variations in weightings over time for each individual, future problems generally will not be exactly the same as present ones (Sec. 3.2.4). Thus, the planner or engineer probably should design for an average (over time) level of importance for each objective if the solution is to be successful.[7]

The final and perhaps greatest hindrance to the systematizing of problems, goals, and objectives is that there is rarely any explicit indication given as to the relative priorities to be attached to the desires of one individual or group of individuals as

---

[6] Anderson [4.23] has some interesting thoughts on planning for flexibility and diversity.
[7] Actually, little is known about predicting variations in weightings of importance, so that this last statement can only be viewed as tentative.

opposed to another. In fact, up to this point in this chapter we have temporarily ignored a question central to study in the fields of welfare economics and political science: that of which individuals or groups *should* benefit (or be disbenefited) from various public "improvements" such as transportation facilities. In welfare economics, for example, a commonly held objective known as the *Pareto optimum* [4.24] is "the condition in which it is impossible to make some individual 'better off' without making any other 'worse off'."

Similarly, political science often has been referred to as "the study of who gets what," so that, again, gains and losses to individuals are significant matters. What is important at this point, however, is that in contrast to welfare economics and political studies, overall transportation studies rarely delineate the degree of emphasis given to one segment of the population as opposed to another. The result for the study is that the relative priorities are implied and thus not open to scrutiny and easy change. The result for the practice of problem and goal formulation is vagueness and confusion. In any case, the question of who should gain (or who does gain inadvertently)—the rich or the poor or the handicapped or the trucker or the downtown merchant, and so forth—is one that should nag at the conscience of the planner in attempting to create solutions for present and anticipated transportation problems. More will be said about this question in the discussion of evaluation and decision making in Chap. 11.

BIBLIOGRAPHY

4.1   Automotive Safety Foundation: *Urban Highways in Perspective*, Washington, D.C., 1968.
4.2   U.S. Department of Transportation: *The Freeway in the City*, Publ. No. TD2.102: F 87, U.S. Government Printing Office, Washington, D.C., 1968.
4.3   Schimpeler, C. C., and W. L. Grecco: "System Evaluation: An Approach Based on Community Structure and Values," *Highway Research Board Record* 238, 1968.
4.4   Falk, E. L.: "Measurement of Community Values: The Spokane Experiment," *Highway Research Board Record* 229, 1968.
4.5   Von Neumann, J., and O. Morganstern: *Theory of Games and Economic Behavior*, (3rd ed.), Princeton Univ. Press, Princeton, N.J., 1953.
4.6   Fishburn, P. C.: *Decision and Value Theory*, Wiley, New York, 1964.
4.7   Churchman, C. W., and R. L. Ackoff: "An Approximate Measure of Value," *Operations Research*, vol. 2, 1954.
4.8   Manheim, M. L.: *Highway Route Location as a Hierarchically-Structured Sequential Decision Process*, Dept. of Civil Engineering, M.I.T., Cambridge, 1964.
4.9   L. C. Fitch and Associates: *Urban Transportation and Public Policy*, Chandler Press, San Francisco, 1964.
4.10 Wohl, M., and B. V. Martin: *Traffic System Analysis for Engineers and Planners*, McGraw-Hill, New York, 1967.
4.11 American Association of State Highway and Transportation Officials: *A Manual on User Benefit Analysis of Highway and Bus Transit Improvements: 1977*, Washington, D.C., 1978.
4.12 Mohring, H., and M. Harwitz: *Highway Benefits: An Analytical Framework*, Northwestern University Press, Evanston, Ill., 1962.
4.13 Wachs, M.: "Relationships Between Drivers' Attitudes Toward Alternate Route and Driver and Route Characteristics," *Highway Research Board Record* 197, 1967.
4.14 Gerlough, D. L., and F. A. Wagner: *Improved Criteria for Traffic Signals at Individual Intersections*, NCHRP Report 32, Highway Research Board, Washington, D.C., 1967.
4.15 Hitch, C. J., and R. N. McKean: *The Economics of Defense in the Nuclear Age*, Atheneum, New York, 1966.
4.16 Chapin, F. S. Jr., and S. F. Weiss (eds.): *Urban Growth Dynamics in a Regional Cluster of Cities*, Wiley, New York, 1962.
4.17 Chapin, F. S. Jr., and E. Kaiser: *Urban Land Use Planning* (3rd ed.), University of Illinois Press, Champaign-Urbana, Ill., 1979.
4.18 Kent, R. J. Jr.: *The Urban General Plan*, Chandler, San Francisco, 1964.
4.19 Haworth, L.: *The Good City*, University of Indiana Press, Bloomington, 1963.

4.20 Berry, B. J. L., et al.: *A Goal Achievement Framework for Comprehensive Social-Physical Planning in the City of Chicago*, Center for Urban Studies, The University of Chicago, Chicago, June, 1968.

4.21 Fishburn, P. C.: "A Note on Recent Developments in Additive Utility Theories for Multiple Factor Situations," *Operations Research*, vol. 14, no. 6, Nov.–Dec., 1966.

4.22 Hille, S. J., and T. K. Martin: "Consumer Preference in Transportation," *Highway Research Board Record 197*, 1967.

4.23 Anderson, S. (ed.): *Planning for Diversity and Choice: Possible Futures and Their Relations to the Man-Controlled Environment*, M.I.T. Press, Cambridge, Mass., 1968.

4.24 Little, I. M. D.: *A Critique of Welfare Economies*, Oxford University Press, London, 1950.

4.25 Guilford, J. P.: *Fundamental Statistics in Psychology and Education* (4th ed.), McGraw-Hill, New York, 1965.

4.26 Osgood, C. E., G. J. Suci, and P. H. Tannenbaum: *The Measurement of Meaning*, University of Illinois Press, Urbana, 1957.

4.27 Barton-Aschman Associates, Inc.: *Guidelines for New Systems of Urban Transportation*, vol. 2, Chicago, April 1968.

4.28 American Association of State Highway Officials: *A Policy on Geometric Design of Rural Highways*, Washington, D.C., 1965.

4.29 American Association of State Highway and Transportation Officials: *A Policy on the Design of Urban Highways and Arterial Streets*, Washington, D.C., 1973.

4.30 Pardee, F. S., et al.: *Measurement and Evaluation of Transportation System Effectiveness*, Memorandum RM-3869-DOT, The Rand Corporation, Santa Monica, Calif., September 1969.

4.31 Dickey, J. W., and J. P. Broderick: "Toward a Technique for a More Exhaustive Evaluation of Urban Area Performance," *Environment and Planning*, vol. 6, no. 1, March 1972.

4.32 Winfrey, R., and C. Zellner: *Summary and Evaluation of Economic Consequences of Highway Improvements*, NCHRP Report 122, Highway Research Board, Washington, D.C., 1971.

4.33 Golob, T. F., E. T. Canty, R. L. Gustafson, and J. E. Vitt: "An Analysis of Consumer Preferences for a Public Transportation System," *Transportation Research*, vol. 6, no. 1, March 1972.

4.34 Zettle, R. M.: "On Studying the Impact of Rapid Transit in the San Francisco Bay Area," *Highway Research Board Special Report 111*, Washington, D.C., 1970.

4.35 Roget, P. M.: *Thesaurus of English Words and Phrases*, Longmans, Green and Co., Ltd., London, 1936.

4.36 Goals for Dallas: *Goals for Dallas: Submitted for Consideration by Dallas Citizens*, Graduate Research Center of the Southwest, Dallas, 1966.

4.37 Ockert, C. W., and C. E. Pixton: *A Strategy for Evaluating a Regional Highway-Transit Network*, Regional Planning Council, Baltimore, 1968.

4.38 Thurstone, L. L.: *The Measurement of Values*, 4th Impression, University of Chicago Press, Chicago, 1967.

4.39 De Chiara, J., and L. Koppelman: *Planning Design Criteria*, Van Nostrand Reinhold, New York, 1969.

4.40 Highway Research Board: "Measures of the Quality of Traffic Service," *Special Report 130*, Washington, D.C., 1972.

4.41 Benjamin, J., and L. Sen: "Comparison of Four Attitudinal Measurement Techniques," Paper presented at the Transportation Research Board Meeting, Washington, D.C., January 1980.

EXERCISES

4.1 Pick a transportation problem of interest and use the Dickey–Broderick technique to identify and describe briefly some important impacts related to the goal of reducing the problem. (Note: This procedure can get quite tedious, so limit the impacts considered to, say, 30.)

4.2 For the goal of increasing the accessibility of the handicapped to job opportunities, develop and explain briefly four possible objectives and associated criteria.

4.3 For the goal of reducing transport energy consumption, identify and explain briefly four possible standards and/or constraints.

4.4 Use the Wilson technique in Sec. 4.6 on a sample of 10 people you know. Describe and discuss in about 500 words differences between your results and those found by Wilson.

4.5 Gather and present evidence (studies, articles, quotes, etc.) about the preferences of a particular group (e.g., the elderly, construction workers, truck drivers, etc.) for different transportation characteristics such as travel time, cost, and the like. Summarize in about 400 words.

5 Models I: Transportation System Characteristics and Interrelationships

In Chap. 2, reference was made to models and their use in the transportation planning process. In that chapter a model was defined as "something which in some respects resembles or describes the structure and/or behavior of a real life counterpart." A model was also said to consist of five basic elements:

1. Variables over which the planner has complete control (X_i's)
2. Variables over which the planner has no control (Z_j's)
3. Variables over which the planner has indirect control (Y_k's)
4. Relationships between variables (R_m's)
5. Parameters (specific coefficients, constants, exponents, etc.) (P_n's)

In this and the following three chapters, we explore several of the various types of models relevant to transportation planning.

In addition to identifying model types, these chapters also will deal with the calibration and use of models, since these two aspects are so closely related. In other words, the discussion will cover both the third and fifth stages of the transportation planning process as described in Chap. 2.

This particular chapter will focus on models of transportation system characteristics and their interrelationships. In Chap. 6 we will discuss user costs, as well as land use and external factors. This will be followed in Chap. 7 by models of travel behavior and in Chap. 8 by impact models.

5.1 MODELS IN THE TRANSPORTATION PLANNING PROCESS

The modeling stage of the transportation planning process has been divided into six phases in Chap. 2. These phases and their interconnections are diagrammed in

Fig. 5.1. It is perhaps easiest to understand this diagram by first looking at the inputs and outputs of the calibration and model-use stage (stage V) itself and then looking at its internal workings.

The major inputs to all phases of stage V come from data collection, stage IV of the transportation planning process. In addition, phase (c) of stage V—estimation of transportation system performance characteristics—can only be accomplished through knowledge of the type of transportation system being considered as a solution. This input would come from stage VII (generation of alternative solutions stage), in which new solutions are formulated. This connection is represented by the dashed vertical arrow on the right-hand side of Fig. 5.1. The outputs of the calibration and model use stage serve as inputs to the evaluation and decision-making stage (VI).

Looking now at the internal workings of stage V, we see that the prediction of the location and extent of human activities [phase (d)] requires informational inputs from three of the other five phases and, as such, provides a good basis for explaining the interaction between the phases. To elaborate further, an activity model is one that is used to estimate the future (zonal) locations and intensities of population, employment, income, and land areas.[1] The factors that seem to affect the level of activities in any section of a metropolitan region are: (1) the projected overall influence of external factors (e.g., national economic conditions), (2) the past and present intensity of activities in each zone, and (3) the past, present, and future levels of certain other variables, primarily transportation, water supply, and sewage-disposal facilities, zoning, and the quality of schools and other municipal services. Inputs from phases (a), (b), and (c) of Fig. 5.1 thus can be identified directly, with the projected levels of external factors such as regional growth of population and employment coming from phase (a), information on present and future municipal services other than transportation coming from (b), and, of course, the transportation system performance characteristics coming from (c). Information on present zonal activities is drawn from the output of the data-collection stage (IV) of the planning process.

Looking again at Fig. 5.1, we see that both the level of future zonal activities and the performance characteristics of the proposed transportation system are the principle factors which influence future travel patterns. This relationship, to be discussed in more detail in Chap. 7, is a natural one since population and employment centers are the focus of most travel and since this travel cannot be done without transportation facilities.

The final relationship in Fig. 5.1 involves other affected factors such as noise levels, appearance, smell, community disruption, and so forth. Most of these impacts stem directly from the transportation system, travel on it, and land use. Noise, for example, might be created by traffic on a street (transportation system) and is heard by residents of abutting buildings (land use).

After estimating these additional impacts, we would proceed to the evaluation stage where all future impacts—land use, travel, noise, relocation, and so on—would be examined and compared to established goals and objectives. If deemed unsatisfactory, these impacts could be altered through modifications of the transportation system. These changes, which would be made in the generation stage (stage VII) of the planning process (see dashed arrow in Fig. 5.1), would be fed back to the transportation system in phase (c), which in turn would be fed into and influence subsequent changes in zonal land use, travel, and the other affected factors. After several iterations through this feedback process, (hopefully) desirable patterns of impacts would evolve,

---

[1] In Chap. 2, many of these activities came under the heading of "land use."

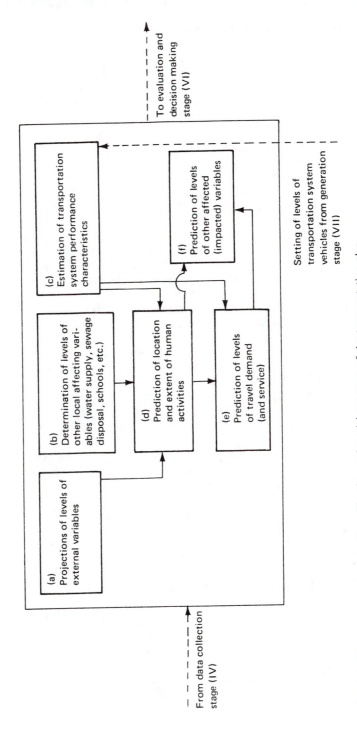

Fig. 5.1 The six major phases in the calibration and model use stage of the transportation planning process.

93

at which time the corresponding transportation system modifications would be specified in more detail (stage VIII), implemented (stage IX), and then operated and maintained (stage X).

From the preceding description of the phases of the calibration and model-use stage it should be apparent that all five elements of a model are involved. First, it is generally assumed that the transportation planner has complete control of the transportation system and can mold its component parts—vehicles, networks, terminals, and controls—to any type of transportation solution that may seem desirable.[2] The transportation system thus is considered to be an X variable.

Other factors cannot be controlled by the transportation planner, however. The growth of regional population and employment usually is assumed to be independent of the planner's actions, as is the quality, location, and timing of construction of water supply and sewage disposal facilities, schools, and other public services. These generally are under the control of other agencies and firms and thus must be considered as uncontrollable or Z variables. Even some transportation system performance characteristics are affected by factors (not shown in Fig. 5.1), such as climate and geologic conditions, that have to be accepted as uncontrollables by the planner. The only action that can be taken with respect to these factors is to estimate their future levels and use these estimates to help predict zonal activities, travel demand (and service), and other relevant impacts.

We see from the above discussion that zonal activities, travel demands, and other impacts are indirectly controllable variables (Y's). Transportation system changes (X's) can be created that can affect, say, subregional land use growth. The construction of radial freeways, for example, might encourage the flight of higher income families to the suburbs. Yet this migration probably would also be aided by changes in school systems, existing land use patterns, regional growth and many other factors (Z's) outside the control of the transportation planner. Hence, transportation plays only a partial role in land development, which thus must be considered a Y variable.

The relationships and parameters in models connecting the X, Y, and Z variables will not be discussed in detail at this point. One example may prove helpful, however. In the *Chicago Area Transportation Study* [5.19], the percent of urban trips by transit was found to correlate with the population per 1,000 sq ft of residential land (considered here to be a Z variable) and the number of autos per 1,000 people in a zone (considered here to be an X variable). The associated equation was (from Ref. [5.20])

$$Y = 15.5 + 21.745 \log Z - 16.72 \log X \qquad (5.1)$$

The relationship between variables is a linear one except that a logarithm is taken of each independent variable before the additions are made. The parameters are the constant 15.5 and the coefficients 21.745 and 16.72. In other cities these parameters may vary, but they are treated as constants in Chicago.

Additional relationships will be displayed throughout this and the succeeding chapters. In many cases it will be difficult to specify controllable, uncontrollable, and indirectly controllable variables. This will be left to the reader. Nonetheless, the distinction should be kept in mind since it is desirable to foresee which variables have to

---

[2] In reality the planner obviously does not have complete control over the entire transportation system. However, this role generally is assumed temporarily so that *recommendations* can be made concerning the system.

be forecasted before using a particular model, which are forecast within the model, and which can be varied by the planner.

5.2 TRANSPORTATION PERFORMANCE CHARACTERISTICS

In Chap. 2 transportation systems were described as being composed of four major components—vehicles, networks, terminals, and controls. Each of these components has certain performance characteristics, some of which are listed in Table 5.1. In this table, the characteristics are divided into two categories: (*a*) mechanics and (*b*) construction and operation. The former category contains those characteristics commonly associated with the strength and motion of a transportation system, while those in the latter category are more akin to the physical development of a system and its use. As an example, costs, while related to the characteristics of strength and power, fall more reasonably under the heading of a "construction and operation" characteristic.

Each of the four transportation system components generally has each of the characteristics listed in Table 5.1, although there are many exceptions. A good illustration is that of capacity. Vehicles have capacities—50 seated persons per bus; networks have capacities—2,000 vehicles per hour per lane on a freeway; and terminals and control systems have capacities—a Greyhound Bus terminal built in Los Angeles can handle 500 buses and 14,000 passengers per hour [5.21], while a fixed-time traffic signal, under ideal conditions, can pass from 1,200 to 1,500 vehicles per hour of green time. Table 5.1 thus contains a list useful for identifying and modeling many of the important performance characteristics of most transportation system components.

While each of these characteristics is important to someone who deals with particular aspects of transportation (e.g., strength and weight of a bridge to the structural engineer), not all are of concern to the planner. Those that we have picked out for further discussion in this regard include speed and acceleration in the first set and

Table 5.1 Some Performance Characteristics of Transportation Systems

I   Mechanics (statics, kinematics, and dynamics)

1. Strength	5. Speed-velocity	9. Lift
2. Weight	6. Acceleration	10. Drag
3. Stress	7. Power	11. Friction
4. Strain	8. Stability	12. Heat-temperature

II   Construction and operation

1. Volume	9. Cleanliness	17. Service life
2. Density	10. Sanitariness	18. Costs
3. Headway	11. Appearance	(a) Construction or purchase
4. Spacing	12. Privacy	(b) Land
5. Capacity	13. Smell	(c) Labor
6. Safety	14. Comfort	(d) Fuel
7. Flexibility	15. Light	(e) Operation and maintenance
8. Reliability	16. Noise	(f) Finance
		(g) Engineering and management

volume, density, capacity, and costs in the second. Some other characteristics, such as noise and appearance, are treated in detail in Chap. 8 on "impacts."

5.3 VOLUME, SPEED, ACCELERATION, AND DENSITY

Perhaps the most discussed characteristics of highway traffic are those of volume, speed, and capacity. Volume, as defined in the *Highway Capacity Manual* [5.15], *is a flow*, not a static quantity or amount. Properly expressed, volume is "the number of vehicles that pass over a given section of a lane or a roadway during a time period of one hour or more." Oftentimes it is necessary to refer to various averages and peaks of volumes occurring during different periods of time. The most common definitions are:

Average annual daily traffic (AADT)—The total yearly volume divided by the number of days in the year.

Average daily traffic (ADT)—The total volume during a given time period in whole days greater than 1 day and less than 1 year divided by the number of days in that time period.

Tenth, twentieth, thirtieth, etc. highest annual hourly volume (10HV, 20HV, 30HV, etc.)—The hourly volume on a given roadway that is exceeded by 9, 19, 29, etc. hourly volumes, respectively, during a designated year.

Peak-hour traffic—The highest number of vehicles found to be passing over a section of a lane or a roadway during 60 consecutive minutes.

Rate of flow—The hourly representation of the number of vehicles that pass over a given section of a lane or a roadway for some period less than 1 hour.

This list of definitions starts with the longest time period usually considered, the year, and ends with the shortest, some period less than an hour. Each attempts to indicate expected variations in traffic volumes. The last one corresponds to the shortest time period. A flow of, say, 600 vehicles in 15 min would be presented as an hourly volume of 2,400 vph *if the flow were to continue at its present rate*.

The concept of speed,[3] like volume, also can be expressed in a variety of ways. The most important distinction arises from the manner in which speed is measured— either in terms of time taken to pass a given (small) length of traveled way or in terms of distance covered in a very small interval of time. The former of these two concepts leads to the idea of "spot speed":

Spot speed—The time a vehicle takes to cover a specified (small) length on a roadway divided into that length. The averaging of spot speeds creates an "average spot speed" or, as it is better known:

Space mean speed $(\bar{U}_s)$—The total of the small, given distances on a roadway traveled by a set of vehicles divided by the sum of the times all vehicles take to traverse that distance. Stated symbolically this definition becomes

$$\bar{U}_s = \frac{n\,\Delta x}{\displaystyle\sum_{i=1}^{n} \Delta t_i} \tag{5.2}$$

[3]The word "speed" usually is used in Traffic Engineering instead of "velocity" since the direction of the motion of the vehicle is apparent.

where $\bar{U}_s$ = space mean speed for a set of n vehicles

Δx = a fixed, small distance on a roadway

Δt_i = time vehicle i takes to traverse distance Δx

The second means of measurement of average speed, involving a fixed time interval instead of a fixed distance, is

Time mean speed $(\bar{U}_t)$—The sum of the distances on a roadway traversed by a set of vehicles divided by the total (over all vehicles) of the small, given intervals of time needed by each vehicle to traverse the corresponding distance. Put in a form comparable to that of Eq. (5.2), the time mean speed is

$$\bar{U}_t = \sum_{i=1}^{n} \Delta x_i / n \, \Delta t \tag{5.3}$$

where Δx_i = the distance traveled by vehicle i in fixed time interval Δt.

The actual difference between these two definitions of speed may not be obvious until one considers two hypothetical situations involving the use of a series of time-coded movie films of traffic flow along a freeway lane on which stripes are painted at 1-ft intervals. In both situations, two drivers are given instructions to drive at a speed (car speedometer) of 20 and 40 ft/sec (13.5 and 27 mph), respectively, but in the first situation we determine how much time it takes for each car to cover an 80-ft section of the freeway lane, whereas, in the second, we determine the distance which each car travels in 1 sec. Thus, $\Delta x = 80$ ft, $\Delta t_1 = 80$ ft/20 fps = 4 sec, $\Delta t_2 = 80$ ft/40 fps = 2 sec, and $\Delta t = 1$ sec, $\Delta x_1 = 20$ fps (1 sec) = 20 ft, and $\Delta x_2 = 40$ fps (1 sec) = 40 ft. Then $\bar{U}_s$ can be calculated as

$$\bar{U}_s = \frac{2 \, (80 \text{ ft})}{4 \text{ sec} + 2 \text{ sec}} = \frac{160}{6} = 26.67 \text{ ft/sec}$$

while

$$\bar{U}_t = \frac{20 \text{ ft} + 40 \text{ ft}}{2 \, (1 \text{ sec})} = \frac{60}{2} = 30.00 \text{ ft/sec}$$

An obvious difference exists, indicating that any further description of the average speed of a transportation mode ought to be referenced to the type of measurement performed. Otherwise some discrepancies might arise. Unfortunately, the means of measurement usually is not specified in many studies, so that the reader must judge from the general context of the presented information whether time or a space mean speed is employed.

For example, if the average speed of vehicles on a transit facility were given, one might suspect that it was calculated by using various time measurements over the prescribed transit route, thereby leading to a space mean speed. On the other hand, if the average speed of automobiles on a highway were computed using readings from the speedometer of each car, the average speed in this case probably would be a time mean speed since speedometers generally measure the number of revolutions of the wheel, and therefore the distance traveled, per unit of time[4] (note that in the previous example the average of the speedometer speeds was $(40 + 20)/2 = 30$ fps $= \bar{U}_t$).

[4] The unit of time is implicit in the mechanical operation of the speedometer.

A final illustration concerning the definition and measurement of speed regards the situation in which neither the distance nor time intervals are fixed for measurement. In this case an average equal to neither $\bar{U}_t$ or $\bar{U}_s$ might be found. If, for instance, the 20 and 40 ft/sec speedometer measurement were made for time a duration of 3 and 5 sec, respectively (implying that different distances were covered), the average speed would be

$$\frac{(20 \text{ ft/sec})(3 \text{ sec}) + (40 \text{ ft/sec})(5 \text{ sec})}{3 \text{ sec} + 5 \text{ sec}} = \frac{60 + 200}{8} = \frac{260}{8} = 32.50 \text{ fps}$$

The result is a mean speed differing from both of the previous two. No specific name has been given to this, perhaps because it is hoped that no one will be tempted to calculate an average speed in this manner. However, measurements of this sort can be made inadvertently as, for example, by summing bus travel times over routes of varying lengths and then making the division. To eliminate such occurrences, the analyst must keep constant watch on the manner in which various travel time and distance data are combined.

Other speed definitions of relevance to the specification of transportation systems are:

Design speed—A speed selected for purposes of design and correlation of those features of a traveled way, such as curvature, superelevation, and sight distance, upon which the safe operation of vehicles is dependent.

Average highway speed (AHS)—The weighted (by length of subsection) average of the design speeds within a section of a traveled way, when each subsection within the section is considered to have an individual design speed.

Operating speed—The highest overall speed at which a vehicle can travel on a given traveled way under favorable weather conditions and under prevailing traffic conditions without at any time exceeding the safe speed as determined by the design speed on a section-by-section basis.

The definitions are almost self-explanatory.

Density was the third characteristic mentioned earlier in conjunction with volume and speed. Also referred to as *concentration*, it can be defined as:

Density (D)—The number of vehicles occupying a unit length of the through lanes of a traveled way in any period (very small) of time.

The easiest way to visualize what is meant by density is to consider an aerial photograph of a highway at some "instant" (actually the small time interval needed to take the picture). A count could be made of the number of vehicles along, say, a measured mile of the highway, giving the density in vehicles per mile. Again, as was true of speed, we could fix either the time period or distance in defining an average density. However, the usual definition is that where the latter entity is held constant. Volume, too, usually is "averaged" over time so that it would be consistent to use figures for volume measured over a fixed small distance in conjunction with space mean speeds and densities.

Several other characteristics of a traffic stream should be defined explicitly. Their meanings are fairly straightforward:

Delay—The time consumed while traffic or a specified component of traffic is impeded in its movement by some element over which it has no control.

Vehicular gap—The interval in time or distance between individual vehicles measured from the rear of one vehicle to the head of the following vehicle.

Spacing—The interval in *distance* from head to head of successive vehicles.
Headway—The interval in *time* from head to head of successive vehicles as they pass a given point.

These characteristics will be alluded to throughout the remainder of the text.

Some of the relationships between volume, speed, density, spacing, headway, and unit (per distance) travel time can be stated quite simply. For instance,

$$V = \bar{U}_s D \qquad (5.4)$$

$$\bar{U}_s = Vs \qquad (5.5)$$

$$D = Vm \qquad (5.6)$$

$$s = \bar{U}_s h \qquad (5.7)$$

$$h = ms \qquad (5.8)$$

$$m = Dh \qquad (5.9)$$

where V = volume
$\bar{U}_s$ = space mean speed[5]
D = density
s = spacing
m = unit travel time
h = headway

The first equation, while not having an obvious interpretation, perhaps can be understood by imagining a paper with a given density of dots on it being slid past a point (represented by the arrow). The greater the density of points on the paper

and the greater the speed at which it is pulled, the greater the number of dots that will pass the point per unit of time—a greater volume. An analogous relationship would exist in freeway flow, with the density and speed variables being multiplied together in order to obtain consistent units

$$\left(\frac{veh}{mi}\right) \times \left(\frac{mi}{hr}\right) = \left(\frac{veh}{hr}\right)$$

Similar analogies can be developed to explain the remaining relationships, yet the "consistent units" idea is one of the easiest ways to remember and understand each equation. For example, if we know the average headway (sec/veh) and wanted to determine the average spacing (ft/veh), we would have to multiply the headway by ft/sec (the speed $\bar{U}_s$) to get the correct units. This would mean that

---

[5] For the most part we will be dealing with averages over time (with distance fixed) so that the space mean speed is the proper measure to be employed.

$$s \quad = \quad h \qquad \bar{U}_s$$

$$\left(\frac{ft}{veh}\right) = \left(\frac{sec}{veh}\right) \times \left(\frac{ft}{sec}\right)$$

which is equivalent to Eq. (5.7).

5.3.1 Acceleration, Deceleration, and Cruise Speed

Primary constraints on top speed for any passenger carrying vehicle are its acceleration and deceleration characteristics. At a given acceleration it takes a finite time and distance for a vehicle to reach its top speed. The same is true for decelerating to a stop. The distance remaining between two terminals after acceleration and deceleration would be traveled at the cruise speed.

For constant acceleration and starting at zero initial velocity,

$$S_A = \tfrac{1}{2} A t^2 \tag{5.10}$$

and since

$$t = U/A \tag{5.11}$$

then

$$S_A = \tfrac{1}{2} U^2/A \tag{5.12}$$

where U = cruise speed
A = rate of acceleration
S_A = distance covered during acceleration
t = time

If there were a situation requiring a low initial velocity, such as maintaining a low constant speed until the vehicle was a certain distance from the terminal, then

$$S_A = \tfrac{1}{2} A t^2 + U_0 t_0 \tag{5.13}$$

where U_0 is the initial speed and t_0 is the time at the initial speed. Or

$$S = \tfrac{1}{2} U^2/A + U_0 t_0 \tag{5.14}$$

So it can be seen that the distance required for a vehicle to reach its cruise speed is primarily dependent upon acceleration. The maximum acceleration possible depends upon the design capability of the vehicle, the track or roadway, and the passenger. For a non-air-cushion vehicle, the amount of friction available between the wheels and the roadway becomes a determining factor. With a flat grade and a maximum coefficient of friction of 0.4 (a not uncommon value for rubber wheels on a wet pavement), the maximum value of A would be about 8.8 mph/sec. An acceleration of this magnitude would not be too severe providing the passengers were either facing forward or were adequately strapped in their seats. It would be much too high if doing other activities (such as serving meals) were desired. If 8.8 mph/sec could be tolerated, and if a high-speed vehicle were desired at about 200 mph top speed, then by Eq. (5.12) and converting mph to ft/sec,

$$S_A = \tfrac{1}{2} (200 \times 1.47)^2 /(8.8 \times 1.47) = 3,333 \text{ ft}$$

or about two-thirds of a mile. If a more comfortable acceleration were used, say on the order of 2 mph/sec, the distance to reach top speed would be

$$\frac{8.8}{2} \times 3,333 = 14,665 \text{ ft}$$

or about $2\tfrac{3}{4}$ miles. An uphill gradient would, of course, increase this value.

On the deceleration phase everything is reversed but about the same numbers result. At high deceleration rates, passengers sitting face-forward must be strapped in their seats. Maximum deceleration would be 8.8 mph/sec. About 2 mph/sec would again represent a slow, comfortable deceleration, similar to that of an automobile slowing down in gear with only slight brake application.

From this discussion it can be seen that where comfort is to be maximized (low acceleration and deceleration rates) and a high cruise speed is to be used, long distances between stops are required to take advantage of the high-speed capabilities of the equipment. Where short distances between stops are required, the high-speed capability would be wasted.

5.3.2 Freeway Speed–Volume–Density Relationships

The relationship between certain characteristics many times cannot be established as easily as was done in the previous paragraph. Often observations must be obtained from the field and curves must be "fitted" to these observations either by hand or through some technique such as regression. Figures 5.2 through 5.4 exhibit three pairs of empirically established relationships between volume, density, and speed for uninterrupted flow [5.23]. It has been found that the data points can be approximated best with two-part logarithmic functions suggested by Edie, with the breakpoint coming, interestingly enough, at the point where the volume of traffic is approximately equal to the lane capacity.[6] Actually, this occurrence would come as no surprise to anyone who has observed flow on a heavily traveled freeway for any period of time. In order to get capacity flow, traffic must proceed in an extremely orderly fashion. Otherwise, any small disturbance, such as a quick acceleration or deceleration, would create large headways between vehicles and subsequent smaller volumes. With capacity flow being so sensitive to small interruptions, a rather unstable condition at capacity can be expected, a situation reflected in the breakpoints in the volume–density–speed relationships.

While it is difficult to remember the values of each parameter in the mathematical equations indicated in Figs. 5.2 through 5.4, the reader should be able to recall the *general form* of each by thinking of them as being essentially continuous in nature and by bearing in mind the boundary conditions for each characteristic. If, for instance, vehicles averaging 20 ft in length were lined up solidly along a 1-mile stretch of roadway lane, the density would be (5,280 ft mile)/(20 ft/veh) = 264 veh/mile, and we could anticipate that the speed as well as the volume would be zero[7] since no vehicles would be moving in this situation.

At the other extreme of density—only a single vehicle on the 1-mile stretch—the

---

[6] See Sec. 5.4 for definitions and models of capacity.

[7] The volume would be zero since no vehicles could move in order to pass a point on the roadway.

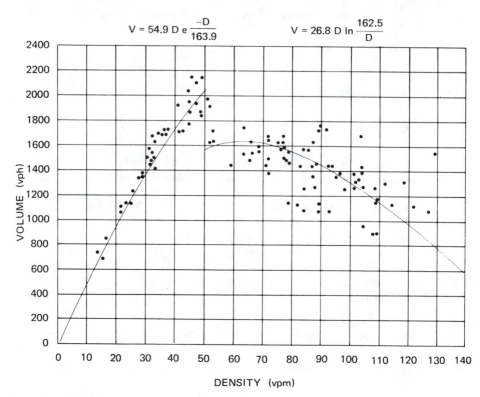

Fig. 5.2   Volume–density relationship—Edie hypothesis [5.23].

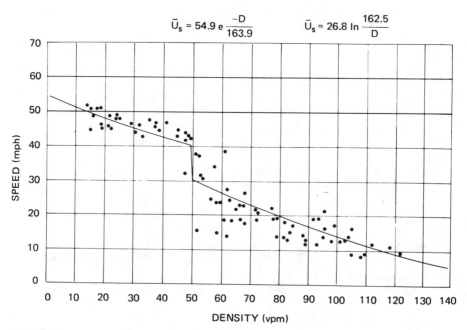

Fig. 5.3   Speed–density relationship—Edie hypothesis [5.23].

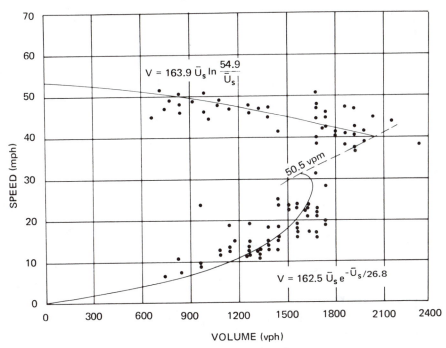

Fig. 5.4 Speed–volume relationship—Edie hypothesis [5.23].

speed probably would be very high because of the lack of interference, whereas the volume, as in the preceding case, would be very low (almost 0). These conditions, along with the constraint that the volume cannot exceed the capacity (which occurs at some intermediate density and speed), would lead to the type of diagrams in Fig. 5.5. The average speed of vehicles on a section of road is called the *mean free speed* and is denoted by $\bar{U}_f$.

Returning to the Edie models of density, volume, and speed, we find that the model relating the latter two characteristics is fairly reliable, with a correlation coefficient of 0.83 and a standard error of regression of 3.55 mph. The analyst can therefore place some amount of faith in the relationship although, as is demonstrated by the scatter of observation points in Fig. 5.4 as well as by the correlation coefficient, *complete* trust is not warranted.

5.3.3 Speeds on Other Types of Highways

The preceding discussion has brought out the general point that highway speeds decrease as the volume of traffic approaches the capacity. This is shown for a variety of highway settings in Table 5.2. On a two-way arterial without parking, as an example, the capacity in a fringe area location is approximately 800 vph per lane. When traffic is light (volume to capacity ratios less than 0.25), average speeds of 25 mph are obtained (29 mph if there is full traffic-signal progression). On the other hand, as the volume gets close to the capacity, average speeds drop to about 15 mph, even if traffic signals are set for progression. Congestion simply is so great that the interference keeps down speeds and prevents vehicles from attaining the consistent flow needed for progression.

Comparisons within Table 5.2 also show that speeds will increase roughly 30% in

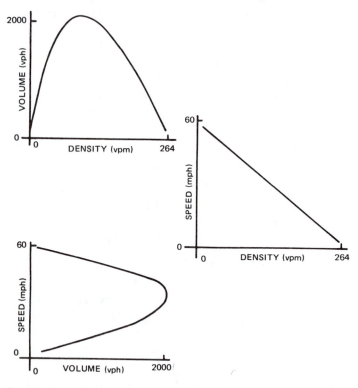

Fig. 5.5 General relationships between speed, volume, and density.

going from CBD to CBD fringe locations and 50% from CBD to residential (the lowest density areas). By contrast, moving up from the lowest type of facility (arterial two-way, with parking) to the highest (freeway) results in speed increases of approximately 150%.

5.3.4 Transit Speeds

The speeds of various transit operations depend on several factors. All forms require stopping to load and unload passengers, so that travel time is added for deceleration, actual loading/unloading (dwell time), and acceleration back to the maximum speed possible. For buses, trolley buses, and other modes usually caught in the motor-vehicle traffic stream, the speed of that stream conditions the maximum possible. For transit having its own right-of-way, the technological capabilities of the vehicle, along with passenger comfort and safety, determine the maximum possible speed. On the other hand, if stations are spaced too closely together, the train or individual vehicle may not be able to reach its cruise speed before it has to decelerate again.

The equations employed to estimate average speed (or mph) for a transit vehicle on its own right-of-way are based on Eqs. (5.10) to (5.12). When station spacing is sufficient to reach cruise speed,

$$\bar{U} = \frac{3600S}{(U/2A) + (U/2B) + (3600S/U) + t_D} \tag{5.15}$$

When station spacing is insufficient,

$$\bar{U} = \frac{3600}{[7200(A + B)/(SAB)]^{1/2} + t_D/S} \tag{5.16}$$

where A = (constant) acceleration rate (mph per sec)
B = (constant) deceleration rate (mph per sec)
U = cruise speed (mph)
S = station spacing (miles)
t_D = dwell time (sec)
$\bar{U}$ = average speed (mph)

**Table 5.2 Per Lane Capacity and Average Speed (mph)
on Various Roadways at Different Locations**

Location	Capacity* (vph)	Speed (mph) at different percent ratios			
		0	0.50	0.75	1.00
Freeway					
Central business district	1750	48	38	33	28
Fringe	1750	48	38	33	28
Residential	1750	67	57	50	34
Outlying business district	1750	58	48	41	30
Expressway					
Central business district	800	37	34	33	31
Fringe	1000	44	38	35	32
Residential	1100	47	44	41	38
Outlying business district	1000	37	34	33	31
Two-Way Arterial with Parking					
Central business district	400	17-22†	17-20	15-15	12-12
Fringe	550	25-29	20-27	18-25	15-15
Residential	550	28-32	25-30	23-28	15-15
Outlying business district	550	22-24	20-22	18-18	13-13
Two-Way Arterial without Parking					
Central business district	600	17-22	17-20	15-15	12-12
Fringe	800	25-29	20-27	18-25	15-15
Residential	800	28-32	25-30	23-28	15-15
Outlying business district	800	22-24	20-22	18-18	13-13
One-Way Arterial					
Central business district	700	17-22	17-20	15-15	12-12
Fringe	550	25-29	20-27	18-25	15-15
Residential	900	28-32	25-30	23-28	15-15
Outlying business district	650	22-24	20-22	18-18	13-13

*Capacity calculated at level of service E of Fig. 5.6, absolute capacity.
†First value shows speed assuming lack of coordinated signal progression; second value shows speed assuming full signal progression.
Source: [5.1, p. iv-5 to iv-6; as based on 5.15].

To check on the station spacing, we calculate

$$S \leqslant (U^2/7200A) + (U^2/7200B) \quad ? \tag{5.17}$$

To illustrate the utilization of these equations, suppose that a station spacing of 0.30 miles is proposed, with a rail vehicle capable of a cruise speed of 50 mph and acceleration and deceleration rates each of 2 mph/sec. The dwell time will average 15 sec. First, use is made of Eq. (5.17):

$$[50^2/7200(2)] + [50^2/7200(2)] = 0.35 \text{ miles}$$

This is greater than the proposed spacing, so Eq. (5.16) is employed next:

$$\bar{U} = \frac{3600}{[7200(2 + 2)/0.30(2)(2)]^{1/2} + (15/0.30)} = 17.6 \text{ mph}$$

If the spacing had been greater than 0.35, Eq. (5.15) would have been used.

Equations like those shown can help in forecasting average transit system speeds for a variety of alternate station spacings and passenger loads (which affect dwell times). Tradeoffs between speed and station accessibility then can be made under more informed circumstances. Actual speeds may vary from those predicted, however. Table 5.3 gives some typical values for selected rail systems in the United States and Europe. Those presented for light rail depend heavily on the type of right-of-way and signal control. The Stadtbahn Linie A, opened in Hamburg in 1978, for example, runs in part through tunnels and thus can obtain an operating speed of 23 mph. And the Market St. tunnel line in San Francisco is due to have operating speeds of 35 mph [5.2, p. 573].

Bus speeds also depend on surrounding traffic. Express buses, particularly on a separate freeway line, can obtain operating speeds of 45 mph. On the other hand, local buses on collector streets in the peak hour may average only 5 mph. This certainly leaves much to be desired. Dedicated bus lanes in the CBD help. A weighted average speed in nine major city CBD's for such facilities was 9.4 mph, with the Fifth Avenue line in New York being about 11.6 mph [5.6, Table C-1].

At the pedestrian level, walking speeds vary from 1.36 to 4.43 mph, with an average being about 2.98 [5.6, p. II-1]. This speed depends on age, of course. Fruin

Table 5.3 Typical Operating Speeds for Rail Facilities

Station spacing (miles)	Speed (mph)		
	Rail rapid transit	Commuter rail	Light rail
0–0.25	20–25	20–30	9.9–14.3
0.25–0.50	20–25	20–30	9.3–18.6
0.50–1.00	20–25	20–30	
1.00–2.00	35–40	20–30	
2.00–3.00	45–50	28–35	
3.00–5.00	50–55	33–40	
5.00–6.00	50–55	38–45	

Source: [5.2; various pages, adapted by permission of Prentice-Hall, Inc., Englewood Cliffs, N.J.].

Table 5.5 Passenger Car Equivalents of Trucks on Freeways and Expressways, on Specific Individual Subsections or Grades

Grade (%)	Length of grade (mi)	Passenger car equivalent, E_T				
		3% Trucks	5% Trucks	10% Trucks	15% Trucks	20% Trucks
0-1	All	2	2	2	2	2
2	$\frac{1}{4}$ to $\frac{1}{2}$	5	4	4	3	3
	$\frac{3}{4}$ to 1	7	5	5	4	4
	$1\frac{1}{2}$ to 2	7	6	6	6	6
	3 to 4	7	7	8	8	8
3	$\frac{1}{4}$	10	8	5	4	3
	$\frac{1}{2}$	10	8	5	4	4
	$\frac{3}{4}$	10	8	5	4	5
	1	10	8	6	5	6
	$1\frac{1}{2}$	10	9	7	7	7
	2	10	9	8	8	8
	3	10	10	10	10	10
	4	10	10	11	11	11
4	$\frac{1}{4}$	13	9	5	4	3
	$\frac{1}{2}$	13	9	5	5	5
	$\frac{3}{4}$	13	9	7	7	7
	1	13	10	8	8	8
	$1\frac{1}{2}$	13	11	10	10	10
	2	13	12	11	11	11
	3	13	13	14	14	14
	4	13	14	16	16	15
5	$\frac{1}{4}$	14	10	6	4	3
	$\frac{1}{2}$	14	11	7	7	7
	$\frac{3}{4}$	14	11	9	8	8
	1	14	13	10	10	10
	$1\frac{1}{2}$	14	14	13	13	13
	2	14	15	15	15	15
	3	14	17	17	17	17
	4	16	19	22	21	19
6	$\frac{1}{4}$	15	10	6	4	3
	$\frac{1}{2}$	15	11	8	8	8
	$\frac{3}{4}$	15	12	10	10	10
	1	15	14	13	13	11
	$1\frac{1}{2}$	15	16	15	15	14
	2	15	18	18	18	16
	3	15	20	20	20	19
	4	20	23	23	23	23

Source: [5.15, p. 258].

Table 5.6 Passenger Car Equivalents of Intercity Buses on Freeways and Expressways, on Specific Individual Subsections or Grades

Grade* (%)	Passenger car equivalent,† E_B
0–4	1.6
5‡	2
6‡	4
7‡	10

*All lengths.
†For all percentages of buses.
‡Use generally restricted to grades over .5 mile long.
Source: [5.15, p. 260].

the first three. In this case, however, a transformation first is made of trucks and intercity buses driving on a given length and percent of grade to an equivalent number of passenger cars (E_T and E_B, respectively). Then the adjustment factors, designated T for trucks and B for intercity buses, are determined based on the previously found equivalency values. The reason for this intermediate transformation lies in the theory that a truck or bus, because of its slowness in climbing various lengths and steepnesses of grades, takes up a space or creates a headway in the traffic stream that could be filled by some much greater, equivalent number of passenger cars. In any real situation the equivalency will vary according to the speed of the truck or bus at the bottom of the hill, the number of similar vehicles directly following each other, the horsepower and torque of the vehicles, and so forth, so that the equivalency number actually pertains to the "average" influence of a truck or bus.

The charts for determining the passenger car equivalents and corresponding adjustment factors are displayed in Tables 5.5 and 5.6. The second of these is for bus equivalents and, as can be seen, gives numbers that generally are lower than for trucks. The bus or truck equivalents then are entered into the left-hand column of Table 5.7, and the adjustment factor is found under the respective percentage column. By way of illustration, suppose that there were 10 percent trucks (and city buses) and 5 percent intercity buses on a 3 percent grade 1 mile long. What would be the values for T and B? We find from Table 5.5 that $E_T = 6$ while from Table 5.6 that $E_B = 1.6$. Entering Table 5.7, we then determine that the adjustment factor for trucks (T) is 0.67 and for buses (B), through interpolation, 0.94.

The calculation of the actual capacity of a section of a freeway now can be found by multiplying the 2,000 vph/lane figure by the number of lanes and the three adjustment factors, W, T, and B. Thus

$$C = 2,000\,NWTB \tag{5.18}$$

where C is the capacity of a given section of freeway and N is the number of lanes. Using the preceding example situation and adjustment factors, we would arrive at a (one-directional) capacity of

$$C = 2,000(3)(0.77)(0.67)(0.94) = 2,910 \text{ vph}$$

Table 5.7 Adjustment Factors* for Trucks and Buses on Individual Roadway Subsections or Grades on Freeways and Expressways (Incorporating Passenger Car Equivalent and Percentage of Trucks or Buses)†

Passenger car equivalent, E_T or E_B	Truck adjustment factor T, (B for buses)‡ Percentage of trucks, P_T (or of buses, P_B) of:														
	1	2	3	4	5	6	7	8	9	10	12	14	16	18	20
2	0.99	0.98	0.97	0.96	0.95	0.94	0.93	0.93	0.92	0.91	0.89	0.88	0.86	0.85	0.83
3	0.98	0.96	0.94	0.93	0.91	0.89	0.88	0.86	0.85	0.83	0.81	0.78	0.76	0.74	0.71
4	0.97	0.94	0.92	0.89	0.87	0.85	0.83	0.81	0.79	0.77	0.74	0.70	0.68	0.65	0.63
5	0.96	0.93	0.89	0.86	0.83	0.81	0.78	0.76	0.74	0.71	0.68	0.64	0.61	0.58	0.56
6	0.95	0.91	0.87	0.83	0.80	0.77	0.74	0.71	0.69	0.67	0.63	0.59	0.56	0.53	0.50
7	0.94	0.89	0.85	0.81	0.77	0.74	0.70	0.68	0.65	0.63	0.58	0.54	0.51	0.48	0.45
8	0.93	0.88	0.83	0.78	0.74	0.70	0.67	0.64	0.61	0.59	0.54	0.51	0.47	0.44	0.42
9	0.93	0.86	0.81	0.76	0.71	0.68	0.64	0.61	0.58	0.56	0.51	0.47	0.44	0.41	0.38
10	0.92	0.85	0.79	0.74	0.69	0.65	0.61	0.58	0.55	0.53	0.48	0.44	0.41	0.38	0.36
11	0.91	0.83	0.77	0.71	0.67	0.63	0.59	0.56	0.53	0.50	0.45	0.42	0.38	0.36	0.33
12	0.90	0.82	0.75	0.69	0.65	0.60	0.57	0.53	0.50	0.48	0.43	0.39	0.36	0.34	0.31
13	0.89	0.81	0.74	0.68	0.63	0.58	0.54	0.51	0.48	0.45	0.41	0.37	0.34	0.32	0.29
14	0.88	0.79	0.72	0.66	0.61	0.56	0.52	0.49	0.46	0.43	0.39	0.35	0.32	0.30	0.28
15	0.88	0.78	0.70	0.64	0.59	0.54	0.51	0.47	0.44	0.42	0.37	0.34	0.31	0.28	0.26
16	0.87	0.77	0.69	0.63	0.57	0.53	0.49	0.45	0.43	0.40	0.36	0.32	0.29	0.27	0.25
17	0.86	0.76	0.68	0.61	0.56	0.51	0.47	0.44	0.41	0.38	0.34	0.31	0.28	0.26	0.24
18	0.85	0.75	0.66	0.60	0.54	0.49	0.46	0.42	0.40	0.37	0.33	0.30	0.27	0.25	0.23
19	0.85	0.74	0.65	0.58	0.53	0.48	0.44	0.41	0.38	0.36	0.32	0.28	0.26	0.24	0.22
20	0.84	0.72	0.64	0.57	0.51	0.47	0.42	0.40	0.37	0.34	0.30	0.27	0.25	0.23	0.21
21	0.83	0.71	0.63	0.56	0.50	0.45	0.41	0.38	0.36	0.33	0.29	0.26	0.24	0.22	0.20
22	0.83	0.70	0.61	0.54	0.49	0.44	0.40	0.37	0.35	0.32	0.28	0.25	0.23	0.21	0.19
23	0.82	0.69	0.60	0.53	0.48	0.43	0.39	0.36	0.34	0.31	0.27	0.25	0.22	0.20	0.19
24	0.81	0.68	0.59	0.52	0.47	0.42	0.38	0.35	0.33	0.30	0.27	0.24	0.21	0.19	0.18
25	0.80	0.67	0.58	0.51	0.46	0.41	0.37	0.34	0.32	0.29	0.26	0.23	0.20	0.18	0.17

*Computed by $100/(100 - P_T + E_T P_T)$, or $100/(100 - P_B + E_B P_B)$. Use this formula for larger percentages.
†Used to convert equivalent passenger car volumes to actual mixed traffic; use reciprocal of these values to convert mixed traffic to equivalent passenger cars.
‡Trucks and buses should not be combined in entering this table where separate consideration of buses has been established as required, because passenger car equivalents differ.
Source: [5,15, p. 261].

The preceding discussion has been directed toward the estimation of freeway capacities. What should be emphasized at this point is that, as proposed earlier, we are attempting to relate one characteristic of a transportation system to several others. In this case capacity is taken as the dependent variable, whereas the dimensions of the roadway (number and widths of lanes, lateral clearance) and the mechanics of vehicle operation (acceleration and power of different types of vehicles) are the independent variables, the latter type being implicit in the passenger car equivalency factor. The estimation of freeway capacity thus demonstrates the interrelationship of characteristics involved.

Still lacking in the freeway capacity model, however, is some statistic, such as a correlation coefficient, that would indicate the reliability and possible estimation accuracy of the model. Unfortunately, such a measure or index is not given in the *Highway Capacity Manual*, with the result being that the analyst is likely to place undue faith in the estimates, having no reason to think otherwise. To provide some perspective on this question, one study of *intersection* capacities [5.14] showed that about 67 percent of the estimates were within ±406 vph of the actual capacity. Obviously this is not a good record, and despite the fact that freeway traffic is subject to fewer interferences than that at intersections, we still should maintain some skepticism of freeway capacity estimates.

5.4.2 Levels of Service

The capacity of a facility sometimes is not the only volume of concern. In fact, as mentioned, it is rare for vehicles to flow continuously at capacity rates because such flow must be very smooth and uniform.

Since no facility would be designed for full capacity, some intermediate level is desirable. In highway planning, several "levels of service" actually have been designated. These are shown in the speed–volume relationship for a freeway lane in Fig. 5.6. The service levels are lettered from that with the least volume-to-capacity ratio (A), and therefore the freest flow, to capacity itself (E). Level F, on the lower part of the

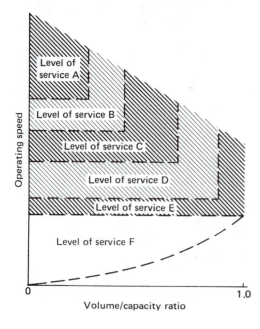

Level of service A

Level of service B

Level of service C

Level of service D

Level of service E

Level of service F

Operating speed

0   1.0

Volume/capacity ratio

Fig. 5.6 Generalized relationships among speed, level of service, and volume-to-capacity ratio for roads. Level of service and characteristics. A: Free flow; low volumes and high speeds; most drivers can select own speed. B: Stable flow; speeds restricted slightly by traffic; service volume used for design of rural highways. C: Stable flow; speed controlled by traffic; service volume used for design of urban highways. D: Approaching unstable flow; low speed. E: Unstable flow; low, varying speeds; volumes at or near capacity. F: Forced flow; low speed; volume below capacity; stoppages [5.15, pp. 80-81].

curve, represents a relatively unusual situation where, upstream from a bottleneck, volumes are high but vehicles have to slow down considerably in anticipation of the congestion. All levels are shown as areas on the curve since there is likely to be some variation in speeds for given volumes, as demonstrated by the spread of data points in Fig. 5.6.

Similar levels of service indices have been developed for intersections, arterials, and CBD streets (although these generally are governed by the signal systems). The planner and/or decision-maker then has the choice of level of service. The tradeoff essentially is between extent of congestion and cost to alleviate it. The free flow of traffic inherent in level of service A could be achieved, as an example, through extensive construction of freeways. Most decision-makers would not want the increased speed and minimal congestion inherent in level A, given the high costs involved. Still, there may be some circumstances where such is desired—thus the choice of service levels.

5.4.3 Capacities of Other Modes

In estimating auto-highway systems capacity, it usually is not necessary to investigate the capacity of the vehicle. Highway planning and design is based on the number of vehicles, not people. The same cannot be said for transit planning, however, since it is important to determine how many vehicles should be purchased to handle a given travel volume, and this depends on their capacity as well as that of the line and station. There is, in addition, a desire to compare capacities among different modes. This requires consistency of definition.

"Capacity" can be measured in several different ways. For highways, as noted, it is in vehicles per time unit (usually an hour). For transit it could be in terms of seats, spaces, or likely passengers per time unit (ranging from 15 min to an hour). The advantage of using "seats" is that they are easy to count. Yet in many systems riders often stand, and in fact the vehicles are designed with fewer seats and more standing room in mind. The extreme measure of capacity under these conditions is the so-called "crush" capacity, which needs little explanation.

A "spaces" definition of capacity takes into account comfort standards. The area allocated per seat usually is between 3.23 sq ft (0.30 m^2) and 5.92 sq ft (0.15 m^2). For standees the range is between 1.62 sq ft (0.15 m^2) and 2.69 sq ft (0.25 m^2) per person [5.2, p. 527]. The former value would approximate "crush" capacity, while a more usual value would be about 2.15 sq ft per standee. Table 5.8 shows actual vehicle capacities, in terms of spaces per meter length of vehicle, for typical auto and transit modes. The auto, at its maximum, can handle about 0.8 to 1.1 spaces per meter of length. Two-level regional (commuter) cars can handle up to 11.0.

Neither in the case of the auto nor in that of transit vehicles do passenger loads actually equal spaces available, except in very short time spaces. Hence the third definition of capacity attempts to show capacity in terms of likely utilization or occupancy. For autos this would be in the range of 1.2 to 2.2 people per vehicle (see Table 5.8). For rail rapid transit, Vuchic has found averages on 15 North American, European, and Japanese systems of from 12.7 passenger · km/vehicle · km (in London) to 69.7 (in Tokyo). Philadelphia had the highest U.S. figure of 31.0 [5.2, p. 449].

The capacity (C_w) of a transit guideway is governed by the capacity of the vehicles (C_v), the number of vehicles in a train unit (n), and the headways between train units required for safe stopping (h_{min}).[8] An equation for establishing this capacity,

---

[8] Assuming that one train cannot go around another while the second is in a station.

Table 5.8 Typical Linear Vehicle Capacity of Different Modes

Mode	Total capacity (spaces)	Length over bumpers/couplers (m)	Vehicle capacity (spaces/m)
Passenger auto	1.2–2.2	4–6	0.2–0.6
Maximum usable (average occupancy)	4–6	4–6	0.8–1.1
Theoretical maximum capacity	4–6	4–6	0.8–1.1
Van	8–12	4–6	1.5–2.2
Transit bus			
Minibus	20–30	6–7	3.5–5.0
Standard	60–80	10–12	5.0–7.0
Articulated	100–120	16–18	5.5–7.0
Light rail transit (LRT)			
Single body	100–120	14–16	7.0–8.0
Articulated	160–250	18–30	7.0–8.0*
Rapid transit	140–280	15–23	8.0–10.0
Regional rail			
Standard	140–210	18–26	6.0–9.0
Bilevel	200–280	22–26	9.0–11.0

*Articulated LRT cars often have wider bodies than do single-body cars.
Source: [5.2, p. 531, adapted by permission of Prentice-Hall, Englewood Cliffs, N.J.].

on a spaces per hour basis, is

$$C_w = 3600nC_v/h_{min} \qquad (5.18)$$

Assume, for example, that a rail rapid transit operation involves six car train units, with each car having a capacity of 140 spaces. The safety headway is 18 sec. Under these conditions, the lane capacity is

$$3600(6)(140)/18 = 168,000 \text{ spaces/hr}$$

The minimum headway, or distance between the front of the leading train and the front of the following, in this case was computed as the sum of the times needed for:

1. Operator reaction
2. Deceleration to a stop
3. Maintaining a stopped distance equal to the length of the leading train plus a suitable gap

Thus

$$h_{min} = t_r + \frac{nL + S_0}{5280U} + \frac{U}{B} \qquad (5.19)$$

where t_r = reaction time (sec)
    n = number of vehicles in train unit
    L = length of each vehicle (ft)

S_0 = extra separation distance between stopped trains (ft)
U = cruise speed (mph)
B = deceleration rate (mph/sec)

Hence, for the example, with a reaction time of 3.0 sec, a 60-ft length of vehicle, a separation of 30 ft, a cruise speed of 45 mph, and a comfortable deceleration of 3 mph/sec,

$$h_{min} = 3 + \frac{6(60) + 30}{5280(45)} + \frac{45}{3} = 18 \text{ sec}$$

The calculated capacity of 168,000 spaces/hr is quite high, and in fact is greater than actual volumes found in any corridor in any U.S. city. The largeness of the figure can be seen by comparison with that for 35-ft, 70-space buses on a dedicated busway, where for comparable input values

$$h_{min} = 3 + \frac{1(35) + 30}{5280(45)} + \frac{45}{3} = 18 \text{ sec}$$

and

$$C_w = 3600(70)(1)/18 = 14,000 \text{ spaces/hr}$$

Actual flows vary substantially from theoretical high values like those calculated above. Vuchic indicates that the maximum spaces per hour utilized for 12 rapid rail systems around the world was 65,340 on the Queens-53rd St. Tunnel in New York. The lowest was 18,450 for the S-Bahn in Munich [5.2, pp. 574–575]. For buses the maximum utilization was 27,600 spaces/hr, with regular buses on reserved lanes being more typically in the range of 3000 to 5000 [5.2, p. 57] (see Table 5.9). Streetcar and light rail lines have been known to carry up to 26,000.

For pedestrians, maximum stairway capacities of 10.9 people/minute · ft of stair width have been noted. This is for the upward direction. The figure for downward travel is 20.0 [5.6, p. 4]. Escalators have nominal capacities of between 3750 and 8025 people/hr, depending on the tread width (24 to 40 in) and incline speed (90 to 120 ft/min). Moving walkways, at a horizontal level, possibly could handle 10,000 people per hour. Some activity center systems have theoretical line capacities of up to 96,000 passengers per hour, but reported or installed capacities rarely are above 10,000 [5.6, various pages].

5.4.4 Modal Speed and Capacity Comparisons

A summary comparison between different modes is particularly difficult to make. One can pick circumstances that favor one mode over another, but it is almost impossible to specify "average" circumstances for each mode. Table 5.10 is one attempt at comparison. It is based on a spaces/hr capacity definition, and as such probably favors the auto (with 4 spaces/veh) over the bus, especially since no "bus on dedicated freeway lane" is presented. Rail systems generally come out most favorably.

A combined measure of "productive capacity" is presented. This is obtained by multiplying the space capacity and the speed of travel. In other words, if a particular mode results in a great number of spaces being moved at rapid speeds, the productive capacity is high. By this measure, rapid and commuter rail have the highest productive

Table 5.9 Actual Capacities of Bus Lines on Streets with at Least 2 Lanes per Direction*

Type of operation	City	Facility	Frequency (veh/hr)	Line capacity			Operating speed (km/hr)	Productive capacity (10³ spaces · km/hr²)
				Max. utilized (people/hr)	Peak 15–20 min. rate (people/hr)	Max. offered (spaces/hr)		
Regular buses on streets, bypassing possible	New York	Hillside Ave.	150	10,251	10,824	11,394	9	102
	San Francisco	Market St.	130	7,553	8,500	8,947	8	72
	Cleveland	Euclid Ave.	90	4,316	5,600	6,222	10	62
Regular buses on reserved lanes	Rochester	Main St.	93	4,982	5,978	6,333	10	63
	Atlanta	Peachtree St.	67	2,807	3,504	3,893	9	35
	Chicago	Washington Blvd.	66	3,235	3,600	4,000	10	40
Express buses on streets	St. Louis	Gravois St.	66	2,918	4,185	4,650	32	149
	Cleveland	Clifton Blvd.	32	1,872	2,700	3,176	18	57
	Chicago	Archer Ave.	29	1,896	2,500	2,941	21	62
Way capacity, single contraflow freeway lane	New Jersey, access to New York	I-495 toward Lincoln Tunnel	500	23,000	—	26,500	48	1400
			600	27,600		31,800		

*For qualifications, see the source.
Source: [5.2, p. 572].

116

Table 5.10 Sample Summary Characteristics of Various Models

Mode	Actual minimum headway (sec)	Line capacity (veh/m · lane)	Typical vehicle capacity (spaces/veh)	Maximum offered capacity (spaces/hr)	Operating speed of capacity (mph)	Productive capacity (10^3 spaces · mi/hr^2)
Auto	2	1800	4	7,200	35	252
Van	2.5	1440	10	14,400	35	504
Express bus on streets	30	90	75	6,750	12	81
Articulated bus on streets	33	110	110	12,100	9	109
Streetcar on streets	33	220	100	22,000	8	198
Light rail on private ROW	60	120	180	21,600	24	518
Rail rapid transit	100	340	175	63,000	31	1953
Commuter rail	120	270	180	48,600	35	1701

Source: Adapted from [5.2, p. 579].

capacity, being about 7 times that of the auto and 3 times that of the next best mode—light rail on its own right-of-way. Again, these conclusions will depend on local circumstances.

5.5 TRANSPORTATION SUPPLY COSTS

One of the most important characteristics of any proposed solution to a transportation problem is its cost. Not only must there be a concern for the amount of funds required to purchase and construct or install the proposed system but also for the amount to operate and maintain it at an appropriate level. To be taken into account in addition are the funds needed to plan and design the changes from the present situation.

In this section several sets of cost data from previously built or anticipated transportation systems will be presented. Most of the figures will not be in a form detailed enough for final cost estimates. Nor are the figures recent enough for such an endeavor. Instead the emphasis will be on presenting costs that show in rough terms the dollar magnitudes that may be involved. Some figures have been derived from rather crude estimates, others from slightly more detailed studies. As a consequence, it should be stressed that most of the costs should not be taken as fixed and final.

The discussion of costs is broken into several categories. First there is a division according to vehicles, networks, terminals, or control systems. Then there is a secondary division into purchase, construction, or installation; operation; and maintenance costs. Thereafter follows a section about the effects of inflation and rising prices on yearly cost levels.

5.5.1 Vehicle Costs

Most of us are familiar with the costs of various types and makes of automobiles so that it will not be necessary to go into this cost item here. Table 5.11 shows costs for different kinds of urban transit vehicles. As can be seen, the average for a new bus in New York City in 1979 was $100,000, which comes out to about $10/sq ft of vehicle floor space. In another case an advanced design bus being considered for use in Florida was estimated to cost $135,000 in 1980. Rail rapid transit vehicles are more expensive than buses on a per seat basis, with the cost in 1975 being roughly $5500 versus $1400. Some units, like double-deck rail cars, are not self-propelled and thus require an external motive power source such as a diesel-electric locomotive. Cost comparisons between vehicles are not easy since they generally have different expected service lives, different numbers of seats, and oftentimes serve different purposes.

5.5.2 Network Costs

Costs for transportation networks include those for guideways, intersections or interchanges, and bridges. Construction costs are, of course, a major consideration for networks on the ground. The cost of land or right-of-way for transportation also becomes a significant component in this instance.

For highways, cost estimates for design purposes generally are obtained by using unit prices. As an example, average contract prices in urban areas in the fourth quarter of 1980 for Federal-aid highway construction were [5.12, various pages]:

Common excavation: $2.11/cu yd
Portland cement concrete surface: $15.60/sq yd
Bituminous concrete surface: $28.30/ton

Table 5.11   Mass Transit Vehicle Costs

Vehicle type	Year	Cost per unit	Notes
Bus			
Bus, 40 ft	1976	70,000	Various U.S. cities
Bus, 35 ft	1976	63,000	Various U.S. cities
Bus, 40 seats	1979	99,646	New York City bid
Bus, advanced design	1980	135,000	
Jitney, 5 seats	1976	8,000	
Bus wagon, 10 seats	1976	16,000	
Minibus, 19 seats	1976	28,000	
Articulated bus, 70 seats	1976	167,000	
Trolley bus	—	50,000	Not made in United States
Rail			
Rapid rail, 50 seats	1974	316,000	Boeing Vertol for CTA
Rapid rail, 72 seats	1972	370,000	Rohr for BART
Commuter rail, 129 seats	1975	728,000	GE for Reading
Commuter rail, 94 seats	1975	241,152	Hawker-Siddeley
Light rail, 68 seats	1973	328,000	U.S. Standard
Activity-center systems			
Dashaveyor I, 40 seats	1973	100,000	
Morgantown PRT, 8 seats	1973	150,000	
Skybus, 12 seats	1968	250,000	
Carveyor, 6 seats	1968	6,000	
Jetrail, 6 seats	1968	35,000	
Rohr M–N	1972	120–160,000	

Sources: All are from [5.6] except the 35- and 40-ft bus [5.9, p. 247], the seat bus [5.5, p. 22], and the advanced-design bus [5.11, p. 10].

Structural reinforcing steel: $0.485/lb
Structural steel: $0.932/lb
Structural concrete: $246.67/cu yd

These vary, of course, by locality and state. For instance, a composite price index of the above in 1980 had a low of 115.0 in Alaska to a high of 273.6 in Connecticut. The U.S. average was 163.0 [5.12].

When unit costs such as these are multiplied by the corresponding amounts of land, labor, and capital involved, the guideway construction, station construction, and land purchase costs on a per lane mile basis come out to be at about the levels indicated in Table 5.12. Note that the variations by population, facility type, and location can be quite substantial. These should be taken into account in assessing the per lane or single track mile costs in Table 5.13. These have been quoted for those conditions thought by the authors to be most common (e.g., new arterial road construction for a metropolitan area population of 500,000 to 1 million).

A side-by-side comparison of the figures in Table 5.13, recognizing the major variations above, indicates that rapid rail transit is highly expensive on a per single track mile basis. Of course, rail rapid transit can carry as much as 10 times as many passengers per lane as a freeway, so capacity is an obvious point to be considered in a comparison. Still, rail rapid transit costs generally are high, especially compared to a busway, which can carry almost the same number of people under ideal conditions.

Table 5.12 Arterial Construction Costs*
(Million $ per Lane Mile)

Facility type	Location	\multicolumn{6}{c}{Population groups (1000s)}

Facility type	Location	0–50	50–100	100–250	250–500	500–1000	Over 1000
New roads	CBD	0.43	0.43	0.47	0.51	0.60	0.73
	Fringe	0.38	0.38	0.39	0.43	0.47	0.60
	Residential	0.34	0.34	0.35	0.38	0.43	0.50
Reconstruction[†]	CBD	0.43	0.43	0.46	0.47	0.52	0.64
	Fringe	0.39	0.39	0.41	0.43	0.46	0.51
	Residential	0.37	0.37	0.38	0.38	0.41	0.43
Major widening	CBD	0.43	0.43	0.47	0.51	0.55	0.72
	Fringe	0.41	0.41	0.43	0.45	0.51	0.59
	Residential	0.41	0.41	0.43	0.45	0.51	0.59

*Note: Costs projected from a 1973 base to a 1976 level using the FHWA Federal Aid Highway Construction Index. The above data reflect a per lane cost.
†Costs of periodic resurfacing are included in these figures.
Source: [5.6, p. IV-20, which comes from: K. Bhatt and M. Olsson, *Capacity and Cost Inputs for Community Aggregate Planning Model* (CAPM), working paper 5002-3, The Urban Institute, Washington, D.C., December, 1973].

Table 5.13 Transportation Guideway Capital
Costs (1976 Million $ per Lane Mile)

Item	Construction	Per station	Land
Rapid rail	37.7	7.4	3.24
Light rail	1.2	0.1	1.26
Busway	3.89		2.29
Expressway			
New	1.57		0.94
Reconstruction	0.93		0.47
Major widening	1.74		0.45
Arterial			
New	0.60		0.64
Reconstruction	0.51		0.18
Major widening	0.59		0.42
ACT	3.00		
Morgantown PRT	11.00		
Jetrail	0.40		
Skybus	2.60		
Carveyor	2.60		
Escalator	0.080		
Elevator	0.038		

Source: [5.6, various pages, which in turn are based on other sources].

Individual large-scale bridges and tunnels are perhaps the most expensive items of all. A good example is the Straits of Bosphorus Bridge in Turkey, which is a suspension structure almost 1 mile long and is expected to cost $185 million. Costs for similar kinds of projects can be noted in Table 5.14.

5.5.3 Terminal Costs

Another source of capital expenditure is for terminal and maintenance facilities. For the automobile, this means primarily parking lots and structures. For transit these are vehicle storage facilities, inspection garages, and maintenance facilities.

The costs of surface parking naturally depend quite heavily on the price of land. An average 330 sq ft stall, for instance, could cost anywhere from $1/sq ft to 15 times that amount, according to 1976 dollar figures generated by Parking Design Standards Associates.[9] Construction costs meanwhile are about $430 per stall. It is conceivable, then, that land could represent 95% of overall costs. Figures for multilevel parking structures are presented in Table 5.15. Land could go to $150/sq ft since structures, with their greater capacity (and thus greater potential fiscal return per land area), might be built in more valuable (and accessible) locations. The cost per stall naturally goes down as the number of floors increases, although there is a corresponding disadvantage in ingress and egress time for the parkers.

Bus facility costs, as shown in Table 5.16, also can vary widely, but such facilities

[9] Updated prices from their *Parking Standards Report*, Los Angeles, 1971.

Table 5.14 Approximate Costs and Dates of Construction of Various Bridges and Tunnels Around the World

Cost (in million $)	Date	Description	Reference
70	1970	Eight-lane sunken-tube tunnel under Thames River in London; $\frac{1}{2}$ mile long	*ENR*[*] March 12, 1970
67	1970	Four-lane tube tunnel, Mersey River in Liverpool, England; 1.5 miles long	*ENR* March 12, 1970
185	1969	5,118-ft Turkish bridge over the Straits of Bosphorus	*ENR* Dec. 4, 1969
69.5	1969	3,500-ft Four-track tunnel under the East River and Welfare Island in New York City	*Civil Engineering* Dec. 1969
45	1969	Kneibrucke (bridge) across the Rhine River at Dusseldorf, Germany; 1,800-ft span, tunnel approach and 4,300-ft underwater section, all 92 ft wide	*ENR* Nov. 26, 1969
57.6	1970	9,000-ft Mersey River Bridge in Liverpool, England	*ENR* Oct. 16, 1969
10.1	1969	Silver Bridge, Pt. Pleasant, West Virginia; 1,800 ft long with four lanes	*ENR* Nov. 6, 1969
23.7	1970	1.072-Mile-long tunnel in Virginia (two tunnels each 26 ft wide and 16.5 ft high)	*ENR* Sept. 18, 1969
180	1970	San Francisco Bay Area Rapid Transit (BART) tube across bay—3.6 miles long, 48 ft wide and 24 ft high	A BART publication

[*]*Engineering News Record.*

Table 5.15 Multilevel Structure Parking Costs*

Land cost ($/sq ft)	Total cost per stall[†]				Annual operating[‡] cost per stall
	3 Levels	5 Levels	7 Levels	9 Levels	
150	$18,150	$12,050	$9,720	$8,650	$505
125	15,400	10,400	8,540	7,730	460
100	12,650	8,750	7,360	6,820	410
80	10,450	7,430	6,420	6,080	375
60	8,250	6,110	5,480	5,350	335
40	6,050	4,790	4,535	4,620	295
20	3,850	3,470	3,590	3,880	260
10	2,750	2,810	3,120	3,520	240
5	2,200	2,480	2,890	3,330	230
2	1,870	2,280	2,740	3,220	225

*Data projected from 1970 base to 1976, using *ENR* construction index for construction expenditures and using consumer price index for operating costs.
[†]Costs include construction and prorated land costs based on a 330 sq ft stall.
[‡]Includes property taxes.
Source: [5.6, pp. 11–24, which in turn is based on Parking Standards Design Associates, *Parking Standards Report*, Los Angeles, 1971].

can be located on much less valuable land than auto parking garages. Interestingly, the total cost per bus for storage, inspection, and maintenance could reach about $50,000 (in 1975), which was about the cost of a new bus then. Wise transit planning thus would not overlook this substantial capital expenditure.

5.5.4 Control System Capital Costs

Another element of transport system capital costs is that for controls. Automotive traffic controls would consist of markings, signs, signals, and the like. Table 5.17 shows unit costs for several of these items in 1976–1977. To the uninitiated, what is most surprising is that a simple item such as a traffic sign with pole costs as much as $75 to purchase and install (at that time). These costs are not high, however, in comparison to traffic signals. A sophisticated eight-phase solid-state traffic controller for one intersection cost $12,000, and a 30-ft mast arm pole to which the traffic signal heads would be attached cost $500. In more sophisticated systems—those for freeway surveillance and control—the average cost (1980 dollars) of implementation for four

Table 5.16 Bus Facility Costs (March 1975)

Facility type	Unit cost ($/sq ft)	Unit cost ($/veh)
Storage facilities	16–28	5,250–15,000
Inspection garages	23–33	13,050–27,800
Main maintenance facilities	32–48	5,400–8,500

Source: Bus Maintenance Facilities: A Transit Management Handbook, report no. UMTA-VA-08-004-75-5, prepared for the U.S. Department of Transportation, Urban Mass Transportation Administration, Office of Transit Management, by the Mitre Corporation, November 1975.

Table 5.17   Installed Costs for Control Services and Lighting

Item	Unit*	1976–1977 Cost per unit, $ (installed)
Electrical service (240 V)	L.S.	2,500.00
Underground conduit	L.F.	6.00
Pull box	Each	120.00
Signal controller with cabinet		
8-Phase actuated	Each	12,000.00
2-Phase actuated	Each	2,500.00
Fixed time	Each	1,500.00
Signal head		
3-Section, 8 inch	Each	200.00
5-Section, 12 inch	Each	400.00
Pedestrian	Each	300.00
3-Section, programmed visibility	Each	700.00
Loop detectors		
6 Feet by 6 feet	Each	100.00
6 Feet by 60 feet	Each	350.00
Wiring, AWG 10	L.F.	0.40
Wiring, loop lead-in	L.F.	0.50
Poles		
Steel, 40 ft	Each	500.00
Steel, 10 ft	Each	150.00
Mast arm	Each	100.00
Pole foundation	Each	200.00
Luminaires		
High-pressure sodium, 150 W	Each	200.00
High-pressure sodium, 400 W	Each	300.00
Traffic sign, with pole	Each	75.00
Sign panel (large overhead)	S.F.	10.00
Pavement stripe, reflective	L.F.	0.10
Raised pavement marker, reflective	Each	3.00

*L.S. denotes lump sum; L.F. denotes linear foot; S.F. denotes square foot.
Source: [5.16, pp. 22–24].

sites in an equivalent number of cities in the United States was $273,000 [5.16, p. 44]. The range was wide, from $101,000 to $506,000, and in some cases did not include expenditures for bus bypass ramps and control-center buildings.

The costs of operation and maintenance of control systems are not insignificant. The Federal Highway Administration states that for advanced computer-based control systems these can range from $1,100 to $2,000 per signal per year. Even if the only effort made were simply to improve the settings of previously interconnected signals, the per signal cost still would run between $300 and $400 annually [5.17, p. 41].

5.5.5   Transit Operating and Maintenance Costs

Transit operating and maintenance costs usually are divided into five major categories:

Table 5.18 Annual Operating and Maintenance Costs ($1976 per Car · Mile)

Item	Transportation	Maintenance of ways and structures	Maintenance of vehicles	Power	General and administrative	Total
Rapid rail	1.00	0.37	0.31	0.32	0.40	2.40
Commuter rail	3.10	0.44	0.88	*	0.27	4.69
Light rail	0.97	0.51	0.42	0.30	0.62	2.82

*Included in "Transportation."
Source: [5.6, various pages, which in turn are based on other sources].

1. Transportation
2. Maintenance of ways and structures
3. Maintenance of vehicles
4. Power
5. General and administration

The averages for each of these on a per car · mile basis are indicated for rail rapid transit, commuter rail, and light rail for 1975 (in 1976 dollars) in Table 5.18. As can be seen, commuter rail tends to be almost twice as expensive as the others in these terms. It should be recognized, however, that the commuter rail cars can range in size from about 118 seats to 168, while rapid rail cars contain only between 40 to 84. These differences account in great measure for the variations in cost per *car* · mile.

Costs also tend to vary substantially by locale, as brought out in Table 5.19. There the lowest is less than half of the highest—from $1.88 per car · mile in Chicago to $3.90 in Philadelphia. Not shown explicitly in any of these figures is the cost of labor, which would include that for operators, supervisors, fare collectors, repair and

Table 5.19 Rail Rapid Transit System Operating Costs*

Cost item	New York	Chicago	Philadelphia	PATH	PATCO	Average
Maintenance of ways and structures	38	19	68	92	26	37
	(16)	(10)	(17)	(24)	(8)	(15)
Maintenance of vehicles	31	35	54	40	26	31
	(13)	(19)	(14)	(10)	(8)	(13)
Power	34	14	40	29	35	32
	(14)	(7)	(10)	(7)	(10)	(13)
Transportation	104	72	136	110	53	100
	(43)	(38)	(35)	(28)	(16)	(42)
General and administrative	33	49	92	117	193	40
	(14)	(26)	(24)	(31)	(58)	(17)
Total	239	188	390	388	333	240
	(100)	(100)	(100)	(100)	(100)	(100)
Car · miles (in thousands)	305,458	49,343	14,560	10,657	4,193	
Annual passengers (in thousands)	1,077,595	—	54,757	38,340	11,120	

*Expressed in terms of 1976 costs. Numbers in parentheses are percent of total cost by specific category.
Source: [5.6, p. B-10, taken in part from American Public Transit Association, *Transit Operating Reports*, 1976].

maintenance personnel, and general officers and staff. Their wages and salaries usually are about 75% of all expenses. Transit thus is highly labor-intensive. Surprisingly, rail transit tends to be even more labor-intensive, on an employee per vehicle basis, than does bus.

Bus operation and maintenance in 1975 (in 1976 dollars) ranged on a total cost per bus mile from about $0.75 to $2.95 [5.6, p. III-7]. As might be expected, the higher costs came in metropolitan areas with the largest populations (over 2,500,000). There the average cost was $2.09 per bus · mile, compared to $0.88 for systems in Standard Metropolitan Statistical Areas (SMSAs) of under 100,000 population.

Average highway and bus-lane operation and maintenance costs are presented in Table 5.20. These cover a wide variety of efforts. Under "operations," for instance, are included:

Cutting and clearing vegetation   Upkeep of guardrail
Snow removal and ice control   Subway and drainage pumping
Cleaning dirt and debris   Electric lighting
Traffic operation   Rest-area operation

Generally not included is the cost of police surveillance and protection. This can be a significant expense item for both highways and rail rapid transit systems.

Operations costs for multistoried parking garages were displayed in Table 5.15. They can range annually anywhere from $225 to $505 per stall. Underground operating expenses have been found to vary from $280 to $360, with the former being for self-parking and the latter for attendant parking. Costs for surface lots can be as low as $155 annually per stall, but with a high just slightly less than that for underground operations.

5.5.6 Cost Trends

The cost figures shown in the previous sections naturally cannot be employed directly in planning analyses, since inflation has acted to make most of them obsolete. Figures 5.7 and 5.8 and Table 5.21 show some of the changes in costs over time. These usually are measured by means of various price indices. For those that relate mainly to purchases by private individuals and firms, the measure is called the *consumer price index* for the particular good or service. One year is selected as the base year, with each index arbitrarily set to 100 in that year. In most of the figures to be shown, that year is either 1967 or 1977.

The average annual compound percentage rate of change (r) in a price index over a

Table 5.20 Annual Cost of Maintenance (1976 $/Lane Mile)

Item	Type of maintenance		
	General	Lighting	Total
Expressway	2860	2490	5350
Arterial	1470	580	2050
Residential/CBD streets	1100	1050	2150
Reserved freeway bus lanes			16,000–262,000*

*Several lanes may be involved.
Source: [5.6, various pages, which in turn are based on other sources].

1977 = 100

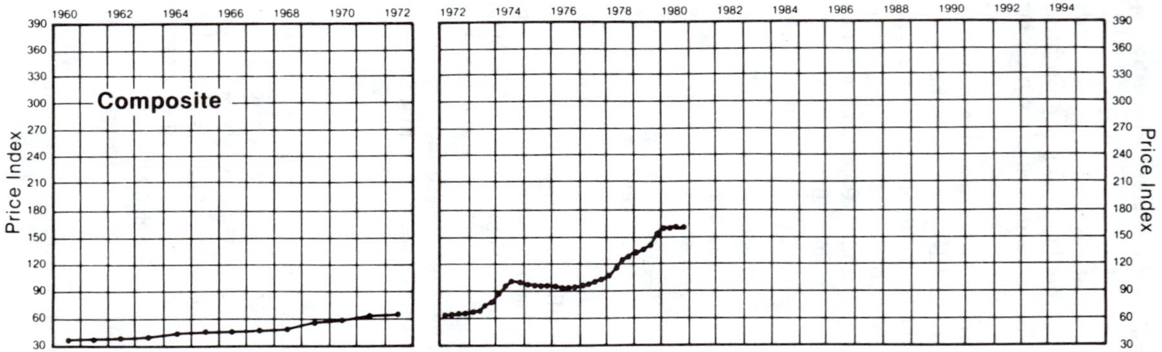

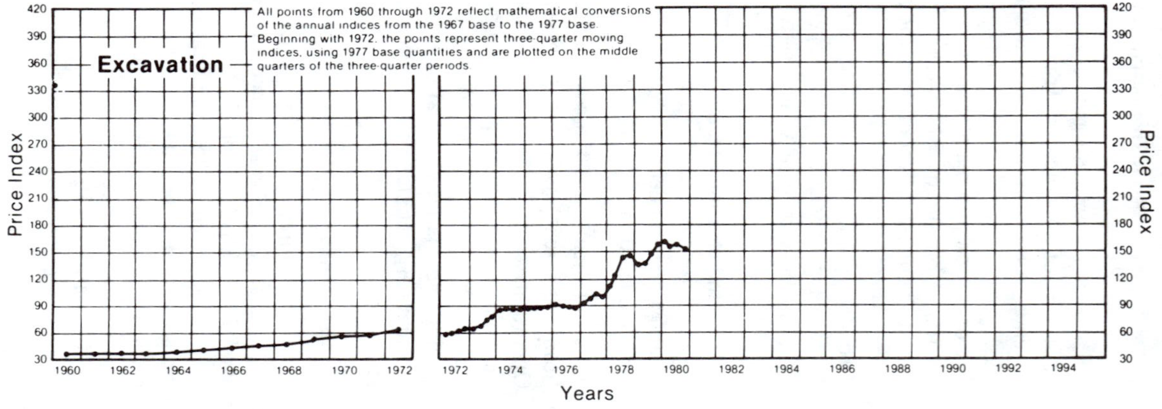

Price Index

Composite

Excavation

Years

All points from 1960 through 1972 reflect mathematical conversions of the annual indices from the 1967 base to the 1977 base. Beginning with 1972, the points represent three quarter moving indices, using 1977 base quantities and are plotted on the middle quarters of the three-quarter periods.

126

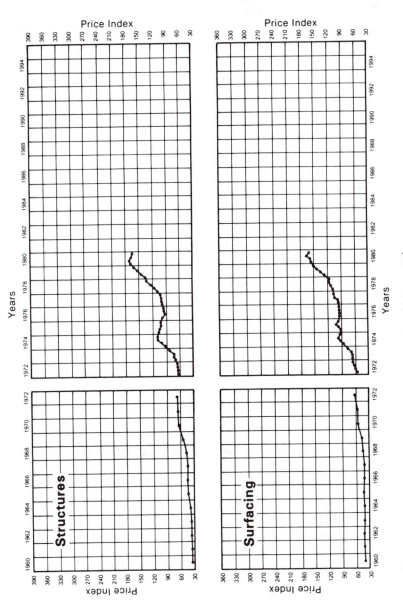

Fig. 5.7 Price trends for Federal-aid highway construction [5.12, p. 57].

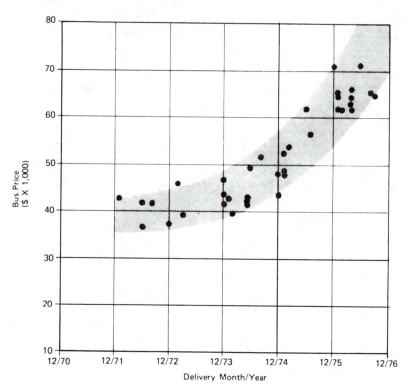

Fig. 5.8 Price trend of 12-m (40-ft) buses, 1971–1976 [5.9, p. 246].

Table 5.21 Cost Trends—Highway Maintenance and Operation*
(1977 = 100)

Year	Labor	Material	Equipment	Overhead	Total
1960	33.52	46.72	44.80	53.11	38.61
1965	40.44	48.96	48.53	58.05	44.18
1970	57.59	54.79	54.88	73.69	57.55
1971	61.20	57.91	55.59	77.45	60.46
1972	65.23	61.37	61.79	81.20	64.89
1973	69.87	64.35	68.86	84.95	69.86
1974	75.83	84.09	79.05	88.71	78.18
1975	81.72	95.60	87.85	92.46	85.24
1976	91.08	95.11	94.95	96.21	92.69
1977	100.00	100.00	100.00	100.00	100.00
1978	106.99	115.17	107.45	103.72	107.83
1979	114.51	136.26	120.84	107.48	118.17

*These data were prepared from the unit cost information submitted each year by state highway departments, and cover both physical maintenance and major traffic service items, including snow and ice control.
   Source: [5.13, various pages].

period of years (n) is found by, first, taking the ratio of the price in the later year, CPI(h), to that in the earlier year, CPI(o); second, taking the nth root of that ratio; third, subtracting 1 from the result; and, fourth, multiplying by 100 to get a percentage:

$$r = 100\{[CPI(h)/CPI(o)]^{1/n} - 1\} \qquad (5.20)$$

The compound growth rate in the cost index for maintenance labor in Table 5.21 from 1960 to 1979 thus is found to be

$$r = [100(114.51/33.52)^{1/19} - 1] = 6.7\%$$

If we now want to assume that this rate will hold in the future from 1979 until, say, 1983, the price index in that year will be

$$CPI(1983) = (1 + r/100)^4 \, CPI(1979) \qquad (5.21)$$

where 4 is the number of years between 1979 and 1983. This would give

$$CPI(1983) = (1 + 6.7/100)^4 (114.51) = 148.42$$

Now if the actual price in 1979 were \$10.00/hr, the 1983 price would be estimated as

$$(\$10.00)148.42/114.51 = \$12.96/hr$$

The rise in the price indices for various Federal-aid highway construction elements, and the composite, are displayed in Fig. 5.7 (1967 = 100). That for highway maintenance and operation is in Table 5.21 (in which 1977 = 100). The composite index has risen more than has that for all costs and services. This has put many transportation agencies in a tight financial bind—revenues have been leveling or dropping off, particularly due to decreased fuel tax receipts,[10] while construction costs have gone up rapidly. The result, consequently, has been much less construction in the late 1970s than in previous years.

The trends for transit have been analogous. Figures from the American Public Transit Association show that overall transit operating expense has risen 136% between 1970 and 1978 ([5.18, p. 20], while fares have gone up more slowly—62% (see Table 6.14). The result has been the need for much increased subsidies or else reductions in services.

The trends for individual transit components are similar. Gasoline, diesel, and electricity all rose rapidly in the 1970s (see Table 6.14 for gasoline), as did transit wage rates and bus prices (Fig. 5.8). Since transit is primarily labor, but also fuel intensive, the above increases have been particularly devastating.

5.6 SUMMARY

Urban transportation characteristics can be specified in many ways. In this chapter we have looked at speed, acceleration, volume, density, capacity, and costs. The interrelationships between some of these characteristics can be quite complex.

---

[10] See Chap. 14.

Capacity, for instance, depends on the speeds of traffic, the widths of lanes, the types of vehicles, gradients, and many other factors (including weather).

Costs are an important aspect of any transportation system. They generally can be divided according to capital, maintenance, and operation. On another dimension, we might be concerned with costs for vehicles, guideways, terminals, and controls. In all cases, future cost estimation is difficult because of the uncertainty inherent in inflation.

BIBLIOGRAPHY

5.1   Newell, G. F.: *Traffic Flow on Transportation Networks*, MIT Press, Cambridge, 1981.

5.2   Vuchic, V. R.: *Urban Public Transportation: Systems and Technology*, Prentice-Hall, Englewood Cliffs, N.J., 1981.

5.3   Homburger, W. S. (ed.): *Transportation and Traffic Engineering Handbook* (2d ed.), Institute of Transportation Engineers, Arlington, Va., 1982.

5.4   Fruin, J. J.: *Pedestrian Planning and Design*, Metropolitan Association of Urban Designers and Environmental Planners, Inc., Churchill, N.Y., 1971.

5.5   Brown-West, O. G.: "The Life Cycle Cost of a Bus—A New York City Case Study," *Transit Journal*, vol. 7, no. 1, Winter 1981.

5.6   U.S. Dept. of Transportation: *Characteristics of Urban Transportation Systems—A Handbook for Transportation Planners*, Washington, D.C., July 1977.

5.7   Holthoff, W. C.: *Cost Increases, Cost Differences, and Productivity of Transit Operations in New York State*, Preliminary Research Report 94, Planning Research Unit, New York State Dept. of Transportation, Albany, N.Y., Oct. 1975.

5.8   Morlok, E. K.: *Introduction to Transportation Engineering and Planning*, McGraw-Hill, New York, 1978.

5.9   Ferreri, M. G.: "Comparative Costs of Transit Modes," in G. E. Gray and L. A. Hoel (eds.), *Public Transportation: Planning, Operations and Management*, Prentice-Hall, Englewood Cliffs, N.J., 1979.

5.10   Gray, G. E., and L. A. Hoel (eds.): *Public Transportation: Planning, Operations and Management*, Prentice-Hall, Englewood Cliffs, N.J., 1979.

5.11   MacLeod, D. S.: "The Economics of Bus Rehabilitation," *Transit Journal*, vol. 7, no. 1, Winter 1981.

5.12   U.S. Dept. of Transportation, Federal Highway Administration: "Price Trends for Federal Aid Highway Construction: 1977 Base, First Quarter, 1981," Department of Transportation, Washington, D.C., 1981.

5.13   U.S. Dept. of Transportation, Federal Highway Administration: *Highway Statistics 1979*, Government Printing Office, Washington, D.C., 1980.

5.14   Wohl, M., and B. V. Martin: *Traffic System Analysis for Engineers and Planners*, McGraw-Hill, New York, 1967.

5.15   Highway Research Board: *Highway Capacity Manual—1965*, Special Report 87, Washington, D.C., 1965.

5.16   U.S. Department of Transportation, Federal Highway Administration: *Design of Urban Streets*, Government Printing Office, Washington, D.C., Jan. 1980.

5.17   U.S. Department of Transportation, Federal Highway Administration: *Traffic Control System Improvements: Impacts and Costs*, Washington, D.C., March 1980.

5.18   American Public Transit Association: *Transit Facts, '78–'79*, Washington, D.C., Dec. 1979.

5.19   *Chicago Area Transportation Study* (3 vols.): Chicago, 1959, 1960, 1962.

5.20   Baerwald, J. E.: *Traffic Engineering Handbook*, Institute of Traffic Engineers, Washington, D.C., 1965.

5.21   American Society of Civil Engineers, Committee on Terminals: "Terminal Planning for Future Highways," *Journal of the Highway Division*, ASCE, vol. 92, no. HW2, Oct. 1966.

5.22   U.S. Department of Labor, Bureau of Labor Statistics: *Monthly Consumer Price Indices Report*, Washington, D.C. (monthly).

5.23   Drake, J. S., J. S. Schofer, and A. D. May Jr.: "A Statistical Analysis of Speed Density Hypotheses," *Highway Research Record 154*, Washington, D.C., 1967.

5.24   C.E.I.R., Inc.: *Final Report on Intersection Traffic Flow*, Arlington, Va., 1960.

EXERCISES

5.1   Do the equations represented in Fig. 5.2 constitute a "model"? If so, what are the five basic elements?

5.2   If the density of vehicles on an *automated* highway were 200 veh/mile (in one lane) and the unit travel time were 2.0 min/mile, what would be the volume (in veh/hr)?

5.3   A new rail vehicle takes 1,000 ft and 15.06 sec to decelerate at a constant rate to a stop. What is the constant deceleration rate, in mph/sec?

5.4   According to the hypothesized Edie relationships, if the volume on a limited-access freeway lane were 1,200 vehicles per lane per hour, what would be the speed under level of service F?

5.5   If, on a limited-access busway, stations were spaced 0.6 mile apart, the buses could accelerate and decelerate comfortably at 2 mph/sec, dwell times averaged 10 sec, and the cruise speed were 55 mph, what would be the average speed?

5.6   A divided freeway has three 10-ft lanes in each direction, obstructions on one side within 4 ft, and 5% trucks and no buses on a 2% grade, 3 miles long. What is the one-way capacity (level of service E) of the freeway?

5.7   An express bus lane in a city is utilized by 40 buses/hr, offering 70 spaces/bus, at a speed of 20 km/hr. What is its productive capacity (as defined by Vuchic)?

5.8   Through the library or personal contacts, find the latest price for any kind of bus. What is included in that price (e.g., air conditioning)?

5.9   Contact the appropriate department in your locality and find the cost to purchase, install, and maintain a stop sign.

5.10   For the "total" highway maintenance and operation cost index in Table 5.21, find the annual (compound) rate of change from 1970 to 1975. If that rate continued to 1979, what would the index be in that year?

6 Models II: User Costs and Human Activities

As brought out in the beginning of the preceding chapter, transportation system performance has an immediate effect on service, which in turn influences costs to the user. Increased speed, for instance, generally leads to reduced travel-time costs and possibly rising fuel costs. The correlations are not complete, however. New land-use development, as a major related aspect of human activity, may be so spread out as to negate the potential travel-time gain. Or change in an external factor, such as a leap in gasoline prices, may add even further to fuel costs to the user.

In this chapter we first explore the different types of user costs. We then turn to the influence on these of relevant human activities, with particular emphasis on land-use changes (since these can both affect, and be affected by transportation). Finally, attention turns to a small number of relevant external factors, although we cannot possibly give this topic anywhere near the attention it might deserve.

6.1 USER COSTS

There are many costs that accrue directly to the user of transportation systems. The main ones, along with the approximate level of 1978 nationwide personal consumption expenditures, are given in billions of dollars by [6.1] and [6.2]:

New autos	50.3
New purchases of used autos	17.7
Other motor vehicles	12.8
Tires, tubes, accessories, other parts	10.4
Repair, greasing, washing, parking	29.5

Gasoline and oil	50.9
Bridge, tunnel, ferry, road tolls	0.9
Automobile insurance premiums paid[1]	32.3
Local transit systems	2.2
Local taxicab	1.2

These, together with outlays for intercity transit (airline, bus, railway, and other), made up about 14.2% of all personal consumption expenditures in 1978. This increased from 13.1% in 1960 [6.1].

Not shown above is the major user cost—the value of time spent in travel. This can be estimated only roughly, using the following somewhat conservative assumptions for 1978:

1. Trip time = 15 min (0.25 hr)
2. Trips per capita per day = 2.0
3. Hourly value of time = $\frac{1}{4}$ wage rate = 0.25 ($227/wk for 40 hr/wk) = $1.42/hr
4. U.S. population 218.5 million
5. Applies 365 days/year

So:

$$0.25(2.0)(\$218.5 \text{ million})(1.42)(365) = \$56.6 \text{ billion}$$

This, as can be noted, is more than for any of the other user cost items above.

Looking more closely at the cost of purchasing and operating automobiles and vans, we find the distribution of costs as displayed in Table 6.1. This is based on a 100,000-mile estimated service life over 10 years for each vehicle type. Hence the figures shown also can be interpreted on a cents-per-mile basis. For a standard auto, the total turns out to be $24,600 (1979 dollars) or $0.246/mile, with depreciation being the largest component, followed by gasoline and repairs/maintenance. Tires and tax costs are lowest per mile, with insurance and parking in the middle range.

Transportation improvements by the government might be expected to have relatively little impact on depreciation, somewhat more on repairs/maintenance, tires, and indirectly on insurance, and most directly on fuel consumption, parking (at least that under government control), and taxes (particularly as related to fuel consumption). The most important impact will be on travel time, which is directly affected by guideway, vehicle, and control improvements. Impacts in these latter two categories thus will receive the most attention here. Taxes are discussed in Chap. 14.

6.1.1 Travel Time

Travel-time savings are estimated on the basis of speed and distance changes inherent in the transportation/land-use plans being proposed. As a simple example, if signal-system improvements along with construction of a less circuitous route reduce the trip length between two points from 10 to 6 miles and at the same time increase speeds from 35 to 45 mph, the resulting travel-time reduction can be computed as:

$$(10 \text{ miles}/35 \text{ mph}) - (6 \text{ miles}/45 \text{ mph}) = 0.153 \text{ hr} = 9.18 \text{ min}$$

The estimation process thus is relatively straightforward, although often it is not

---

[1] Computed from the National Underwriter Co., "Argus F.C.&S. Chart," Cincinnati, Ohio (annual).

Table 6.1 10-Year Cost ($1,000) of Operating an Automobile, by Size of Car: 1979

Item	Total	Costs (excluding) taxes)	Depreciation	Repairs/ maintenance	Tires	Gasoline	Insurance	Parking	Taxes and fees
Standard	24.6	23.0	6.3	4.8	0.6	5.4	2.5	3.2	1.6
Compact	21.7	20.4	5.2	4.2	0.5	4.8	2.3	3.2	1.3
Subcompact	18.5	17.3	3.8	3.4	0.5	4.0	2.2	3.2	1.1
Passenger van	36.2	34.1	10.2	5.3	0.6	7.2	7.2	3.2	2.1

Source: U.S. Department of Transportation, Federal Highway Administration, "Cost of Operating an Automobile" (periodic).

possible to determine exactly the resultant speed on the improved sections. A likely range in the above example, for instance, might be from 41 to 48 mph. It also must be assumed that *all* traffic will travel the estimated speeds, which obviously is an average figure for autos, trucks, buses, motorcycles, mopeds, and the like.[2]

The more difficult task is that of establishing a monetary value for travel time. This can be expected to vary with characteristics of the traveler, the purpose for which the trip is being made, and the actual amount of time saved.

There are several ways to obtain monetary values for travel time [6.3, pp. 30-41]:

1. Study of actual situations where a tradeoff was made between time and money (e.g., a fast toll road versus a slower, free road)
2. Development of travel demand functions (see Chap. 7), where time and money are both included
3. Estimation of the monetary gains in private and governmental productivity if lower travel times occur
4. Development of housing demand functions, where the price of housing is related to travel times (especially to work)

The latter two approaches still are experimental. The first one suffers because it provides only a lower limit. For instance, if a driver selects the faster toll road, this means that the time saved is worth *at least* the toll (but possibly more). Still, this approach probably is the one employed most often.

Figure 6.1 gives 1975 values of travel-time savings based on the first approach.[3] These are a function of the time saved per trip, both for different trip types and the average for all trip types, for travelers in average income levels (originally $10,000 to $12,000 per year; in 1975 equivalent to about $14,000 to $17,000 per year). No significant variation in travel-time values as a function of the time or length of individual trips has yet been found.

---

[2] Of course, separate speed categories can be made for each class of vehicle. This adds considerably to the computations, and still leaves some variations within each category.
[3] Much of the next six paragraphs is adopted from the AASHTO "Redbook." [6.4, pp. 14-19].

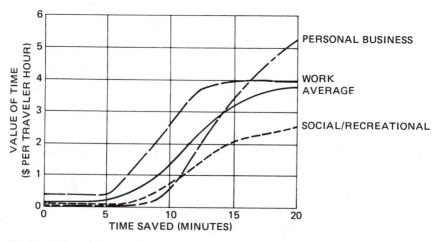

Fig. 6.1 Value of time as a function of time saved, by trip type [6.4, p. 16].

Table 6.2   Value of Time as a Function of Time Saved and Trip Type

	Value* of time per traveler hour	Percentage of average hourly family income
For low time savings (0–5 min)		
Average trips	$0.21	2.8%
Work trips	0.48	6.4
For medium time savings (5–15 min)		
Average trips	1.80	24.2
Work trips	2.40	32.2
For high time savings (over 15 min)		
Average trips	3.90	52.3
Work trips	3.90	52.3

*In 1975 dollars.
Source: [6.4, p. 17].

In principle, Fig. 6.1 can be used to obtain estimates of travel-time value once the time saving per trip for a particular highway or transit improvement is known. However, many such improvements may only be parts of longer term, more extensive improvements that *cumulatively* save much more time. Moreover, uncertainties over the exact time saved by cumulative improvements may be such that it becomes a needless refinement to select the exact point on the curve that corresponds to the estimated cumulative time savings for each case. Accordingly, planners may wish to (1) use cumulative time savings for all planned improvements to a route or network in selecting the appropriate time savings per trip, and (2) establish standard values of time for several general categories of time saved. Table 6.2 is an example of tabulated standard values of time that could be used in place of Fig. 6.1. The average values in Table 6.2, as in Fig. 6.1, are unweighted arithmetic means for the different trip types shown.

The "percentage of average family income" figures shown in Table 6.2 are based on 2,080 working hours per year for a $15,500 average family income in the $14,000 to $17,000 range, or $7.45 per hour. These percentages can be used to adjust time value factors proportionally when associated incomes are outside that range.

The per-person time figures shown in Fig. 6.1 and Table 6.2 can be converted to average values per *vehicle* by multiplying by the vehicle occupancy factor (Table 6.3). Note that such values vary considerably from place to place and over time.

For transit improvements, it is advisable to specify different values for time spent in the transit vehicle and that spent out of the vehicle, either walking or waiting. It

Table 6.3   Vehicle Occupancy Factors

Trip type	Adults per vehicle
Work	1.22
Social-recreational	1.98
Personal business	1.64
Average	1.56

generally is conceded that the latter has a higher value to the user than time spent riding. For average conditions, waiting and/or walking time is probably worth 1.5 times the in-vehicle travel time—for high time savings (over 15 min). This could be adjusted upward to 2.0 or more if out-of-vehicle comfort or safety conditions are below average. (Unlike travel time in the vehicle, even small improvements in waiting or walking time are perceived as important to travelers.) Where data permit, a higher ratio (such as 2.0) could also be used for initial waiting time (versus transfer and walking), which tends to be less well tolerated than delays in the transit journey. A final qualification on the use of a value for automobile and travel-time savings is the indication from recent studies that total average personal daily travel time has, over many years, been extremely invariant in different urban areas, at about 1.1 hr per capita per day. This means that the long-term—or in some cases the short-term—results of reductions in travel time caused by improved personal transportation facilities or operations usually show up in two ways:

Longer trips—the tendency in large urban areas to increase spatial opportunities and decrease residential density

More frequent trips—such as increases in trips for cultural and social purposes

Travel-time savings thus might be regarded as a surrogate for other values that travelers seek.

For trucks, travel-time savings represent market costs rather than the value of personal user time commitments. The AASHTO "Redbook" [6.4] recommends time values of about $7.00/hr (or $0.115/min) for single-unit trucks and $8.00/hr (or $0.13/min) for semi-trailer diesel trucks, based on 1975 driver wages and fringe benefits.

In addition to the truck driver's time, there are other time-related costs that may be affected by highway improvements. These include:

Shipper inventory costs in transit

Time-related deterioration of cargo

Receiver inventory costs

In general these costs are small relative to driver wages and vehicle operating costs [6.4, p. 19].

6.1.2 Fuel Consumption

The 1970s were renowned for the leaps in petroleum prices that took place. The consumer price index for U.S. gasoline (excluding taxes) stood at 105.6 (1967 = 100) in 1970 and jumped drastically to 369.1 in 1980 (see Table 6.17). This rapid rise had many effects, the one of most concern here being in auto and truck fuel efficiencies. In the early 1970s, nationwide, the auto fleet was averaging around 13.5 mpg and the truck fleet about 8.3 mpg. By 1979, as displayed in Fig. 6.2, these had risen to around 14.3 and 9.0, respectively, with the trends heading upward rapidly. Congressional mandates on new car sales weighted miles per gallon (SWMPG) are aiding these trends significantly. The current standard for autos, for instance, is an SWMPG of 27.5 by 1985. This means that the mean mpg of all American-made autos sold in the United States must be at least 27.5. The mixture of these into the fleet naturally will increase its overall efficiency.

Many factors can influence the rate of energy consumption. These include characteristics of the driver or operator, type of propulsion system, weight of the vehicle,

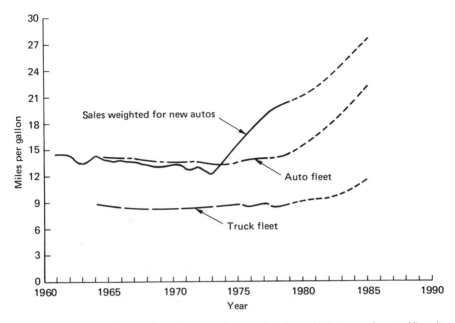

Fig. 6.2 Nationwide auto and truck mpg and new auto sales weighted mpg. *Source:* Historic Data from U.S. Dept. of Transportation, Federal Highway Administration, *Highway Statistics* (various years). Forecasts by the author.

weather, guideway geometrics and conditions, and the like. Since metropolitan governments have little control over most of the above except the guideway geometrics, transportation planners tend to concentrate on these factors.

For automotive fuel consumption, the first two affecting factors considered usually are speed and vertical grade. Variations in fuel consumption for each are shown in Fig. 6.3 (for positive grades only) and in Table 6.4. Consumption in both is measured in gallons per mile (the reciprocal of mpg). A "grade" of a certain percentage is defined as the vertical rise or fall divided by the associated horizontal distance. A vertical rise of 1 ft over a horizontal distance of 100 ft thus is a 1% grade.

Consider two relevant features of Fig. 6.3 and Table 6.4. First, minimum fuel consumption for most automobiles is reached at a uniform speed of about 35 mph. This means that from the strict perspective of energy savings, traffic flow on arterials is to be preferred to that on freeways and certainly to that on slow-moving downtown or residential streets. Second, at 40 mph, an 8% grade almost triples auto fuel consumption over that on level sections, while an 8% downgrade cuts consumption by three-quarters. A combined trip of 1 mile up such a grade and one mile down thus results in a total consumption of 0.124 + 0.012 = 0.136 gallons, while a corresponding trip on level terrain consumes only 0.046 + 0.046 = 0.092 gallons. Steep grades therefore are to be avoided.[4]

Other factors that can be taken into account using data developed by Claffey [6.7, pp. 16–20] include wearing surface roughness, horizontal curvature, traffic volume, type of roadway (freeway, arterial, CBD street), stop/go speed changes, street

---

[4] Note that this does not hold at 4%, where the hypothetical 2-mile trip total is 0.078 + 0.014 = 0.092 gallons, which equals that on level terrain.

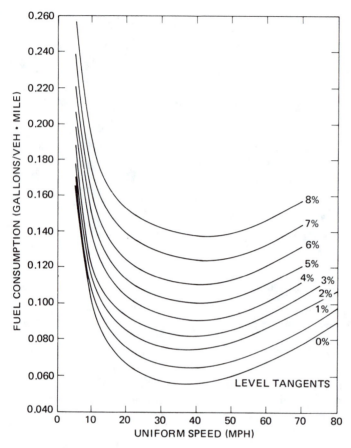

Fig. 6.3 Total automobile fuel consumption for various plus grades [6.4, p. 135].

parking, and stops per mile. Several of these are highlighted in Table 6.5. Suppose, for illustration, that a six-lane CBD street with a 1% upgrade causes 5 stops per mile with a one-way traffic volume of 170 vph. From Table 6.4, the fuel consumption on the 1% upgrade at an attempted speed of 25 mph (assumed in Table 6.5) is (by interpolation) about 0.054 gallons/mile. The correction factor from Table 6.5 is 2.04, which leads to an overall consumption rate of

$$0.054(2.04) = 0.110 \text{ gallons/mile}$$

The corresponding charts for a two-axle, six-tire truck (average of 6 ton gross vehicle weight) are displayed in Tables 6.6 and 6.7.

The cost of fuel now can be determined by multiplying the gpm by the cost per gallon. The latter might be known for a given year, say 1970, and subsequently have to be updated given the corresponding consumer price index (CPI). As an example, the 1970 nationwide price, excluding taxes, was about $0.30 per gallon. For the indices in Table 6.17, the price in 1980 would be

$$\$0.30(369.1/105.6) = \$1.05/\text{gallon}$$

Table 6.4 Automobile Fuel Consumption as Affected by Speed and Gradient—Straight High-Type Pavement and Free-Flowing Traffic*

Uniform speed (mph)	Gasoline consumption (gpm) on grades of										
	Level	1%	2%	3%	4%	5%	6%	7%	8%	9%	10%
Plus grades											
10	0.072	0.080	0.087	0.096	0.103	0.112	0.121	0.132	0.143	0.160	0.179
20	0.050	0.058	0.070	0.076	0.086	0.094	0.104	0.116	0.128	0.144	0.160
30	0.044	0.051	0.060	0.068	0.078	0.087	0.096	0.110	0.124	0.138	0.154
40	0.046	0.054	0.062	0.070	0.078	0.087	0.096	0.111	0.124	0.140	0.156
50	0.052	0.059	0.070	0.076	0.083	0.093	0.104	0.118	0.130	0.145	0.162
60	0.058	0.067	0.076	0.084	0.093	0.102	0.112	0.126	0.138	0.152	0.170
70	0.067	0.075	0.084	0.093	0.102	0.111	0.122	0.135	0.148	0.162	0.180
Minus grades											
10	0.072	0.060	0.045	0.040	0.040	0.040	0.040	0.040	0.040	0.040	0.040
20	0.050	0.040	0.027	0.022	0.021	0.021	0.021	0.021	0.021	0.021	0.021
30	0.044	0.033	0.022	0.016	0.014	0.013	0.013	0.013	0.013	0.013	0.013
40	0.046	0.035	0.025	0.018	0.014	0.013	0.012	0.012	0.012	0.012	0.012
50	0.052	0.041	0.030	0.025	0.021	0.018	0.014	0.013	0.010	0.010	0.008
60	0.058	0.048	0.036	0.037	0.030	0.027	0.022	0.018	0.014	0.011	0.008
70	0.067	0.058	0.048	0.043	0.039	0.036	0.031	0.027	0.022	0.016	0.013

*The composite passenger car represented here reflects the following vehicle distribution: large cars, 20 percent; standard cars, 65 percent; compact cars, 10 percent; small cars, 5 percent.

Source: [6.7, p. 17].

Table 6.5 Correction Factors to Adjust the Values of Table 6.4 for Traffic Volume—Six-Lane CBD Streets with Parking in Both Curb Lanes*

One-way traffic volume (vph)	Correction factors by frequency of stops per mile[†]										
	0	1	2	3	4	5	6	7	8	9	10
0–40	1.00	1.18	1.37	1.55	1.73	1.91	2.12	2.30	2.40	2.65	2.80
40–80	1.00	1.18	1.37	1.55	1.73	1.91	2.12	2.30	2.40	2.65	2.80
80–120	1.00	1.21	1.39	1.58	1.76	1.95	2.14	2.33	2.50	2.70	2.89
120–160	1.00	1.22	1.41	1.60	1.78	1.97	2.20	2.37	2.56	2.75	2.94
160–200	1.01	1.23	1.42	1.64	1.85	2.04	2.24	2.42	2.61	2.81	3.00
200–240	1.02	1.24	1.43	1.69	1.89	2.08	2.31	2.49	2.68	2.86	3.06
240–280	–	1.26	1.47	1.74	1.95	2.16	2.39	2.55	2.70	2.91	3.17
280–320	–	–	1.49	1.80	2.02	2.23	2.45	2.64	2.80	3.00	3.26
320–360	–	–	–	1.84	2.11	2.33	2.52	2.73	2.93	3.10	3.28
360–400	–	–	–	–	2.21	2.44	2.68	2.85	2.98	3.18	3.39
400+	(Level of service E = unstable flow)										

*Automobiles are assumed to attempt to travel at 25 mph with traffic signals set approximately 500 ft apart. Average stopped delay when stopped is 30 sec.
†Traffic volume corrections determined for the standard-size U.S. car (65 percent of vehicle population) represented by a Chevrolet sedan at 4,400 lb gross vehicle weight. When vehicle stops are involved, this table should not be used for grades greater than 1.5 percent.
Source: [6.7, p. 18].

Thus, if, as above, 0.110 gallons/mile were consumed, the cost would be

$1.05(0.11) = $0.115/mile

Procedures for *forecasting* future costs are presented in Chap. 8 in the section on energy consumption impacts.

6.1.3 Parking

When costs for parking are considered, questions have to be answered first about ownership and subsidy. If parking is provided for the public at large by a governmental agency, there may be a subsidy involved. If there is, the price charged for parking will be less than the costs, and the user will gain accordingly. The extreme case of this is free parking—especially for government's own employees.

The subsidy question also is important with respect to private enterprises. Some offer parking services at a price to the general public. Presumably they make a profit over and above their costs. Others offer free or low-cost parking (e.g., shopping centers), particularly to entice prospective customers. The cost of parking then shows up as a business expense, which is passed on to the customer in the form of higher prices for goods and services.

In practice there are no good ways to take all these issues into account. The resultant approach usually is to take the prices as stated. As brought out in Table 6.1, parking costs average about $0.032/mile, although this will vary according to location. A person parking in downtown Washington, D.C., in 1981, for instance,

Table 6.6 Two-Axle Six-Tire Truck Fuel Consumption as Affected by Speed and Gradient— Straight High-Type Pavement and Free-Flowing Traffic*

Uniform speed (mph)	Gasoline consumption (gpm) on grades of										
	Level	1%	2%	3%	4%	5%	6%	7%	8%	9%	10%
	Plus grades†										
10	0.074	0.094	0.120	0.143	0.175	0.195	0.225	0.255	0.289	0.324	0.357
20	0.059	0.080	0.112	0.140	0.167	0.190	0.214	0.254	0.295	0.344	0.394
30	0.067	0.094	0.121	0.150	0.181	0.206	0.232	0.268	0.305	—	—
40	0.082	0.112	0.141	0.173	0.210	0.228	—	—	—	—	—
50	0.101	0.130	0.159	0.194	—	—	—	—	—	—	—
60	0.122	0.150	—	—	—	—	—	—	—	—	—
	Minus grades										
10	0.074	0.064	0.055	0.053	0.051	0.051	0.051	0.051	0.051	0.051	0.051
20	0.059	0.049	0.039	0.034	0.030	0.030	0.030	0.030	0.030	0.030	0.030
30	0.067	0.054	0.041	0.034	0.027	0.026	0.025	0.025	0.024	0.024	0.024
40	0.082	0.071	0.051	0.041	0.032	0.029	0.025	0.023	0.021	0.020	0.020
50	0.101	0.090	0.072	0.058	0.045	0.038	0.031	0.025	0.020	0.020	0.020
60	0.122	0.110	0.090	0.075	0.062	0.052	0.043	0.035	0.025	0.020	0.020

*The composite two-axle six-tire truck represented here reflects the following vehicle distribution:

Two-axle trucks at 8,000 lb gross vehicle weight: 50 percent
Two-axle trucks at 16,000 lb gross vehicle weight: 50 percent

†Operation is in the highest gear possible for the grade and speed (no. 4, no. 3, or no. 2). When vehicle approach speed exceeds the maximum sustainable speed on plus grades, speed is reduced to this maximum as soon as the vehicle gets on the grade.
Source: [6.7, p. 24].

143

**Table 6.7   Correction Factors to Adjust the Values of Table 6.6
for Traffic Volume—Six-Lane CBD Streets with Parking
Both Curb Lanes***

One-way traffic volume (vph)	Correction factors by frequency of stops per mile[†]										
	0	1	2	3	4	5	6	7	8	9	10
0–40	1.00	1.30	1.60	1.84	2.20	2.50	2.80	3.10	3.40	3.70	4.00
40–80	1.00	1.30	1.60	1.84	2.20	2.50	2.80	3.10	3.40	3.70	4.00
80–120	1.00	1.30	1.60	1.84	2.20	2.50	2.80	3.10	3.40	3.70	4.00
120–160	1.00	1.30	1.62	1.84	2.21	2.50	2.80	3.10	3.41	3.70	4.00
160–200	1.00	1.31	1.63	1.86	2.22	2.51	2.80	3.11	3.42	3.70	4.00
200–240	—	1.33	1.65	1.88	2.24	2.53	2.81	3.12	3.43	3.72	4.00
240–280	—	1.34	1.66	1.89	2.25	2.54	2.83	3.13	3.43	3.74	4.00
280–320	—	—	—	1.90	2.27	2.55	2.86	3.16	3.47	3.76	4.04
320–360	—	—	—	1.91	2.29	2.58	2.88	3.18	3.47	3.77	4.08
360–400	—	—	—	1.93	2.33	2.63	2.91	3.19	3.50	3.81	4.12
400+	(Level of service E = unstable flow)										

*Traffic volume correction factors determined for the two-axle six-tire truck at 24,000 lb gross vehicle weight.

[†]Vehicle operation at 25 mph attempted. Traffic signals are approximately 500 ft apart. Average stopped delay when stopped is 30 sec. When vehicle stops are involved, this table should not be used for grades greater than 1 percent. If grades exceed 1 percent, basic data on fuel consumption due to stops and traffic conditions in [6.7] should be used to adjust the values of Table 6.6 for traffic volume.

Source: [6.7, p. 26].

might pay up to $7.00/day, which probably would come out to be much more than $0.032/mile traveled.

6.1.4   Accidents

It is questionable whether individuals consider safety in making or not making a trip on a given transport facility. Exceptions may be those circumstances where people ride transit instead of being in an automobile to reduce the chance of an accident. Still, the costs of accidents to users can be quite significant, and usually are not ignored in overall family expenditure considerations (e.g., in determining whether to buy auto insurance for a young, male driver).

A recent study in Maricopa County, Arizona (which includes the cities of Phoenix, Tempe, Mesa, Scottsdale, and Glendale) helps to illustrate the safety record for private motor vehicles [6.5]. Accidents were analyzed according to the extent of injuries and property damage (including to the vehicles themselves).

The overall accident rate per 100 million (100m) vehicle miles of travel (VMT) in Maricopa County was almost twice as much as the U.S. average in 1979 of 2.8. A similar proportion holds when fatalities are considered on a basis of per 100m VMT, per 10,000 population, or per 10,000 registered vehicles. The rates for all of these generally have increased in that county since 1960, while the opposite holds true for U.S. cities as a whole (Table 6.8).

Accident and death rates (versus VMT) vary considerably over time. The death rate is highest in December, indicating some unhappy holidays. January, by contrast,

Table 6.8 Accident Rates

	Maricopa County	U.S. cities
Accidents per m VMT	5.7	2.8
Deaths per 100m VMT	4.7	2.3
Deaths per 10,000 population	3.0	1.1
Deaths per 10,000 vehicles	4.3	2.1

has the lowest death rate. Weekends bring the highest fatalities, with the worst times in the early morning hours (Fig. 6.4).

Almost 40% of all deaths and injuries occurred to those in the 15–24-year age bracket. Second was the 25–34-year bracket, with about half the percentage of the younger group. About 33% of all accidents took place at intersections, and, of the overall accidents, 1.7% involved pedestrians, another 1.7% pedacyclists, 84.7% two vehicles, and the remaining 11.9% only one motor vehicle.

The timing of accidents suggests that alcohol is a major factor. In fact, at the national level the general rule is that alcohol is involved in about half of all traffic fatalities. By "involvement" is meant significant bloodstream alcohol on the part of one or more drivers and/or pedestrians. The situation in Maricopa County appears somewhat better in that in only 32% of the accidents in which deaths occurred had at least one driver been drinking.

Putting a monetary value on losses from accidents naturally is a difficult task. Property values perhaps are easiest to assess since bids can be obtained for auto repairs, damages to houses, etc. Even hospital costs for certain types of injuries can be

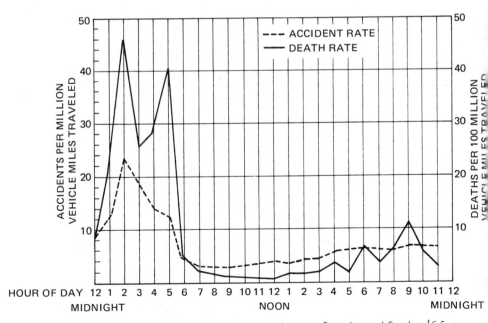

Fig. 6.4 Vehicular travel accident rate and death rate by hour on Saturday and Sunday [6.5, p. 22].

estimated with some degree of assurance. The difficult part comes in those cases with permanent injury or death.

The AASHTO "Redbook" [6.4, p. 64] reports costs per fatal accident in a wide range from $18,800 to $307,210. The low figure results from consideration only of out-of-pocket costs (e.g., funerals) while the highest figure takes into account:

1. Discounted value of estimated future earnings
2. Costs of $13,000 (1975 dollars) for pain and suffering in partial disability accidents
3. Higher wage losses for all disabling accidents
4. A cost of $11,800 for loss of services to home and family in partial disability accidents

Adjustments also are made for underreporting of accidents, since the California Department of Transportation has found that only about 93% of injury accidents and 52% of property damage (only) accidents are reported.[5]

That such monetary judgments are made at all by planners and other transportation analysts may seem to the reader to be somewhat unnecessary. On the other hand, society, through the courts (and juries) and through other projects that involve potential life savings, makes these judgments all the time. In fact, many of the numbers suggested above are based on jury decisions.

Table 6.9 gives one version of cost figures that may be employed for analysis purposes. This shows a breakdown by degree of road access, number of lanes, and urban/suburban location.

While accidents involving transit passengers are much fewer on a per million passenger mile basis, they still happen. The *Characteristics of Urban Transportation Systems* [6.6] manual reports the ranges shown in Table 6.10. Accident costs for transit systems usually are included in insurance costs of the transit operator and accordingly are not treated in a separate cost analysis.

6.1.5 Tire Wear

Tire wear is a small part of overall vehicle user costs. It can vary substantially, however, by speed, type of roadway surface, and degree of horizontal curve.[6] Some of these variations for an automobile are displayed in Tables 6.11 and 6.12. Note that asphalt pavement is about 50% more wearing than high-type concrete, and gravel about three times as much. Curvature adjustments can be substantial. For example, travel around an 8-degree curve at 40 mph adds 12.50 times as much wear as on a straight road.

The costs in Table 6.11 are in 1969 dollars. Although not mentioned by Claffey [6.7], the tires tested probably were not radial, and thus did not have the added wear of those at present.

6.1.6 Motor Oil

Vehicle oil consumption is another relatively minor cost component. It does increase, however, with speed, as indicated in Table 6.13. At 60 mph the additional oil needed is about four times that at 30 mph.

---

[5] California Department of Transportation, Division of Highways, memo to district directors entitled "Accident Costs," from J. E. Peddy, Chief, Office of Systems and Program Development, Sacramento, Calif., Oct. 18, 1974.

[6] The "degree of a curve" is the central angle subtended by a 100-ft length of curve.

Table 6.9 Accident Rates and Costs by Road Type

Road type	Fatal accidents			Injury accidents			Property damage only accidents			Total accidents	
	Number per MVM*	Cost ($/KVM)*	Fatal % of total	Number per MVM	Cost ($/KVM)	Injury % of total	Number per MVM	Cost ($/KVM)	PDO % of total	Number per MVM	Cost ($/KVM)
Urban											
No access control											
2 Lanes	0.045	5.06	37	1.51	5.30	39	3.38	3.38	25	4.94	13.73
4 Or more lanes, undivided	0.040	4.52	28	2.12	7.43	45	4.49	4.89	27	6.65	16.44
4 Or more lanes, divided	0.027	3.06	25	1.65	5.76	48	3.19	3.19	27	4.86	12.01
Subtotal	0.032	3.54	27	1.71	5.99	46	3.43	3.43	26	5.17	12.95
Partial access control											
2-Lane expressway	0.033	3.73	53	0.65	2.28	32	1.05	1.05	15	1.73	7.06
Divided expressway	0.022	2.51	30	1.08	3.76	45	2.04	2.04	25	3.14	8.32
Total nonfreeway	0.031	3.45	28	1.65	5.77	46	3.29	3.29	26	4.97	12.51
Freeway	0.012	1.39	37	0.40	1.39	37	1.01	1.01	27	1.43	3.79
Total urban	0.016	1.78	33	0.64	2.22	41	1.45	1.44	26	2.10	5.45
Suburban											
No access control											
2 Lanes	0.048	6.12	43	1.26	5.05	35	2.56	3.08	22	3.88	14.25
4 Or more lanes, undivided	0.037	4.67	31	1.58	6.33	42	3.31	3.98	27	4.93	14.98
4 Or more lanes, divided	0.030	3.78	35	1.10	4.42	41	2.24	2.69	34	3.37	10.88
Subtotal	0.039	4.89	38	1.26	5.02	39	2.57	3.08	24	3.86	13.00
Partial access control											
2-Lane expressway	0.096	12.18	71	0.82	3.29	19	1.42	1.71	10	2.34	17.17
Divided expressway	0.060	6.37	57	0.82	3.29	29	1.29	1.54	14	2.16	11.20
Total nonfreeway	0.043	5.42	42	1.16	4.65	36	2.30	2.75	21	3.50	12.82
Freeway	0.015	1.91	47	0.32	1.27	31	0.74	0.89	22	1.07	4.07
Total suburban	0.021	2.67	45	0.50	2.00	34	1.08	1.29	22	1.60	5.97

*MVM = million vehicle miles; KVM = thousand vehicle miles.
Source: [6.4, pp. 66–67]

Table 6.10   Transit System Passenger Accident Rates

Mode	Injuries	Fatalities	Rate per million
Bus	14–21	NA*	Bus miles
Rail rapid	164	0.70	Train miles
Commuter rail	0.0591	0.0026	Passenger miles
Light rail	39.2	NA	Train miles

*NA = not available.
Source: [6.6, various pages].

6.1.7   Maintenance/Repair

It is difficult to separate the costs of maintenance due to calendar time of operation of a vehicle and those due to characteristics of the roadway system. Obviously, if many stop/go cycles are needed because of traffic signs and signals, for instance, then the brake system is more likely to need maintenance and/or repairs. Claffey has estimated an average maintenance/repair cost, affected by travel distance, for most automobile components of 1.15 cents per mile per vehicle [6.7]. Presumably this is in 1971 dollars.

6.1.8   Depreciation

The total depreciable cost of a vehicle is that for a new vehicle minus its scrap value when removed from highway service. Part of this cost is attributable to calendar time, as manifested by body corrosion, interior wear, consumer preference changes, and the like. The remaining depreciation is due to highway use, and might be expected to vary according to roadway and traffic conditions.

Table 6.11   Automobile Tire Cost as Affected by Speed and Type of Surface—Straight Road and Free-Flowing Traffic*

Uniform speed (mph)	Cost of four tires (cents/mile)[†]		
	High-type concrete	High-type asphalt	Dry well-packed gravel
20	0.09	0.27	1.03
30	0.19	0.36	1.05
40	0.29	0.43	1.07
50	0.32	0.45	1.10
60	0.31	0.46	—
70	0.30	0.44	—
80	0.27	0.43	—

*The composite passenger car represented here reflects the following vehicle distribution: large cars, 20 percent; standard-size cars, 65 percent; compact cars, 10 percent; and small cars, 5 percent.
[†]Tire costs were computed using a weighted average cost of $119 for a set of four new medium-quality tires based on the following unit tire costs by vehicle type (as noted in the northeastern states in 1969): Large cars, $35 per tire; standard-size cars, $30 per tire; compact cars, $25 per tire; and small cars, $15 per tire. There are approximately 1,500 g usable tire tread in 80 percent of passenger car tires. This weight of usable tire tread was also recorded for the tires used in the tire wear test.
Source: [6.7, p. 31].

Table 6.12 Correction Factors to Adjust the Values in Table 6.11 for Curvature*

Degree of curve	Correction factors by uniform speed of automobiles (mph)					
	20	30	40	50	60	70
0	1.00	1.00	1.00	1.00	1.00	1.00
2	1.02	1.53	3.06	5.27	9.21	15.70
4	1.10	2.00	6.11	10.67	19.05	29.58
6	1.30	2.56	8.88	17.11	30.28	–
8	1.60	3.33	12.50	28.40	–	–
10	1.90	4.33	16.66	44.80	–	–
12	2.10	5.33	20.44	89.40	–	–
14	4.00	8.50	–	–	–	–
16	6.10	12.70	–	–	–	–
30	10.83	–	–	–	–	–

*Correction factors apply on concrete, asphalt, and gravel surfaces. Test operations were also carried out on a 90° curve at a speed of 20 mph with stops at $\frac{1}{2}$-mile intervals. Tire wear was found to be approximately 1,000 times that on tangent.

Source: [6.7, p. 31].

Winfrey [6.9] and others have made some attempts to measure and price the latter. There appears to be a great amount of uncertainty in (a) estimating the depreciation costs due to highway use, and (b) attributing these costs to different roadway and traffic characteristics. For these reasons, and because the costs per mile appear to be relatively low, they are ignored here.

Table 6.13 Engine Oil Additions between Oil Changes to Make Up That Lost by Leakage and Combustion*

Speed in free-flowing traffic (mph)	Quarts/1,000 miles traveled by		
	Composite passenger car[†]	Pick-up truck	Two-axle six-tire truck[‡]
30	0.27	0.27	0.27
35	0.27	0.27	0.27
40	0.42	0.42	0.34
45	0.58	0.58	0.50
50	0.75	0.75	0.66
55	0.94	0.94	0.83
60	1.08	1.08	1.00

*Minimum trip length = 10 miles.
[†]Both the composite passenger car and the pick-up truck are represented by an eight-cylinder Chevrolet sedan (engine displacement = 283 cu in).
[‡]The two-axle six-tire truck is represented by a two-axle six-tire truck with six cylinders (engine displacement 351 cu in) weighing 16,000 lb.
Source: [6.7, p. 37].

Table 6.14 Average Fares

Mode	Cost (cents/trip)	Increase since 1960 (%)
Motor bus	35.75	99
Light rail	33.40	51
Heavy rail	46.09	186
Trolley coach	28.12	55

6.1.9 Transit Fares

American users of transit in 1978 faced average fares calculated by the American Public Transit Association [6.13] which are shown in Table 6.14. Since 1960 these have gone up at a much lower rate than overall inflation (see Table 6.17).

Transit systems offer a variety of fare schemes that depend on frequency of use, length of trip, time of day, age, special service provided (e.g., charter), etc. Perhaps one of the most interesting examples is the scheme put in effect by the Port Authority of Allegheny County (Pittsburgh area) in 1975, as displayed in Table 6.15. Special fares were created for purchase of weekly, monthly, or annual permits, a special express service, early trips, transfers, and even a free "wild card" for one morning trip daily.

Most transit systems have special rates for senior citizens, students, and children. The fare ranges for these groups on 36 systems in 1972 are shown in Table 6.16. On average, students and children seem to have received somewhat greater reductions than senior citizens.

Table 6.15 Special-Price Packages Offered to Commuters
by the Port Authority of Allegheny County

Description of service	Special fare	Regular one-trip fare
Reduced weekly permit	$2.60/week	$.40
Reduced monthly permit	$10.00/month	$.40
Reduced annual permit	$100.00/year	$.40
Ten trip tickets in outer zones for 49 Red Flyer Express and other multizone routes	$4.05	$.45 (express)
Early-bird special (before 7 a.m.)	$.25	$.40
Wild-card bus (one morning trip daily)	Free	
Stop-over transfer for 1 hr	$.10	

Other provisions
  Payroll deduction and annual subscription programs
  Credit card charge for permit purchases
  Free outlying parking (3,600 spaces)
  Twenty-trip downtowner zone ticket for close-in park-n-ride parking for $4.00
    (2,000 spaces)

Source: [6.12, p. II.20, as based upon "A Price Package for Every Rider's Pocketbook," Port Authority of Allegheny County, June 6, 1975, and "Fare Reports," American Public Transit Association, December 1973].

Table 6.16 Fare Level by Fare Type (36 Urbanized Areas)*

Fare (cents)	Basic adult	Senior citizen	Student	Child
0–24	3	16†	26	24
25–29	10	11	7	7
30–34	5	3	3	5
35–39	8	1		
40–44	5	3		
45–49	3	1		
50+	2	1		
Total number of areas analyzed	36	36	36	36

*All information as of 1972. Information obtained from transit operator statistics submitted to the American Transit Association.
†Six transit firms offer free senior-citizen fares.
Source: [6.12, p. II.6, based upon U.S. Department of Transportation, *A Study of Urban Mass Transportation Needs and Financing,* Washington, D.C., 1974, p. V-10].

6.1.10 Taxi Fares

Taxi revenue passengers in 1970 surprisingly numbered about 40% of those on mass transit. The average paid length was 2.95 miles and the corresponding receipts were $1.95 [6.10, p. 2-2]. In 1976 the median fare for a 3-mile trip was $2.45, excluding traffic delays and extras. Flag-drop rates (the initial meter charge) centered between $.70 and $.80. The typical rate for waiting time and traffic delay was $6/hr ($.10/min) [6.10, pp. 6-9 to 6-11].

6.1.11 Cost Changes Over Time

None of the costs quoted in the preceding sections have remained constant. Gasoline, as noted, rose rapidly—at an annual rate of 13.33% (compounded) between 1970 and 1980. In about the same period, trucking labor rose 9% and auto maintenance 8.3%, each above the national rate for all goods and services of 7.8%. In contrast, gasoline taxes went up very little, and new autos just somewhat more. The consumer price indices for these and several other user costs items are presented in Table 6.17.

6.2 AFFECTING FACTORS

The basic relationships shown in Fig. 5.1 indicate that direct transportation service impacts (that is, user costs, as discussed in the preceding section) are influenced not only by transportation systems but also by a variety of affecting or external factors. These can be exceedingly diverse, encompassing everything from international events such as raising of prices by petroleum-exporting countries to metropolitan events like urban-to-suburban migration of high-income families.

In between, and of probable great importance, are national events, particularly those created by the Federal government in its roles related to expenditure, taxation, and regulation. A prime example is the Federal law requiring the sales weighted miles per gallon (SWMPG; see Chap. 8, Sec. 8.3) of new cars sold each year to rise to certain levels.[7] This has a direct effect on the costs of fuel consumption and subsequently on

---

[7] The level in 1985, for instance, is 27.5 mpg.

Table 6.17 Consumer Price Indices for Selected User Cost Items
(Goods or Services)

Year	Gasoline	New auto	Auto insurance	Gasoline taxes*	Trucking labor	Auto maintenance	Parking	Intracity mass transit fares	All goods and services
1970	105.6	107.6	126.7	100.0	118.8	120.6	124.0	134.5	116.3
1971	106.3	112.0	141.1	100.0	134.7	129.2	135.3	143.4	121.3
1972	107.6	111.0	140.5	109.1	149.5	135.1	144.5	150.1	125.3
1973	118.3	111.1	138.0	118.2	163.8	142.2	152.8	150.1	133.1
1974	159.9	117.5	138.1	118.2	176.0	156.8	158.9	147.6	147.7
1975	170.8	127.6	145.9	118.2	188.1	176.6	172.1	170.3	161.2
1976	180.8	135.7	187.7	118.2	199.7	189.7	183.9	175.8	170.5
1977	189.1	139.0	200.0	118.2	214.0	203.7	195.4	178.5	181.2
1978	196.3	153.8	216.6	118.2	238.7	220.6	206.6	181.8	195.4
1979	265.6	165.9	228.7	118.2	259.3	242.6	226.5†	189.8	217.4
1980	369.1	179.3	247.4	127.3	—	268.3	—	217.6	246.8

*Virginia only.
†Estimated by author.
Source: U.S. Department of Labor, Bureau of Labor Statistics, Monthly Consumer Price Indices Reports.

travel. Many states also enhance or detract from transportation services and their impacts through relevant legislation and regulatory actions.

Another way of viewing affecting or external factors is through changes in man (and groups), elements of the natural and manmade environment, and activity agents and their behavior. This is the construct presented in Tables 4.1 to 4.4 in Chap. 4. Transportation technology,[8] as one example element, can change significantly over time. This might be influenced, for instance, by alterations in fuel types, as might happen if, say, a synthetic diesel fuel (containing no petroleum products) were developed.

Metropolitan transportation planners can have little effect on any of these external factors, of course. The best they can do is forecast likely upcoming changes and vary proposed plans accordingly. Even this approach involves many severe difficulties.

One area in which metropolitan planners have played a role, discussed briefly here, is in predicting urban area or region-wide population and economic development. This is important because it provides "control totals" for estimates of a variety of human activities in particular parts of the city. As an illustration, regional population figures sometimes are "allocated" to zones within the region, and these zonal figures in turn are utilized as a basis for estimating travel to and from each zone.

The discussions here will be short, simply to introduce models for making forecasts of regional-level activities.

6.2.1 Regional Population Models

Population projections have been the center of attention of researchers for many years. This emphasis has been due to the integral part such projections play in any

[8] See Sec. 12.1.2.

planning endeavor, be it urban, corporate, or even defense. Prime among the techniques available are the three to be discussed here. They are:

1. Extrapolation
2. Gompertz or logistics curves
3. Growth composition analyses[9]

6.2.1.1 Extrapolation of population The extrapolation technique, simply stated, is one in which a line is extended from past data points into the future and the population for the future period is found by "reading" the value for the particular year in question. Figure 6.5 indicates such a procedure. Data have been taken from past *Census* studies, a hand-drawn smooth line has been created and extrapolated into the future, and the population for 1990 has been calculated. This line need not be linear nor must it always be hand-drawn. Regression often has been used to establish the relationship between population and time.

Difficulties with the extrapolation technique are many. Most important, underlying causal factors, essentially covered over by this crude model, may change considerably, thereby creating entirely different growth rates. An example of this kind of situation would be the construction of a large manufacturing plant in an area previously lacking substantial employment opportunities. People would tend to accumulate in nearby residential areas and thus increase the population of these areas.

6.2.1.2 Gompertz or logistics curves Another difficulty with extrapolation is that the potential maximum growth limit of a region is not taken into account. The Gompertz curve technique eliminates this problem by considering population growth to start out slowly, building cautiously on a small base, then increasing fairly rapidly, and finally slowing down again as the maximum population that a region can "hold"

---

[9] See [6.14] and [6.15] for detailed discussions of these and other techniques.

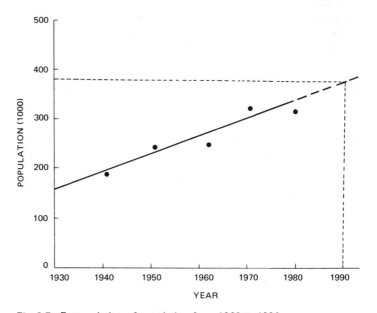

Fig. 6.5 Extrapolation of population from 1980 to 1990.

is approached. The result is an S-shaped curve of population growth over time whose general form is

$$P_{t+\theta} = ka^{b^{\theta}}$$ (6.1)

where $P_{t+\theta}$ is the population at a time θ years from the present time (t) and k, a, and b all are parameters whose values are established through some kind of curve-fitting procedure (like regression) utilizing past data.

A similar model of population growth is that of the logistic curve, whose formula is

$$P_{t+\theta} = \frac{k}{1 + e^{a+b\theta}}$$ (6.2)

where $P_{t+\theta}$, k, a, and b are defined in the same terms as in Eq. (6.1) and e is the base of the natural logarithms. Both of these models have been used quite extensively, but they suffer from the same drawback as the extrapolation technique—no underlying causal factors are identified, and these, if significant, may lead to large, unexpected changes.

6.2.1.3 Growth composition analysis To help reduce the above-mentioned deficiencies in population growth models, several analysts have employed models that include four important variables: births, deaths, in-migration, and out-migration. The equation for this model is

$$P_{t+\theta} = P_t + B_\theta - D_\theta + I_\theta - O_\theta$$ (6.3)

where P_t = the present (time t) population
B_θ = the births during period θ
D_θ = the deaths during period θ
I_θ = the in-migration during period θ
O_θ = the out-migration during period θ

A basic need for a model such as Eq. (6.3) is to have estimates of B_θ, D_θ, I_θ, and O_θ for the upcoming period θ. To obtain such information requires that projections of these values be available and also reliable, a situation that brings us back to the extrapolation technique as a necessary tool for all predictions. In other words, somewhere in the prediction process we eventually must rely on some sort of extrapolation into the future.

6.2.2 Regional Employment Models

Also required as input to most land-use models is the future number of persons falling in certain employment categories. Usually, although not always, these inputs are obtained through models that relate future employment to future population, present employment, and interindustry production and consumption relations, which are assumed to hold constant into the future. The input–output model [6.15] is representative of this kind of concept.

6.2.2.1 The regional input–output model This type of model relies heavily on data concerning the amount of money spent by an industry in a region for goods and services from other industries in the region. As an example, consider the purchases and sales in Table 6.18 by two industries that, for simplicity, are assumed to be the only ones in a region—that is, with the exception of the households in the region—that

Table 6.18 Input-Output Matrix (Dollar's Worth of Products)

| From | To | | | |
	Agri-culture	Manu-facturing	Final demand	Total
Agriculture	5	3	2	10
Manufacturing	4	1	4	9
External	1	5	0	6
Total	10	9	6	25

also purchase industry goods, like food and furniture, for final consumption. There is also an "external" industry that makes up the difference between supply and demand for each of the other industries. The values in each row of Table 6.18 actually correspond to an equation of the form

$$X_i - x_{i1} - x_{i2} = Y_i \quad \text{(all } i) \tag{6.4}$$

where X_i = the total (or gross) output of industry i
  x_{i1} = sales of industry i to industry 1 (agriculture)
  x_{i2} = sales of industry i to industry 2 (manufacturing)
  Y_i = final (household) demand for outputs of industry i
Thus, for row one we have equalities $10 - 5 - 3 = 2$, and for row two $9 - 4 - 1 = 4$. Now, excluding the final demands, let us divide each x_{ij} by X_j to determine the *fraction* of goods purchased by industry i from industry j, a_{ij}. Thus

$$x_{ij} = a_{ij} X_j \tag{6.5}$$

which leads to the set of production coefficients displayed in Table 6.19 (excluding the external industry).

Now assume that the final demand by households for goods and services is known, estimated perhaps on the basis of future population and average household income projections. All other variables in Eq. (6.4) are to be predicted. Thus, there are six unknowns $(x_{11}, x_{12}, X_1, x_{21}, x_{22},$ and $X_2)$ and there also are six equations:

$$X_1 - x_{11} - x_{12} = Y_1 \tag{6.6}$$

$$X_2 - x_{21} - x_{22} = Y_2 \tag{6.7}$$

$$x_{11} = a_{11} X_1 \tag{6.8}$$

Table 6.19 Input-Output Production Coefficient Matrix

| From | To | |
	Agri-culture	Manu-facturing
Agriculture	0.50	0.33
Manufacturing	0.40	0.11

$$x_{12} = a_{12}X_2 \tag{6.9}$$

$$x_{21} = a_{21}X_1 \tag{6.10}$$

$$x_{22} = a_{22}X_2 \tag{6.11}$$

If $Y_1 = \$4 \times 10^6$, $Y_2 = \$6 \times 10^6$, and the a_{ij}'s are as in Table 6.19, we can solve Eqs. (6.6) to (6.11) simultaneously by inserting the equalities (6.8) through (6.11) where needed in the first two equations and get

$$X_1 - 0.50X_1 - 0.33X_2 = 4 \tag{6.12}$$

and

$$X_2 - 0.40X_1 - 0.11X_2 = 6 \tag{6.13}$$

which lead to $X_1 = \$17.82 \times 10^6$, $X_2 = \$14.74 \times 10^6$, $x_{11} = \$8.91 \times 10^6$, $x_{12} = \$4.91 \times 10^6$, $x_{21} = \$7.12 \times 10^6$, and $x_{22} = \$1.62 \times 10^6$.

We thus have found the gross output of each industry at some given future point in time. What remains to be done is to use either some present or extrapolated ratio of employees/dollars of gross output, r_i, to obtain the number of employees. For instance, if at present $r_1 = 2 \times 10^{-4}$ employees/dollar of gross output and $r_2 = 1 \times 10^{-4}$ employees/dollar of gross output, the future number of employees in agriculture will be $r_1 X_1 = (2 \times 10^{-4})(17.82 \times 10^6) = 3,564$ and in manufacturing $(1 \times 10^{-4})(14.74 \times 10^6) = 1,474$.

6.2.2.2 Comments on input–output models Several comments should be made about the advantages and disadvantages of an input–output model:

1. It can be expanded to take care of any number or different categories of industry so that it can be easily adapted to producing employment inputs for other purposes.
2. Data on sales and purchases of many industries often are difficult to find.
3. Production coefficients and ratios of employment to gross output may not be constant over time due to technological, union, management, and governmental changes.
4. The complexity of the model is such that most studies probably would be willing to sacrifice reliability to get predictions in a quicker and less expensive manner.

6.3 HUMAN ACTIVITIES (LAND USE)

Figure 5.1 shows that human characteristics and activities are major contributors to travel, and thus to mobility. As will be brought out in Chap. 8, they also can be impacted greatly by the transportation system and resultant travel. There are, of course, an almost infinite variety of activities and characteristics that might be considered. Activities of the handicapped, for instance, can vary from almost zero for those people who are bedridden to near normal for, say, people with a slightly sprained ankle. The resultant travel patterns thus can also vary substantially.

If transportation plans are to be made for such people, it is necessary to know the number of likely users with different handicaps and the activities they would undertake (e.g., work, shopping, and medical visits). An estimate then would have to be made of the origins, destinations, times, etc., of the subsequent travel patterns (Chap. 7). This highlights the point that the demand for different kinds of transportation depends on the type, amount, and location of human characteristics and activities.

There is, of course, far too much variety in human characteristics and activities to discuss here. An important subgroup, however, is "land use." As pointed out in Chap. 2, this can refer not only to the type, but also to the *intensity* of use. This could be measured by acres of ground area, square feet of floor area, number of employees, sales, and even household income and auto ownership. The definition thus is very broad.

There are two main ways in which land use can be employed as an input to the transportation planning process. First, the land use *plan* (if such exists) in a locality can be assumed to hold and thus taken as a basis for travel prediction. Alternately, where no plan exists or where it is unlikely to be implemented as proposed, forecasts can be made of land-use development in different sections of the urban area. Naturally, there also can be combinations of the above two approaches. And both may take into account the influence of the transportation system on land use as well as the reverse (see Chap. 8).

The development of land use plans is discussed in Chap. 13. Here we will describe methods for *forecasting* land-use changes.

6.3.1 Land-Use Models

Prediction of zonal land-use development is difficult, especially since it is not easy to see that a relationship with transportation and land use exists. Certainly this reaction is to be expected since the causal influence is a gradual one, appearing only after 5, 10, or an even greater number of years after changes in the transportation system and other facilities. Nevertheless, as can be seen in Fig. 6.6, the actual changes can be dramatic. In this figure, the growth in various parts of the Chicago metropolitan area is shown, and while the development that appears cannot all be attributed to the transportation system itself, the fact that most of it is within a short distance of these facilities provides a strong indication of the importance of the transportation system, and possibly other public and private facilities, in development.

6.3.1.1 Basic land-use development models There is a wide variety of urban models available, and development and improvement of such has continued from the late 1950s. Unlike the travel models to be discussed in Chap. 7, however, no single land-use forecasting scheme has been incorporated as part of a widely accepted computer program package. Thus, in the late 1960s the EMPIRIC land-use model [6.24] was applied in several large urban areas, while lately EMPAL DRAM [6.21] has been most frequently employed, although not widely.

In general, most planning agencies attempting to forecast land-use development in different parts of a metropolitan area seem to follow a process, either implicitly or explicitly, similar to that in Fig. 6.7. In the upper left-hand side we see the estimation of areawide population (3) and economic growth (or decline) (4) using techniques like those described in Sec. 6.2. Combination of these lead to a forecast of the total number of jobs to be located in the metropolis (5).

Actual job location (6) within parts (zones) of the city may be affected by several factors:

Job location policies
Available land within each zone (9)
Travel time from other zones (8)
Residential population within other zones (11)

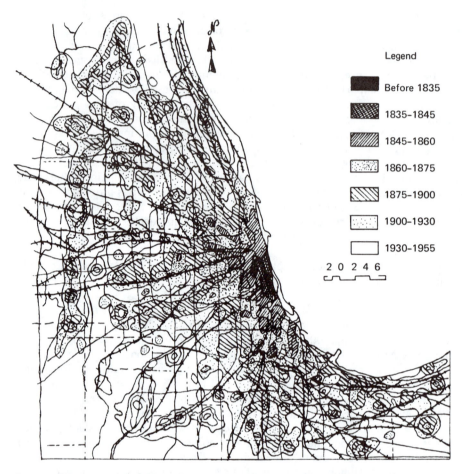

Legend

███	Before 1835
▨▨▨	1835–1845
▨▨▨	1845–1860
░░░	1860–1875
▧▧▧	1875–1900
⬚	1900–1930
☐	1930–1955

2 0 2 4 6

Fig. 6.6 Chicago growth patterns. The urbanized area has spread out in waves from the site of the original settlement. The finger development has followed the suburban railroad line [6.17, vol. 1, p. 15].

For example, an entrepreneur looking for a place for an industrial park might first determine which sites were permitted this type of use (job location policy), and then whether the land available were sufficient. He subsequently would see whether enough potential employers lived within a certain travel-time radius (e.g., 40 min) from the prospective site.

The other side of the ledger is housing location. Population forecasts would indicate the number of households to be located (10). People in these households probably would want to live in somewhat close proximity to work, schools, shopping areas, and the like. The choice of location (11) thus would depend on:

House location policies (e.g., types of housing required)
Availability of land within each zone
Travel time to work, shops, schools, etc.

These point to a joint determination of housing and job locations.

To complete the cycle it may be necessary to recalculate travel times, since greater

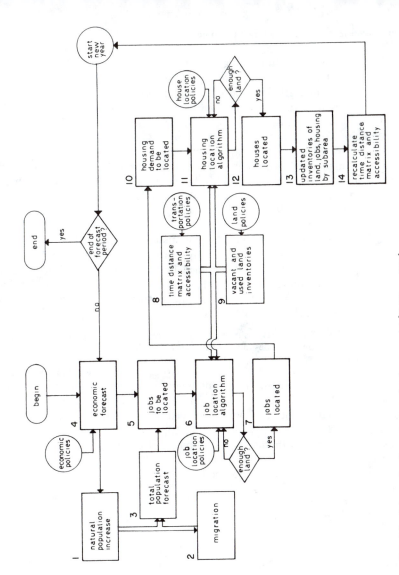

Fig. 6.7 Schematic of a prototype land use model [6.18, p. 152].

concentration of jobs and housing in certain zones may overload the nearby links on the transportation network and thus raise the times (14). This in turn may lead to some associated relocation of housing and employment.

Models of the type suggested by Fig. 6.7 can become very complicated. Employment and housing demand may be divided into several categories, with the location of one being dependent on several of the others. Transportation may be classified by mode with, say, transit having a different impact on office employment location than it does on high income housing location. The EMPIRIC model, for instance, divides land use into four types of households (by income class) and five types of employment (by standard industrial classification). As shown in Fig. 6.8, future levels of each of these in each zone are influenced in part by the levels of the others in that zone and by seven other factors.

6.3.1.2 A simple example A simplified illustration of a residential location allocation model has been developed by Putman [6.21]. A hypothesis is made that people's place of residence is determined by their place of work, and by the desirability of making work trips involving certain specified "costs" (which include the value of time spent in travel; costs for fuel, parking, and auto maintenance; and the like). Thus the model may be stated as:

> The number of people choosing to reside in zone i equals the sum, over all zones, of the employees in j times the desirability of making a work trip to j.

This can be envisioned in Fig. 6.9 where there are five zones, with the employment levels indicated in the boxed numbers and the "costs" of travel shown on the associated links. We imagine a group of people trying to decide whether to live in zone 1. They survey the employment opportunities there, as in other zones, and weigh them against the desirability of making trips there, as represented in part by the "costs." That zone with the highest product of employees and desirability of travel attracts the largest proportion of dwellers.

An equation summarizing the models is:

$$N_i = \sum_{j=1}^{5} E_j \, d_{ij} \qquad (6.14)$$

where N_i = population (in terms of employed residents) in zone i
E_j = employment (at place of work) in zone j
d_{ij} = desirability of work trip between zones i and j
Analyses of travel patterns (see Chap. 7) give some clues on determination of d_{ij}. These have shown that the perceived worth of a trip, w_{ij}, depends on "costs," and falls off in a form like that in Fig. 6.10 as "costs," c_{ij}, increase. This leads to an equation like

$$w_{ij} = c_{ij}^{-2} \qquad (6.15)$$

It is now assumed that the trip desirability to potential home location seekers is not simply the worth number in Fig. 6.10 but its value *relative* to that in other zones. In other words, the trip desirability is judged by comparing the worth of work trip costs to zone j to that from all other zones. In equation form this becomes

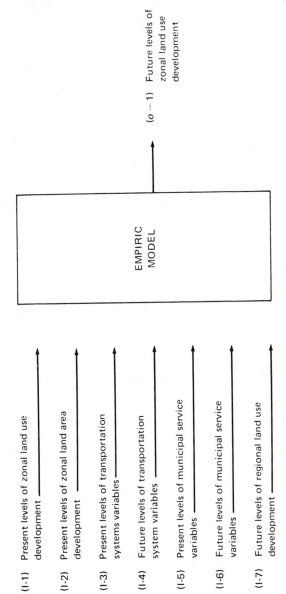

Period t inputs

(I-1) Present levels of zonal land use development ———→

(I-2) Present levels of zonal land area development ———→

(I-3) Present levels of transportation systems variables ———→

(I-4) Future levels of transportation system variables ———→

(I-5) Present levels of municipal service variables ———→

(I-6) Future levels of municipal service variables ———→

(I-7) Future levels of regional land use development ———→

EMPIRIC MODEL

Period $(t + 1)$ outputs

$(o - 1)$ Future levels of zonal land use development

Fig. 6.8 Major inputs and outputs for the EMPIRIC land use model.

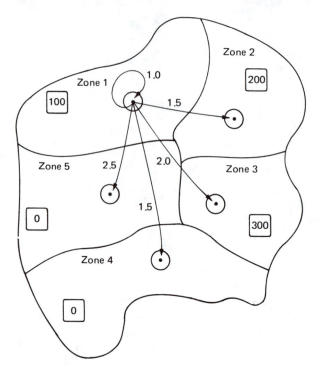

Fig. 6.9 Schematic of urban area zonal employment and travel "costs." Boxes show number of employees in each zone. Zone centers (centroids) are marked with circles. "Cost" of travel between respective zone centroids is shown next to arrows.

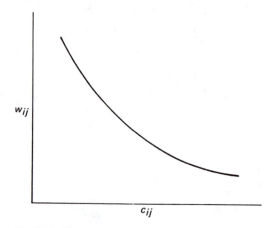

Fig. 6.10 Hypothetical decline in work trip worth (w_{ij}) with "costs" (c_{ij}).

$$d_{ij} = \frac{w_{ij}}{\sum\limits_{k=1}^{5} w_{kj}} \tag{6.16}$$

or

$$d_{ij} = \frac{c_{ij}^{-2}}{\sum\limits_{k=1}^{5} c_{kj}^{-2}} \tag{6.17}$$

which leads Eq. (6.14) to become

$$N_i = \sum\limits_{j=1}^{5} \frac{E_j c_{ij}^{-2}}{\sum\limits_{k=1}^{5} c_{kj}^{-2}} \tag{6.18}$$

As a numerical sample, suppose we have the urban area shown in Fig. 6.9 along with the zone-to-zone travel "costs" in Table 6.20. From Eq. (6.17), the desirability of, say, travel from zone 1 to zone 2 is

$$d_{12} = (1.5)^{-2}/[(1.5)^{-2} + (1.0)^{-2} + (1.5)^{-2} + (2.0)^{-2} + (2.0)^{-2}] = 0.186$$

The other figures for zone 1 can be computed as

$$d_{11} = 0.437 \quad d_{13} = 0.119 \quad d_{14} = 0.186 \quad d_{15} = 0.080$$

Equation (6.14) subsequently gives the number of residents of zone 1 as

$$N_1 = 100(0.437) + 200(0.186) + 300(0.119) + 0(0.186) + 0(0.080) = 116$$

For the other zones the resultant residents are, respectively, 166, 191, 76, and 51.

As an example of the use of this model, suppose that the travel "costs" between zones 3 and 5 are reduced from 2.5 to 1.5. This results in the figures in Table 6.21.

Table 6.20 Interzonal Travel "Costs"
for Hypothetical Example

Zone	1	2	3	4	5
1	1.0	1.5	2.0	1.5	2.5
2	1.5	1.0	1.5	2.0	2.0
3	2.0	1.5	1.0	2.0	2.5
4	1.5	2.0	2.0	1.0	1.5
5	2.5	2.0	2.5	1.5	1.0

Source: [6.21, p. 26].

Table 6.21 Number of Employed Residents in Each Zone Under Two Cases

Zone	1	2	3	4	5
Base case	116	166	191	76	51
Policy test 1	112	159	174	72	83

Source: [6.21, p. 28].

The transportation improvement thus allows more employees from zone 3 to live in zone 5, as well as producing minor changes in other zones.

6.3.1.3 DRAM The model in the previous section is a simplified version of DRAM, the disaggregated residential allocation model. There are two principal differences between the two:

1. DRAM can locate several (usually four) types of population simultaneously.
2. The term (d_{ij}) in DRAM contains two components, a trip desirability component and an attractiveness component. The trip-worth component is a two-parameter function of travel time or cost. The attractiveness measure is a multivariate function of several variables which describe each zone. Each such variable has an associated parameter. These parameters must be estimated, empirically, for each urban area in which the model is to be applied.

These extensions to the simplified model are developed in a few simple steps. First, the d_{ij} term is expanded to include, separately, an attractiveness measure A_i at the residence zone i and the travel-worth function. So we get

$$d_{ij} = \frac{A_i w_{ij}}{\sum\limits_{k=1}^{n} A_k w_{kj}} \tag{6.19}$$

In DRAM, the worth function used is

$$w_{ij} = c_{ij}^a \exp (bc_{ij}) \tag{6.20}$$

where both a and b are empirically derived parameters. Further, in most of those model applications A_i has been a single variable such as population or floor area of zone i. A multivariate attractiveness measure is used in DRAM,

$$A_i = V_i^a P_i^b R_i^d N_{1,i}^q N_{2,i}^r N_{3,i}^s N_{4,i}^t \tag{6.21}$$

where V_i = vacant, buildable land in zone i
P_i = 1.00 plus the percentage of buildable land in zone i that has already been built upon
R_i = residential land in zone i
$N_{1,i}$ = 1 plus the percentage of employed residents of zone i who are in the lowest income quartile
$N_{2,i}$ = 1 plus the percentage of employed residents of zone i who are in the lower middle income quartile

$N_{3,j}$ and $N_{4,j}$ = upper middle and upper income quartiles
a, b, d, q, r, s, t = empirically derived parameters
Thus the full equation structure for DRAM is, for each household type:

$$N_i = \sum_j E_j \ \frac{A_i c_{ij}^a \exp\left(bc_{ij}\right)}{\sum_k A_k c_{kj}^a \exp\left(bc_{kj}\right)} \tag{6.22}$$

Use of DRAM requires the estimation of nine parameters. As the model directly forecasts the location of employed residents in each of the four income quartiles, it is therefore necessary to estimate four sets of nine parameters each.

6.3.1.4 Comments on land-use models As can be seen from the preceding example, the DRAM can be quite complicated. This is not unexpected, however, seeing that the urban area or region being modeled also is extremely complicated in nature. But the major problem still remains—most people and many professionals do not have the mathematical capabilities to understand, much less apply, a model like DRAM.

Despite its complexity, DRAM, as a representative of almost all land-use models, still has many theoretical difficulties, which in comparison to the ideal leave something to be desired. First is the question of the time period involved. DRAM usually would be calibrated at one point in time and then employed recursively for forecasting up to 20 years into the future. Such forecasts should be suspect since the one calibration point may not be representative. Then too, one must always be suspicious of the input data to the calibration phase since they are quite numerous, somewhat expensive to collect, and subject to error of measurement and/or interpretation.

Another theoretical question concerns the form of the relationships. In the attractiveness measure, Eq. (6.21), for instance, all the variables have been multiplied together. There really is no theoretical justification for this type of relationship, but easy-to-use methods for other types of multivariate analysis are not presently available. This situation leads us to be suspicious of the parameters in DRAM, especially if (1) they are assumed to hold constant for 20 years into the future, and (2) they are employed in a situation where the model might inadvertently be utilized outside of the range of the variables with which it was calibrated.

A final difficulty with DRAM and similar models is the uncertainty associated with the input variables, including those not actually employed in the model. As far as the *actual* inputs are concerned, many have to be forecast themselves: future highway and transit travel times, percentage of buildable land in a zone, and regional population and employment. There thus is some question about possible forecast errors in the input variables that may be reflected in errors in the model outputs.

The second part of the difficulty—uncertainty with variables *not* employed in DRAM—is, we feel, more serious. There are so many factors that can affect land-use development that it is almost impossible to include all of them in *any* model. These factors range from "too small closets in a house" (thereby inducing people to move) to "excessive national capital gains taxes" for industry. Between are a variety of factors including changes in attitudes toward environmental preservation, development, and subsidy of complete new towns in a region, and changes in sewage treatment technology (e.g., complete in-house recycling of wastes).

DRAM, like any land-use model, thus must be utilized with significant amount of judgment on the part of many persons, including those familiar both with land-use modeling and with the geographic area being modeled.

6.4 SUMMARY

In this chapter we have explored various kinds of user costs. These included the value of time spent in travel, fuel consumption, parking, accidents, tire wear, motor oil, and transit and taxi fares. Data on some of these actually are somewhat old, the latest being based on road tests made in the early 1970s. In any case, they usually have to be updated anyway because of inflation and technological change. The future unit cost of tires, for example, will not be the same as in the past due to inflation and improvements in tire performance (e.g., mileage).

The next major section of this chapter dealt with those factors affecting general change in a region. These included population and economic growth (or decline). Relatively simple descriptions were given of models often used for estimating these changes. Attention then turned to human activities (and, more specifically, land-use development) within various parts (zones) in an urban area. A simplified example was given of DRAM (disaggregated residential allocation model) and its use to predict household location relative to employment opportunities. Generally speaking, land-use models are not a standard part of any transportation planning package, although EMPAL–DRAM seems to be most widely used at this time.

BIBLIOGRAPHY

6.1   U.S. Bureau of Economic Analysis: *The National Income and Product Accounts of the United States*, 1929–1974, Government Printing Office, Washington, D.C., 1977.

6.2   U.S. Bureau of Economic Analysis: *Survey of Current Business*, July issues (monthly).

6.3   Hensher, D. A.: "Review of Studies Leading to Existing Values of Travel Time," *Transportation Research Record 587*, 1976.

6.4   American Association of State Highway and Transportation Officials: *A Manual on User Benefit Analysis of Highway and Bus Transit Improvements: 1977*, Washington, D.C., 1978.

6.5   Maricopa Association of Governments, Transportation and Planning Office: *Accident Study for the Maricopa County Area*, Phoenix, April 1981.

6.6   U.S. Department of Transportation: *Characteristics of Urban Transportation Systems*, Washington, D.C., July 1977.

6.7   Claffey, P. J.: *Running Costs of Motor Vehicles as Affected by Road Design and Traffic*, NCHRP Report 111, Washington, D.C., 1971.

6.8   Curry, D. A., and D. G. Anderson: *Procedures for Estimating Highway User Costs, Air Pollution, and Noise Effects*, NCHRP Report 133, Washington, D.C., 1972.

6.9   Winfrey, R.: *Economic Analysis for Highways*, International, Scranton, Pa., 1968.

6.10 U.S. Department of Transportation: *The Role of Taxicabs in Urban Transportation*, Washington, D.C., Dec. 1974.

6.11 U.S. Department of Transportation: *Taxicab Operating Characteristics*, Washington, D.C., March 1977.

6.12 U.S. Department of Transportation: *Public Transportation Fare Policy*, Washington, D.C., May 1977.

6.13 American Public Transit Association: *Transit Fact Book '78–'79*, Washington, D.C., Dec. 1979.

6.14 U.S. Department of Commerce, Bureau of Public Roads, Urban Planning Division: *The Role of Economic Studies in Urban Transportation Planning*, Government Printing Office, Washington, D.C., Aug. 1965.

6.15 Isard, W.: *Methods of Regional Analysis: An Introduction to Regional Science*, M.I.T. Press, Cambridge, Mass., 1961.

6.16 Berry, B. J. L., and F. E. Horton: *Geographic Perspectives on Urban Systems*, Prentice-Hall, Englewood Cliffs, N.J., 1970.

6.17 *Chicago Area Transportation Study* (3 vols.): Chicago, 1959, 1960, 1962.

6.18 Goldberg, M. A.: "Simulating Cities: Process, Product, and Prognosis," *American Institute of Planners Journal*, April 1977.

6.19 Batty, M.: "Urban Models," in N. Wrigley and R. J. Bennett (eds.), *Quantitative Geography: A British View*, Routledge and Kegan Paul, London, 1981.

6.20 Batty, M.: *Urban Modelling: Algorithms, Calibrations, Predictions*, Cambridge University Press, Cambridge, England, 1976.

6.21 Putman, S. H.: *Integrated Policy Analysis of Metropolitan Transportation and Location*, Urban Simulation Laboratory, University of Pennsylvania, Philadelphia, Feb. 1980.

6.22 Hamburg, J. R., and R. H. Sharkey: *Land Use Forecast*, Chicago Area Transportation Study, Chicago, 1961.

6.23 Rosenthal, S. R., et al.: *Projective Land Use Model—PLUM* (3 vols.), Institute of Transportation and Traffic Engineering, University of California, Berkeley, 1972.

6.24 Hill, D. M., D. Brand, and W. B. Hansen: "Prototype Development of Statistical Land Use Prediction Model for the Greater Boston Region," *Highway Research Board Record* 114, 1966.

6.25 Putman, S., and F. Ducca: "Calibrating Urban Models 2: Empirical Results," *Environment and Planning A*, vol. 10, 1978.

6.26 U.S. Department of Transportation, Federal Highway Administration: *An Introduction to Urban Development Models and Guidelines for Their Use in Urban Transportation Planning*, Government Printing Office, Washington, D.C., Oct. 1975.

EXERCISES

6.1 A new bus route increases average trip speed from 20 to 24 mph and cuts the route length from 8 miles to 6 miles. What is the saving in travel time (in min)?

6.2 Pretend the year is 1985 and the construction of a new subway has saved 1,000,000 work travelers, on average, 10 min time. Using the information in Table 6.2 and assuming average hourly family income has increased since 1975 at 9% per annum (compounded), what is the value (in 1985 dollars) of all time saved?

6.3 What is the fuel consumption (gpm) for a two-axle, six-tire truck on a six-lane CBD street with parking in both curb lanes? The truck is attempting to run up the street on a 0.5% grade at 25 mph, but has to make six stops per mile in a one-way traffic volume of 100 vehicles per hour. Assume that the year is the same for the data in Tables 6.6 and 6.7.

6.4 For a selected locality, find the number of accidents and fatalities in the preceding year. Determine a unit dollar value for each, and justify those numbers. Update them to the preceding years (if they are not so already). Compute the total cost of all the accidents and fatalities.

6.5 According to Table 6.11, what would be the cost of auto tire wear (in 1969) for travel on a high-type asphalt road at a uniform speed of 40 mph around a 6-degree curve?

6.6 If you live in a city with a transit system, describe the fare structure (for every type of trip and traveler).

6.7 If you live in a city with at least one taxi company, describe the fare structure for the selected company.

6.8 In the library, find the latest consumer price index for new autos. What has been the annual (compound) rate of change since 1970?

6.9 In a given metropolitan area, the current population is 1 million. If (a) net migration (in minus out) can be found from a logistic curve where $k = 838,906$, $a = 1.5$, and $b = 0.05$; (b) the birth rate over a decade is 20 per 1000 existing population; and (c) the death rate over a decade is 5 per 1000, what will the population be 10 years from now?

6.10 A region has only two industries—manufacturing and retailing. Their sales to each other and to households (final demand) in billions of dollars are:

	To		
From	Manufacturing	Retail	Households
Manufacturing	3	6	1
Retail	6	8	6
External	1	6	0

If final demand increases in 20 years to $3 and $21 billion, respectively, what will be the gross output of the manufacturing and retail sectors (assuming all production coefficients stay the same)?

6.11 Suppose in the DRAM example in Fig. 6.11 and Table 6.20 that 100 employees (jobs) are relocated from zone 2 to zone 5. What would be the resultant distribution of residences? How does this compare to the other two cases in Table 6.21?

7 Models III: Travel Demand

The estimation of future demand for travel under a specific set of forecasted land-use developments and an anticipated transportation plan is considered the backbone of the transportation planning process. Planners and engineers have realized through their previous efforts that planning for an individual segment or link of a metropolitan transportation system is not adequate, and is most often misleading when viewed in the context of long-range planning efforts of transportation facilities. A systems approach to transportation planning and development, where flows in transport networks are analyzed in an integrated manner, is deemed necessary for a proper comprehensive and coordinated transportation plan. This approach required new demand estimation techniques.

The projection of past trends in constant increments into the future was often treacherous, as pointed out at the time by Kanwit, Steel, and Todd [7.30]. Perhaps the major reason has been that projections could not be made for completely new facilities or for facilities passing through areas yet to be developed. Furthermore, trending did not relate travel to any causal variables in the urban setting.

In formulating travel demand forecasting techniques, transportation planners had to base their theories on some assumptions or hypotheses. Among the important ones were that human travel behavior is orderly, rational, and measurable, and that the amount of traffic is a function of human characteristics and activities and, more specifically, land use. These assumptions are brought out explicitly in the report of the Chicago Area Transportation Study (CATS), in which it is stated [7.1, pp. 5, 6, 11]: "It is a basic theory of the study that there is an order in human travel behavior in urban areas which can be measured and described." The study also states that: "... the study worked on the hypothesis that there is a measurable relationship

169

between land use and the amount and distribution of traffic. . . . Since land use can be predicted with some assurance, future traffic demands can also be predicted."

Figure 7.1 shows diagrams of daily trip ends and square feet of floor area by zone for the Chicago region, aptly demonstrating the existence of correlation between the two variables.

To translate these theories and hypotheses into a working travel estimation process is not an easy task. A look at the amount and variation of passenger travel in metropolitan areas illustrates the difficulties inherent in the problem.

7.1 METROPOLITAN PASSENGER TRAVEL

To describe the various passenger travel patterns in a metropolitan area requires some kind of stratification. A stratification by tripmaker, trip, mode, and route characteristics is most often used. An attempt to classify metropolitan travel using these four basic vectors is illustrated in Table 7.1.

Each vector can be divided into many components, as shown in the table. Furthermore, each component can be subdivided into a number of groups to highlight the relationship between specific subgroup characteristics and their effects on travel. Distinct arrangements and combinations among the different vector components would result in a large number of alternative travel desire patterns.

It is not our intention to discuss in detail here all of the various components of travel. Instead we want to bring to the reader's attention the complexity of travel behavior and the difficulty associated in modeling it for a metropolitan area.

7.1.1 Peaking

One component of the trip vector that necessitates immediate discussion is the variation of tripmaking over time, that is, peaking characteristics. The temporal distribution of trips and their peaks is significant to the modeling as well as overall planning process for transportation facilities.

Probably the most significant temporal variation from a planning standpoint is that within the 24-hr period. In most urban areas the peaks come at two periods— the morning and afternoon rush hours—with a larger volume usually occurring in the latter period. The series of diagrams in Figs. 3.5 (Chapter 3), 7.2, and 7.4 show rather clearly the hourly variations that may be anticipated, first by purpose of trip and then by volume. Figure 7.3 indicates the variations in the proportion of the afternoon peak hour to daily volume at different locations and for different days in the Nashville urban area. As noted above, the afternoon peak stands out fairly distinctly, representing anywhere from 8 to 14 percent of the total daily traffic. Trips to home comprise most of this peak, with those from work combined with those of family members returning from school, shopping, and personal business visits. Travel by modes other than automobile are particularly peaked, pointing to a major difficulty that faces planners and managers of mass transit systems.

It has generally been found in most transportation planning studies that one weekday's travel is similar to any other weekday's. The Saturday and Sunday hourly maximums are not as great percentage-wise as those during the week: 8 percent versus approximately 11 percent (Fig. 7.4). In fact, this difference is the basis for the usual planning procedure of concentrating on weekday travel as the major source of traffic problems. Nevertheless, it often turns out that weekend traffic is greater in magnitude than weekday, so that the peaks almost are equivalent in terms of volume, with the

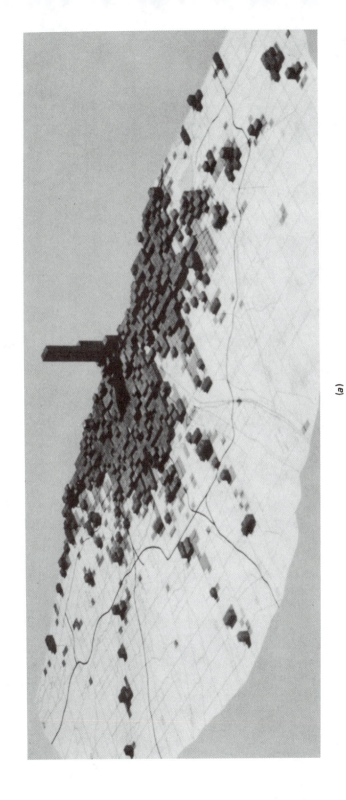

(a)

Fig. 7.1 Total person trip destinations and floor area models, Chicago, 1956. (a) The highest block represents 144,000 trip destinations per quarter square mile [7.1, pp. 23, 25].

171

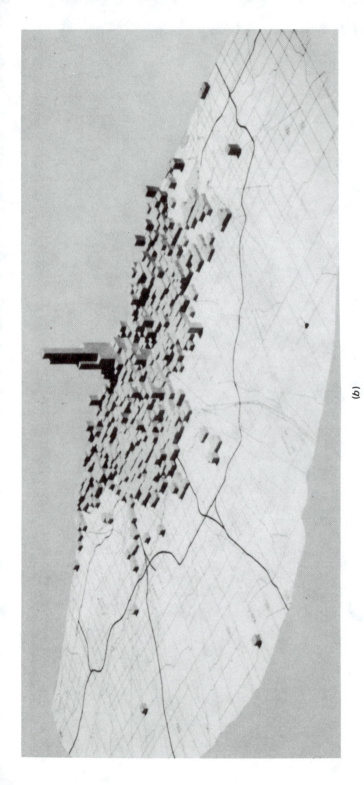

(b)

Fig. 7.1 Total person trip destinations and floor area models, Chicago, 1956. (b) The highest block represents 32,000,000 sq ft per quarter square mile [7.1, pp. 23, 25].

Table 7.1 Metropolitan Passenger Travel Classification

Trip-maker vector	Trip vector	Mode vector	Route vector
Age	Purpose	Auto	Highway routes
Sex	Time	Taxi and limousine	Transit routes
Income	Orientation	Transit buses	
Family size	(origin–destination)	School buses	
Auto ownership		Rail transit	
Occupation		Commuter rail	
Household head		Air	
Other		Miscellaneous	
		Bicycles	
		Walkways	

added characteristic that little of the weekend travel is by transit. This situation would be particularly noticeable on routes leading to large recreational areas.

Traffic volumes also vary by the month of the year, being higher in the summer months and, as expected, lower during the winter. Again, recreation routes have much greater peaks than do in-city routes.

The easiest way engineers have found to summarize peaking characteristics is by means of a chart similar to that in Fig. 7.5 where the volumes (taken as a percentage of the annual average daily traffic) in each of the 8,760 hr in the year are arranged in descending order, with the hour corresponding to the highest hourly volume being called the "first highest hour," the hour corresponding to the second highest volume called the "second highest hour," and so forth. The 30th and 50th highest hourly

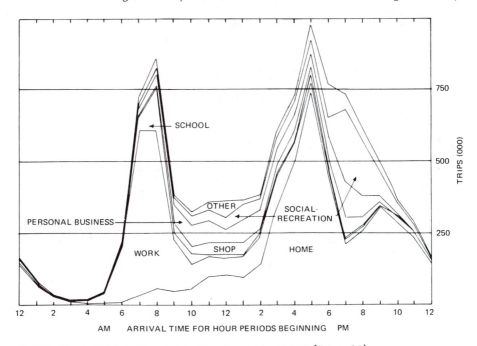

Fig. 7.2 Hourly division of internal person trips by trip purpose [7.1, p. 35].

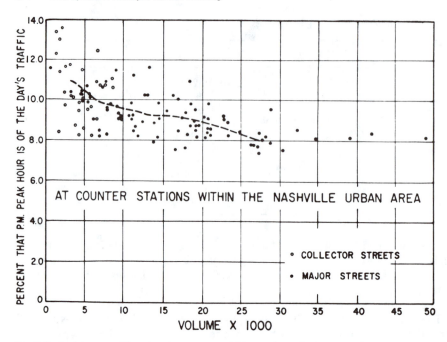

Fig. 7.3 Peak-hour traffic volume as percent of 24-hr volume [7.27, p. 147].

volumes are employed extensively in many design considerations, especially in regard to setting the number of lanes and other geometric features. The reason for this, as can be perceived in Fig. 7.5, is that most of the hourly volumes after the 30th or 50th highest hour stay relatively constant (in percentage of AADT), while those before rise rather abruptly, thus implying that much more money be spent to provide adequate

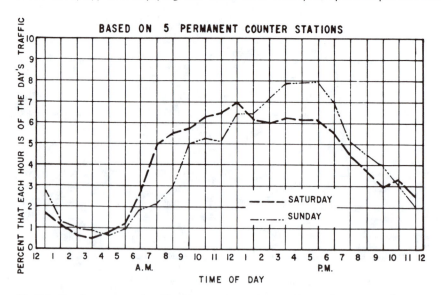

Fig. 7.4 Hourly patterns of traffic-volume variation by day of week [7.27, p. 146].

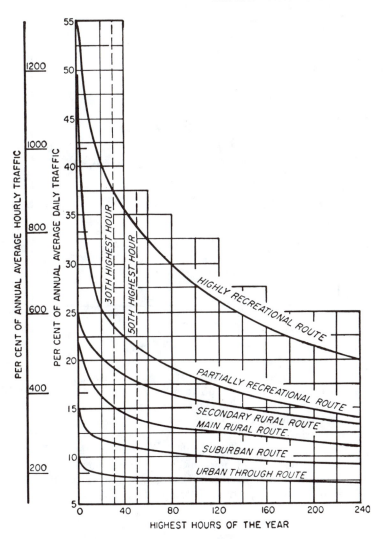

Fig. 7.5 Hourly volumes expressed as percentages of average daily traffic, for typical types of highway [7.27, p. 15].

capacity for each hour above the 30th or 50th than beyond them. This generally is true, but economy may not be the sole reason for arriving at system modification schemes.

7.1.2 Peaking at Special Generators

Certain generators, like airports, shopping centers, and manufacturing plants, are unique in that (1) they have a great influence on travel and congestion in neighboring areas, and (2) they attract trips at times that do not correspond necessarily to the peak periods for the urban areas as a whole. Having these two characteristics, they deserve further analysis at this point, and, while no elaborate detailing of these characteristics will be presented, at least the reader will become familiar with the general peaking features portrayed in Figs. 7.6 to 7.8.

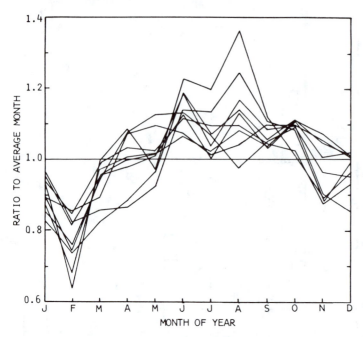

Fig. 7.6 Monthly variation in air passengers [7.29, p. 10].

For airports, perhaps the most important peaking characteristics are those by month of the year. In Fig. 7.6 it can be seen that the variation is quite marked, with increased trip-making in the summer months and a slow decrease to a low point in February. Hourly trips, on the other hand, follow the same general pattern as overall urban-area trips [7.29].

Shopping-center traffic generally peaks during the day at a time different from that of the general morning and afternoon peaks found in most other parts of a city, with the maximum number of arrivals coming around noontime and, for those centers with evening openings, around 7:00 p.m. (Fig. 7.7). Shopping at Christmas time also

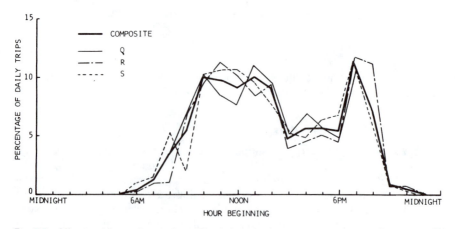

Fig. 7.7 All auto drivers trips to three Miami shopping centers, by arrival hour [7.29, p. 43].

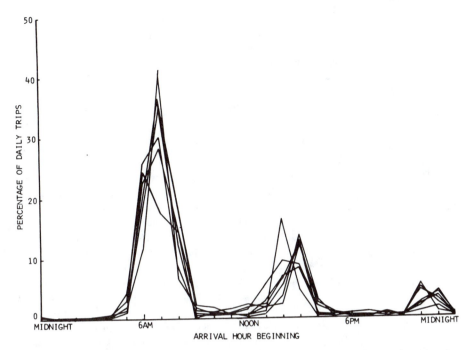

Fig. 7.8 Home interview auto driver trips to work, at selected plants, by employment size groupings [7.29, p. 71].

leads to significant increases in arrivals, with the week after Thanksgiving being the busiest.

As should be expected, work trips, which usually occur in the morning rush hour, are the major contributor to travel to manufacturing plants. Consequently, we see in Fig. 7.8 a sharp peak for most plants at about 6:30 a.m. While there is nothing startling about this characteristic, we should be aware that travel in the vicinity of a very large plant may be extremely peaked, thereby requiring additional transportation facility capacity that under usual circumstances would not be necessary.

7.2 THE TRAVEL ESTIMATION PROCESS

Early transportation planners recognized the large task of modeling the trip actions of individuals in an urban area and resorted to a variety of methods. One was to aggregate the action of individual trip-makers to a geographical area, such as a traffic zone, and then to further aggregate according to trip purpose and orientation. Another method was to divide the large task into smaller ones, where each could be modeled separately, and then grouped and sequenced in such a way that the output of one model would be the input to the next. This resulted in the well-known four-step procedure of trip generation, trip distribution, modal split, and traffic assignment. To repeat the definitions of Chap. 2, *trip generation* is concerned with the number of trips per time period made to or from a given areal unit or zone (regardless of the trips' origins or destinations), *trip distribution* with finding the zones to or from which the generated trips are directed, *mode choice* with the determination of the particular mode of transportation used for the zone-to-zone trips, and *trip assignment* with the particular route selected by travelers going between each pair of zones on each mode of transport.

Figure 7.9 gives a schematic portrayal of the division of generated (produced or attracted) trips into subcategories according to zone origin and destination, mode, and route. This figure also shows the type of results desired from travel estimation models and, in a sense, provides a definition of travel or travel patterns. By knowing the values of all T_{ijmr} and the time period to which they correspond, we can obtain a fairly complete description of a trip as:

A one-way journey made within a given time period between two places (usually two areal zones, i and j) on a certain route (r) of a certain mode of transport (m).

Travel would be simply the overlay of all component types defined above. Of special interest in Fig. 7.9 is the fact that the number of trips from a given zone (i) to any other zone (j), T_{ij}, is not necessarily equal to T_{ji}. Nor for that matter is the trip production of any zone i, P_i, necessarily equal to the trip attraction, A_i, of that zone. Only over a 24-hr period is $T_{ij} = T_{ji}$ and $P_i = A_i$, and even for this interval of time, the use of equalities may not be exactly correct because of long weekend and holiday exoduses from certain localities.

In general, then the absence of these equalities is an indication of variations and peaks in tripmaking, the most familiar of these being in the morning and evening rush hours. Since these two peaks are so distinct and sizable, they have been the source of most of the attention of transportation planners. The usual approach to predicting the amount and pattern of travel within these periods has been first to calculate average 24-hr trip volumes through the four-step procedure shown in Fig. 7.9 on the assumption that $T_{ij} = T_{ji}$ and $P_j = A_j$ and then to multiply the resulting values by suitable peak to average ratios to obtain the peak volumes. In other words, to complete the picture of the overall travel estimation modeling process, we would need to add a further step to the form shown in Fig. 7.9. In this step, the 24-hr T_{ijmr} variable would be divided into various time periods.

7.3 TRIP GENERATION

Trip generation is concerned with the number of trips produced and attracted to a parcel of land or a zone. The number of trips entering or leaving a zone is considered dependent on the characteristics of people and land use in that zone. Forecasts of these characteristics (such as population, employment, and land areas) for the horizon year are employed to estimate the zonal trip productions and attractions for that year. The basic assumption is that the relationships between the trip rates and the zonal characteristics are stable over time. The major drawback of this assumption is, however, that trips are generated independently of transportation systems characteristics. In other words, the demand for travel is assumed independent of supply.

Most of the trips generated in an urban area start or end at home. Studies [7.1, 7.2] show that above 80 percent of urban trips fall into that category. They are referred to as "home-based." Those that start at home are called "residential." The home end of the home-based trip, whether it is the origin or destination, is called a *trip production* and the non-home end a *trip attraction*. Both are often divided into two major groups: home-based work and home-based other. The home-based work trips are of special interest to engineers and planners, since they constitute a good part of residential trips and are concentrated in a few rush hours, thereby creating peak demands on transportation facilities. Non-home-based trips, known also as

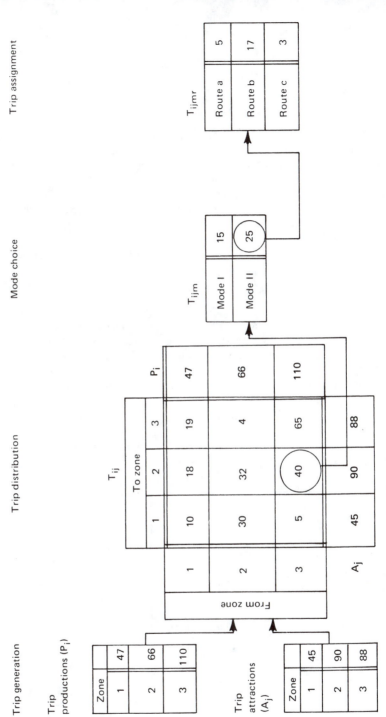

Fig. 7.9 Example of steps in the travel estimation process.

nonresidential trips, have their origins designated as productions and their destinations as attractions. Nonresidential trips have either one or both of their ends at other than a residence.

7.3.1 Trip Generation From Residences

The determinants of trips from residences have received considerable attention in transportation studies. Variables such as household size, car ownership, and residential density in a zone have been studied for their impacts on trip generation. Figures 7.10 and 7.11 show some of the relationships that have been found between these types of variables and number of trips per household. The indicated relationships are non-linear—that is, a unit increase in the amount of explanatory variable does not produce a constant increase in the number of generated trips, a phenomenon often neglected in the modeling methods used.

The general methods used to model trip generation from residences are of two types:

Linear regression
Cross classification

7.3.1.1 Linear regression This method gained wide application in most transportation studies in the 1960s because it was easy to use and relevant "canned" computer programs were available. The method involves a linear equation between some independent, explanatory variables $(x_1, x_2, \ldots, x_k)$ and a dependent variable (y), in this case the trips generated from residences. The general form of the model is:

$$y_i = b_0 + b_1 x_{1i} + b_2 x_{2i} + \ldots + b_k x_{ki} + e_i \tag{7.1}$$

where i stands for a single observation and e_i is the error term.

Among the explanatory variables usually considered are:

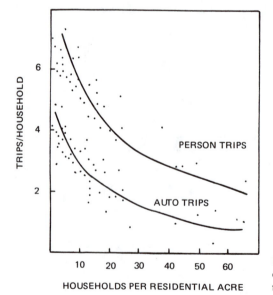

HOUSEHOLDS PER RESIDENTIAL ACRE

Fig. 7.10 Effect of residential density on trip production by districts in Pittsburgh [7.8, p. 8].

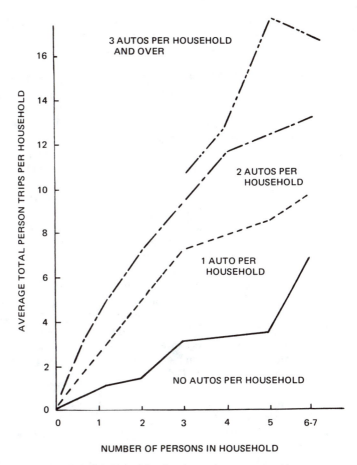

Fig. 7.11 Relationship of family size and auto ownership to average total person trips per dwelling unit [7.8, p. 20].

1. Car ownership
2. Family income
3. Household size
4. Number of persons 5 years old and over in the household
5. Number of persons 16 years old and over who drive
6. Occupation of head of household
7. Distance from the central business district (CBD)
8. Residential density
9. Structure type of house (single detached, single attached, combination, apartments, etc.)
10. Appraised house value and/or median rent or value

The selection of variables to be used in the analysis is based on their strong, logical relationship to trip-making, both causally and statistically. They should be easy to forecast and limited in number so that the analysis is simple and is not distorted with interrelated factors.

The equation is calibrated on base-year data. Mathematical formulation and calibration of such models are discussed in App. A. Although the regression relationship in Eq. (7.1) is strictly linear, nonlinear ones can be obtained, as shown in App. A.

The validity of the model is tested through several statistical measures. The important ones are (1) the coefficient of determination, which indicates the variation explained by the regression out of the total variation of the dependent variable (y), and (2) the standard error of the estimate, which measures the average deviation of observation points from the established regression line.

The regression method used in trip generation has been applied on two levels of data—aggregate zonal level and disaggregate dwelling unit or household level. The former can involve aggregate totals or, alternatively, rates. An example where aggregate totals are used is in estimating trips per zone and average automobiles per zone, while aggregate rates would be average trips per household per zone and average automobiles per household per zone. Aggregate totals, unlike rates, are dependent on the size of unit of aggregation. Some studies have mixed totals and rates in the same relationship, a practice with no logical basis, which may lead to erroneous results.

Kassoff and Deutschman [7.3] compared aggregate-rate and aggregate-total trip generation equations based on data obtained from home-interview surveys conducted by the Tri-State Transportation Commission in 1963 and 1964. The two equations with their statistical measures are shown in Table 7.2. For comparison, the rate equation was multiplied by the number of households in the zone to yield zone totals. The aggregate total equation had a slight advantage over the aggregate-rate in terms of statistical measures (for example, $r^2 = 0.929$ vs. 0.888). The authors still recommended the use of aggregate rates, however, because they offer more flexibility and efficiency in analyzing the data.

A common assumption in trip generation analysis is that households within a zone are much more similar to each other than to those outside a zone. McCarthy in his analysis of travel data from Raleigh, North Carolina refuted the validity of zonal homogeneity [7.4]. The within-zones variation contributed 87.74 percent to the total variation of trips per household, in comparison to 12.26 percent between zones variation. Trip generation equations based on the same aggregated and disaggregated data are shown in Table 7.3. Interestingly, based on the coefficient of determination (r^2) and on the standard error of the estimate (S_E), the equation based on zonal averages appears superior to that based on individual household data. However, this is not the case. The aggregate equation actually explains only 61 percent of 12.26 percent or 7.5 percent of total variation, whereas the disaggregate explains 38 percent of total variation. The equation based on aggregated data thus shows a deceptively high value for r^2. A similar conclusion has been reached in the works of other researchers such as Fleet and Robertson [7.10], Fleet, Stowers, and Swerdloff [7.9], Wooten and Pick [7.12], and Robinson [7.11].

The suppression of within-zone variation in aggregate trip generation models may also lead to the decrease of significance of some variables, even to the extent of their exclusion from the model. These might otherwise prove to be significant and desirable at the disaggregate level. Besides, aggregate equations are area-specific, that is, they depend on the size and number of zones in the study area. They do not directly relate to the socioeconomic characteristics of the household units that are the causal base for all trip-making characteristics. These effects have been considered among the major reasons for the failure of trip-generation equations to exhibit applicability in more than one city.

Table 7.2 Comparison of Aggregate-Rate and Aggregate-Total Trip-Generation Equations

Analysis unit	Number of observations	Dependent variable Y	Independent variables		Equation	Constant/ mean	Standard error/ mean	r^2
			X_1	X_2				
Zone	305	Trips per zone	Automobiles per zone	Persons per zone	$Y = 4.343X_1$ $+ 0.758X_2$ $- 66$	0.001	0.181	0.930
		Avg. trips per household per zone	Avg. automobiles per household per zone	Avg. person per household per zone	$Y = 3.458X_1$ $+ 2.054X_2$ $- 2.94$	0.418	0.209	0.714

Source: [7.3].

Table 7.3 Comparison of Basic Trip-Generation Equations Developed from Nonaggregated versus Aggregated Data

Level of aggregation	Trip-generation equation*	Number of observations	r^2	S_E	$\bar{Y}$	Percent $S_E/\bar{Y}$
Household	Home-based trips per household = (−1.0889 + 0.63941IL + 1.8328PFO + 1.4401AO − 0.1774FS)	4,158	0.38	4.18	7.49	55.81
Zonal average	Home-based trips per household = (−1.4169 + 0.74851IL + 1.6482PFO + 1.6849AO − 0.1846FS)	184	0.61	1.38	7.43	17.23

* r^2 = coefficient of determination, S_E = standard error of estimate, $\bar{Y}$ = mean of the dependent variable, Percent $S_E/\bar{Y}$ = percentage of standard error of the estimate, IL = household income level, PFO = number of persons aged 5 and over, AO = household auto ownership, and FS = household family size.
Source: [7.4].

In summary, the recommended procedure for developing multiple-regression equations in trip generations is to use individual household data, and then to aggregate the results of this analysis on the zonal level for inputs into the trip distribution and modal split models.

7.3.1.2 Cross classification Cross classification, referred to as *category analysis*, is also used for trip-generation analysis. No assumptions about linearity and normality are needed, as required by linear regression. The dependent variable, such as trips per household, is cross-tabulated against two or more explanatory variables, such as income and auto ownership per household. The explanatory variables, whether discrete or continuous, are divided into distinct classes, and the mean values of the observations on the dependent variable are then allocated to the appropriate cells within the established classes. An example of trips per household cross-classified against car ownership and number of persons per dwelling unit is shown in Table 7.4.

In that table we can examine the conditional variation of total trip rates per dwelling unit for each class of household size. This procedure facilitates the understanding of the impact of different factors on, and their relationship to, trip generation. This is in contrast to the regression approach where we must try to understand mathematical interrelationships, constants, and coefficients.

Forecasts are usually made by locating the horizon year number of persons (row) and vehicles (column) per dwelling unit for a specific zone in the appropriate cell of the matrix. The trip rates for that cell then are multiplied by the number of dwelling units to produce the total trips generated. In those cases where detailed extrapolation is needed, the data from the matrix are plotted graphically. Smooth curves are fitted to the data and extended out, based on the shape of curves and logic. The curve values are then used to develop a completed matrix, which is used for future trip forecasts. Extrapolation in this fashion should be only used for approximations, because the validity of the obtained trip rates is not supported by real data.

Zonal aggregate and disaggregate household data are both used in cross-classification tables. However, the problems with zonal aggregation hold in cross classification as in regression.

Recent publications [7.7, 7.13] recommend only the cross-classification technique for household-level trip-generation analysis. This recommendation is based on the belief that the approach is straightforward in nature, easy to develop and apply in

Table 7.4 Mean Number of Total Trips per Dwelling Unit, Classified by Car Ownership and Household Size (Detroit, 1955)

Number of persons per dwelling unit	Number of vehicles per dwelling unit			
	0	1	2 or more	Total
1 or 2	1.71	5.09	6.68	4.00
3	3.32	6.92	8.82	6.93
4	3.40	7.63	11.28	7.91
5 or more	4.12	9.05	13.15	9.55
Total	2.40	6.93	10.58	6.64

Source: [7.5, p. 90].

a short time period, and lower in cost than regression analysis and other previously applied methods. Other advantages as stated are [7.7]:

Ease of understanding
Efficient use of data
Ease of monitoring and updating
Accuracy in forecasting as well as in base year
Policy sensitive through introduction of factors representing the relevant issues
Applicable at differing study levels—since the approach is based on household level data for residential generation, the rates developed should be applicable to any areal level of study
Transferable to other areas—since the analysis is based on household data at the production end, similar variables can be used for stratification in other areas

An application of the graphic cross-classification procedure is shown in Fig. 7.12. The trips per dwelling unit data obtained from an origin-destination survey (see Chap. 9) or "borrowed" from other areas are cross-classified utilizing two household characteristics, income, and auto ownership. The data are further cross-tabulated by percent of dwelling units and by trip purpose, using income and auto ownership as controlling variables. The matrix tables then are graphically plotted and fitted with smooth curves, as shown in the three sections of Fig. 7.12.

As a hypothetical example [7.7] for residential trip generation, suppose a zone has:

Total number of dwelling units (DU) = 1,000

Zonal average income/DU = $12,000

To find the number of trips generated:
1. Enter Curve A with zonal income/DU to determine car ownership level by household.

2% "0" auto households =	20 DU	
32% "1" auto households =	320 DU	
52% "2" auto households =	520 DU	
14% "3" auto households =	140 DU	
	1,000 DU	

2. Enter Curve B with income to determine the total person trip production from each household.

Trips from "0" auto households = 5.5 trips/DU X 20 DU = 110 trips

Trips from "1" auto households = 12.0 trips/DU X 320 DU = 3840 trips

Trips from "2" auto households = 15.5 trips/DU X 520 DU = 8060 trips

Trips from "3" auto households = 17.2 trips/DU X 140 DU = 2408 trips

Total trips = 14,418

Average trips/DU = 14.4

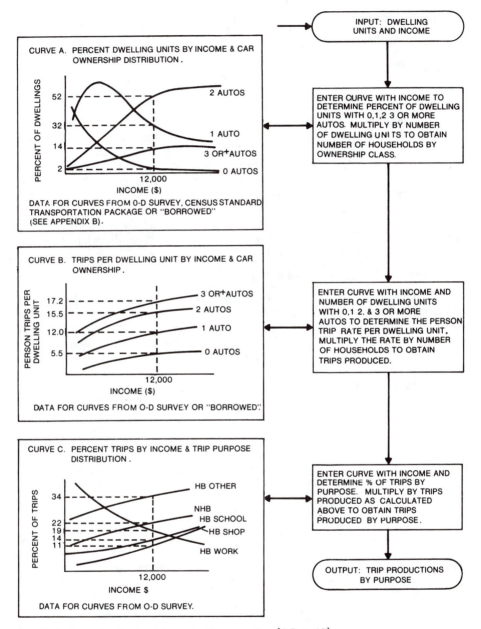

Fig. 7.12 Example of urban trip production procedure [7.7, p. 45].

3. Enter Curve C with income to determine the percent trips produced by purpose.

Home to work trips = 19% × 14,418 = 2739 trips

Home to shop trips = 11% × 14,418 = 1586 trips

Home to school trips = 14% X 14,418 = 2018 trips

Home to other trips = 34% X 14,418 = 4903 trips

Non-home based = 22% X 14,418 = 3172 trips

14,418 trips

The cross-classification method, as others, has drawbacks. The mean trip rates in the matrix do not usually represent the "average" of all household trip values in that cell. The distribution of cell values may be skewed and have a wide dispersion.[1] In this case the analyst should consider further subdivision of initial variables, adding other stratified variables, and/or reevaluating the initial choice of variables. The addition of variables, however, increases geometrically the number of cell entries. Hence the sample sizes must be large to obtain a value of reasonable accuracy in every cell. The trip rates at the extreme of the matrix are unreliable because of the relatively low number of observations found there. In addition, the variances between households in a specific cell are totally suppressed.

7.3.2 Trip Generation of Nonresidences

The importance of the interrelationship between trip-making and land use becomes most apparent when considering nonresidential trips. In fact, the models used for estimation of nonresidential trip generation often specify the classification and intensity of land use activities as causal factors. Consequently, the most prevalent models are simply cross-classification summaries of rates of trip-making (per acre or per square foot) for the various land use activity classes. These rates are assumed to hold constant into the future.

An example of cross-classification scheme is given in Table 7.5 where the number

[1] Because of the skew in the distributions, the median might be a better measure of the "average" (central tendency).

Table 7.5 1956 Person · trip Generation Rates in Chicago

Ring	Average distance from Loop (miles)	Person · trip destinations/acre				
		Manufacturing	Transportation	Commercial	Public buildings	Public open space
0	0.0	3,544.7	273.1	2,132.2	2,013.8	98.5
1	1.5	243.2	36.9	188.7	255.5	28.8
2	3.5	80.0	15.9	122.1	123.5	26.5
3	5.5	86.9	10.8	143.3	100.7	27.8
4	8.5	50.9	12.8	212.4	77.7	13.5
5	12.5	26.8	5.8	178.7	58.1	6.1
6	16.0	15.7	2.6	132.5	46.6	2.5
7	24.0	18.2	6.4	131.9	14.4	1.5
Average for study area		49.4	8.6	181.4	52.8	4.2

Source: [7.8, p. 16].

Table 7.6 Total Person Trips versus Floor Area and Employment*

Land use (j)	a_j	Floor area (b_{1j})	Employment (b_{2j})	r^2
Food and kindred products	17	−0.003	1.24	0.959
Tobacco manufacturers	−5	−0.001	1.14	0.998
Textile mill products	53	0.005	1.05	0.959
Apparel	−8	0.003	1.12	0.994
Lumber	5	0.009	1.21	0.997
Furniture and fixtures	303	0.034	0.77	0.889
Paper	63	0.220	1.08	0.996
Printing	53	−0.001	1.08	0.961
Chemicals	−101	0.209	0.50	0.822
Petroleum refining	51	0.019	1.04	0.993
Rubber and plastics	47	−0.021	1.24	0.986
Leather products	18	0.053	0.57	0.914
Stone, clay, and brass	−3	0.000	1.25	0.972
Primary metals	48	0.044	0.96	0.999
Fabricated metals	−35	0.017	1.09	0.998
Mechanical machinery	−23	0.012	1.17	0.997
Electrical machinery	−47	0.038	1.00	0.989
Transportation equipment	13	0.002	1.37	0.947
Scientific instruments	−1	0.031	1.10	0.990
Miscellaneous manufacturing	67	0.023	1.01	0.913

*Data derived from *Chicago Area Transportation Study*.
Source: Table partially restructured from [7.31, p. 121].

of daily person trips made in Chicago in 1956 is divided into different categories depending on the type of land use—manufacturing, transportation, commercial, public buildings, and public open space—and distance from the Loop, the central business district (CBD) of Chicago. Each trip total within the table is subsequently divided by the number of acres of ground area of the respective land use in the ring of territory whose average distance from the CBD is as indicated. The result of this procedure is a rate that in each case is specified in terms of daily person trip destinations per acre of ground area. When assumed to hold constant into the future, each of these rates can be employed in conjunction wih exogeneously predicted land acreage figures (at the given distance from the CBD) to give the future quantity of trip destinations.

As an example of the use of these models for prediction of nonresidential trip generation, consider the rate in Table 7.5 of 132.5 person trips per acre for commercial land uses lying at a distance of between 14 and 20 miles from the CBD. It may be that, for 20 years hence, land-use predictions may indicate a growth from 150 to 350 acres in this category, thereby giving an expected increase in trip generation, from 132.5(150) = 19,875 daily person trips to 132.5(300) = 39,750.

Linear regression equations also are used in determining nonresidential trips. In some cases they are combined with cross-classification to obtain greater reliability: that is, by establishing a regression equation within each and every class of the matrix. Illustrations of this technique can be seen in Table 7.6, where regression equations have been developed within each of several manufacturing subclasses (regardless of distance from CBD). The numbers in the second, third, and fourth columns represent the parameters a_j, b_{1j}, b_{2j} for the general model

$$Y_{ij} = b_{1j}X_{1ij} + b_{2j}X_{2ij} + a_j \tag{7.2}$$

where Y_{ij} = total daily person trips to land use j in zone i
$\quad X_{1ij}$ = floor area of land use j in zone i
$\quad X_{2ij}$ = number of employees associated with land use j in zone i
The correlation coefficients corresponding to each subclass model are presented along with the parameters and appear to be fairly high, thus providing the analyst with the satisfaction of knowing that the models are reliable, at least if used to estimate present trip generation. For prediction of trips to be made at some future time, possibly 20 years hence, the assumption will have to be made, as in residential trip-generation models, that the parameters will be the same at that date. If this is accepted, then the number of daily person trips to, say, apparel manufacturing plants in a zone with 50 apparel employees and 40,000 sq ft of apparel floor area would be predicted as

$$Y = 0.003(40,000) + 1.12(50) - 8 = 168 \text{ person trips/day}$$

Table 7.7 shows variables used in modeling efforts in several other planning studies where the classification was based on the *purpose* of the nonresidential trips rather than on the land-use activity category and distance from the CBD. The variety of entries in this table emphasizes the differences in trip-making between cities: differences which, as already mentioned with residential generation, point to the need for distinct models for each individual study area.

7.3.3 Major Generators

While models similar to those presented in the previous two subsections provide the engineer or planner with much information on future trip generation, there are several types of urban facilities which because of their large size and innate attractiveness to tripmakers should be handled on a separate basis. The three to be examined here are airports, shopping centers, and large industrial plants. It should be noted that we probably will be concerned with travel to and from these kinds of facilities in the more immediate future (1–5 years), especially for the purposes of gauging parking and traffic control equipment needs. Consequently, we can use independent variables that are much more precise in nature than those employed in, say, 20-year forecast models. The designers of these soon-to-be-constructed major generators often will have detailed data on the characteristics of the development.

7.3.3.1 Airports Figure 7.13 shows data on the number of annual air passengers originating at 180 Standard Metropolitan Statistical Area (SMSA)[2] airports as related to SMSA population. The relationship appears to be reliable except, perhaps, at the points most distant from the origin. Air passenger trips are not the only source of travel to airports, however, as is indicated in Tables 7.8 and 7.9, where social recreation (serve passenger and sight-seeing) trips and especially work trips are seen to be numerous, often outnumbering the actual passenger trips.

Further study of trip generation of airports by Keefer [7.29] has led to the three regression equations in Table 7.10. The dependent variables are expressed in terms of passenger originations while the independent variables cover a wide range of

---

[2] For simplicity, an SMSA can be thought of as any metropolitan area (city plus suburbs) associated with a city with a population greater than 50,000 people.

Table 7.7 Factors Used to Estimate Trip Attractions

Study Year-pop.	Trip purpose category					Non-home based	Trucks
	Home-based						
	Work	Shop	Soc.-rec.	Other	Special		
Washington, D.C. 1963-2,900,000	E	S_R	S_R,DU	E,S_R,DU	[1]A_{SC}, DU	E,S_R DU	
New Orleans 1960-645,000	NRD,DU,S_R E,A_C,I,DRD	NRD,DU,C S_R,DRD	DU,P,C P_s,S_R	[2]C,E DRD	[3]NRD,P/C,S_R E,I,H,DRD	NRD,DU,C S_R,E,A_C,SC	DU,D,E A_C,SC
Kansas City, Kan.-Mo. 1959-643,000	[4]E,D	[4]S_R,D	[5]P,DU		[5][6]P	[5]D	
Ft. Worth, Tex. 1964-540,000	E	[7][8]E_C		P,E_C E_M,E_O		P,E_C E_M,E_O	P,E_C E_M,E_O
Charleston, W. Va. 1965-250,000	[9]E,E_V	E_R,S_C,S_P		[9]S_V,E_V,DU A_R,SC_V,I	[10]SC_V	[9]E,E_V,SC_V S_C,DU,A_R,I	[9]E_V,T S_V,I
Nashville, Tenn. 1961-250,000	[11]E,E_W,E_B	[11]A_C,D	[11][12]P,I	[13]A_C	[14]SC,I	A_C,D	
Chattanooga, Tenn. 1962-240,000	E	[15]A_C	[15][16]P,I	[16][17]A_C	[15][18]SC,I	A_C	
Waterbury, Conn. 1963-190,000	E	E_R		P,E_C E_M,E_O		P,E_C E_M,E_O	P,E_C E_M,E_O
Erie, Pa. 1963-140,000	E	[19]E_R	P,E_R,E_O			P,E_R E_M,E_O	[20]P,E_R E_M,E_O
Greensboro, N.C. 1964-130,000	E	E_R		P,E_C E_M,E_O		P,E_C E_M,E_O	P,E_C E_M,E_O
Fargo, N.D. 1965-70,000	E,E_R	[21]P,E_R,E_O		P,DU C,E	[22]DU,E_R,E_O	DU,E,E_R	C,E,E_R
Appleton, Wis. 1965-55,000	[23]E,D A_{PU},A_I			[23]DU,A_{PU},A_I A_C,E		D,E,A_C A_I,A_{PU},DU	DU,E A_C,A_I

Comments

[1] School trips
[2] Personal business
[3] School trips
[4] Different procedures used for stable and unstable zones
[5] Different procedures used depending on type of zone
[6] School trips
[7] Includes related business, eat, and convenience and shopping goods
[8] Gross sales and floor area suggested as possible alternatives
[9] Different factors used to estimate AM and PM peaks
[10] School trips
[11] Different procedures used for stable and unstable zones
[12] Recreation trips computed by uniform factor expansion
[13] Business trips
[14] School trips
[15] Different procedures used for stable and unstable zones
[16] Recreation trips distributed in proportion to surveyed recreation trips
[17] Business trips
[18] School trips
[19] Special adjustments made for shopping centers
[20] Special adjustments made for areas adjacent to major railroads
[21] Retail employment alone used for CBD and outlying areas
[22] Personal business trips
[23] Different factors used to estimate origins and destinations

Key to entries

Employment: E = Total employment; E_R = Retail employment; E_M = Manufacturing employment; E_C = Commercial employment; E_O = Employment other than retail and manufacturing; E_W = White collar employment; E_B = Blue collar employment; E_V = Various specialized employment.

Sales: S_R = Retail sales; S_C = Convenience goods retail sales; S_P = Personal service sales; S_V = Retail sales by various specialized categories.

Area: A_{PU} = Acres of public and semipublic land; A_I = Acres of industrial land; A_C = Acres of commercial land; A_R = Acres of residential land; A_{SC} = Acres of school land.

School Enrollment: SC = Total school enrollment; SC_V = School enrollment by various grade levels.

Household Characteristics: P = Population; P_s = Persons five years of age or older; H = Persons per dwelling unit; DU = Number of dwelling units; NRD = Persons per net residential acre; DRD = Dwelling units per net residential acre; I = Income; C = Number of automobiles.

Miscellaneous: D = Distance from CBD; T = Truck ownership.

Source: [7.32, p. 77].

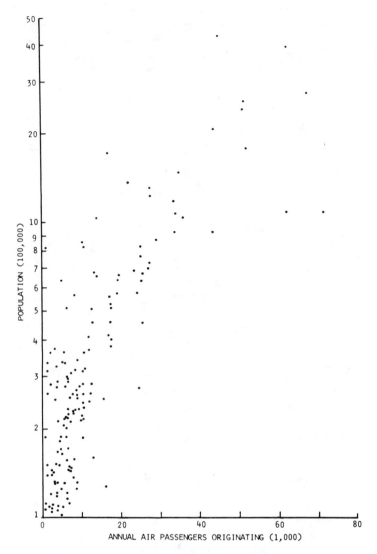

Fig. 7.13 Annual air passengers originating per 100,000 population, 180 standard metropolitan statistical areas, 1960 [7.29, p. 22].

characteristics of the SMSA such as population per square mile (X_2), and so forth. Correlation coefficients are high in the cases of Y_1 and Y_2 but do not appear to be very substantial for Y_3. Besides, some of the independent variables display negative coefficients, which would indicate improbable relationships with the dependent variables. This is most probably caused by the high collinearity among the independent variables. Increasing the number of independent variables does not necessarily improve the rationale of the model.

7.3.3.2 Shopping centers The number of shopping centers in the United States has risen remarkably. These types of commercial areas attract unusually high numbers

Table 7.8 Purpose Distribution of Person Trips to and from Selected Airports, All Travel Modes*

Airport	Trips to airport (%)			Trips from airport (%)		
	To work	To soc.-recr.	To air travel	To home	To pers. business	To other
Atlanta	67.8	5.8	26.4	--	--	--
Buffalo	23.3	33.7	43.0	55.7	14.1	30.2
Chicago (Midway)	34.7	25.7	39.5	82.6	6.0	11.4
Minneapolis–St. Paul	46.8	19.7	33.6	80.3	7.1	12.6
Philadelphia	24.2	32.8	43.1	70.0	9.7	20.3
Pittsburgh	43.0	20.6	36.5	85.9	4.7	9.3
Providence	39.8	37.7	22.5	--	--	--
San Diego	45.9	21.6	32.4	--	--	--
Seattle–Tacoma	35.0	24.2	40.8	81.3	12.4	6.3
Washington (National)	69.8	15.8	14.4	80.1	9.9	10.0

*From transportation study data (home interviews) for the various cities.
Source: [7.29, p. 8].

Table 7.9 Definition of Variables Used in Multiple Regression Equations Displayed in Table 7.10

Variable type	Symbol	Definition
Dependent	Y_1	Total aircraft departures performed in scheduled service, calendar year 1960
	Y_2	On-line revenue passenger originations in scheduled service, calendar year 1960
	Y_3	$Y_2/100,000$ SMSA population
Independent	X_1	Population, thousands
	X_2	Population, per square mile
	X_3	Nonwhite population, 0.1%
	X_4	Median age, 0.1 year
	X_5	Families with incomes over $10,000, 0.1%
	X_6	High school graduates, 0.1%
	X_7	Manufacturing employment, hundreds
	X_8	Transportation employment, hundreds
	X_9	Trade employment, hundreds
	X_{10}	Institutional employment, hundreds
	X_{11}	Manufacturing establishments with at least 100 employees
	X_{12}	Services receipts, million
	X_{13}	Unemployment rate

Source: [7.29, p. 106].

Table 7.10   Selected Multiple Regression Results, Airports ($N = 176$)

Equation	r^2	Standard Error*
$Y_1 = -63 + 0.17X_3 + 0.78X_5 - 0.06X_7 + 0.76X_8$ $+ 0.19X_9 - 0.43X_{12} - 0.62X_{13}$	0.86	60
$Y_2 = -2,977 - 3.36X_1 + 2.50X_3 + 16.43X_5 - 2.27X_7$ $+ 12.32X_8 + 13.06X_9 - 2.09X_{10} - 3.09X_{12}$	0.92	73
$Y_3 = -944 - 0.06X_2 + 0.69X_3 + 1.49X_4 + 3.01X_5$ $+ 0.58X_6 - 0.48X_7 + 0.95X_9 - 0.41X_{10} + 0.33X_{11}$ $+ 0.30X_{12} + 1.51X_{13}$	0.35	83

*As percentage of $\bar{Y}$.
Source: [7.29, p. 107].

of trips as compared to other uses, so that it is important to gauge the impact of such centers on travel in surrounding territories. Of course, many factors can be easily recognized as influencing trip-making to shopping centers—the amount and quality of merchandise available there, the distance or travel time to the center, the number of parking spaces, and perhaps even the weather (Table 7.11). While it is not possible to take all these factors into account, Keefer [7.29] employs many of them in determining relationships relevant to trip generation. A set of three multiple-regression models developed in the report are displayed in Table 7.12. It should be noted that the independent variables utilized in the regression models are restricted almost entirely to characteristics of the center, not to the population in the surrounding area. This restriction could be harmful in that an urban center and a rural center with similar characteristics would be predicted to have the same number of attracted auto driver trips. Quite obviously this would not be realistic, so that a warning must be

Table 7.11   Definition of Variables Used in Multiple Regression
Equations Displayed in Table 7.12

Variable type	Symbol	Definition
Dependent	Y_1	All auto driver trips
	Y_2	Auto driver trips to shop
	Y_3	Other auto driver trips
Independent	X_1	Number of parking spaces
	X_2	Total person work trips
	X_3	Distance from major competition, 0.1 mile
	X_4	Age of study data, years
	X_5	Age of center at time of study, years
	X_6	Reported travel speed of tripmakers, mph
	X_7	Floor space for convenience goods, 1,000 sq ft
	X_8	Floor space for shopping goods, 1,000 sq ft
	X_9	Floor space for other uses, 1,000 sq ft

Source: [7.29, p. 108].

Table 7.12 Selected Multiple Regression Results, Shopping Centers

Equation	r^2	Standard error*
$Y_1 = 3{,}875 + 5.35X_2 + 291.9X_3 - 578.5X_4 - 0.65X_6$ $- 22.31X_9$	0.920	21
$Y_2 = 2{,}841 + 3.23X_2 + 241.4X_3 - 410.8X_4 - 0.34X_6$ $- 10.45X_7 + 4.32X_8 - 25.70X_9$	0.892	25
$Y_3 = 801 + 0.06X_1 + 0.90X_2 + 31.2X_3 - 108.0X_4$ $+ 35.7X_5 - 0.18X_6 - 1.47X_7 - 2.36X_8$ $+ 1.67X_9$	0.985	14

*As percentage of $\bar{Y}$.
Source: [7.29, p. 109].

issued concerning the injudicious insertion of values into the model equations in a strictly "mechanical" manner.

7.3.3.3 Manufacturing plants Major industrial plants many times are the producers of significant traffic congestion, the problem being doubly acute when an entire shift of plant employees is released at the same time that other, smaller establishments are closing for the day. In small towns, this concurrent influx of traffic onto the transportation system can be particularly bothersome and not susceptible to much relief, since it usually would be impractical and costly to provide sufficient capacity for these large but not very lengthy peaks.

Keefer also has produced multiple-regression-equation models for determining the quantity and percentage of various types of trips (Table 7.13) made to manufacturing plants. These equations, shown in Table 7.14, deal with daily automobile-driver and transit-passenger trips based on characteristics of the plant site, the employees, and the region within a three mile radius of the plant. In using these and other regression models, the analyst should attempt to gather as much relevant additional information as possible in order to make independent checks of model-calculated trip-generation figures.

7.3.4 Travel by the Elderly and Handicapped

Of particular concern in recent years has been the mobility of the elderly and handicapped. Grey Advertising did a national survey of current travel patterns for these groups. Looking at the number of trips per day by people 16 years and older, they found, in comparison to the overall U.S. average [7.38]:

All persons:	total United States	: 2.32
	private vehicles	: 2.20
	public transport	: 0.12
All persons 65+	total United States	: 1.03
All persons 65+	private vehicles	: 0.96
All persons 65+	public transportation	: 0.07
All persons, areas served by transit		: 2.00
All handicapped persons, areas served by transit		: 1.00
All handicapped persons, urban areas		: 1.00

Table 7.13 Definition of Variables Used in Multiple Regression Equations Displayed in Table 7.14

Type	Symbol	Definition
Dependent	Y_1	Auto driver trips
	Y_2	Transit passenger trips
Independent	X_1	Population within 5-mile radius, thousands
	X_2	Automobiles within 5-mile radius, thousands
	X_3	Residential land within 5-mile radius, 0.1 acre
	X_4	Net residential density in plant zone, persons/acre
	X_5	Net manufacturing density in plant zone, persons/acre
	X_6	Plant site area, acres
	X_7	Prime shift percentage, three highest morning hours
	X_8	Employees from car-owning households
	X_9	Employees not licensed to drive
	X_{10}	White-collar employees
	X_{11}	Male employees
	X_{12}	CBD-plant distance/CBD cordon line distance
	X_{13}	Average distance, home to work, 0.01 miles
	X_{14}	Total work trips to plant
	X_{15}	Total manufacturing work trips, plant zone

Source: [7.29, p. 109].

From these figures it is obvious that elderly and handicapped people have trip-making rates less than half the average for the U.S. Interestingly, the rate for the handicapped is about the same as for the elderly (although many people fall into both categories).

The distribution of trips by the handicapped by mode, as displayed in Table 7.15, highlights the fact that a much greater number are auto passengers than drivers. It also shows that the proportion going by transit (14%) is not significantly larger than for the average trip-maker (12%).

As pointed out aptly by Helen Meier [7.37], these figures do not indicate how much the elderly and handicapped *would* travel if transportation systems and programs (as well as, say, building systems) were focused more on accessibility for these groups. This *induced trip-making* is difficult to estimate, particularly when there has been little experience with substantially more accessible systems.

Table 7.14 Selected Multiple Regression Results, Manufacturing Plants

Equation	r^2	Standard error*
$Y_1 = 1,449 - 3.02X_4 + 1.34X_5 + 1.18X_6 - 9.46X_7$ $- 0.97X_8 - 1.58X_9 - 0.79X_{10} - 0.62X_{11}$ $- 0.64X_{12} - 1.22X_{13} + 2.32X_{14} - 0.01X_{15}$	0.98	13
$Y_2 = -287 + 0.78X_9 + 0.43X_{10}$	0.82	64

*As percentage of $\bar{Y}$.
Source: [7.29, p. 110].

Table 7.15 Mode Choice of Trip-makers in Mass Transit Areas

	Percent trips* by	
Mode	Transportation handicapped	Non-transportation handicapped
Auto	71	78
Driver	38	62
Passenger	33	16
Transit	14	12
Bus	12	9
Subway	2	3
Walking	6	3
Taxi	6	3
Personally owned van	1	1
Association van	1	—
Other modes (e.g., school bus)	1	3
Total	100	100

*Total monthly trips by persons 16 years old and over.
Source: Adapted from [7.38, p. 38].

7.4 TRIP-DISTRIBUTION MODELS

After having predicted the number of trips produced by a zone or attracted to it, the next step in the overall travel-estimation process is that of "linking" the productions with the attractions, that is, determining how the trips produced in a zone are distributed among all zones. Stated in a behavioral context, we are trying to predict how people who are about to make trips decide on their possible destinations. Obviously there are a multitude of reasons why one destination would be chosen over another: lack of jobs in certain zones, better highways between certain points, dangerous neighborhoods that must be traversed, and so forth. Generally speaking, then, traffic distribution can be considered a function of [7.33, p. I-1]:

1. The type and extent of transportation facilities available in an area
2. The pattern of land use in an area, including the location and intensity of land use
3. The various social and economic characteristics of the population of an urban area

Many mathematical models have been developed to describe and forecast the distribution of traffic. They are generally divided into two groups: (1) growth-factor methods and (2) theoretically based methods. In the first group, four basic types of models are the *uniform*, the *average-factor*, the *Fratar*, and *Detroit*. In the second group, the two types of models most well known are the *gravity model* and the *intervening opportunities model*. It is not possible to discuss adequately all these models in this text, so we have chosen to present in some detail the Detroit model as well as the gravity model, since they are the most widely used.

7.4.1 Growth-Factor Methods

Growth-factor methods work on the premise that the future number of trips between a pair of zones can be found by proportioning the relative increases (growth) in trip ends in those zones. This proportioning process, as will be shown, is iterative in nature: that is, a first proportion is worked out based on initial conditions, new trip end totals are computed, a new proportion established, and so on until some stable numbers are obtained. To show this process mathematically, we first must introduce some appropriate notation:

T_i = actual trip ends in zone i in base year

T_i^* = forecasted trip ends in zone i in horizon year

t_i^k = computed trip ends in zone i in iteration k

t_{ij} = actual trips between zones i and j in base year

t_{ij}^k = computed trips between zones i and j in iteration k

To start, the initial growth factor for zone i, F_i, is computed by dividing the forecasted trips by actual trip ends:

$$F_i = \frac{T_i^*}{T_i} \tag{7.3}$$

More generally, for the kth iteration:

$$F_i^k = \frac{T_i^*}{t_i^k} \tag{7.4}$$

For the whole study area, we sum trip ends over all zones to get the corresponding average area-wide growth factors, F and F^k,

$$F = \frac{\sum_i T_i^*}{\sum_i T_i} \tag{7.5}$$

and

$$F^k = \frac{\sum_i T_i^*}{\sum_i t_i^k} \tag{7.6}$$

Total trip ends in zone i are obtained through

$$T_i = \sum_j t_{ij} \quad i \neq j \tag{7.7}$$

and

$$t_i^k = \sum_j t_{ij}^k \quad i \neq j \tag{7.8}$$

The mathematical expressions used to find the future interzonal trips can be quite complicated. Those for the first and kth iteration of the Detroit model are

$$t'_{ij} = t_{ij} \frac{F_i F_j}{F} \tag{7.9}$$

$$t^k_{ij} = t^{k-1}_{ij} \frac{F_i^{k-1} F_j^{k-1}}{F^{k-1}} \tag{7.10}$$

In words, the number of trips between zones i and j increases in proportion to (1) the growth of trip ends in the origin zone (i); and (2) the growth of trip ends in the destination zone (j) in proportion to those in the overall area.

The iterations in the above model end when the values of the F factors approach unity. Typical closure criteria are

$$0.95 \leqslant F_i^k \leqslant 1.05 \tag{7.11}$$

and for the whole area

$$0.95 < F^k < 1.05 \tag{7.12}$$

An example showing the application of the Detroit model will illustrate the computational procedure involved. Consider the four-zone system displayed in Fig. 7.14. Part (a) shows the existing travel patterns among the different zones and the forecasted trip ends in each zone.

First we compute the overall growth factor, F,

$$F = \frac{T_i^*}{T_i} = \frac{1,200 + 2,000 + 2,400 + 600}{600 + 400 + 800 + 600} = 2.6$$

and initial growth factors for each zone:

$$F_1 = \frac{1,200}{600} = 2 \quad F_2 = \frac{2,000}{400} = 5 \quad F_3 = \frac{2,400}{800} = 3.0 \quad F_4 = \frac{600}{600} = 1.0$$

Using Eq. (7.9) for the first iteration,

$$t'_{12} = 100 \frac{(2)(5)}{2.6} = 385$$

$$t'_{23} = 200 \frac{(5)(3)}{2.6} = 1,153$$

$$\ldots$$

$$t'_{34} = 300 \frac{(3)(1)}{2.6} = 346$$

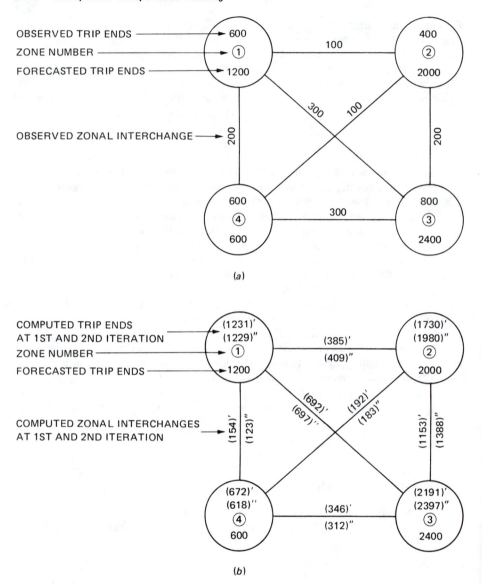

OBSERVED TRIP ENDS

ZONE NUMBER

FORECASTED TRIP ENDS

OBSERVED ZONAL INTERCHANGE

COMPUTED TRIP ENDS
AT 1ST AND 2ND ITERATION

ZONE NUMBER

FORECASTED TRIP ENDS

COMPUTED ZONAL INTERCHANGES
AT 1ST AND 2ND ITERATION

Fig. 7.14 Distribution of future trips by Detroit method. (*a*) Existing travel pattern. (*b*) Travel pattern after first and second iteration.

with $t'_{13} = 692$, $t'_{14} = 154$, and $t'_{24} = 192$. Total trip ends in each zone then can be computed from Eq. (7.7) as

$$t'_1 = 385 + 692 + 154 = 1,231 \qquad t'_2 = 385 + 1,153 + 192 = 1,730$$

$$t'_3 = 692 + 1,153 + 346 = 2,191 \qquad t'_4 = 154 + 192 + 346 = 692$$

From Eq. (7.4) we then obtain F factors for each zone,

$$F_1' = \frac{1,200}{1,231} = 0.97 \quad F_2' = \frac{2,000}{1,730} = 1.16 \quad F_3' = \frac{2,400}{2,191} = 1.10$$

$$F_4' = \frac{600}{692} = 0.87$$

and an areawide factor from Eq. (7.6) of

$$F' = \frac{1,200 + 2,000 + 2,400 + 600}{1,231 + 1,730 + 2,191 + 692} = 1.06$$

In the second iteration in Eq. (7.10),

$$t_{12}'' = 385 \frac{0.97(1.16)}{1.06} = 409$$

$$t_{23}'' = 1,153 \frac{1.16(1.10)}{1.06} = 1,388$$

$$t_{34}'' = 346 \frac{1.10(0.87)}{1.06} = 312$$

with $t_{13}'' = 697$, $t_{14}'' = 123$, and $t_{24}'' = 183$. Computed zonal trip ends from Eq. (7.8) are

$$t_1'' = 409 + 697 + 123 = 1,229 \quad t_2'' = 409 + 183 + 1,388 = 1,980$$
$$t_3'' = 1,388 + 697 + 312 = 2,397 \quad t_4'' = 312 + 123 + 183 = 618$$

Computed growth factors for each zone [Eq. (7.4)] then are

$$F_1'' = \frac{1,200}{1,229} = 0.98 \quad F_2'' = \frac{2,000}{1,980} = 1.01 \quad F_3'' = \frac{2,400}{2,397} = 1.00$$

$$F_4'' = \frac{600}{618} = 0.97$$

and the computed overall growth factor is

$$F'' = \frac{1,200 + 2,000 + 2,400 + 600}{1,229 + 1,980 + 2,397 + 618} = 1.00$$

Because the growth-factor (F) values have reached a satisfactory closure, as specified in Eqs. (7.11) and (7.12), the distribution of zonal trip ends by the Detroit method is completed in the second iteration.

The Fratar and Detroit models are considered to have better mathematical expressions and to be computationally more efficient than the uniform growth and average factor models. In any case, the growth-factor models find most application in estimating trips from external to internal or other external zones since there are no land-use data available for the external areas outside the study region.

These models are advantageous in that they:

1. Are simple, inexpensive, and easy to apply
2. Are well-tested
3. Require no distance variables
4. Need no calibration
5. Can be applied to peak directional flows
6. Are useful in updating origin–destination surveys (see Chap. 9)

The disadvantages are:

1. Only a single growth factor for each zone, and assumed stable to the horizon year
2. Inability to account adequately for major changes in land use or interzonal activity
3. No explicit term relating to any form of travel cost, time, or other impedances
4. Zones having zero interchanges in the base will show zero interchanges in the horizon year
5. Errors in original distribution due to sampling or other factors will be carried forward and magnified

7.4.2 Gravity Model

The gravity model, as its name implies, is adopted from the "law" of gravity as advanced by Newton in 1686 to explain the force between (and consequent motion of) the planets and stars in the universe. As originally proposed, the model equation for the force between two bodies was

$$F_{12} = G \frac{M_1 M_2}{d_{12}^2} \tag{7.13}$$

where F_{12} = the gravitational force between bodies 1 and 2
M_1 = mass of body 1
M_2 = mass of body 2
d_{12} = distance between bodies 1 and 2
G = a constant

In viewing this model, travel researchers noted an interesting analog, especially in regard to shopping travel: M_1 might represent the "mass" of trips available at, say, a residential area; M_2 the "mass" or attractiveness of a shopping area; d_{12} the distance between the two areas; and F_{12} the number of trips between the two areas. These interpretations would imply through the gravity model that the greater the size or attractiveness of the two areas (masses) and the less the distance between them, the more would be the number of interarea trips. This was found to resemble many real world situations.

When the effect of several competing attraction areas (i.e., multiple masses) was taken into account, the gravity trip-distribution model became

$$T_{ij} = P_i \frac{A_j/d_{ij}^b}{(A_1/d_{i1}^b) + (A_2/d_{i2}^b) + \ldots + (A_j/d_{ij}^b) + \ldots + (A_n/d_{in}^b)} \tag{7.14}$$

where T_{ij} = number of trips produced in zone i and attracted to zone j

$\quad\quad P_i$ = number of trips produced by zone i

$\quad\quad A_j$ = number of trips attracted to zone j

$\quad\quad d_{ij}$ = distance between zone i and zone j, generally expressed as the total *travel time* (t_{ij}) between i and j

$\quad\quad b$ = an empirically determined exponent that expresses the average area-wide effect of spatial separation between zones on trip interchange

What this formula states in essence is that the percentage of the P_i trips produced by zone i allocated to destination zone j is dependent upon both the attractiveness (A_j) of and travel time to that zone *relative to* the same features of all other attracting zones. Thus a zone in which a new shopping center is built (increased A_j), or to which a new transportation facility is constructed (decreased d_{ij}), increases its relative "pull" on the P_i trip productions and subsequently draws a greater proportion of these productions to itself.

The gravity model has undergone an additional change in form in order to make it more general in concept. As used in most applications, it now appears as

$$T_{ij} = P_i \frac{A_j F_{ij} K_{ij}}{\sum\limits_{j=1}^{n} A_j F_{ij} K_{ij}} \tag{7.15}$$

where F_{ij} is called the *travel-time factor* [specified as $1/(t_{ij}^b)$ above] and where K_{ij} is a specific zone-to-zone adjustment factor for taking into account the effect on travel patterns of defined social or economic linkages not otherwise incorporated in the gravity model formulation (somewhat equivalent to the "G's" in Newton's gravity model). One major change to be noted here is that we no longer are required to have a travel time function of the form $1/(t_{ij}^b)$ but could have others such as

$$F = a t_{ij}^{-b} \tag{7.16}$$

$$F = a t e^{-bt^2 + ct + d} \tag{7.17}$$

$$F = \frac{1}{a + bt} \tag{7.18}$$

All of these express the general idea of a drop in F as t increases, but one may lead to more reliable trip distribution estimates than the others under certain circumstances. The second major change is, of course, in the inclusion of the K_{ij} factors, these being added because of some unusual differences in travel distributions noted in some cities and attributed to the social and economic makeup of the travelers and zones. For example, higher income residents have a greater tendency to make longer work trips than their lower income counterparts [7.33]. Basically the use of K_{ij} factors stems from the inability of the F_{ij} factors alone to replicate the base-year trip interchanges in many zones.

7.4.2.1 Calibration of the gravity model   After establishing the minimum time paths from each zone centroid (center) to all others to obtain the current zone to zone travel times (see App. A), the engineer or planner must calibrate the gravity model to be able to make forecasts of future trip distributions using the total trip attractions and trip productions of each zone obtained from trip-generation analysis. Calibration for this model involves determination of the values of the travel-time factors (F_{ij}) and zone-to-zone adjustment factors (K_{ij}) that will produce the zone-to-zone trip tables of the base year from the trip ends (productions and attractions) observed in that year. These factors, F_{ij} and K_{ij}, are then assumed constant over time, and by applying them to the trip ends computed for the forecasted year, the future trip interchanges from zone to zone can be computed. Because of the nature of the gravity model shown in Eq. (7.15), there is no estimation procedure available for determining F_{ij} and K_{ij} factors directly from the known T_{ij}'s, P_i's, and A_j's. An initial value of 1.0 is assumed for all K_{ij} factors and then a trial-and-error process is used to estimate the values of F_{ij} factors, subject to the conditions that $\Sigma P_i = \Sigma A_j$; $\Sigma_j T_{ij} = P_i$; $\Sigma_i T_{ij} = A_j$.

The following steps are used to estimate the F_{ij} factors:

1. Assume all $K_{ij} = 1.0$.
2. Assume initial values for travel-time factors F_{ij} (usually 1.0) at 5-min (or other) time intervals.
3. Group the zones within the predetermined time intervals.
4. Use these assumed travel time factors to distribute the trip ends among zones, employing Eq. (7.15).
5. Check the number of trips obtained by the gravity model versus those observed in the base year for each of the time interval groups.
6. Adjust the travel time factors for each time interval as follows:

$$F(\text{adj}) = F(\text{old}) \frac{\text{total trips observed}}{\text{total trips calculated}}$$

7. Stop calibrating for F_{ij} factors when the ratio for the trips observed to trips calculated is close to 1.0—in other words, when no improvements in F_{ij} factors are realized from the old ones.

To understand the calibration method better, let us take the following hypothetical example (Table 7.16) of three zones with the corresponding travel time and trip interchange matrices. We assume all K_{ij} values are 1.0, and group the zones within 5-min time intervals.

First trial Assume $F_{ij} = 1.0$, calculate T_{ij} values from the gravity model, and compare with the observed values. The model under these circumstances has the form

Table 7.16 Example Base Year Travel Time and Trip Figures

Travel-time matrix (min)				Trip-interchange matrix				
	To zone				To zone			
From zone	1	2	3	From zone	1	2	3	P_i
1	2	9	13	1	100	350	100	550
2	10	4	11	2	240	150	210	600
3	14	12	3	3	60	120	200	380
				A_j	400	620	510	1,530

Table 7.17   First-Trial Trip-Interchange Matrix

From zone	To zone			P_i
	1	2	3	
1	144	223	183	550
2	157	243	200	600
3	99	154	127	380
A_j	400	620	510	1,530

$$T_{ij} = P_i \frac{A_j}{\Sigma A_j} \tag{7.19}$$

which leads to the calculations

$$T_{11} = 550 \frac{400}{1,530} = 144 \qquad T_{12} = 550 \frac{620}{1,530} = 223 \quad \cdots$$

$$T_{23} = 380 \frac{510}{1,530} = 127$$

The other results are presented in Table 7.17.

Grouping the results in zone pairs according to 5-min travel-time intervals and using the adjustment process in step 6 above, we establish the first-trial F_{ij} factors as indicated in Table 7.18. The matrix of F_{ij} values is displayed in Table 7.19.

Second trial   Using the observed trip ends (P_i's and A_j's) and the F_{ij} factors from the first trial, distribute the trips using the gravity model:

$$T''_{ij} = P_i \frac{A_j F'_{ij}}{\Sigma A_j F'_{ij}} \tag{7.20}$$

This leads to the calculations

$$T_{11} = \frac{550(400)(0.875)}{(400)(0.875) + (620)(1.533) + (510)(0.77)} = 113$$

Table 7.18   First-Trial F_{ij} Factors

Travel-time interval	Zone pairs in intervals	Total observed trips	Assumed F_{ij}	First trial calculated trips	First-trial F_{ij} factors
0.1–5.0	1–1, 2–2, 3–3	450	1.0	514	$(450/514)1.0 = 0.875$
5.1–10.0	1–2, 2–1	590	1.0	380	$(590/380)1.0 = 1.553$
10.1–15.0	1–3, 2–3, 3–1, 3–2	490	1.0	636	$(490/636)1.0 = 0.770$

Table 7.19   First-Trial F_{ij} Matrix

From zone	To zone 1	2	3	P_i
1	0.875	1.553	0.77	550
2	1.553	0.875	0.77	600
3	0.77	0.77	0.875	380
A_j	400	620	510	1,530

$$T_{12} = \frac{550(620)(1.533)}{(400)(0.875) + (620)(1.533) + (510)(0.77)} = 310$$

. . .

$$T_{33} = \frac{380(510)(0.875)}{(400)(0.77) + (620)(0.77) + (510)(0.875)} = 138$$

and $T_{13} = 127$, $T_{21} = 240$, $T_{22} = 209$, $T_{23} = 151$, $T_{31} = 95$, and $T_{32} = 147$.

Grouping the results in the same form as in Table 7.18, we find the second-trial F_{ij}'' factors as in Table 7.20.

Third trial   Similarly to second trial, we use the gravity model to distribute the observed trip ends utilizing the second-trial F factors. We then obtain

$$T_{11} = \frac{550(400)(0.856)}{(400)(0.856) + (620)(1.669) + (510)(0.726)} = 107$$

$$T_{12} = \frac{550(620)(1.669)}{(400)(0.856) + (620)(1.669) + (510)(0.726)} = 326$$

. . .

$$T_{33} = \frac{380(510)(0.856)}{(400)(0.726) + (620)(0.726) + (510)(0.856)} = 141$$

and $T_{13} = 117$, $T_{21} = 255$, $T_{22} = 203$, $T_{23} = 142$, $T_{31} = 94$, and $T_{32} = 145$.

Table 7.20   Second- and Third-Trial Trips and F_{ij} Factors

Travel-time interval	Second-trial calculated trips	Second-trial F_{ij} factors	Third-trial calculated trips	Third-trial F_{ij} factors
0.1–5.0	460	(450/460)0.875 = 0.856	452	(450/452)0.956 = 0.852
5.1–10.0	550	(590/550)1.553 = 1.669	581	(590/581)1.669 = 1.695
10.1–15.0	520	(490/520)0.77 = 0.726	498	(490/498)0.726 = 0.714

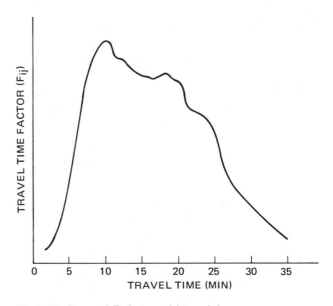

Fig. 7.15 Decay of F_{ij} factors with travel time.

Since the calculated trips are within 0.95 and 1.05 of observed trips for the same travel-time intervals, the iteration process is terminated and the third travel-time factors, F_{ij}''', are adopted for future forecasts. Before their adoption is made final, the travel-time factors should be checked to see whether they approximate some reasonably regular decay with travel time. In other words, if they are plotted against travel time, they should produce a graph similar to the one shown in Fig. 7.15. "Badly behaved" travel-time factors will create considerable difficulty in balancing future zonal trip interchanges with the observed ones.

The third-trial trip-interchange matrix now is as shown in Table 7.21. The table shows that the calculated trip productions, P_i, are equal to the observed values. However, this is not true for the trip attractions, A_j. The trial-and-error procedure used to determine the F_{ij} factors does not guarantee that the zonal trip attractions estimated by the model will be equal to those found in the observed data. So it is common to use a successive column-and-row factoring process in which the zonal trip attractions

Table 7.21 Third-Trial Trip Interchanges

From zone	To zone			
	1	2	3	P_i
1	107	326	117	550
2	255	203	142	600
3	94	145	141	380
Calculated A_j	456	674	400	1,530
Observed A_j	400	620	510	
Column factors	0.875	0.92	1.275	

Table 7.22 Column-Factored Trip Interchanges

From zone	To zone 1	To zone 2	To zone 3	Calculated P_i	Observed P_i	Row factors
1	95	300	149	544	550	1.011
2	223	187	181	591	600	1.015
3	82	133	180	395	380	0.962
A_j	400	620	510	1,530		

and productions are adjusted to make them equal to their observed values. This process is carried out as follows.

1. All interzonal trips are factored by the ratio of the observed to calculated zonal attractions (column factors in Table 7.21).
2. All interzonal trips are factored by the ratio of observed to calculated zonal productions (row factors in Table 7.22).
3. The first two steps are repeated sequentially until *both* the calculated P_i's and A_j's are sufficiently close to the corresponding observed values.

The final matrix that results from our example (after two iterations) is displayed in Table 7.23.

Determining the K_{ij} factors The K_{ij} factors are established from the final balanced matrix using the relationship

$$K_{ij} = \frac{T_{ij} \text{ (observed)}}{T_{ij} \text{ (calculated)}} \tag{7.21}$$

For example, for the trips from zone 1 to 2,

$$K_{12} = \frac{350}{302} = 1.159$$

7.4.2.2 Forecasting zonal trip interchanges using a calibrated gravity model
After the gravity model shown in Eq. (7.15) has been fully calibrated, the user is in a

Table 7.23 Final Trip-Interchange Matrix

From zone	To zone 1	To zone 2	To zone 3	P_i
1	96	302	152	550
2	225	190	185	600
3	79	128	173	380
A_j	400	620	510	1,530

Table 7.24 Horizon-Year Travel-Time Matrix (min)

From zone	To zone		
	1	2	3
1	3	9	14
2	10	6	12
3	15	13	5

position to make forecasts of horizon-year trip interchanges. As an example of this process, let us employ the preceding three-zone example model (including F_{ij}'s and K_{ij}'s) with the new travel-time matrix and horizon-year trip productions and attractions shown in Tables 7.24 and 7.25. Now, let

$$Z_i = \sum_j A_j F_{ij} K_{ij} \tag{7.22}$$

The Z_i terms are the denominators of Eq. (7.15). Then

$$Z_1 = 600(0.852)(1.042) + 1,100(1.695)(1.159) + 950(0.714)(0.658) = 3,139.95$$

with $Z_2 = 3,326.10$ and $Z_3 = 1,997.53$. Consequently, from the gravity model equation,

$$T_{11} = \frac{750(600)(0.852)(1.042)}{3,139.95} = 127$$

$$T_{12} = \frac{750(1,100)(1.695)(1.159)}{3,139.95} = 516$$

$$\dots$$

$$T_{33} = \frac{900(950)(0.852)(1.156)}{1,997.53} = 422$$

Table 7.25 Forecasted Horizon-Year Trip Attractions and Productions

	Zone		
	1	2	3
Productions	750	1,000	900
Attractions	600	1,100	950

Table 7.26   Final Forecasted Horizon-Year Trip-Interchange Matrix

From zone	To zone			
	1	2	3	P_i
1	136	466	148	750
2	328	374	298	1,000
3	136	260	504	900
A_j	600	1,100	950	2,650

After these trip interchanges have been estimated, the column-and-row factoring procedure used in the calibration is applied. This again is needed because the total of the trips into a zone does not necessarily equal the forecasted trip attractions. The result, after three iterations, is the trip distribution presented in Table 7.26.[3]

7.4.2.3   Comments on the gravity model   The popularity and common use of the gravity model stems from several advantages.

1. It accounts for competition of trips between land uses by highlighting trip attractions versus productions.
2. It is sensitive to changes in travel times between zones.
3. It recognizes trip purposes as affecting zonal interchanges.
4. It is easy to understand (at least at first glance) and therefore easy to apply in particular areas.

Some of the disadvantages are, however:

1. It is very unlikely that the travel-time factors by trip purpose would remain constant throughout the urban area to the horizon period.
2. The changing nature of travel times between zones with time of day makes questionable the use of single values for the travel time factors.
3. It tends to overestimate near trips and underestimate far trips.
4. It only shows approximate agreement with field data, and considerable "playing" with K_{ij} factors is needed to balance the total number of productions and attractions.

7.5   MODE CHOICE

At this point in the travel estimation process we have first predicted the number of trips produced or attracted by each zone in a region and then, through the use of trip distribution models, attempted to determine how these generated trips will spread themselves among the zones. The next step, in referring back to Fig. 7.11, is to find the percentage of the inter- and intrazonal trips taking some kind of transit as opposed to going by private automobile.

Many mathematical models have been developed for this purpose. They generally are divided into two types: aggregate and disaggregate. The aggregate versions include

---

[3] Space does not permit elaboration of all the calculations here. Readers may want to verify these on their own.

the trip-end and the trip-interchange modal-split models [7.14]. The methods commonly used in these models are multiple regression, cross-classification, and diversion curves. The disaggregate types, referred to as individual mode-choice models, are probabilistic and behavioral in nature and are based on the theory of utility or disutility of a certain mode to a particular traveler. The methods generally used are the probit or logit functions.

All these models include some factors that relate to the three broad categories of characteristics of the trip, tripmaker, and transportation system. In this section we discuss one model of each type.

7.5.1 A Trip-End Modal-Split Model

The trip-end modal-split model described here [7.14] is the one used in the Pittsburgh Area Transportation Study (PATS). A trip-end model is one that splits total person-trip origins and destinations to the alternate modes of transportation prior to trip distribution.

The split in this model is between two modes: transit and auto. The trip purposes are for CBD, school, and other. The main variables are auto ownership (autos per 1,000 persons), net residential density (persons per acre) (NRD), and distance from CBD.

The percent of CBD trips by transit for the base year are cross-classified with autos per household, distance from CBD, and NRD, as shown in Table 7.27. The 1980 forecast of the proportion of all CBD person trips using transit was made from the data in that table. The inputs were the forecasted autos per household per zone, and the forecasted NRD for zones 1 mile from CBD and over 1 mile from CBD. The assumptions were that there would be no improvement in transit service, and the existing service would be adequate to meet the anticipated demand. This assumption implicitly indicates that transit improvements have no effect on total transit ridership, and that transit users continue to be mainly captives.

School transit trips were estimated from the following logarithmic regression equation [7.14], developed from the base-year data:

$$\log Y = 3.30 - 0.91 \log X \qquad (7.23)$$

where Y = school transit trips per 1,000 persons
X = net residential density (persons per acre)
This had a correlation of -0.748 and a standard error of the estimate[4] of 0.32.

[4] See App. A.3 for an explanation.

Table 7.27 Percent of CBD Trips by Transit

Autos per household	Zones under 1 mile from CBD	Zones over 1 mile from CBD	
		NRD 12 or more	NRD under 12
0	77.0	88.0	61.0
1	35.5	53.5	36.0
2 or more	13.5	31.5	20.5

Source: [7.14, p. 20].

In forecasting 1980 school transit trips, planners assumed that the number of schools in areas of similar density would be relatively the same in 1980 as in the base year. Similarly, "other" transit trips were estimated by using linear regression equations developed from the base-year data for different classes of car ownership, as shown in Table 7.28.

The number of transit trips for all purposes for 1980 was summed and subtracted from total person trips. The resulting auto person trips subsequently were divided by the 1980 auto-occupancy rates to provide the vehicle trips. The distribution of auto drivers and transit trips was then made.

A potential theoretical problem associated with trip-end modal-split models is that tripmakers might not choose their mode before knowing the destination. In fact, there might not be any transit serving their destination. Later trip-end models overcame this problem by including characteristics of the transportation system, such as the relative accessibility of the remaining zones to the origin zone for the transport mode under consideration.

In general the trip-end models have the same problems associated with aggregate models, such as loss of within-zone variation and lack of transferability from one area to another.

7.5.2   The Trip-Interchange Modal-Split Model

The choice or "split" between two modes is seen in this model [7.34] as a function of four variables that describe both the transportation alternatives between each pair of zones and the socioeconomic characteristics of the people who avail themselves of the alternatives. Since transportation supply characteristics are included in these models, it is employed after the distribution of trips. The considered variables are: (1) relative travel time, TTR; (2) relative travel cost, CR; (3) economic status of the tripmaker, EC; and (4) relative travel service, L.

Several other variables, such as trip length, population density, employment density, and so forth were investigated for their possible relevance to modal split. However, they were found to be highly correlated with the four listed above, so that their influence already was accounted for through the TTR, CR, EC, and L variables.

The first of the four independent variables is defined as follows:

$$TTR = \frac{X_1 + X_2 + X_3 + X_4 + X_5}{X_6 + X_7 + X_8} \tag{7.24}$$

Table 7.28   Correlation Summary of "Other" Transit Trips[*]

Class	Variables	Equations	r	s	$\bar{Y}$
0-Car household	Y: Other transit trips/1,000 (D) X: Net residential density (D)	$Y = 84.02 + 8.9X - 0.094X^2$	0.519	68	212
1-Car household	Y: Other transit trips/1,000 (Z) X: Net residential density (Z)	$Y = 3.04 + 3.20X - 0.026X^2$	0.748	22	71
2-Car household	Y: Other transit trips/1,000 (Z) X: Net residential density (Z)	$Y = 16.4 + 3.6X - 0.0334X^2$	0.832	10	56

[*]Used for 1980 estimates: (Z) = formulated by zone (s, $\bar{Y}$ based on trip rate per 1,000 population); (D) = formulated by district.
Source: [7.14, p. 24].

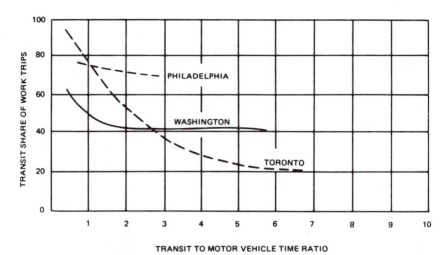

Fig. 7.16 Travel time ratio diversion curve for work trips in peak periods [7.14, p. 279].

where X_1 = time spent in transit vehicle

X_2 = transfer time between transit vehicles

X_3 = time spent waiting for a transit vehicle

X_4 = walking time to transit vehicle

X_5 = walking time from transit vehicle

X_6 = auto driving time

X_7 = parking delay at destination

X_8 = walking time from parking place to destination

TTR represents the ratio of the *door-to-door* travel time by transit to that by automobile. The effect of TTR on the transit share (percentage) of *work* trips can be seen in Fig. 7.16, where, for the three cities studied, it appears that door-to-door travel by transit must take at least 2.5 times that by auto before the transit share drops below 50 percent.

The second variable is that of relative travel cost CR,

$$CR = \frac{X_9}{(X_{10} + X_{11} + 0.5X_{12})/X_{13}} \tag{7.25}$$

where X_9 = transit fare

X_{10} = cost of gasoline

X_{11} = cost of oil change and lubrication

X_{12} = parking cost at destination

X_{13} = average car occupancy

The significance of the division of the denominator by average car occupancy, X_{13}, and the halving of the parking cost (X_{12}), is that auto costs must be put on a per person per one-way trip basis in order to be comparable to the costs for transit. Moreover, the only auto costs considered are those which a *person must take out of his pocket.* The underlying reason for this assumption is that these costs, and not those for such long term items as tires and car insurance, are the ones that apparently influence a person's day-to-day decision as to whether or not to take the car to work.

The third variable, the economic status of the tripmaker (EC), is defined strictly in terms of the median income per worker in the zone of trip production. The last variable, travel service (L), is difficult to quantify because it pertains to such intangible factors as the atmosphere within the vehicle, comfort, appearance, ride smoothness, availability of seats, and convenience of transfer. In the trip-interchange modal-split model, L is designated somewhat arbitrarily as the amount of time spent other than in actual travel in the transit vehicle or automobile, giving the relationship

$$L = \frac{X_2 + X_3 + X_4 + X_5}{X_7 + X_8} \qquad (7.26)$$

where each of the variables have been defined previously.

Having interpreted these variables in a definite manner, most analysts probably would proceed by utilizing a statistical procedure such as multiple regression to establish a connection between the independent and dependent variables. This was done in a partial manner for the trip-interchange modal-split model, but, for graphic simplicity, the CR, L, and EC variables were divided into discrete stratifications, and then regression equations relating the two variables—transit share of trips, TS, and relative travel time, TTR—were constituted within each stratification. This procedure led to a set of two-dimensional plots as presented in Fig. 7.17. The categories for the three stratified variables in this figure are:

$CR_1 = 0.0$ to 0.5

$CR_2 = 0.5$ to 1.0

$CR_3 = 1.0$ to 1.5

$CR_4 = 1.5$ and over

$EC_1 = \$0$ to $\$3,100$ per annum

$EC_2 = \$3,100$ to $\$4,700$ per annum

$EC_3 = \$4,700$ to $\$6,200$ per annum

$EC_4 = \$6,200$ to $\$7,500$ per annum

$EC_5 = \$7,500$ per annum and over

$L_1 = 0.0$ to 1.5

$L_2 = 1.5$ to 3.5

$L_3 = 3.5$ to 5.5

$L_4 = 5.5$ and over

Each separate chart in Fig. 7.17 has four curves corresponding to the four stratifications of L. Several charts have relationships derived from studies not only in Washington, D.C., but also in Philadelphia and Toronto. The equations are linear except in the shaded portion of each chart where manual adjustments have been made to approximate more closely actual conditions in the study area. Most correlation coefficients for the linear portion of the curves are over 0.85, but should not be accepted injudiciously since they are based on only a small number of data points (a situation which generally leads to higher correlation coefficients).

7.5.2.1 Use of the trip-interchange modal-split model The prediction of the share (percentage) of work trips going by transit at some future point in time would involve the straightforward use of the charts presented in Fig. 7.17. The difficult part of the process, however, would be to forecast the values for each of the 14 factors used to calculate TTR, CR, EC, and L. The estimation of future costs of gasoline, oil change and lubrication, and parking would be particularly risky. For the Washington study [7.34], the following assumptions about these factors were made to guide the prediction process:

1. Auto and transit travel times and distances were calculated along minimum time paths.
2. Transit transfer time was taken as half of the time between arrivals of transit vehicles.

3. Walking times to transit stops were established by observation of zone size, transit location, and average walking speeds.
4. Waiting times for transit were equated to half the average time between arrivals of transit vehicles.
5. Parking delay generally was assumed as 1 min.
6. Walking time from parking was taken as 1 min except in downtown areas where 2 to 5 min was used.
7. Parking costs were assumed to be 1.6 times the zonal *average* of all parking costs at the present time.
8. No increases from present zonal median worker incomes were assumed, but at the same time no increases in car operating and transit fare costs were projected, the feeling being that the relative increases in these cost factors would cancel each other out.

In employing the trip-interchange modal-split model in a region other than Washington, D.C., the analyst probably would be wise to recalibrate the regression equations, that is, find new sets of coefficients and constants based on present data. This would be desirable because the variations between Washington, D.C., Philadelphia, and Toronto (shown in Fig. 7.17) are large enough to make us suspect similar differences in other regions. Mode choice relations also should be formulated for trip purposes other than work. In Washington, the three purposes of work, school, and nonwork-nonschool were utilized.

7.5.2.2 Comments on the trip-interchange modal-split model In many studies, mode choice relationships are derived through the straightforward use of regression. The trip-interchange model goes a step beyond this theoretically by considering a variety of categories of characteristics of users and the transportation system and *then* employing regression. The model has been calibrated with fairly good results in three large urban areas, a fact that gives us an honest basis for comparison. On the negative side, however, we find:

1. Only two "modes" are considered—highway and transit. There is no way to differentiate between bus, streetcar, and commuter rail. This also means that if a new transit technology were introduced (e.g., gravity vacuum tube), it would be difficult if not impossible to determine the proportion of the travel market it would capture.
2. The model is used after all trips have been distributed. It may be that decisions about mode choice are made before the traveler decides where he will go (or even before he decides if he will go).
3. In predicting future mode choice, the model may be employed outside the ranges in which the original variables were measured. This could lead to significant errors.

7.5.3 An Individual Mode Choice Model

The individual mode choice model to be described was developed by Peat, Marwick, Mitchell, and Co. for the Comprehensive Planning Organization of San Diego County, California, in 1972. Such models, also referred to as disaggregate travel-behavior models, are employed to determine the probability that a particular individual will choose a given mode among the available alternatives based on behavioral variables describing the choice situation. The predominant mathematical formulation used in individual choice modeling is that of the logit function

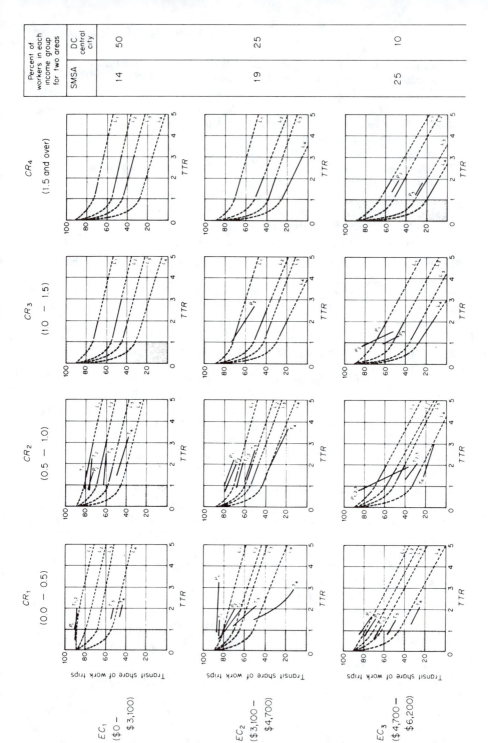

Percent of workers in each income group for two areas

	DC central city	SMSA
	50	14
	25	19
	10	25

216

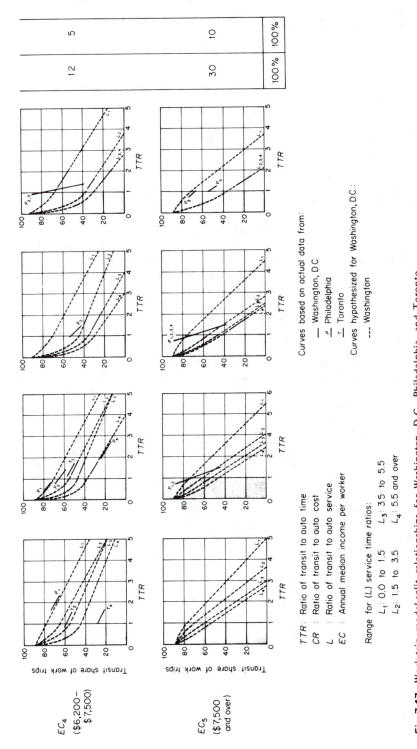

12		5
30		10
100%		100%

Fig. 7.17 Work-trip modal-split relationships for Washington, D.C., Philadelphia, and Toronto.
Source: Traffic Systems Analysis for Engineers and Planners by M. Wohl and B. V. Martin, copyright 1967, McGraw-Hill Book Co. Used with permission of McGraw-Hill Book Co.

TTR: Ratio of transit to auto time
CR : Ratio of transit to auto cost
L : Ratio of transit to auto service
EC : Annual median income per worker

Range for (L) service time ratios:

L₁: 0.0 to 1.5   L₃: 3.5 to 5.5
L₂: 1.5 to 3.5   L₄: 5.5 and over

Curves based on actual data from:
— Washington, D.C.
ᵖ Philadelphia
ᵀ Toronto

Curves hypothesized for Washington, D.C.:
--- Washington

217

$$P_i = \frac{e^{U_i}}{\sum\limits_{j=1}^{N} e^{U_j}} \qquad (7.27)$$

where P_i = the probability of an individual choosing alternative i
U_i = the utility function of mode i
N = set of modal alternatives
e = base of natural logarithms

The U_i terms used in Eq. (7.27) generally are linear functions of the variables describing the choice situation

$$U_i = \sum_{\alpha} \theta_\alpha X_{i\alpha} \qquad (7.28)$$

where θ_α and $X_{i\alpha}$ are the coefficients and independent variables, respectively, describing alternative i. The variables are transportation system characteristics as well as socioeconomic characteristics of the tripmaker, possibly stratified by trip purpose. Examples for home-to-work and "other" trips are travel cost, travel time, comfort, and household income.

The coefficients (θ_α) are most commonly estimated by the method of maximum likelihood, which will not be discussed here. Linear regression analysis does not apply because the dependent variable of an individual choice model takes the values of either zero or one and the independent variables (X's) can be either continuous and/or discrete. The characteristics and calibration of logit models are discussed by McFadden [7.19] and Stopher [7.20].

The alternative travel modes specified in the San Diego models are transit passenger (TP), auto driver (AD), and auto passenger (AP). Individual mode choice models were developed for home-to-work trips during the morning peak (5 a.m. to 10 a.m.). Separate versions were created for CBD and non-CBD trips. For the sake of brevity, only home-to-work trip models will be discussed here. Interested readers are referred to [7.16].

The logit model formulas used in the study were

$$P_{AP} = \frac{1}{1 + e^{U_{AD}} + e^{U_{TP}}} \qquad (7.29)$$

$$P_{AD} = \frac{e^{U_{AD}}}{1 + e^{U_{AD}} + e^{U_{TP}}} \qquad (7.30)$$

$$P_{TP} = \frac{e^{U_{TP}}}{1 + e^{U_{AD}} + e^{U_{TP}}} \qquad (7.31)$$

where P_{AP} = probability of a traveler being driven to work
P_{AD} = probability of a traveler driving to work
P_{TP} = probability of a traveler using transit to work
U_{AD}, U_{TP} = linear utility expressions for auto driver and transit passenger, respectively

The utility expression for auto passenger could be set to zero, making the exponential expression $e^{U_{AP}}$ equal to one, since the other two modes were compared to that for the auto passenger.

The linear utility expressions for the two modes, auto drivers and transit, for CBD trips were

$$U_{AD} = -1.4809 + 1.9500 \, (\text{TI35}) \tag{7.32}$$

$$U_{TP} = 1.1636 + 0.0916 \, (\text{DX3}) + 0.0563 \, (\text{DL3}) + 0.0106 \, (\text{DCH}) \tag{7.33}$$

and for non-CBD trips,

$$U_{AD} = -0.5441 + 2.6800 \, (\text{TI35}) \tag{7.34}$$

$$U_{TP} = 1.6600 + 0.1314 \, (\text{DX3}) + 0.0192 \, (\text{DL3}) + 0.0184 \, (\text{DCH}) \tag{7.35}$$

where TI35 = transformed household income variable
       = $1 - e^{-0.035 \, (I)}$, where I is annual household income (in \$1,000)
DX3 = difference in excess time
     = (auto terminal time at origin) + (auto terminal time at destination) − (walk to transit time) − (transit wait time) − (walk from transit time)
DL3 = difference in line haul time
     = (auto travel time) + (auto access time) − (vehicle in transit time) − (transit transfer time)
DCH = difference in travel cost
     = (5 cents/mile × auto distance) + (auto parking cost/2) − (transit fare)

Differences in line-haul time, excess time, and travel cost turned out to be the most appropriate transportation system variables, while income was the only socio-economic variable that was statistically significant. The transportation system variables were created on the basis of the following assumptions:

1. Auto and transit vehicle times and distances were calculated along zone-to-zone peak-hour minimum time paths on the highway and transit networks, respectively.
2. Auto access times—those taken by the traveler while using local streets to access the principal highways—ranged from 0 min in the CBD, to 1 to 2 in the densely settled areas, to 5 and above in the outlying zones.
3. Auto terminal times—taken for (1) walking between the parking lot and the actual origin or destination of the trip, and (2) waiting or walking within the parking facility—were calculated as a function of the availability of parking spaces in a zone.
4. Auto parking costs were computed as the average daily values for each zone.
5. Walk-to-transit times were based on the average walking times to transit networks for each zone.
6. Transit waiting times were equated to half the transit headway times, up to a maximum of 15 min.
7. Transit transfer times were taken as half the transit headways at the transfer points.
8. Transit fares were developed from a fare matrix based on the most probable zone-to-zone transit routes.

The models were able to reproduce the observed mode-split distributions for various incomes and travel times with reasonable accuracy. In addition, to test the sensitivity to different transit service conditions, the models were run on data obtained from Boston and San Francisco. Both of these areas had better transit service than San Diego. The models closely approximated the observed data set in both places, a result that shows both their broad range of applicability to various public transportation service options and their transferability from one area to another.

7.5.3.1 Example As an example of the use of the model, suppose an individual belongs to a household with an annual income of $5,000. The probability of, say, making a CBD trip by transit can be determined from Eq. (7.31), which in turn depends on Eqs. (7.32) and (7.33). As input to the latter two, let us assume that difference in excess time (DX3) is −12 min; in line haul time (DL3), is −10 min; and in cost (DCH), +$1.60. The transformed household income variable first is calculated as

$$TI35 = 1 - e^{-0.035\,(5)} = 0.16$$

and, from Eq. (7.32), the utility associated with auto driving is

$$U_{AD} = -1.4809 + 1.9500(0.16) = -1.169$$

and, for transit travel,

$$U_{TP} = 1.1636 + 1.0916(-12) + 0.0563(-10) + 0.0106(1.60) = -0.482$$

Then, from Eq. (7.31), the probability of taking transit is

$$P_{TP} = \frac{e^{-0.482}}{1 + e^{-1.169} + e^{-0.482}} = 0.321$$

In other words, there is a 32.1% chance that the individual will take transit.

7.5.3.2 Comments on the model Use of the San Diego model to forecast the future share of work trips by transit, auto driver, and auto passenger depends on the ability to predict individual income as well as the transportation system variables. As with the trip-interchange model, the estimation of future travel costs would be particularly difficult and risky.

The aggregation of the individual choices to the zonal level for travel-demand forecasting in the traditional four-step procedure is generally considered difficult and sometimes impossible. The summation of individual choices—that is, of individual probabilities for taking a certain mode (for different socioeconomic characteristics)— is computationally impossible in a large data set, and the information rarely is available in most cases.

Certain procedures [7.21, 7.22] have been developed to estimate group behavior from individual choice models, following some adjustments and corrections. The San Diego individual-choice models, however, are dependent on one variable computed at the household level: income. The system variables all were derived from zonal travel-time matrices. It was suggested that separate choice models be computed for each income group found in the zone-to-zone interchanges. The mode split between any zone pair could then be computed by multiplying the number of person trips between the zones for each income group, by the mode-choice probabilities for those groups obtained from the models. It has also been suggested that reasonable approximations could be made by stratifying income on the basis of the origin zone only [7.18].

Another shortcoming of individual-choice models, and particularly the San Diego versions, is the use of zone-to-zone travel times as proxies for point-to-point. The zonal-level system variables represent mean values for travel along the "best path" between two zone centroids. These may not necessarily represent the actual route taken by a traveler.

A more general inherent disadvantage to the individual-choice models is their property of the independence of irrelevant alternatives (IAA). All share models work on the assumption that the ratio of the market proportion of two alternatives is independent of the set of considered alternatives. This property is valid as long as any new alternative competes equally with each existing one. This is unrealistic in most transportation planning applications, however. It is well known, for instance, that most rapid transit systems divert former bus passengers. They compete more heavily for the bus than for the auto market. A model having the IIA property would not be able to account entirely for these differences in competitiveness.

On the other hand, individual-choice models have a number of advantages. They are directly sensitive to transportation policy issues, such as parking charges, gasoline prices, transit fares, and improvements in the transit service. They incorporate the total variation in the data set, in contrast to the aggregate zonal models. They require a small data base for calibration, and they can be employed to compare several alternatives in a simple model. They may be transferable from one urban area to another. Finally, they try to explain the rationale (cause) for the mode-choice behavior of individuals.

7.6 TRIP ASSIGNMENT

The final phase of the travel estimation process deals with the assignment of the interzonal, modal trips to the various routes of each mode. Rephrased, this means that we are going to be concerned with why tripmakers choose one route over another. The reader also should expect that we will concentrate only on automobile travel since, for the most part, modes other than auto do not provide a multiplicity of routes between zonal pairs.

The question that naturally arises at this point is, "What are the factors that lead people to choose one route over another?" Generally speaking, researchers have identified at least four: (1) travel times, (2) travel costs, (3) comfort, and (4) levels of service (volume/capacity).

While all of these are considered important, the first is used almost exclusively in all models of route choice or trip assignment, the main reason being the relative ease by which travel time as opposed to the other three variables can be measured. In addition, all four variables are somewhat interrelated, so that one often can be used to represent the whole group.

If travel time is utilized as the major factor in trip assignment, a desirable feature in a trip assignment model would be the incorporation of a tradeoff between travel time and trip volume, because, as can be readily observed on any street or highway, the greater the volume of traffic (as compared to the capacity) the greater the travel time of any one vehicle traversing the facility. On the other hand, the greater the travel time, the fewer the number of people who will take the particular route, a fact that indicates that, after a period of time, an equilibrium on a set of routes should be reached in which any person by switching his route could only *increase* his individual travel time.

Early trip-assignment techniques, referred to in the literature as noncapacity restraint methods, did not consider the relationship between trip volume and travel time. The "all-or-nothing" procedure, for instance, assigns the total volume of trip interchanges to the minimum time path between the appropriate zonal centroids. The travel time on the best path is considered independent of the volume. Trip assignments with this technique often give unrealistic traffic volumes. This also is due to failure to take into account the preferences of some drivers for using arterial streets rather than freeways, especially when the time difference is small.

In response to this problem, "diversion" curves were developed, such as shown for California in Fig. 7.18, to determine the percentage of freeway usage in comparison to other major streets based on two parameters—time and distance savings. As an illustration, in Fig. 7.18 if the travel time on a freeway and an arterial were equal, only about 48% would use the freeway. And if the same comparison were made between an arterial and a major street (lower level than an arterial), only 35% would use the latter.

Capacity restraint methods, such as the TRC trip-assignment model to be discussed next, attempt to go one step further by showing how volumes (relative to capacities) affect travel times, and these in turn influence volumes.

7.6.1 The TRC Trip-Assignment Model

This model involves two travel-time versus volume relationships used iteratively to arrive at predictions of volumes on up to four separate[5] routes between any two zones [7.36]. The first equation, utilizing route volume as the dependent variable, was developed from observations from radar detectors mounted at the approaches to eight

[5] A modified Moore's algorithm technique ([7.35] and App. A) can be utilized to find the second, third, and fourth shortest routes between a pair of zones.

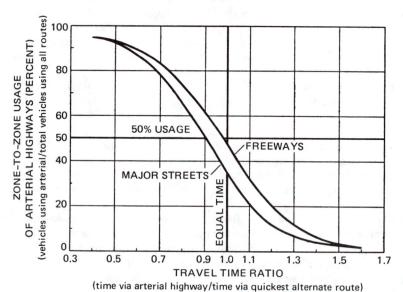

Fig. 7.18 Time ratio diversion curve [7.15, p. 31].

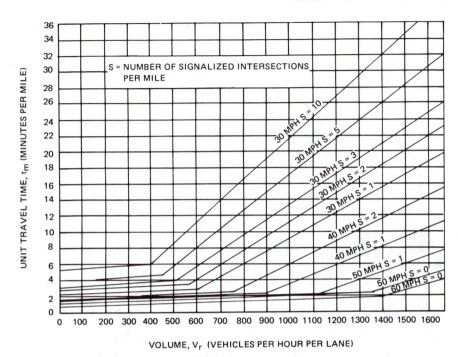

Fig. 7.19 Unit travel time versus volume on various types of roadways [*E. R. Ruiter, ICES–TRANSNET Procedures Manual, M.I.T., Civil Engineering Dept., Cambridge, Mass., 1968, p. 20*].

signalized intersections in Toronto. As seen in Fig. 7.19, 10 separate curves, each composed of two connected linear pieces, have been produced, with each one representing a route with a given speed limit and a given number of signalized intersections per mile. For simplicity it can be assumed that the slope of the low-volume linear portions is 0.5 and of the high volume portion 10.0, irrespective of the particular category to which the curve corresponds. In addition to its graphical portrayal, each curve can be represented symbolically as

$$t_r = t_{rc} + \frac{d(V_r - V_{rc})}{V_{rc}} L_r \qquad (7.36)$$

where t_r = travel time on route r (minutes)
V_r = volume of traffic on route r (vehicles/hr · lane)
V_{rc} = critical volume[6] for route r (vehicles/hr · lane)
t_{rc} = unit travel time at the critical volume (min/mile)
L_r = length of route r (miles)
d = delay parameter (min/mile)
The parameter d takes on the aforementioned values of 0.5 for $V_r < V_{rc}$ and 10.0 for $V_r \geqslant V_{rc}$.

The second relationship alluded to above is the converse of Eq. (7.36), used for

---

[6] See Fig. 7.19 where the critical volume is that corresponding to the break point between the two linear portions of the curve for each category of street.

predicting the volume on route r given the travel time,

$$V_r = \frac{1/t_r}{\sum\limits_{r=1}^{m} 1/t_r} V \tag{7.37}$$

where V is the total volume of trips from zone i to j on all m routes and V_r and t_r are as in Eq. (7.36). In effect, Eq. (7.37) divides up the volume of trips from zone i to j among the various routes in accordance with the reciprocal of the travel times. A decrease in t_r thus leads to an increase in t_r^{-1} and a corresponding increase in the proportion of the V trips assigned to route r. The value for t_r is found from Eq. (7.36), inserted into Eq. (7.37), and the resulting V_r is entered back into Eq. (7.36). This cyclical procedure continues until the changes in volumes or travel times become negligible.

7.6.1.1 Example of the TRC trip-assignment model The interplay between the two TRC trip-assignment equations can be demonstrated through an example in which a pair of zones, with the vehicular interchanges of 1,200 vph, is connected by two routes whose characteristics are indicated in Table 7.29. Travel on route 1 is slower, but the route is shorter in distance. Critical volumes, corresponding critical travel times, and travel times when no traffic is on each route (the "ideal" travel times) are found from Fig. 7.19 based on the characteristics presented in Table 7.29.

The assignment procedure starts by using Eq. (7.37) in conjunction with the ideal travel times for the entire length of each route. Thus if no traffic were on route 1, the travel time would equal 2.5 min/mile $\times$ 3 miles = 7.5 min, while that for route 2 would be 1.5(4) = 6.0 min, leading to

$$V_1 = \frac{1/7.5}{1/7.5 + 1/6.0} (1,200) = 533 \text{ vph/lane}$$

and

$$V_2 = \frac{1/6.0}{1/7.5 + 1/6.0} (1,200) = 667 \text{ vph/lane}$$

Switching to Eq. (7.36) and using the above values, we find that

$$t_1 = \left[3.0 + \frac{0.5(533 - 600)}{600} \right] 3 = 8.83 \text{ min}$$

Table 7.29 Example Route Characteristics

Route no.	No. of lanes	Speed limit	Signals/ mile	Length	Critical volume	Critical travel time	Ideal travel times
1	1	30 mph	1	3 miles	600 vph/lane	3.0 min/mile	2.5 min/mile
2	1	50 mph	1	4 miles	1100 vph/lane	2.0 min/mile	1.5 min/mile

and

$$t_2 = \left[2.0 + \frac{0.5(667 - 1,100)}{1,100} \right] 4 = 7.21 \text{ min}$$

where in both cases $d = 0.5$ since both route volumes are less than their respective critical volumes. Going back to the previous equations for determining route volumes, and given the above travel times, we obtain

$$V_1 = \frac{1/8.83}{1/8.83 + 1/7.21} (1,200) = 539 \text{ vph/lane}$$

and

$$V_2 = \frac{1/7.21}{1/8.83 + 1/7.21} (1,200) = 661 \text{ vph/lane}$$

Using these results as inputs to the next set of iterations, we discover that

$$t_1 = \left[3.0 + \frac{0.5(539 - 600)}{600} \right] 3 = 8.85 \text{ min}$$

and

$$t_2 = \left[2.0 + \frac{0.5(661 - 1,100)}{1,100} \right] 4 = 7.20 \text{ min}$$

where, again, $d = 0.5$ because the V_r's are less than their respective V_{rc}'s. The last set of travel times do not differ significantly from the previously calculated ones, so that the procedure can be terminated at this point. The final results are: $V_1 = 536$ vph/lane, $V_2 = 661$ vph/lane, $t_1 = 8.85$ min, and $t_2 = 7.20$ min.

The example calculations closed on the accepted values rather quickly. This may not always be the case, especially if the interzonal volume falls within a range of, say, ±300 vph/lane of the sum of the critical volumes, which in this example is 600 + 1,100 = 1,700 vph/lane. The reason for this statement is that, at the indicated values, the d parameter will jump from 0.5 to 10.0 and back, causing corresponding fluctuations in the related travel times.

Another interesting point to note is that the travel times on the two routes in the example are not equal, although they are fairly close. This circumstance, which is a general situation, may seem to contradict the previously stated concept that a driver cannot improve travel time by changing route. However, realistically, most drivers probably do not know exactly what the minimum time path for their trip may be, especially since this path probably changes from moment to moment. Moreover, as brought out before, there are other factors besides travel time that affect route choice, so that a discrepancy in travel times resulting from the use of the two equations, if in the right direction, probably leads to more realistic representations of route choices. As it turns out, tests of the TRC trip-assignment model generally have shown it to be fairly reliable.

The adjustment of interzonal travel times by trip assignment will change the trip distribution base, which uses assumed travel times as input for distributing trips between zones. Therefore, logically, a feedback loop should exist between the end of trip-assignment phase and trip-distribution phase, until the travel times between zones remain stable. This technique was adopted by TRC [7.36].

7.7 THE URBAN TRANSPORTATION PLANNING SYSTEM (UTPS) PACKAGE

The Urban Transportation Planning System (UTPS) is a tool useful for multi-modal transportation planning in a variety of contexts. Particularly valuable for systems analysis, it has been used for both long- and short-range planning. UTPS is comprised of computer programs, attendant documentation, user guides, and manuals covering both computerized and manual planning methods. It supports many types of analyses from detailed network design to highly simplified sketch planning and ad hoc problem solving.

The computer programs are very flexible, user-oriented, and adhere to uniform high quality standards. The Urban Mass Transportation Administration (UMTA) and the Federal Highway Administration (FHWA) have developed this package and are committed to distribute, maintain, and upgrade UTPS on a continuing basis. These standards and this commitment are what make UTPS unique as a source of technical support to transportation planners in the United States and around the world.

Each UTPS computer program relates to one or more of five analytical categories. While each program performs different functions in different categories, they are all compatible and communicate readily through common data bases. The analytical categories are:

A. Highway network analysis   D. Data capture and manipulation
B. Transit network analysis   E. Sketch planning
C. Demand estimation

Highway and transit network analysis pertains to network preparation (coding) for computer input, system costing, and other network evaluation procedures. Demand forecasting subsumes trip generation, trip distribution, and mode split—including both model calibration and forecasting in an aggregate or disaggregate manner. Passenger loading associates the cell values of a matrix of interzonal travel demand (trip table) with the appropriate components of a transportation network.

Additionally, network analysis and demand forecasting require a great deal of data manipulation, which entails calculation, restructuring, modification, and disclosure of the contents of the many household, zonal, link, line, or matrix data sets used in transportation planning, e.g., interviews, link data, land-use data, zone-to-zone trip tables, travel-time matrices, etc. All of these data types, and others, are readily manipulated using UTPS.

The demand estimation category of the UTPS package is composed of five programs: UMODEL, UFIT, UREGRE, AGM, and ULOGIT. *UMODEL* is a software program that accommodates virtually any demand or modal-choice model. It contains a "default" demand model for users who require a first-cut demand estimate without spending the money and time to construct and calibrate their own model. Alternatively, users may insert their own model. It also reads relevant land-use (zonal) data and transportation system characteristics (matrix data), and outputs origin/destination demand matrices with observed trip matrices or disaggregate data and generates statistics useful in model calibration. *UFIT* is a linear least-squares regression program that allows the user to formulate, test, and apply multivariate linear models easily and in a structured manner. *UREGRE* is also a regression program that records either raw data or the condensed output of UMODEL, along with a user-provided set of regression-model specifications, and calculates the linear multiple-regression parameters and goodness of fit estimates. A residual analysis is also available. UREGRE is included in the UTPS only to support users wishing to delay their use of UFIT, which replaces it. *AGM* is the gravity-model calibration and application program. It assists in the "manual calibration" of gravity models, while also providing an automatic calibration feature. *ULOGIT* calibrates models (usually modal-choice models) of the linear logit form. It uses a maximum-likelihood estimation technique so the disaggregate (trip-based) data can be used as input.

To obtain a copy of UTPS, inquire by writing to the Office of Planning Methods and Support, UPM–20, Urban Mass Transportation Administration, U.S. Department of Transportation, Washington, D.C. 20590.

7.8 COMMENTS ON THE FOUR-STEP TRAVEL ESTIMATION PROCEDURE

Several criticisms have been levied against the traditional four-step travel-demand estimation process. Among them are that it is cumbersome, expensive, and requires a large amount of data. It is based on cross-sectional data obtained at one point in time, and the results are reached through iterative processes rather than through direct optimal solutions. The errors at one step are likely to magnify at other steps, since the inputs to one model are the outputs of another. The generation of trips is independent of the transportation supply characteristics and possible technological improvements, and the models are generally site-specific—that is, they are not transferable from one urban area to another.

Despite all those criticisms, the process is still the most used, primarily because it has been well tested and is completely operational. Several modifications and refinements have been carried out on this procedure, such as the inclusion of transit assignment and disaggregate behavioral modal-split models. The transit-assignment techniques are not discussed in this chapter because of their variety and their dependence on the transit network under consideration. Quick-response travel-demand estimation [7.23] is gaining popularity. However, it is not designed or meant to replace the four-step process. Simplified procedures, mainly based on the four-step sequence, have been developed and used in small urban areas to avoid the costs and the time needed to run the large models [7.24]. A fair amount of research work has been conducted in the travel forecasting field, but so far has failed to come up with any radical departure from the set of four models that have been described in this chapter.

7.9 AGGREGATE MODELS

In contrast to the four-step process and to disaggregate models are those that are highly aggregate in nature. These are relatively new and not widely understood or utilized yet. Still, they provide some interesting insights, and so will be discussed here briefly.

One of the first efforts in aggregate modeling has been to investigate average overall travel times, distances, and expenditures [7.26]. As an illustration, Zahavi has gathered a wide variety of evidence to find that average daily time spent in travel (per traveler) is relatively constant over time, between cities, by distance from CBD, and so on.[7] For the whole United States, for instance, the figure in 1970 was 1.06 hours/day · traveler (for auto). For St. Louis it was 1.04 (1976), Washington, D.C. 1.09 and 1.11 (1955 and 1968), Munich 1.16 (1976), and the Twin Cities 1.14 and 1.13 (1958 and 1970).

Average daily travel distance (per traveler), on the other hand, varies with travel speed and, subsequently, with trip distance. The former relationship is shown for zones in the Washington, D.C., metropolitan area in Fig. 7.20. The interesting conclusion that results from these findings is that trip-makers use the potential savings in trip times brought about by increased speeds (i.e., better transportation) to increase the distance traveled, keeping the same travel time.

In a slightly different vein, Zahavi has identified so-called *alpha* (α) *relationships* between traffic intensity (I, in 1,000 vehicle kilometers driven daily in each square kilometer of a sector of a city) and, for that same sector, the space mean speed (v, in km/hr) and the road density (D, in km/sq km).

The α parameter connects these together in

$$I = \alpha \frac{D}{v} \tag{7.38}$$

The value of α has been found to be fairly constant within cities, as demonstrated by the curves for arterial streets (only) for London and Pittsburgh in Fig. 7.21. Although α values vary somewhat between cities, a value of 400,000 for arterial systems is a good, general approximation.

Equation (7.38) can be employed in a variety of ways. As one example, suppose it is proposed to increase the length of arterials in a 400-sq-km area from 450 to 500 km, with average speed expected to rise from 30 to 32 km/hr (because of the better geometrics of the new arterials). Present traffic intensity can be estimated as

[7] There were some differences between modes, however.

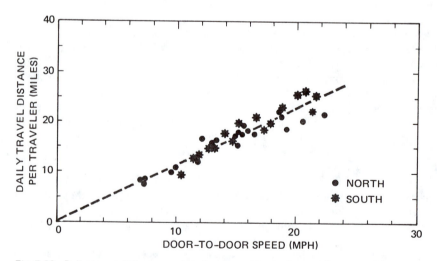

Fig. 7.20 Daily travel distance per traveler, by residence distance from the city center, versus daily mean door-to-door speed, north and south corridors, Washington, D.C., 1968 [7.26, p. 35].

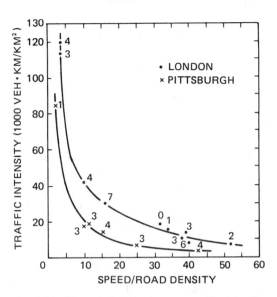

Fig. 7.21 The relationship between traffic intensity and the ratio of speed/road density in London and Pittsburgh [7.39].

$$400,000 \,\frac{450/400}{30} = 15,000 \text{ veh} \cdot \text{km/km}^2/\text{day}$$

This would increase to

$$400,000 \left(\frac{500/400}{32}\right) = 15,625$$

For the 400-sq-km sector, the veh · km of daily traffic thus would change by

$$(15,625 - 15,000)400 = 250,000 \text{ veh} \cdot \text{km/day}$$

The aggregate model in Eq. (7.38) thus presents us with a quick means of estimating gross travel impacts of transportation system (in this case, arterial) changes.

7.10 SUMMARY

The subject of this chapter has been the demand for travel. The most common method for estimating such demand revolves around a four-step process:

1. Trip generation   3. Mode choice
2. Trip distribution   4. Route choice (or assignment)

In the first step, forecasts are made of the number of trips produced by or attracted to each zone. These trips then are "distributed" to other zones based on the relative attractiveness of, and travel times to, those zones. Next, a choice is made of the mode of travel. This choice is seen as a function of the characteristics of the modes (time, distance, cost, etc.) as well as of the travelers (income, purpose, etc.). In the final step the routes between a pair of zones taken by the travelers are identified. At this stage, the relationships between travel volumes and time (i.e., congestion) usually are taken into account.

While new demand-estimation procedures, such as the disaggregate models on the one hand, and on the other the highly aggregated models, have evolved, the four-step process generally has remained the key approach to date.

BIBLIOGRAPHY

7.1   Chicago Area Transportation Study, vol. 1, *Survey Findings*, Chicago, Dec. 1959.

7.2   Pittsburgh Area Transportation Study, vol. 1, *Survey Findings*, Pittsburgh, Nov. 1961.

7.3   Kassoff, H., and H. D. Deutschman: "Trip Generation: A Critical Appraisal," *Highway Research Record* 297, 1969.

7.4   McCarthy, G. M.: "Multiple-Regression Analysis of Household Trip Generation—A Critique," *Highway Research Record* 297, 1969.

7.5   Oi, W. Y., and P. W. Shuldiner: *An Analysis of Urban Travel Demands*, Northwestern University Press, Evanston, Ill., 1962.

7.6   Draper, N. R., and H. Smith: *Applied Regression Analysis*, New York, Wiley, 1968.

7.7   U.S. Department of Transportation, Federal Highway Administration: *Trip Generation Analysis*, Washington, D.C., Aug. 1975.

7.8   U.S. Department of Transportation, Federal Highway Administration: *Guidelines for Trip Generation Analysis*, Washington, D.C., June 1967.

7.9   Fleet, C., J. Stowers, and C. Swerdloff: *Household Trip Production—Results of a Nationwide Survey*, U.S. Bureau of Public Roads, Highway Technical Report 2, Washington, D.C., 1965.

7.10 Fleet, C., and S. Robertson: "Trip Generation in the Transportation Planning Process," *Highway Research Record* 240, 1968.

7.11 Robinson, W. S.: "Ecological Correlations and the Behavior of Individuals," *American Sociological Review*, vol. 15, 1950.

7.12 Wooten, H. J., and G. W. Pick: "A Model for Trips Generated by Households," *Journal of Transport Economics and Policy*, vol. 1, no. 2, May 1967.

7.13 North Atlantic Treaty Organization, Committee on the Challenge of Modern Society: *Travel Forecasting Subproject, Final Report*, Brussels, March 1975.

7.14 U.S. Department of Transportation, Federal Highway Administration: *Modal Split*, Washington, D.C., Oct. 1970.

7.15 U.S. Department of Transportation, Federal Highway Administration: *Traffic Assignment*, Washington, D.C., Aug. 1973.

7.16 Peat, Marwick, Mitchell & Co.: *Implementation of the N-Dimensional Logit Model, Final Report*, Washington, D.C., May 1972.

7.17 Paquette, R. J., N. Ashford, and P. H. Wright: *Transportation Engineering*, Ronald Press, New York, 1972.

7.18 U.S. Department of Transportation, Federal Highway Administration: *Applications of New Travel Demand Forecasting Techniques to Transportation Planning*, Washington, D.C., March 1977.

7.19 McFadden, D.: *The Revealed Preferences of a Government Bureaucracy*, Technical Report No. 17, Institute of International Studies, University of California, Berkeley, 1968.

7.20 Stopher, P. R.: *Transportation Analysis Methods*, Civil Engineering Department, Northwestern University, Evanston, Ill., 1970.

7.21 Koppelman, F. S.: "Prediction with Disaggregate Models: The Aggregation Issue," *Transportation Research Record* 527, 1974.

7.22 Talvitie, A. P.: "Aggregate Travel Demand Analysis with Disaggregate or Aggregate Travel Demand Models," *Transportation Research Forum Proceedings*, vol. 14, no. 1, 1973.

7.23 Sosslau, A. B., et al.: *Quick-Response Urban Travel Estimation Techniques and Transferable Parameters—A User's Guide*, NCHRP Report 187, 1978.

7.24 U.S. Department of Transportation, Federal Highway Administration: *Transportation Planning for Your Community* (series of several manuals), Washington, D.C., 1980.

7.25 U.S. Department of Transportation, Federal Highway Administration: *Traveler Response to Transportation System Changes*, 2nd ed., Washington, D.C., July 1981.

7.26 Zahavi, Y.: *The UMOT Project*, U.S. Department of Transportation, Research and Special Programs Administration, Washington, D.C., 1979.

7.27 Baerwald, J. E. (ed.): *Traffic Engineering Handbook*, 3rd ed., Institute of Traffic Engineers, Washington, D.C., 1965.

7.28 Martin, B. V., F. W. Memmott, and A. J. Bone: *Principles and Techniques for Predicting Future Demand for Urban Area Transportation*, MIT, Department of Civil Engineering, Cambridge, Mass., Jan. 1963.

7.29 Keefer, L. E.: *Urban Travel Patterns for Airports, Shopping Centers, and Industrial Plants*, NCHRP Report 24, Highway Research Board, Washington, D.C., 1966.

7.30 Kanwit, E. L., C. A. Steele, and T. R. Todd: "Need We Fail in Forecasting?", *Highway Research Board Bulletin 257*, 1960.

7.31 Kolifrath, M., and P. Shuldiner: "Covariance Analysis of Manufacturing Trip Generation," *Highway Research Record 165*, 1967.

7.32 Shuldiner, P. W.: "Land Use, Activity and Non-Residential Trip Generation," *Highway Research Board Bulletin 347*, 1966.

7.33 U.S. Department of Commerce, Bureau of Public Roads: *Calibrating and Testing a Gravity Model for Any Size Urban Area*, Government Printing Office, Washington, D.C., Oct. 1965.

7.34 Hill, D. M., and H. G. Von Cube: "Development of a Model for Forecasting Travel Mode Choice in Urban Areas," *Highway Research Record 38*, 1963.

7.35 Dreyfus, S. E.: *An Appraisal of Some Shortest Path Algorithms*, Memorandum RM–5433–PR, The Rand Corporation, Santa Monica, Calif., Oct. 1967.

7.36 Irwin, N. A., and H. G. Von Cube: "Capacity Restraint in Multi-Travel Model Assignment Programs," *Highway Research Board Bulletin 345*, 1962.

7.37 Meier, H.: *Accessible Public Transit*, United Cerebral Palsy Association of San Francisco, San Francisco, 1981.

7.38 Grey Advertising, Inc.: *Summary Report of Data from the National Survey of Transportation for Handicapped People*, U.S. Department of Transportation, Washington, D.C., 1978.

7.39 Zahavi, Y.: *A Method for Rapid Estimation of Urban Transport Needs*, Research Group in Traffic Studies, University College London, London, 1973.

EXERCISES

7.1 Using the data as given in Table 7.30, develop and interpret a regression model for estimating trips per dwelling unit. (Force all the variables into the model.) *Do all calculations necessary for complete interpretation.*

Table 7.30 Data for Trip-Generation Analysis

One	Trip/D.U.	Family income	Single family	Combination	Apartment complex	Car/D.U.	Household size
1	4.2	5,000	0	0	1	1.5	2.7
2	3.8	3,400	0	0	1	1.2	2.3
3	3.9	5,500	0	0	1	1.4	2.3
4	4.5	6,800	0	1	0	1.5	2.8
5	5.6	9,100	0	1	0	1.8	2.5
6	5.8	6,600	1	0	0	1.9	2.7
7	4.4	7,500	0	1	0	1.4	2.3
8	7.0	8,500	1	0	0	2.2	3.1
9	5.9	8,600	1	0	0	1.9	2.8
10	3.1	4,200	0	0	1	1.2	2.3
11	4.6	5,600	0	1	0	1.6	2.4
12	6.6	6,800	1	0	0	2.1	3.0
13	6.1	7,600	1	0	0	1.9	2.8

Residential type (spanning Single family, Combination, Apartment complex)

Table 7.30 Data for Trip-Generation Analysis (*Continued*)

One	Trip/D.U.	Family income	Single family	Combination	Apartment complex	Car/D.U.	Household size
				Residential type			
14	3.5	4,400	0	0	1	1.2	2.3
15	5.0	6,400	0	1	0	1.6	3.0
16	8.1	10,000	1	0	0	2.5	3.8
17	2.5	3,000	0	0	1	1.0	1.5
18	5.7	7,300	0	1	0	1.7	2.5
19	8.5	11,000	1	0	0	2.6	4.2
20	11.0	12,600	1	0	0	3.0	4.8
21	4.9	6,500	0	1	0	1.5	2.7
22	10.5	12,500	1	0	0	2.8	4.5
23	1.5	1,800	0	0	1	0.8	1.6
24	7.2	9,100	1	0	0	2.2	3.3
25	9.6	9,800	1	0	0	2.8	4.3
26	8.3	9,500	1	0	0	2.6	4.0
27	4.4	3,600	0	0	1	1.4	2.4
28	8.2	8,100	1	0	0	2.5	4.0
29	9.3	10,500	1	0	0	2.8	4.3
30	5.0	6,400	0	1	0	1.6	3.0
31	4.5	6,700	0	1	0	1.5	2.3
32	6.0	8,500	1	0	0	1.8	2.7
33	8.5	10,400	1	0	0	2.6	4.0
34	3.6	5,200	0	0	1	1.2	1.8
35	7.8	11,300	1	0	0	2.4	3.8
36	5.8	8,500	0	1	0	1.9	1.9
37	4.7	6,600	0	0	1	1.5	2.7
38	7.6	11,300	1	0	0	2.4	3.8
39	4.2	4,300	0	0	1	1.4	2.1
40	4.3	5,200	0	0	1	1.4	2.1
41	8.5	11,200	1	0	0	2.6	4.0
42	6.5	8,500	1	0	0	2.1	3.4
43	4.7	6,200	0	1	0	1.6	2.5
44	2.7	3,500	0	0	1	1.0	1.5
45	1.6	2,500	0	0	1	0.8	1.5
46	2.7	3,600	0	0	1	1.0	1.5
47	1.8	2,500	0	0	1	0.9	1.5
48	2.9	3,800	0	0	1	1.0	1.6
49	1.5	1,700	0	0	1	0.8	1.5
50	3.0	3,600	0	0	1	1.0	1.8

7.2 Plot family income, cars per dwelling unit, and household size against trips per dwelling unit. Interpret for linearity.

7.3 Review the correlation matrix and check for collinearity (correlation between the independent variables).

7.4 With your knowledge from 7.2 and 7.3, determine the "best" model for estimating trips per dwelling unit. (Do not force any variables into the model.) You may or may not want to use dummy (0–1) variables. You may also desire to regroup into different classes. Your requirement is to determine the "best" model.

7.5 Prepare a write-up for each of the above parts to be handed in with the problem. A brief description of your logical analysis and interpretation of each part will be sufficient.

7.6 Given the data below on three zones, use the gravity model to distribute the forecasted trip productions and attractions among the three zones. Calibrate the model first, using the base-year data.

	Base Year							
Travel-time matrix (min)				Trip-interchange matrix				
	To zone				To zone			
From zone	1	2	3	From zone	1	2	3	P_i
1	3	8	11	1	150	400	250	800
2	7	4	14	2	350	100	250	700
3	12	13	6	3	400	350	200	950
				A_j	900	850	700	

	Horizon Year						
Travel-time matrix (min)				Forecasted trip attractions and productions			
	To zone				Zone		
From zone	1	2	3		1	2	3
1	2	8	12	Productions	1,500	1,000	1,200
2	6	3	11	Attractions	1,300	700	1,700
3	14	10	5				

7.7 From the text references, choose a model-split model or a traffic-assignment model and critique it. Describe the main features of the model (objectives, formulation, inputs, outputs, etc.) and then provide a critical appraisal of the model by discussing its advantages and disadvantages. The critique should not be more than 6 pages.

7.8 Traffic on 300 km of arterials in a 250-sq-km section in an urban area runs at about 35 km/hr. If 50 km of arterials are closed to make pedestrian malls, what will be the resultant decrease in daily vehicle · kilometers of travel in the section? (Assume that $\alpha = 400,000$ and the speeds remain the same.)

8 Models IV: Transportation System Impacts

8.1 OVERVIEW OF IMPACT STUDIES

One of the first steps in any impact analysis is to identify likely impacts and determine whether they are significant. The initial part of this step usually is accomplished using some sort of checklist procedure such as that outlined in Sec. 4.3. There, goals were divided into four major categories:

1. Humans (and groups)
2. Natural environment
3. Artificial environment
4. Activity elements and agents

A set of 49 characteristics then were cross-tabulated with these in an effort to determine relevant "performance characteristics," which here might be interpreted as "impacts." This concept might even be taken a step further by dividing these impacts according to the area (urban or rural), location (corridor, community or system, or region or nation), and timing (before, during, or after construction or implementation).

Most agencies have their own checklists that must be considered in the impact statement. As an illustration, rules promulgated by the Federal Highway Administration in 1974 indicated that the impacts shown in Table 4.8 should be taken into account. The comment also was made that highways may stimulate or induce other *secondary* or *indirect* actions, such as more rapid land development, and that these could turn out to be even more important than the direct impacts listed above.

Several types of impact analysis can be undertaken, depending on the level in the policy chain (strategy, policy, program or system, and project). The latter ones would depend to some extent on the former, of course. In any case it is necessary to specify which impacts are long-term and/or irreversible or unavoidable (at least at any reasonable cost).

Land-use impacts generally would be long-term in nature. A particular concern in the late 1970s, for example, was for the long-term decline of central business district commercial enterprises. For a short time Federal agencies were required to prepare "community impact analyses" if a local official felt that, say, construction of a belt-way might lead to development of large, new shopping malls in the suburbs that would create economic hardships for core-city stores and residents.

"Irreversible" impacts are those whose directions cannot be changed. Several scientists have hypothesized, for example, that the burning of fuel in internal combustion engines has helped to increase the amount of carbon dioxide in the upper atmosphere. This buildup is such that, even if further combustion were halted, the temperature of the earth will continue to rise. This result (if found true) would exemplify an "irreversible" impact, at least in the foreseeable future. Another such impact would be if the sole habitat of an endangered species were wiped out, thereby effectively eliminating that species.

"Unavoidable" or "unsolvable" effects are ones that are built into the technology in such a way that only massive expenditures could keep them from occurring. While it is possible to reduce the noise near a highway through various treatments (e.g., extended grassy plots or baffled walls), the problem cannot be eliminated completely, and so it is an "unavoidable" impact.

Even though certain impacts may create long-term, irreversible, and unavoidable problems, they still may not be significant enough to create concern. The determination of "significance" is far from easy. It requires much professional and political judgment. It also requires an infusion of citizen participation to make local values known. Techniques for helping to gain such participation are discussed in Chap. 11.

This chapter contains a brief overview and survey of impacts and six examples of impact "models" and analyses. The examples have been chosen to represent certain types of economic, man-built (land use), social, and environmental impacts and, to a small degree, to demonstrate some analytic techniques involved. In the final section, BART system impact case studies are employed as a summarizing mechanism.

8.2 SOME TRANSPORTATION SYSTEM IMPACT CONTROVERSIES

One of the easiest ways to identify and anticipate certain transportation system impacts is to locate relevant articles in a good newspaper. Most of the impacts discussed are negative in nature, yet they still have to be considered. Thomas and Schofer [8.43] have gone through newspaper clippings kept by the Bureau of Public Roads (now the Federal Highway Administration) and have produced a summary of many controversial impacts. This is presented in part below:

Highway planners are often tempted to build roads through what often is the only vacant land available—parks and recreation areas. Many groups are legitimately and rightly interested in preserving these open spaces for the use of city dwellers; thus recreation facilities and other natural resources have often become hotly contested issues in transportation disputes. Citizens of Westchester County, N.Y., were concerned about the many acres of wildlife sanctuaries that would be taken in their county during the construction of I-87. In another case, the mayor of Louisville, Ky., in 1966 went to the extreme of blocking construction of I-64 through Seneca and Cherokee parks even though the highway was already under construction on both sides of the parks. . . .

This conflict illustrates well the dilemma of planners who attempt to route a highway through an urban area. Frequently, the alternative to building a road through a park is routing it through a residential or business section. This necessitates the taking of privately owned property, however, and is almost certain to arouse protest. In New York City, the Cross-

Brooklyn Expressway was protested by the residents of Flatbush because it would displace 1,042 families. . . .

In other cases, citizen's groups have protested that land taking would have a deleterious effect on important historical monuments. Local officials and residents of Morristown, N.J., for example, contested the location selected for I-287, because it would be too close to George Washington's Revolutionary War Headquarters. In Philadelphia, the Committee for the Preservation of Philadelphia's Historic Southwark charged that 131 historic homes certified by the Philadelphia Historical Commission would be destroyed unless the route of the Delaware Expressway were changed.

Some cities have attempted to avoid the problem of direct taking of properties for facility construction by building elevated transportation links. This solution usually has not met with unqualified approval either. Protest has usually centered around the question of aesthetics, another aspect of concomitant outputs. Elevated expressways are often considered to be visually unappealing, and the memory of older elevated facilities serves to intensify this feeling. A famous example of this sort of controversy is that of the Riverfront Expressway in New Orleans. The proponents of the plan claimed that elevating the highway would avoid the adverse effect on the economy of the area which taking additional property would produce, and would provide an excellent flood-prevention facility. Opponents countered with the charges that the expressway would be noisy, would shut out light and air, and, most important, would blight and deteriorate the character of the French Quarter, through which it would run. . . .

Some of the most intense recent controversies have been associated with transportation choices where planned facilities were to pass through economically and socially disadvantaged neighborhoods. The principal concern in many of these cases has been the planned destruction of dwelling units and community facilities, although many of the structures were already substandard. The inadequate supply of low cost housing in the affected areas has been more than a contributing factor in these conflicts; it has often been the central problem. The transportation facility, however, has become the focus of controversy, perhaps because of the high visibility of the transportation system and its planning processes. That is, the housing program, if relatively inactive, is not a well-known function that can draw controversy. When transportation plans become known, and when construction begins, considerable attention may result, and the emotions of the community may be directed towards transportation.

8.3 ENERGY CONSUMPTION IMPACTS

Rapid rises in the price of oil during the 1970s have resulted in a great deal of emphasis on fuel conservation. Attention has not been restricted to petroleum-based products, however. This is caused by two major factors—(1) other sources, such as electricity, are generated in part using petroleum, and (2) there have been associated price rises in the other sources, partially because they have been restricted in their expansion. This holds for nuclear power plants, hydroelectric dams, coal from strip-mining, and so on. Much emphasis thus has been given to conservation.

Energy consumption can be measured in several different ways. The three most common are:

Gallon of fuel (gasoline, diesel, etc.). Another measure is a barrel, which contains 42 gallons.

British thermal unit (Btu). The amount of heat needed to raise one pound of water 1°F.

Kilowatt · hour (kWh). Use of 1,000 watts (amps × volts) of power over a 1-hr period.

This last measure is probably the most difficult to understand, but can be exemplified simply as the burning of a 100-W light-bulb for 10 hr.

The second measure is the most general. The other two would be related as

1 gal gasoline = 125,000 Btu   1 kWh = 3,412 Btu

8.3.1 Comparison of Average Modal Consumption Rates

Any attempt to compare different modes in terms of energy consumption rates is difficult, since many factors can affect such rates. This most reasonable comparison, for example, might be on a Btu per passenger • mile basis. But to obtain such a figure for a given mode requires (1) that the amount of energy be translated into Btu, perhaps using the above ratios, and (2) that average passenger • mile data be available, which is not always the case, particularly for transit.

Table 8.1 has been constructed bearing in mind these complications. Note that the figures presented are only for propulsion energy, and then only for the line-haul portion of the trip. A fuller analysis would take into account energy consumption for stations and maintenance, guideway construction, and vehicle manufacturing. The

Table 8.1 Average Propulsion Energy Efficiency for Passenger Transportation Modes

Mode	Years	Vehicle • miles per gallon	Passenger • miles per vehicle • mile	Passenger • miles per gallon	BTU per passenger • mile
Auto					
Overall	1970–1972	10.7–11.1	1.4–1.9	15–28	5,600–8,100
	1979	14.29	1.6*	22.9	5,500*
Small					
Work	1974	13.5	1.6	21.7	5,770
Shop	1974	18.0	2.3	41.4	3,020
Soc-rec	1974	26.8	2.8	74.9	1,670
Standard					
Work	1974	9.8	1.6	15.7	7,970
Shop	1974	9.0	2.3	20.7	6,040
Soc-rec	1974	15.1	2.8	42.2	2,970
Car pool	1976	11.4	3.0	34.2	3,670
Van pool	1976	8.9	9.0	80.4	1,560
Taxi			0.7	8.0	15,600
Truck					
Overall	1979	8.82	1.2*	10.6	11,800*
6-Ton single unit	1974	7	1.2*	8.4	14,810
20-Ton gasoline	1974	2.4	1.2*	2.9	43,103
25-Ton diesel	1974	3.5	1.2*	4.2	29,762
Bus					
Urban	1970–1972	4.0	10–13	48–51	2,680–3,700
School	1970–1972	7.4	22.7*	168	740–1,100
Dial-a-Bus	1976	8.1	1.6	13	9,690
Rail					
Rapid	1970–1972	–	21–24	–	1,650–4,300
Commuter	1972	–	40	–	2,490
Light	1976–1977	–	20	–	3,750
Bicycle	1970–1974	–	1	–	100–200
Walking	1972–1974	–	1	–	300–500

*Estimated by author.
Sources: [8.12, 8.13, 8.14, which also reference various other sources].

Congressional Budget Office (CBO) estimates, for example, that these latter elements amount to about 29% of average auto propulsion energy, 16% of conventional bus, and 27% of new heavy rail [8.13, pp. 32–33]. Also, car pools, van pools, and almost all transit operations involve a certain degree of circuity in trips compared to most auto travel, which usually is fairly direct from origin to destination. For comparison's sake, then, the line-haul propulsion figures in Table 8.1 should be adjusted to take into account this circuity, which the CBO estimates to be almost 1.25 times the average auto trip length for car pools, 1.40 for conventional bus, and 1.20 for new heavy rail [8.13, pp. 32–33].

A third consideration is that many person • trips involve more than one mode. At the least, people have to walk to and from the car or bus. In larger urban areas, trips may also involve transfers from car or bus to rail or another bus. Naturally, it is the energy consumption of actual person travel rather than just the line-haul portion that is of concern to planners and decision makers.

The indications from Table 8.1 are that bicycle trip-making is the least energy-consumptive, with walking being second. Far distant is any motorized mode. The implications thus are that, from a strict energy perspective, emphasis should be given to sidewalks, bikeways, and similar alternatives that encourage trip-making in these fashions.

The worst mode is truck, although most trucks obviously are driven to haul freight rather than passengers. Taxis and dial-a-bus systems are the worst for hauling only passengers. These results are due primarily to their very low occupancy rates (passenger • miles per vehicle • mile). The average dial-a-bus figure of 1.6 may be surprisingly low, but these systems do produce a collection service that is highly personalized, and thus have low average ridership throughout the vehicle trip.

Bus and rail systems have roughly the same ranges of energy efficiencies, which are 2 to 3 times better than the average automobile. Efficiencies for both transit modes would be higher if average occupancies were closer to vehicle seat availabilities. It must be remembered, however, that occupancy, while generally substantial in center-city portions of a transit trip, approaches zero at the other end. The average thus is less than might be assumed from observations of loaded buses in a downtown area.

8.3.2 Trends in Energy Consumption

Changes in auto and truck fuel efficiencies were discussed in Chap. 6, Sec. 6.1.2. These appear to be on the rise (2.5% per year for autos and 0.7% for trucks), with likely faster increases in the 1980s.

The trends for bus and rail are not as distinct, although there was a slight *increase* in energy consumption per bus mile in the late 1970s (Fig. 8.1). This may be due in part to recent requirements for "kneeling" buses, lowered floors, and wheel-chair lifts and special seating for the handicapped. These tend to increase weight and reduce overall passenger space.

8.3.3 Factors Affecting Consumption

Many factors can influence the rate of energy consumption. These include characteristics of the driver or operator, type of propulsion system, weight of the vehicle, weather, guideway geometrics and conditions, and the like. As noted in Chap. 6, metropolitan governments have little control over most of these except the guideway geometrics and types of transit vehicles purchased, so that transportation planners tend to concentrate on these factors.

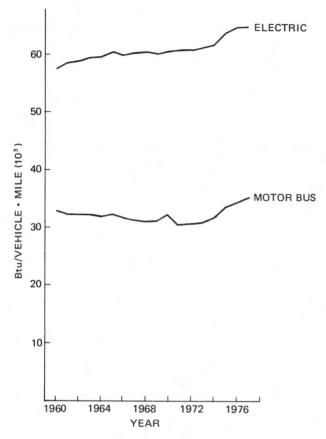

Fig. 8.1 Trends in energy consumption per vehicle · mile on public transportation services [8.14, p. 81].

Because auto and truck efficiencies were discussed in Chap. 6, our attention here will focus on transit. Bus energy efficiencies, like those for autos and trucks, are affected by grades and speeds, but a more useful measure than the latter is stops per mile. The relationship of fuel consumption to this factor is displayed in Fig. 8.2. While there is a range of values for any given number of stops, it can be seen that the gallons per mile at 12 stops is 3 to 4 times larger than for no stops.

Adjustments for buses on grades can be made as follows:

Grade (%)	−6	−4	−2	0	2	4	6	8	10
Adjustment	0.5	0.5	0.5	1.00	1.63	2.75	4.00	5.25	6.38

These should be multiplied by the fuel-consumption figure obtained from Fig. 8.2. For example, if a bus averages 4 stops per mile on a route with a general uphill grade of 2%, the gallons per mile (gpm) fuel consumption would be estimated (taking the middle value in the range in Fig. 8.2) as

$$(0.23 \text{ gpm})(1.63) = 0.37 \text{ gpm}$$

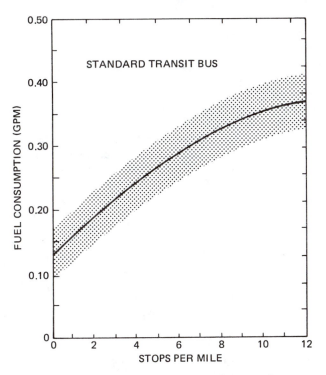

Fig. 8.2 All diesel fuel consumed versus bus stop frequency [8.14, p. A-72].

8.3.4 Forecasting Future Energy Consumption

The information in Tables 6.3 to 6.6 and Fig. 8.2 make it possible to estimate direct fuel consumption for vehicles that are the same as the ones on which the tests were made to obtain the data. The composite vehicles employed by Claffey to get the information had average model years shown in Table 8.2 [8.8, pp. 5–7].

For projections of fuel consumption in the future, a prediction first must be generated of the *fleet* gallons per mile expected at that date. Then, to use the information

Table 8.2 Composite Vehicles: Nationwide Fleet Average Fuel Consumption

Vehicle type*	Model year	gpm
Passenger car	1965	0.072
Two-axle six-tire truck	1964	0.115
Bus (diesel)	—†	0.250

*Data for other types of trucks also can be found in Claffey [8.8].

†It is not clear what model year buses were employed to obtain Fig. 8.2. It has been assumed that their mean gallons per mile was 0.250 (see Table 8.1) in 1975.

in Tables 6.3 to 6.6 and Fig. 8.2, the assumption must be made that the performance characteristics of the future vehicles are proportional to those in the tables and figure. If, for instance, overall auto fuel efficiency in the future is expected to be twice as great as in 1965, then this same ratio is assumed to hold for consumption at different speeds, grades, stops per mile, etc. It also might be necessary to assume that the national efficiency figures hold for the urban area being analyzed, since data specific to individual cities rarely are available.

To illustrate the resulting process for making fuel-consumption forecasts, let us imagine that it is desired to determine the equivalent Btu of fuel consumed by outgoing peak-hour traffic on a single street radiating from the CBD in a medium-sized city in 1990. The street being studied has six lanes, with parking on both sides. Drivers attempt to reach speeds of 25 mph, but autos and trucks average 2 stops per mile over the 2-mile street section. Buses average 8 stops per mile while they pick up and discharge passengers. It has been predicted that there will be 240 autos, 48 trucks, and 10 buses in the outgoing peak hour traffic stream. The street has a 1% uphill grade.

Calculations can be divided into three parts, one for each of the modes—passenger car, truck, and bus. For the first, Table 6.3 indicates that the fuel consumption on a 1% upgrade at a uniform speed of 25 mph can be interpolated to be 0.054 gpm. This, however, is for Claffey's composite 1965 auto. An approximate corresponding figure for an average auto in the 1990 fleet can be obtained by, first, forecasting the overall fleet gpm in that year and, second, assuming a proportionality with the 1965 fleet value.

Extrapolation in Fig. 6.2 would lead to a 1990 auto fleet mpg of about 28.00, or 0.036 gpm. The corresponding figure for 1965, as noted in the chart above, is 0.072 gpm. Since the former is half the latter, it might be assumed that the 0.054-gpm figure above also will be proportionally halved—to 0.027 gpm. As a generality, then, the base consumption, C, in some horizon year, $t + \Delta t$, can be calculated as

$$C(t + \Delta t) = C(t) \frac{\text{gpm}(t + \Delta t)}{\text{gpm}(t)}$$ (8.1)

where $C(t)$ = consumption in year t, for which data are available
     $\text{gpm}(t)$ = average gallons per mile of the fleet in the data year
 $\text{gpm}(t + \Delta t)$ = forecasted average gallons per mile of the fleet in the horizon year, $t + \Delta t$

The last step for passenger cars then is to apply the correction factor for stops and traffic volume, as found in Table 6.4. Note that the volume is composed of the 240 autos + 48 trucks + 10 buses = 298 vehicles per hour. With 2 stops per mile, this leads to a correction factor of 1.49, giving a new gpm of 1.49(0.027) = 0.040. This, when multiplied by the 2 miles of travel and the 240 autos, gives a consumption of

(240 auto trips/hr)(2 miles/trip)(0.040 gal/mile) = 19.20 gal/hr

which equals

(19.20 gal/hr)(125,000 Btu/gal) = 2.4 million (m) Btu

The corresponding values for trucks can be found through the use of Tables 6.5 and 6.6. The extrapolated 1990 fleet gpm value is taken to be 0.063 (mpg = 16.00) compared to the 1964 level of 0.115. Use of Eq. (8.1) with a figure of 0.087 for

25 mph at a 1% grade leads to a base consumption of

$$0.87 \frac{0.063}{0.115} = 0.047$$

with a correction factor of 1.66 from Table 6.6. Final consumption thus is

(0.047 gal/mile)(1.66)(2 miles/trip)(48 truck trips/hr)(125,000 Btu/gal)

= 0.94m Btu/hr

For buses, an optimistic prediction might be that their fleet gpm will decrease about one-third, from 0.250 in 1975 to 0.170 in 1990. At 8 stops per mile, Fig. 8.2 yields a 1975 gpm of 0.30. This will be reduced in 1990 to

(0.30)(0.17)/(0.25) = 0.20 gpm

With a correction factor of about 1.31 (interpolated from the listing in Sec. 8.3.3), the overall consumption becomes

(0.20 gal/mile)(1.31)(2 miles/trip)(10 bus trips/hr)(125,000 Btu/gal)

= 0.66m Btu/hr

The grand total then is

2.40 + 0.94 + 0.66 = 4.00m Btu/hr

If desired for comparison's sake, this value can be expressed on a per passenger basis for the auto and bus trips. Thus, assuming loadings of 1.5 passenger · miles per auto · mile and 12 passenger · miles per bus · mile, the total number of passenger · miles (pm) is, for the auto,

(1.5 auto pm/vm)(240 auto trips/hr)(2 miles/trip) = 720 pm/hr

and for the bus,

(12 bus pm/vm)(10 bus trips/hr)(2 miles/trip) = 240 pm/hr

for a total of 960 pm/hr. For the 2.40 + 0.66 = 3.06m Btu/hr for these two modes,

(3.06m Btu/hr)/(960 pm/hr) = 3,188 Btu/pm

A more comprehensive comparative analysis would take into account vehicle storage and maintenance energy as well as that for vehicle manufacturing and guideway construction. It also would consider trip circuity.

8.4 AIR-POLLUTION IMPACTS

Air pollutants emitted from the internal combustion engine while either mobile or stationary may have significant adverse effects on air quality both on a regional and a local scale. Evaluation of these impacts, particularly from mobile sources, is a complex problem, due largely to the necessity of defining the temporal and spatial distribution of emitted pollutants, and the interactions of these within the atmosphere.

Transportation sources, including aircraft, trucks, buses, trains, automobiles, and other vehicles, are responsible for on the order of 75% of the carbon monoxide (CO), 60% of the hydrocarbons (HC), and 40% of the oxides of nitrogen (NO_x), emitted annually in the United States [8.16]. This amounts to a total of over 100 million tons per year of combined emissions.

By far the greatest concerns for adverse impacts to the general public are associated with emissions from automotive sources. In addition to the burden of these pollutants on air quality directly, some of them serve as the precursors to the generation of photochemical oxidants, the major species of which is ozone (O_3), and other derivatives that make up photochemical "smog."

Particulate matter, including lead and other compounds derived from gasoline additives, and sulfuric acid and other sulfates resulting from the catalytic oxidation of small amounts of sulfur in gasoline by automotive emission control devices, are receiving increasing attention.

In this section we examine methods of assessment of air-quality impacts due to highway vehicular emissions, since these represent a major portion of transportation-related impacts.

8.4.1 Scope of Impact Analysis

Air-quality impact analyses must be performed on temporal and spatial scales consistent with the protection of health and welfare of the general public. The basic guidelines for these determinations are the National Ambient Air Quality Standards (NAAQS), which have been established by the United States Environmental Protection Agency (EPA) [8.17, p. 8187].

These standards are set forth in terms of ambient concentrations of specific pollutants over specified averaging intervals. Table 8.3 lists the major pollutants for which NAAQS have been promulgated to date. The primary standards are based on health-related criteria, while the secondary ones are related to other adverse effects such as material or plant damage.

In the most basic terms, air quality is related to emissions from stationary and mobile sources. Mobile source emissions can be quantified, and mathematical estimates made of their contribution to ambient pollutant concentrations for any given existing or proposed transportation system.

The impact analysis then becomes a matter of relating the resulting estimate of emissions and/or pollutant concentrations to the appropriate NAAQS. Where predicted ambient concentrations for specific averaging intervals exceed these standards, an adverse impact would be indicated. In some instances, even though a change in the transportation system does not result in a violation of the NAAQS, an adverse impact may be inferred. For example, if the change results in a net increase in emissions, or in resultant concentrations of pollutants, the added pollutant burden may be an unacceptable (and thus an adverse) impact.

Because transportation networks or systems frequently involve large areas, and because air pollutants may be dispersed over similarly large areas, the impact-analysis process must include not only local changes in pollutant concentrations, but also in the general regional "background" level which results from many other sources. It thus becomes necessary to consider air-quality impacts in the microscale region (within 100 m or so) of a specific project and in the mesoscale region (within a radius of

Table 8.3 National Ambient-Air Quality Standards

Pollutant	Primary* standard	Secondary* standard
Particulate matter		
Annual geometric mean	75	60
Maximum 24-hr value[†]	260	150
Lead		
Quarterly arithmetic mean	1.5	1.5
Sulfur oxides		
Annual arithmetic mean	80 (0.03 ppm)	60 (0.02 ppm)
Maximum 24-hr value[†]	365 (0.14 ppm)	1,300 (0.5 ppm)
Carbon monoxide		
Maximum 8-hr value[†]	10 (9 ppm)	Same as primary
Maximum 1-hr value[†]	40 (35 ppm)	
Ozone[‡]		
Expected number of days per year when the 1-hr maximum value exceeds the standard (must be ≤ 1)	235 (0.12 ppm)	Same as primary
Hydrocarbons		
Maximum 3-hr (6–9 a.m.) value[†]	160 (0.24 ppm)	Same as primary
Nitrogen oxides		
Annual arithmetic mean	100 (0.05 ppm)	Same as primary

*All measurements in micrograms per cubic meter ($\mu g/m^3$) except for carbon monoxide, which is expressed in milligrams per cubic meter (mg/m^3). Equivalent measurements in parts per million by volume (ppm) are given for the gaseous pollutants.

[†]Not to be exceeded more than once per year.

[‡]Ozone is the surrogate species for photochemical oxidants, and the standard of compliance is statistically evaluated.

Source: [8.17, p. 8187].

several kilometers) of the project or network. Procedures and data requirements for these two types of assessments are similar, but vary in scope, detail, type, and magnitude.

Impact assessment procedures vary also with the specific pollutant of interest. In general, inert or essentially nonreactive pollutants such as carbon monoxide (CO) are analyzed in both the microscale and mesoscale regions, and the results are compared directly with the NAAQS. However, reactive pollutants such as the oxides of nitrogen (NO_x) and hydrocarbons (HC) are not analyzed in the microscale. This is due to their relatively rapid transformations by chemical reactions after release to their more important role in the subsequent formation of photochemical oxidants at some time and distance well removed from their point of emission. Ozone and other photochemically derived species (secondary pollutants) thus are not subjected to microscale impact assessment.

Both reactive precursor and secondary-generated pollutants may be analyzed in the mesoscale region. Procedures range from relatively simple emission "burden" and "trend" analyses to much more complex analytical and numerical modeling procedures for pollutant concentration determinations. The approach depends upon the size of the transportation system and the potential for violations of the NAAQS from these pollutants.

8.4.2 Predicting Air-Quality Impacts

Procedures for predicting changes in pollutant concentrations that would occur for a given set of conditions vary with the magnitude, location, and/or significance of the project or with the stage of network or system planning. Microscale or mesoscale procedures also differ from each other. There are, however, common elements in the methods, and these will be recognized from the following discussions.

8.4.2.1 Input data requirements There are certain basic types of data which will be required to assess potential air quality impacts. Among these are:

Vehicular emissions   Meteorology
Traffic-flow characteristics   Ambient air quality
Topography   Land use
Highway design configurations

The first type describe emission rates for each specific pollutant in terms of mass per unit distance per vehicle, generally in grams per vehicle · mile. These emissions are a function of many parameters, such as type of vehicle, speed, mechanical condition, and type of fuel. The most widely used method for determining vehicle emissions is that developed by the U.S. EPA, based upon standardized test procedures [8.18]. Variables such as vehicle age mix, speed, type mix (automobile, diesel and gasoline truck, etc.), ambient temperature, and fraction of vehicles in warmed-up or cold operating condition are utilized to define an average composite emission factor for each specific situation. The procedures are periodically updated, and projections for future-year emissions are based on current Federal Automotive Emission Standards as prescribed by the Clean Air Act Amendments [8.19].

Table 8.4 contains a list of representative emission rates for CO, HC, and NO_x for a national average vehicle mix for 1972 [8.19]. Note that the use of Federal Motor Vehicle Emission Standards for automotive model years subsequent to 1972 will result in substantial reductions in predicted emissions, particularly of carbon monoxide and hydrocarbons.

Regional and local topography, highway design, and regional and local land-use factors influence air quality. For example, valleys or mountains will influence wind speed and direction. These are reflected in the meteorological data inputs to air pollutant models, as well as background or ambient air-quality levels. Highways may be designed for cut or elevated sections, which in turn exhibit different pollutant dispersion characteristics. Likewise, existing and future land uses will influence the choice

Table 8.4 Average Emission Factors for Highway Vehicles, Calendar Year 1972

Vehicle mix, light/heavy duty	Average route speed	Ambient temperature (°F)	% Cold operation	Emission factor (g/vehicle · mile)		
				CO	HC	NO_x
National	19.6	75	20	76.5	10.8	4.9
Average	45	75	0	29.8	4.7	8.0

Source: [8.19].

of vehicles, operating characteristics, distance traveled, and the like, which in turn will markedly influence air-quality impacts.

Because air pollutants and precursors are emitted from moving and stationary vehicles, it is necessary to define certain traffic flow characteristics in order to estimate air-quality impacts. In particular, the spatial and temporal distribution of traffic throughout the region is required in order to perform a mesoscale analysis. In addition, vehicle speed and other operating parameters are necessary to compute emission factors. Depending on the level of analysis required, the data may range from average speed and daily vehicle · miles traveled in the region, to speed, hourly traffic volume, and operating mode on a particular route segment.

Traffic-flow characteristics and emission factors can be combined to produce a "source strength" or pollutant emission rate for each pollutant. This source strength is an important input parameter, along with meteorological factors, in determining pollutant concentrations that may result from the transportation system or project. These concentrations, when added to existing or future projected background levels of that pollutant from other sources, will provide the necessary measure for comparison with the NAAQS.

The next most important variables that influence pollutant concentrations are meteorological dispersion factors such as wind speed and direction and degree of atmospheric stability. These are difficult to predict, in general, but are estimated on the basis of historical trends.

Winds transport and disperse pollutants and allow chemical and physical interactions to cause changes in concentrations and chemical species. Large-scale or synoptic weather patterns, solar radiation and other phenomena, as well as regional or local-scale patterns and interactions, may have to be determined in detail for a complete air-quality analysis. Again, the type, quantity, and detail of the meteorological input will vary considerably with the level of analysis required for the system or project.

8.4.2.2 Microscale modeling The primary purpose of microscale modeling is to determine the influence of a proposed project (highway section, intersection, bridge, etc.) on the air quality in the area within approximately 0.3 km of the project. Past that point the influence of a single project on the ambient air quality probably will be negligible. Carbon monoxide is the major air pollutant analyzed, although particulate matter such as lead and sulfur compounds recently have been investigated [8.20, 8.21, 8.22]. Models of the Gaussian type are the most widely accepted and best validated currently available. Gaussian equations as described by Turner [8.24] were developed to describe the dispersion of an inert pollutant such as CO from a point source with a constant emission rate. It is assumed that the concentrations of pollutants follow a normal distribution in the horizontal and vertical directions. One of the more widely accepted versions is CALINE 3, developed by the California Department of Transportation. The model is based on the additional assumption that emissions from vehicles on the highway are uniformly mixed in a "mixing zone" surrounding the highway. Empirical data from a number of highway pollutant monitoring studies were used to develop the model and validate predicted results.

CALINE-3 features multiple-link and multiple-source capabilities for up to 20 receptors at varying distances up to 150 meters from the roadway. It can be used for at-grade, cut, and elevated sections, as well as intersections. Provided deposition and settling velocities are known, the model has built-in capabilities for the prediction of particulate-matter dispersion from the line sources. The graphical solution technique [8.26] for CALINE-3, to be illustrated below, is only intended for a rough-cut analysis to provide a quick estimate of CO concentrations. For more complex situations, the computerized or programmable calculator version should be used [8.25]. For complete tables and charts necessary to utilize the graphical solution technique, the reader is referred to the solution manual [8.26].

Assumed initially in the graphical technique are the following:

1. The roadway is straight, level, at-grade, 30 m wide, consists of four lanes of pavement with a median of less than 10 m, and is a segment 0.5 km long.
2. The receptor is located at a height of 1.8 m at a receptor distance measured along a perpendicular line from the roadway centerline.
3. The unadjusted traffic volume is 1,000 vehicles per lane per hour (4,000 vehicles per hour total).
4. Ambient or background CO concentration is 0 ppm, mixing height is 1,000 m, surface roughness is 10 cm, and the averaging time for the calculation is 1 hr.
5. The wind speed is 1 m/sec, with a wind direction oriented at an angle with respect to the roadway such that a crosswind is 90° and a parallel wind is 0°.

The graphical procedure permits limited adjustments to these basic, highly restrictive assumptions to fit the situation under analysis. The actual traffic volume relative to 4,000 vehicles per hour in the assumed case can be used to adjust the corresponding CO concentration linearly. One may also adjust the CO concentration for wind speed using a linear relationship, since concentration is inversely proportional to wind speed. The user must select a stability class and wind angle from the available set in the graphical procedure. Classes D (neutral), E (slightly stable), or F (very stable) are available, and wind-angle charts for these stability categories include those for 0, 10, 20, 30, 45, and 90°.

The procedure is as follows:

1. Select the stability class of atmosphere and the wind angle to be used in the analyses and find the appropriate nomograph.
2. Obtain the appropriate CO exhaust composite vehicle emission factor utilizing the EPA emission-factor estimation procedure.
3. Determine the receptor distance, in meters, from the center line of the roadway, perpendicular to the link.
4. Determine the unadjusted CO concentration for a receptor distance shown in Fig. 8.3 or 8.4 by reading horizontally across from the composite emission factor to that distance and vertically down to the CO concentration. Use linear interpolation for receptor distances not explicitly shown.
5. Adjust the CO concentration for actual traffic conditions:

$$CO_I = \frac{V}{4,000} CO_C \qquad (8.2)$$

where CO_I = intermediate CO concentration
V = actual traffic volume (compared to the assumed 4,000 vehicles/hr)
CO_C = unadjusted CO concentration read from the chart in step 4
6. Adjust the final concentration for wind speed[1] by the ratio

$$CO = (1/WS)CO_I \qquad (8.3)$$

[1] A wind speed of less than 1 m/sec is not allowed.

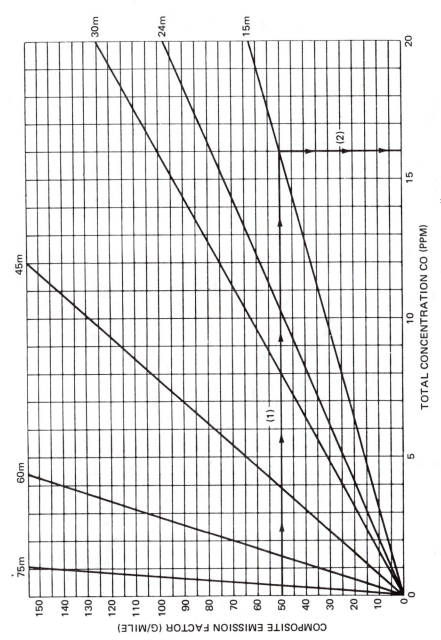

Fig. 8.3 Nomograph for determining unadjusted CO concentration at various receptor distances (m) (stability class F; wind angle at 10°) [8.26].

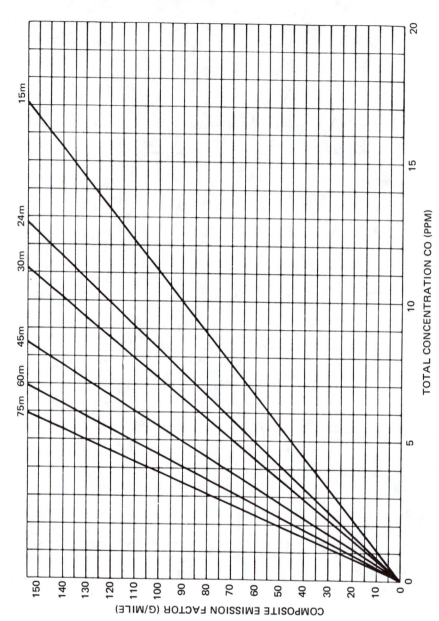

Fig. 8.4 Nomograph for determining unadjusted CO concentration at various receptor distances (m) (stability class D; wind angle of 90°) [8.26].

where CO = final CO concentration
WS = actual wind speed (compared to the assumed 1 m/sec)
CO_1 = intermediate CO value from step 5

To illustrate the process, suppose a potential receptor is located 15 m from a roadway with a traffic volume of 10,000 vehicles per hour. The EPA composite emission factor for CO is 50 g/mile, and the wind speed is 2 m/sec. The wind angle is $10°$ with respect to the roadway, and the stability class is F, very stable.

Step 1. The appropriate graph for a wind angle of $10°$ and a stability class of F is Fig. 8.3.

Step 2. The CO composite emission factor is given as 50 g/mile.

Step 3. The receptor distance is given as 15 m.

Step 4. Reading horizontally across Fig. 8.3 from a composite emission factor of 5 g/mile to the receptor line for 15 m and then down vertically, we obtain the unadjusted CO total concentration of 16 ppm.

Step 5. Adjusting the CO concentration for traffic conditions of 10,000 vehicles per hour via Eq. (8.2) gives

(10,000/4,000) 16 ppm = 40 ppm

Step 6. Adjusting the CO concentration for wind speed of 3 m/sec via Eq. (8.3) yields

(1/3) 40 ppm = 13.3 ppm

For comparison with the NAAQS, it is necessary to assess both the maximum 1-hr CO concentration and the average maximum 8-hr concentration. By assuming a set of "worst case" conditions of low wind speed, limited atmospheric mixing, and peak traffic flow, we can conservatively estimate the "worst case" 1-hr value. However, since steady conditions rarely persist for periods of as much as 8 hr, the average maximum value for the span (CO_8) is more difficult to predict.

If sufficient traffic, meteorological, and air-quality data are available, the preferred and most accurate method for calculating CO_8 would be through an hour-by-hour running average analysis. Since these data are rarely available, a simplified technique may be used. This utilizes the estimated 1-hr maximum concentration, CO, through

$$CO_8 = (V_8/V_1)(CO)(P) \tag{8.4}$$

where V_8 = average hourly traffic volume in both directions during the 8-hr period
V_1 = peak hourly traffic volume in both directions
P = 1-hr to 8-hr meteorological persistence factor for an 8-hr period
The EPA suggests that a P value of 0.6 be used if local data are not available [8.29, p. 32].

8.4.2.3 Mesoscale modeling
Mesoscale analyses generally are carried out in regions from about 0.3 km to up to 20 km or further from the facility or network, depending upon the particular situation. Selection of the procedure for mesoscale modeling varies not only with the location, complexity, and size of the traffic network, but also with the specific pollutants of interest, and whether or not they are reactive.

Table 8.5 is a matrix of mesoscale models for various types of analysis, ranging

Table 8.5 Mesoscale Models

Level of analysis	Pollutant			Program submodels	
	CO	HC, NO$_x$	O$_3$	Input	Output
I Manual	Regional "burden" trends	Regional "burden" trends	Regional by inference	Traffic model Emission model Now-ETC + w/wo proj	T/Day (yr) by region
II Computer	Subarea burden and trend SAPOLLUT	Subarea burden and trend SAPOLLUT	Regional by inference	Traffic model Emission model Now-ETC + w/wo proj	T/Day (yr) by subarea and region
III Computer	Concentration at receptors APRAC 1-A	—	—	Traffic model Emission model Met. model Diffusion model	ppm (1 hr, 8 hr) Freq. dist., hr by hr synoptic, isopleths
IV Computer	Concentrations at receptors SAI	Concentration spatial and temporal SAI, DIFKIN, REM	Concentration spatial and temporal SAI, DIFKIN, REM	All above plus chemical and kinetic model, trajectory, etc.	Spatial and temporal concentration (ppm—isopleths)

from simple ones to complex "state-of-the-art" computer models. The matrix can be used as a "screening" procedure by proceeding first with the simpler models to assess the seriousness of the air pollution problem and continuing with more complex procedures as the need is evidenced.

For mesoscale analyses where few air-quality problems exist, generally in smaller urban areas, simple procedures as illustrated by the first level of analysis (I—Manual) would be used. As shown in the matrix for this level, the requirements are:

1. A traffic model for estimating the number of present and projected vehicular sources. This may simply involve forecasts of total annual vehicle miles traveled in the area.
2. An emission model to provide pollutant emission rates at some average speed. This would be the EPA–AP–42 method described previously.
3. Information relating to changes in expected vehicle miles traveled as a function of time in future, modal split, and various alternative flow patterns.

For carbon monoxide, hydrocarbons, and the oxides of nitrogen, a burden or trend analysis may be undertaken to estimate the added or reduced amounts of pollutants emitted under the various alternatives. These procedures do not give any information relative to *concentrations* of pollutants, and thus do not directly address the issue of the NAAQS. What may be done, if sufficient ambient-air-quality data in the region are available, is to infer relationships between pollutants emitted and present existing ambient pollutant concentrations. Future concentrations may then be estimated based upon these derived inferences.

Since ozone (O_3) is not emitted directly but is generated from precursors present in exhaust gasses, additional procedures have to be used. One developed by the EPA relates measured ambient hydrocarbon concentrations and expected ozone concentrations [8.30, p. 8].

The next level of analysis (II—Computer) requires much the same information and again provides only a burden or trend analysis. The major improvement is in the use of a set of more sophisticated travel models. The whole procedure, called SAPOLLUT, generates inputs by subarea in the mesoscale region. Again, no pollutant concentration data are produced [8.31, p. 66].

The next higher level of analysis (III—Computer) involves the first type of model capable of providing pollutant concentration data for comparison with the NAAQS. The model illustrated is called APRAC and is based on the same Gaussian dispersion concepts discussed previously under microscale models. It also is subject to the same constraints [8.32]. One of the more important limitations is that it may only be used for nonreactive pollutants, namely carbon monoxide. Carbon monoxide, however, may provide an indicator or "tracer" pollutant for other automotive emissions.

APRAC contains submodels for predicting traffic, emissions, meteorological conditions, Gaussian dispersion, and "street canyon" microscale impacts. With appropriate input data, it may be employed to predict carbon monoxide concentrations at several selected receptor points in the mesoscale.

The last level of analysis (IV—Computer) contains representatives of the most complex and advanced modeling procedures used to date. Such models have the potential of providing the spatial and temporal distribution patterns for both reactive and nonreactive air pollutants. Due to substantial and complex input data requirements and computer time involved, models of this type have only been applied in major urban area situations, and at considerable expense of manpower, time, and dollars. Most applications of these procedures have been developmental in nature and have been carried out in California. It is probable that even when the developmental problems have been overcome, the cost and difficulty of applying the models will be prohibitive in all but the most urgent situations.

Input data requirements include initial boundary-condition information. on all pollutant species and on meteorological parameters; complex and extensive surface, upper wind, and other meteorological observations; ambient air quality; vehicular traffic distribution; and complete mobile and stationary emissions.

The submodels included in such photochemical oxidant models as SAI, DIFKIN, REM, and others [8.33, p. 75] are those for traffic, meteorological conditions, air pollutant diffusion, and chemical kinetics. The latter account for the chemical reactivity of the constantly changing, reacting air mass constituents.

A helpful review of both meso- and microscale modeling experience can be found in [8.36].

The following problem will serve to illustrate the methodology for a simplified regional burden or trend analysis. Such procedures, as noted, are utilized in the preliminary screening stage of analysis.

A new freeway is to be built in an urban area. The estimated time of completion of the project is 1977. The planning horizon is 20 years thereafter. The analysis is to be done for conditions with and without the new freeway (Table 8.6).

To illustrate a simple mesoscale model, suppose we need to determine the influence of the proposed construction of a new freeway on the 1977 and 1997 CO burden. We start with the equation

$$CO = (vmt)(EF)(1.1 \times 10^{-6}) \tag{8.5}$$

Table 8.6 Problem Analysis

Year/variable	Freeways	Local streets
Existing highway facilities (no new freeway)		
1977		
Daily vehicle miles (vmt)	810,000	680,000
Average route speed (mph)	45	25
Composite emission factor (g/mile) for carbon monoxide	23	40
1997		
Daily vehicle miles (vmt)	1,840,000	2,000,000
Average route speed (mph)	35	25
Composite emission factor (g/mile) for carbon monoxide	7	7
With construction of new freeway		
1977		
Daily vehicle miles (vmt)	1,200,000	420,000
Average route speed (mph)	50	30
Composite emission factor (g/mile) for carbon monoxide	22	35
1997		
Daily vehicle miles (vmt)	2,800,000	970,000
Average route speed (mph)	45	25
Composite emission factor (g/mile) for carbon monoxide	7	7

Table 8.7 CO Burden

	Year	vmt	EF	CO (tons/day)
		With new freeway		
Freeways	1977	1,200,000	22	29.1
	1997	2,800,000	7	21.6
Local streets	1977	420,000	35	16.2
	1997	970,000	7	7.5
		Without new freeway		
Freeways	1977	810,000	23	20.5
	1997	1,840,000	7	14.2
Local streets	1977	680,000	40	30.0
	1997	2,000,000	7	15.4
	Year	Freeways	Local streets	Grand total
Total (with new freeway)	1977	29.1	16.2	45.3
	1997	21.6	7.5	29.1
Total (without new freeway)	1977	20.5	30.0	50.5
	1997	14.2	15.4	29.6

where   CO = CO emissions (tons/day)
  1.1×10^{-6} = conversion factor from grams to tons
    vmt = daily vehicle miles of travel
    EF = composite emission factor (g/vehicle · mile)
This leads to the calculations of Table 8.7. Note that this analysis indicates an initial benefit in CO burden owing to the construction; however, in 20 years the vmt has increased at a rate sufficient to offset, slightly, the improvements from smoother traffic flow and higher route speeds inherent in the new freeway.

In such cases as this, it may be useful to construct a "critical year" diagram for, say, five-year increments between the base year (1977) and the horizon year (1997). If one were done for the preceding example, a plot such as in Fig. 8.5 might be derived, dependent upon actual traffic and emission factor changes in the interim years [8.34].

8.5 NOISE IMPACTS

Few physical phenomena, particularly among those that surround us constantly, are so commonly misunderstood as is sound. It is measured in units unfamiliar to most people, and even the "experts" disagree about many of the subjective reactions to sound. What is certain, however, is that excessive noise can lower the quality of life for many, and can irritate to the point where people complain. Transportation is a major source of the noise in many urban areas.

8.5.1 Physical Aspects of Sound

The sound we hear consists of very small pressure disturbances traveling through the air. When these strike our ear drums, they cause them to vibrate. This vibration is

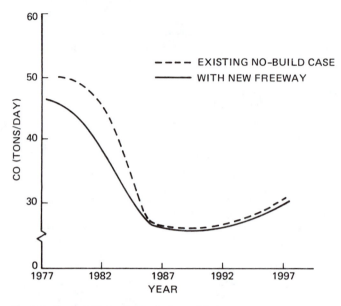

Fig. 8.5 Typical CO burden variations with time.

transmitted through the middle ear and into the inner ear, where it excites nerve cells. What we thus perceive as sound consequently depends on psychological factors and on several characteristics of both the pressure disturbances and our ears.

The pressure disturbances associated with most "real-life" sounds are fairly complex in nature. Thus it turns out to be easiest to discuss sounds in terms of component waves that are simpler in nature. It can be shown that even complex, repetitive sounds can be generated from a properly superimposed set of simple harmonic waves (sine and cosine waves). Consequently, the discussion of sound waves is threaded through with references to the frequency (pitch) and amplitude (size) of simple harmonic waves. Sound waves in the air are compressional in nature, but no great damage is done here if they are visualized as waves traveling across the surface of a deep pond. Both types of waves have many of the same characteristics. They travel at a fixed speed and with (ideally) undiminished amplitude. Sound waves in air spread out from a sound source (particularly as one gets farther away from the source) in a spherical pattern, just as water waves travel outward in circles from the point where a stone is dropped in a pond. The waves have a length (distance between crests for a simple harmonic wave), which is equal to the speed of propagation divided by the frequency (number of disturbances per time period). Sound waves in air in the middle of the human hearing range have wavelengths of about a foot, so very little sound can be blocked by small objects. Sound in air also reflects from hard surfaces, just as water waves do from a wall. In discussing sound and its behavior it is often helpful to come back to these visualizations.

8.5.2 Noise Descriptors

The definition of basic measuring scales begins with the observation that hearing is nearly logarithmic in nature. That is, a wave that is twice as large does not sound twice as loud to an observer. Thus, the measuring scales use the logarithmic unit of the decibel (dB). The sound pressure-level scale is defined by

$$L_p = 20 \log \frac{p}{p_0} \tag{8.6}$$

where log = the logarithm to the base 10
p = the amplitude of the sound wave, in root mean square units
p_0 = an arbitrary reference value
The reference value commonly used for sound waves in air is 20 micropascals (μPa) (or 2×10^{-5} N/m^2, which is the same thing). This is approximately the minimum level for audible sound for young men with good hearing. On this scale, pressure signals standing in a ratio of 1 to 1 million are different by 120 dB.

Closely related to the sound pressure-level scale, and equally important to consider, is the intensity-level scale. This expresses the amount of sound energy falling on a given area, and is defined by

$$L_I = 10 \log \frac{I}{I_0} \tag{8.7}$$

where I = intensity (W/m^2)
I_0 = reference intensity (commonly 10^{-12} W/m^2)
The choice of reference quantities, and the fact that there is a square law relationship between pressure amplitude and intensity, yields pressure-level and intensity-level scales equal in most common situations.

The importance of considering the intensity level scales may be demonstrated by a simple example. Suppose that two independent sound sources, when in operation one at a time, each cause a level of 70 dB at some observer's location. If both are then turned on at the same time, it is clear that twice as much sound energy should be reaching the observer. Manipulation of the intensity-level equation shows that the level then should be

$$L_I = 10 \log \frac{2I}{I_0} = 10 \log \frac{I}{I_0} + 10 \log (2)$$

$$= L_{I_1} + 3 \text{ dB}$$

Thus the two sound sources produce a level only 3 dB greater than either one alone. While the energies have been added (in W/m^2), it is necessary to consider the logarithmic nature of the decibel scale when determining the net level. The results can often be surprising. A little experimentation will show that 3 dB above the *loudest* sound is the limit that can be reached by combining two sounds. If the two sounds, considered separately, have levels that differ by as much as 10 dB, the effect of the lower level sound on the overall level is negligible. Combining of sound pressure levels can be done using the equation

$$L_{p,\text{net}} = 10 \log \left[\sum_{i=1}^{n} 10^{(L_{p_i}/10)} \right] \tag{8.8}$$

where i is the index for any individual sound pressure level.

The next complication in the establishment of a noise scale is the frequency dependence of human hearing. The human hearing range is generally considered to extend from 20 Hertz (Hz) to 20,000 Hz. However, sounds below about 500 Hz are not heard well, and sounds above 12,000 Hz are heard with rapidly decreasing acuity. Sound-level scales that discriminate against these very low and very high frequencies have been found to correlate better with human hearing than scales that treat all frequencies the same. By far the most common of these is the "A-weighted" scale. It is the preferred scale for measuring sound levels in community noise situations. Weighting networks are built into sound-level meters so that levels in dBA may be read directly. Many other scales have been proposed from time to time, but they generally have not been found to be much more effective than the dBA scale and are usually much more difficult to use. Some examples of common noise levels measured in dBA are given in Fig. 8.6.

The remaining major factor that must be brought into community and transportation noise-level description is the variation of level with time. For example, noise levels are usually higher in the daytime than at night. Passage of a car or airplane might cause the noise level in a quiet neighborhood to rise by 10 to 20 dBA for a short period of time. Several measures are currently in use to account for this variation. These include the equivalent noise level, L_{eq}, the day–night noise level, L_{dn}, and the statistical descriptors L_{10}, L_{50}, and L_{90}, the levels exceeded 10, 50, and 90 percent of the time, respectively. All of the noise-level measurements used in these calculations are A-weighted.

The equivalent noise level and the day–night noise level are both 24-hr averages of the noise levels. This averaging must be done on the basis of energy received. Just as in the example for combining two sound sources, therefore, it is necessary to get out of the decibel scale to do the averaging. The definition of L_{eq} is then

$$L_{eq} = 10 \log \left[\frac{1}{T} \sum_{i=1}^{N} 10^{(L_{p_i}/10)} (\Delta t) \right] \tag{8.9}$$

where the time period T is taken to be 24 hr, and Δt is a shorter time period over which the sound pressure level can be assumed to stay constant at L_{p_i}.

The day–night noise level is very similar to L_{eq} in many respects. Its chief difference is that a 10-dB penalty is attached to any noise occurring during the night (10 p.m. to 7 a.m.) before that noise is included in the averaging process.

There is still some debate as to which of these single noise-level descriptors is best to use. Europeans commonly use L_{eq}. In the United States, however, L_{dn} is becoming the recognized measure. The Environmental Protection Agency (EPA), in its "Levels Document" [8.37], chose L_{dn} as the appropriate criterion for community noise specification. The Federal Highway Administration, on the other hand, prefers L_{10} in its design process. Both measures give similar results in many common situations, although it is possible to describe cases in which the results would be different. A comparison of these and other single number noise descriptors can be found in Appendix A-1 of the "Levels Document" [8.37].

The levels recommended by EPA as being acceptable are given in Table 8.8. These are based on experimental results showing that speech interference is the dominant criterion on which people base annoyance. The levels recommended by the Federal Highway Administration (FHWA) are given in Table 8.9. While the numbers appear to

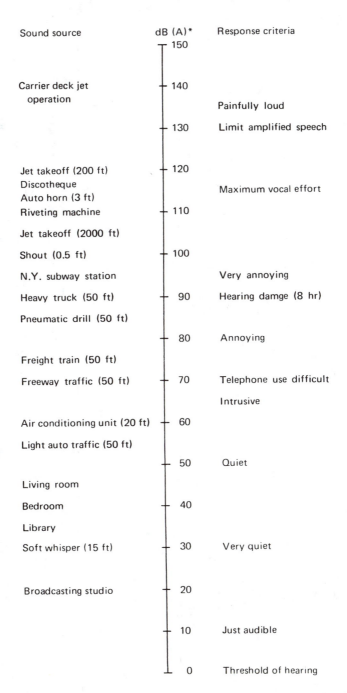

Sound source	dB (A)*	Response criteria
	┬ 150	
Carrier deck jet operation	┼ 140	
		Painfully loud
	┼ 130	Limit amplified speech
Jet takeoff (200 ft)	┼ 120	
Discotheque		
Auto horn (3 ft)		Maximum vocal effort
Riveting machine	┼ 110	
Jet takeoff (2000 ft)		
Shout (0.5 ft)	┼ 100	
N.Y. subway station		Very annoying
Heavy truck (50 ft)	┼ 90	Hearing damge (8 hr)
Pneumatic drill (50 ft)		
	┼ 80	Annoying
Freight train (50 ft)		
Freeway traffic (50 ft)	┼ 70	Telephone use difficult
		Intrusive
Air conditioning unit (20 ft)	┼ 60	
Light auto traffic (50 ft)		
	┼ 50	Quiet
Living room		
Bedroom	┼ 40	
Library		
Soft whisper (15 ft)	┼ 30	Very quiet
Broadcasting studio	┼ 20	
	┼ 10	Just audible
	┴ 0	Threshold of hearing

*Typical A—Weighted sound levels taken with a sound-level meter and expressed as decibels on the scale. The "A" scale approximates the frequency response of the human ear.

Fig. 8.6 Weighted sound levels and human responses [8.54, pp. 62-63].

Table 8.8 EPA Recommendations for Yearly Average (on Energy Basis) Equivalent Sound Levels Identified as Requisite to Protect the Public Health and Welfare with an Adequate Margin of Safety

	Measure	Indoor Activity interference	Indoor Hearing loss consideration	Indoor To protect against both effects†	Outdoor Activity interference	Outdoor Hearing loss consideration	Outdoor To protect against both effects†
Residential with outside space and farm residences	L_{dn}	45		45	55		55
	L_{eq} (24)		70			70	
Residential with no outside space	L_{dn}	45		45			
	L_{eq} (24)		70			70	
Commercial	L_{eq} (24)	*	70	70‡	*	70	70‡
Inside transportation	L_{eq} (24)	*	70	*			
Industrial	L_{eq} (24)§	*	70	70‡	*	70	70‡
Hospitals	L_{dn}	45		45	55		55
	L_{eq} (24)		70			70	
Educational	L_{eq} (24)	45		45	55		55
	L_{eq} (24)§		70			70	
Recreational areas	L_{eq} (24)	*	70	70‡	*	70	70‡
Farm land and general unpopulated land	L_{eq} (24)					70	70‡

*Since different types of activities appear to be associated with different levels, identification of a maximum level for activity interference may be difficult except in those circumstances where speech communication is a critical activity. (See Figure D-2 in [8.37] for noise levels as a function of distance which allow satisfactory communication.)

†Based on lowest level.

‡Based only on hearing loss.

§An L_{eq} (8) of 75 dB may be identified in these situations so long as the exposure over the remaining 16 hours per day is low enough to result in a negligible contribution to the 24-hr average, i.e., no greater than an L_{eq} of 60 dB.

Note: Explanation of identified level for hearing loss. The exposure period which results in hearing loss at the identified level is a period of 40 years.

Source: [8.37].

Table 8.9 Federal Highway Administration Noise Standards

	Design noise level/land-use relationship	
Land-use category	Design noise level L_{dn}	Description of land-use category
A	60 dBA (exterior)	Tracts of land in which serenity and quiet are of extraordinary significance and serve an important public need, and where the preservation of those qualities is essential if the area is to continue to serve its intended purpose. Such areas could include amphitheaters, particular parks or portions of parks, or open spaces that are dedicated or recognized by appropriate local officials for activities requiring special qualities of serenity and quiet.
B	70 dBA (exterior)	Residences, motels, hotels, public meeting rooms, schools, churches, libraries, hospitals, picnic areas, recreation areas, playgrounds, active sports areas, and parks.
C	75 dBA (exterior)	Developed lands, properties, or activities not included in categories A and B.
D		For requirements on undeveloped lands, see *Federal Highway Program Manual.*
F	55 dBA (interior)	Residences, motels, hotels, public meeting rooms, schools, churches, libraries, hospitals, and auditoriums.

Source: [8.38].

be quite different from EPA's, it must be kept in mind that FHWA's are for levels that are more closely related to the average noise level.

8.5.3 Traffic Noise-Level Prediction

The prediction of noise levels caused by automobile and truck traffic is difficult. Space is too limited to allow a complete review of the methods here. Readers with interest in the subject are referred to the *National Highway Cooperative Research Program Report* 174 [8.39], which includes a comprehensive design guide for highway noise computations. Included are both a nomograph-based approach for approximate studies, and a computer-based method for making detailed noise assessments of highways when final alignment and traffic volumes are known. In this section we will give some idea of the important parameters in traffic noise prediction.

The primary parameters are the traffic volume expected (in vehicles per hour), the speed of the traffic (in miles per hour), and the equivalent shortest distance in feet from the point where the noise level is to be calculated to the center of the traffic lane.[2] Automobiles and light trucks must be treated separately from heavy trucks since the noise-generating characteristics of each are quite different.

For automobiles and light trucks the primary radiated noise comes from the tires and pavement. Thus the noise "source" is at ground level. Further, the magnitude of the noise generated increases as the vehicle speed increases. Of course, as the traffic volume rises, so does the noise level. In general, light trucks are noisier than automobiles, so light-truck volumes must be multiplied by 10 before being included in the

---

[2] In some cases several lanes of traffic may be combined in the analysis procedure; hence an equivalent distance is used.

calculations. Because the noise spectra are similar, however, calculations may be based on the sum of the automobile traffic volume and 10 times the light-truck traffic volume. Thus, for example, the approximate method given in [8.39] would yield a prediction at 200 ft from the traffic lane of $L_{10} = 66$ dB for either 3,000 automobiles per hour traveling at 50 mph or 2,000 automobiles and 100 light trucks, all traveling at 50 mph.

The nature of the decibel scale is such that given changes in the sound pressure level, in dB, are associated with fixed ratios of the primary variables used in the calculation. Doubling the traffic volume thus would add 3 dB to the result; doubling the distance from the traffic lane would decrease the level by 3 to 4 dB (for each successive doubling); and halving the speed would drop the noise levels about 6 dB.

The noise radiated by heavy trucks is different from that by automobiles and light trucks. The basic component is usually engine noise, radiated from an exhaust stack that is normally well above the roadway. Truck engines are operated in a relatively narrow speed range, so that the noise tends to be less dependent on vehicle speed than is automobile noise. In fact, the level drops slightly as truck speed increases. The noise levels are higher than for automobiles, however. For example, the approximate method of [8.39] would yield an L_{10} of 68 dB for a truck volume of 100 trucks per hour, all traveling at 50 mph, and measured 200 ft from the center line of the traffic. As before, doubling the truck volume would add 3 dB to the level, and doubling the distance would decrease the level by 3 to 4 dB per doubling. Yet halving the speed of the truck traffic would increase the noise level by about 3 dB.

If both the heavy trucks and the mixture of automobiles and light trucks of the preceding examples were present, an L_{10} of 70 dB would result. Obviously, truck traffic is the major component. In many cases it can be the dominant source. If the heavy truck traffic volume were maintained but the automobile and light truck volume were reduced to 500 vehicles per hour, the truck noise level would be 10 dB above the automobile noise level, and the latter could be ignored completely.

These traffic noise predictions are based on the idea that a uniform line of traffic is proceeding along a level, straight highway segment with no interference to the sound propagation from buildings, barriers, road berms, wind, or other environmental factors. The sound from the line of traffic thus spreads out in a cylindrical fashion from the road. It is evident that corrections to this prediction must be made for all of the factors mentioned, as well as for curves and other changes in the highway alignment that would place more of the vehicles closer to the listener. Furthermore, when the traffic volume falls below about 200 vehicles per hour, the assumption of the Gaussian distribution of vehicles (see Sec. 8.4.2.2) becomes increasingly less valid, and other corrective measures must be taken. All of these factors are taken into account in the computer program for detailed sound pressure-level prediction. It is estimated that this program, when properly used, can predict sound pressure levels to within 2 dB. Obviously, the approximate method gives answers that are less accurate and only useful for separating those areas which need detailed study from those where no noise problem should be anticipated.

8.5.4   Rail-System Noise

Noise caused by rail systems can come from several sources. A diesel locomotive, for example, emits a lot of sound at low frequencies. This is primarily engine exhaust noise. In an electrified line, the primary noise comes from wheel and rail interaction. Particularly in the case of wheel "screech," this noise can be high-pitched and quite disturbing.

Relatively little attention has been paid to the prediction of sound pressure levels caused by rail systems. The exhaust-noise propagation from a diesel engine could be predicted with a good expectation of success using the same basic approach as for a single truck on a highway. Comparatively little is known about wheel screech, and the prediction of sound radiation from this source is complicated. When a rail line is sited in a city, the prediction problem is made more difficult by the presence of buildings and many nearby reflecting surfaces.

Noise in buildings next to rail lines is often caused by the transmission of vibration from the passage of a train. Once the vibration is in the building's structural members, it may be transmitted throughout, often causing a wall or panel to vibrate in turn and to radiate sound unpredictably in different parts of the building. To prevent this problem, new subway and rail lines have had extensive vibration-isolation pads installed under the track in sensitive areas. Of course, these add to the expense of track construction.

8.5.5 Noise-Control Measures

Traditionally, noise-control approaches are divided into three types. These are control at the source, in the sound transmission path, and at the receiver (listener). Control at the source (trucks, automobiles, trains, etc.) is generally not within the province of the urban planner, and control at the receiver is limited. The planner has most control over the transmission path between the source and the receiver.

One of the best noise-control measures, where it can be used, is to keep the sources away from the listener. Thus, whenever possible, noisy transportation systems should be kept away from areas where quiet is important, such as residential areas and parks. Where this is impractical, the next suggestion is usually to employ some sort of a wall or barrier to protect the desired quiet area.

For a barrier to be effective it must block the "line of sight" between the noise source and the listener. In fact, the top of the barrier must extend at least several feet above the line of sight when the source and the receiver are close together. As the distance between the two increases, the height required of the barrier rises proportionally if the same protection level is to be maintained. Furthermore, the barrier must be long, or sound will simply come around the ends instead of over the top. The barrier must be constructed of dense material so that it will block the transmission of sound through itself. Finally, no holes or openings can be left in the barrier. An open area of 10 percent of the wall would nearly destroy its usefulness. Despite these problems, barriers have been used effectively for noise control in some situations.

It must be remembered that the function of a barrier is to reflect sound, placing the listener in the "shadow" of the wall. In an urban situation, however, this sound may be reflected toward another listener across the road. In these situations an additional barrier may be required, turning the roadway into a walled channel. This effect also can be obtained by depressing the roadway. Drivers on such roads naturally will find the noise levels higher than before. In many situations the installation of acoustic absorbing material may also be necessary.

In extreme cases, a depressed roadway in a city may even be decked over to control the noise. This is a very effective technique from a noise-control standpoint if the decking is substantial enough to block the sound and if the openings in the decking are small and properly located.

Trees and shrub plantings are often proposed as control measures. Despite their aesthetic appeal, they provide almost no protection against sound. Very wide bands of trees—typically in excess of 100 ft wide—are required to provide any appreciable noise reduction.

In summary, the best noise control measure available for a road or transit system is still the choice of location. Slight alignment changes often can make a significant reduction. Placement of a roadway or rail line behind a hill or noise-insensitive building can provide substantial shielding, for example. Other noise-control measures are likely to be more expensive and much less successful.

8.5.6 Human Impacts of Noise

Assessment of the impact of noise on humans is by no means an easy task, even after noise-level predictions have been completed. The effects of noise on people are as yet incompletely understood. Extremely loud noises, particularly if experienced over a long period of time, can damage hearing irreparably, of course, but noise levels much below these also can cause trouble. Noise can seriously interfere with sleep. It can also cause stress and, indirectly, stress-related diseases. There is some evidence that circulatory changes can be induced by noise, but the long-range effects of these have not been determined. The area that turns up most frequently when noise complaints are investigated is the interference with speech. This, consequently, is usually employed in predictive methods to determine the impact of noise on community residents.

The relationship between noise levels and the percentage of the population who are highly annoyed can be specified, although the data are scattered, as one would expect. An example of this is shown in Fig. 8.7, which comes from a study on the effects of noise in urban environments [8.40]. Similar correlations between L_{dn} and the percentage highly annoyed have been found by other investigators.

NCHRP Report 174 [8.39] suggests using the percentage highly annoyed, multiplied by the potentially affected population, to obtain an impact number. It envisions a more drastic relationship between the percent highly annoyed (%HA) and the L_{dn} than was found in the urban noise study, however:

$$\%HA = 2(L_{dn} - 50) \tag{8.10}$$

Further, since most noise criteria can be said to have been completely violated when the L_{dn} has risen 20 dB above the criterion level L_c (usually taken as $L_{dn} = 55$ dB, from the "Levels Document"), NCHRP 174 suggests using a fractional impact (FI), which is zero for levels below the criterion level and increases linearly to 1.0 at 20 dB above that. In other words, for $L_{dn} > L_c$,

$$FI = 0.05(L_{dn} - L_c) \tag{8.11}$$

It may be advisable, as NCHRP 174 suggests, to allow the FI to go above 1.0 to penalize flagrant violations of the criterion level. The fractional impact may then be used with population estimates in various ways to find a noise-impact index for a proposed highway or mass-transportation system.

8.6 RETAIL SALES AND LAND VALUE IMPACTS

One type of economic impact relates to retail sales and land values. Horwood et al. analyzed a large number of highway impact studies made in the years up to 1964 [8.42]. These studies focused on many items, but retail sales and land values were prominent. Three categories of investigations were identified: (1) bypass studies,

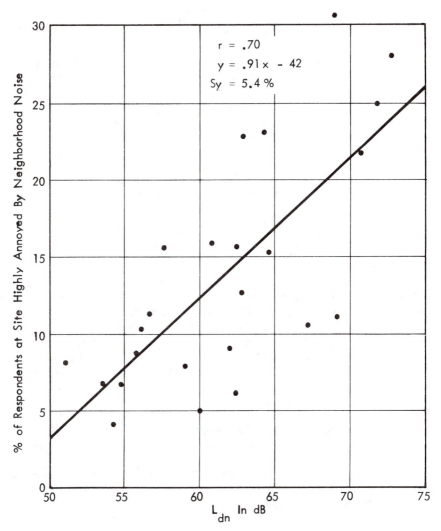

Fig. 8.7 Relationship between noise level and percentage of respondents at each site highly annoyed by neighborhood noise [8.40].

(2) urban circumferential studies, and (3) urban radial freeway studies. These will now be discussed in turn.

8.6.1 Bypass Studies

The bypass studies dealt with sales in highway- and non-highway-oriented retail establishments in bypassed towns falling into various population categories. Data were obtained primarily from sales-tax information provided by appropriate state agencies. In many cases sales comparisons were made with control areas—that is, towns that with the exception of the bypass were as similar as possible to those of main concern to the particular study. In providing a control area, it was possible to determine in a rough way whether gains or losses in sales were due primarily to the bypassing highway or to general economic conditions prevalent in the region.

A summary of total and highway-oriented retail sales changes in bypassed towns is shown in Table 8.10. As can be seen, only 7 of 36 towns showed a sales revenue loss, 28 had a gain, and one had no change. Towns over 5,000 population showed an average increase over twice that of towns under 5,000. In comparison to the control areas, the overall gain of +8.5 percent turns out to be a slight relative loss of 0.3 percent. In other words, the highway bypasses appear to have almost no relative impact on retail sales for the "average" establishment. Towns of different sizes are not affected equally, however; those under 5,000 population lost 3.5 percent more than their corresponding controls, while those over 5,000 averaged gains of 3.7 percent relative to their controls.

Interestingly, highway-oriented sales for fuel, food, and lodging for transients in bypassed towns increased, and at a greater ratio than for total retail sales—21.0 percent versus 8.5 percent. No control areas were established for this set of studies so it is difficult to tell if the highway-oriented sales actually gained relative to other localities. The apparent gains might be suspect, however, since a further analysis of 32 impact studies where service station sales were listed separately, Horwood et al. found that the average change in sales was only +2.3 percent, and −3.2 percent in comparison to control areas (see Table 8.11).

In addition, restaurant sales decreased an average of 13.0 percent (−6.4 percent relative to controls) and motel and hotel sales decreased an average of 23.1 percent (no controls employed). These latter findings are more in accord with expectations, yet there appears to be much variability in impact. In particular, the largest towns almost always appear to gain significantly while the smaller ones lose equivalently. Also, non-highway-oriented sales changes are better in larger cities. There are gains, nonetheless, over all city sizes (averaging +10.3 percent individually and +5.8 percent relative to control areas).

8.6.2 Urban Circumferented Studies

Only six studies found by Horwood et al. dealt with urban circumferentials. A common feature of several of these studies was relatively rapid and intense land-use changes along beltway routes. Commercial and industrial use predominate (in contrast to residential), and land values close by the facility were seen to increase substantially.

8.6.2.1 Land-use changes Lexington and Louisville, Kentucky, found that total agricultural acreage was reduced 35.7 and 29 percent, respectively, after road construction. Louisville had a 23.6 percent increase in commercial and industrial uses and 3.1 percent increase in institutional uses. Minneapolis–St. Paul showed a 60.5 percent increase in commercial and industrial land uses where little had existed previously.

8.6.2.2 Land-value changes Commercial lands showed the greatest average increases in values: 53.5 percent in Louisville and 93 percent in Lexington. Older residential areas were more seriously affected than newer ones in both cities. There was an average 5.5 percent decrease for older areas and 2.0 percent for newer ones. In Baltimore the average land value increase near a circumferential was 10.0 percent.

8.6.3 Urban Radial Freeway Studies

Four studies of the impacts of urban radial freeways were detailed by Horwood et al. [8.42]. These were for the cities of Dallas, Houston, Atlanta, and San Antonio. Unfortunately, time spans in the studies differed somewhat, as did definitions of "land value" and adjustments for inflation and local cost factors. Delineation of study and control areas also varied from area to area.

Table 8.10 Sales Changes in Bypassed Towns

Population category	Avg. change in retail sales (%)	No. of towns with — Gain in sales*	No. of towns with — Loss in sales*	Range (%)	Avg. gain (%)	Avg. loss (%)	No. of towns with — Control area	No. of towns with — More gain or less loss than control*	Avg. gain or loss over control (%)
(a) Total retail sales									
Under 5,000	+5.6	16/20	3/20	−6.4 to +22.5	+8.2	−6.6	18	6/18	−3.5
Over 5,000	+12.2	12/16	4/16	−13.0 to +49.0	+20.4	−7.5	14	11/14	+3.7
5,000–10,000	+16.9	5/6	1/6	−13.0 to +38.0	+17.7	−13.0	6	4/6	+0.85
10,000–25,000	+7.3	5/7	2/7	−3.1 to +40.5	+12.5	−4.8	7	6/7	+4.1
25,000–50,000	−11.4	0/1	1/1	—	—	−11.4	—	—	—
50,000–100,000	—	—	—	—	—	—	—	—	—
100,000 and over	+22.6	2/2	0/2	+4.3 to +49.0	+22.6	—	1	1/1	+19.0
All towns	+8.5	28/36	7/36	−13.0 to +49.0	+12.9	—	32	17/32	−0.30
(b) Highway-oriented sales									
Under 5,000	+20.8	3/6	3/6	−14.7 to +60.9	+51.8	−10.3			
Over 5,000	+21.2	3/4	1/4	−11.8 to +50.4	+32.3	−11.8			
5,000–10,000	+41.5	1/1	0/1	—	+41.5	—			
10,000–25,000	+50.4	1/1	0/1	—	+50.4	—			
25,000–50,000	−11.8	0/1	1/1	—	—	−11.8			
50,000–100,000	—	—	—	—	—	—			
100,000 and over	+4.9	1/1	0/1	—	+4.9	—			
All towns	+21.0	6/10	4/10	−14.7 to +60.9	+42.5	−10.6			

*For example, 16/20 indicates "16 of 20."
Source: [8.42, p. 9].

267

Table 8.11 Service Station Sales Changes in Bypassed Towns

| Population category | Avg. change in retail sales (%) | No. of towns with | | Range (%) | Avg. gain (%) | Avg. loss (%) | No. of towns with | | Avg. gain or loss over control (%) |
		Gain in sales*	Loss in sales*				Control area	More gain or less loss than control*	
Under 5,000	-0.47	8/17	9/17	-33.0 to +39.4	+14.3	-13.6	11	5/11	-0.86
Over 5,000	+5.5	7/15	8/15	-21.0 to +39.0	+20.0	-7.3	10	6/10	-5.75
5,000-10,000	-1.8	2/4	2/4	-21.0 to +17.0	+11.3	-15.0	3	2/3	-20.0
10,000-25,000	-4.2	3/7	4/7	-10.0 to +33.9	+19.1	-7.5	6	3/6	-6.1
25,000-50,000	-4.8	0/2	2/2	-7.0 to -2.5	—	-4.8	—	—	—
50,000-100,000	—	—	—	—	—	—	—	—	—
100,000 and over	+30.3	2/2	0/2	+21.5 to +39.0	+30.3	—	1	1/1	+21.0
All towns	+2.3	15/32	17/32	-33.0 to +39.4	+16.9	-10.6	21	11/21	-3.2

*For example, 9/17 indicates "9 of 17."
Source: [8.42, p. 10].

Table 8.12 displays a comparison between changes in land use values in the four study areas given the limitations brought out in the preceding paragraph. The results that seem most prominent are:

1. In most cases, value of land abutting the radial freeway facility exceeded that of land further removed as well as land in control areas.
2. Unimproved or vacant land appears to receive the largest benefit from the freeway, ranging from two to three times the value increases of improved properties.
3. Land values computed with improvement values deducted doubled to tripled for land including improvements.

Another interesting finding is the rapid decline in land values with distance from the radial freeway. For example, band B in Dallas is quite close to the freeway (next to abutting land), yet values there rose less than one-quarter of those in the abutting band. In fact, the rise in band B is slightly less than for the control area (band D). The rapid fall in land values also is quite noticeable on an overall basis. This means that owners of existing facilities in an area should not expect any great relative change in land values from a new radial freeway unless they happen to be right next to it.

Owners of vacant and unimproved land are likely to gain substantially, however. Even in band B, increases in land values are up to eight times as great as in the control area. The average is two to four times as great. It would seem, then, that the most return on investment can be obtained by purchasing unused land within an approximate 2-block strip from the potential location of a radial expressway.

8.7 RELOCATION IMPACTS

Of all the social impacts of transportation system construction, relocation seems to have been given the most attention. This is probably because it strikes closest to individual citizens and thus exposes many personal problems that otherwise would have laid dormant. Christensen and Jackson express the idea well [8.52, p. 1]:

> When expressways run through a major city, large numbers of people and many businesses are displaced. Unfortunately, the highways are frequently routed through the least desirable sections of the city, and those who are displaced are the poor, the aged, and those who are least able to take care of themselves, and there is little likelihood that many of them will use the expressway that displaces them. . . .
>
> In theory, relocation assistance is simple. In practice, it is difficult, complicated, and time-consuming. Frequently, successful relocation depends on solving personal problems, both financial and social, in addition to finding replacement property.

The authors go on to portray six examples of the unique types of problems that were faced in relocating some families and businesses in Baltimore [8.52, pp. 4, 5], of which some are:

> Mrs. L, 60-year-old widow, lived with her mentally retarded son and daughter, both in their 30's. Conditions were pitiful. They had no furniture and slept on the floor. They had no gas or electricity as these had been shut off in 1960 when they failed to pay a $75 bill. There was no heat. The case looked hopeless when relocation went to work on it. The ideal place for a family of this kind would be public housing, but they refused even to consider it, insisting that they stay in the same general neighborhood. Relocation finally found them a satisfactory apartment nearby at a rent within their welfare allowance. Welfare provided a furniture grant, which was used at Goodwill Industries so as to get the maximum return for each dollar spent. A private charitable organization was found which agreed to pay the back-due gas and electric bill. Finally, relocation provided transportation for the few goods owned, and assisted them in

Table 8.12 Changes in Land Values by Locational Relationship to Freeway (Expressway)*

	Average increase over base year (%)											
	Land with improvements				Land without improvements				Unimproved (vacant) land			
Location	Band A	Band B	Band C	Band D	Band A	Band B	Band C	Band D	Band A	Band B	Band C	Band D
Dallas												
1941 Annexation	431	100	139	106	623	123	185	130	518	383	291	166
1946 Annexation	127	26	(22)	31	1027	538	–	104	1179	766	–	136
Houston												
Unadjusted	250	130	50	90	282	150	38	76	–	–	–	–
Adjusted	245	125	(15)	44	190	96	(70)	(12)	–	–	–	–
Atlanta												
West	99	4	11	102	–	–	–	–	197	12	53	148
East	40	18	(35)	102	–	–	–	–	247	35	(58)	148
San Antonio	251	181	71	(2)	377	264	127	30	–	–	–	–
Overall average	206	83	28	68	500	234	70	66	535	299	95	149
Overall range	40–251	4–181	35–139	-2–106	100–1027	96–538	-70–185	-12–130	197–1179	12–766	-58–291	136–166

*Band A abuts the freeway; band B is next; and so on. Band D is the "control" area.
Source: [8.42, p. 19].

paying the rent deposit. The family is now warm and comfortable, much better than they have been for years. . . .

Mrs. A, a 45-year-old recluse, also has mental problems. Relocation showed her numerous possible locations, yet she refused to move. Because of her very limited income, a charitable landlord was found who agreed to reduce rent to a price she could pay. Still she refused to move. Something had to be done, as the remainder of the block was vacant, and it was dangerous for her to stay in her apartment any longer. Her brother was contacted and asked to assist, but he was unable to change her mind. Finally, with the brother's cooperation, eviction was arranged on court order. As her furniture was moved out of the apartment and onto the sidewalk, her brother arrived with a truck to take her to a new location which he had approved. Mrs. A calls occasionally. It is hard to say whether she is content in her new location or not. Sometimes she says she would like to move, but by the time Relocation reaches her apartment, she has changed her mind and decided that she will stay where she is. Probably this should be rated as a failure, because she had to be evicted; yet she is without any serious problems at her new location, other than those she had before.

Mr. R moved out of a house without telling Relocation. He was traced and visited at his new location which was found to be substandard. He was offered further assistance but refused the offer saying he was satisfied with the new place and would not move again. He had lost ground as a result of his move.

Mr. S operated a small two-chair barber shop. He suffered from cancer and had had a laryngectomy, which left him virtually unable to speak. His attempts to find a new location were met with failure. Even when he finally found a place he thought he could use, his application for a zoning exception was turned down. He was bitter and depressed. Finally, he turned to Relocation, which found him a new location, assisted in processing an application for a permit, assisted him in obtaining credit, and finally arranged for a SBA loan. He is proud of his new, greatly improved shop, and is getting along fine.

These types of problems, although handled here successfully for the most part, are compounded by the size of the relocation program. In Baltimore, the expressway program in the early 1970s was expected to displace 3,800 families (about 15,000 people). Only 20 percent of these families are white; less than 40 percent own their homes; their median income is $4,500. Nearly three-quarters have incomes so low that they qualify neither for public housing nor for other government-subsidized housing programs. A large number are elderly, and many have large families. Some 500 businesses also are expected to be displaced, varying in size from the small neighborhood grocery or barber shop to multi-million-dollar factories [8.52]. On the national level, the Federal–state highway program was expected to be responsible for about 50,000 displacements annually, again in the early 1970s [8.53].

The question now arises as to how the relocatees have fared in general. Were there any groups that were more or less successful than others? What was the overall reaction of relocatees to their forced move? The answers to these questions are not readily available, as relatively little intensive research has been done on the subject of relocation.

The 1968 Federal-Aid Highway Act upgraded the relocation program considerably. Many of the far-reaching provisions of that act were [8.53, p. 13]:

1. New declaration of legislative policy with respect to highway relocation assistance.
2. Provision for assurances to be given by the state highway departments in connection with specific project proposals.
3. Increase in the level of all moving cost payments without a ceiling but with certain limitations.
4. Provision for 100 percent federal share of the first $25,000 of such payments to any person until July 1, 1970.

5. Authorization for an additive to fair market value of property acquired in the form of a replacement housing payment up to $5,000.
6. Provision for a similar additive in the form of a rent supplement for tenants up to $1,500.
7. Sanction of the payment of expenses to the property owner incidental to the transfer of his property to the state.
8. Requirement for an expanded level of relocation assistance services to displacees.
9. Definition of several real property acquisition policies that are mandatory on all Federal-aid highway acquisitions.

The primary effect of this act was to place a significant obligation on highway departments and other state and local agencies to find "decent, safe, and sanitary" housing for those families displaced by highway facilities.

Some statistical information on the impacts of relocation before the 1968 Act are presented by House [8.51]. She studied displacement caused by the North-South Freeway in Milwaukee, which bisected the city's Black area, and concluded that, in general: (a) living conditions were improved for a majority of families, (b) changes in housing induced positive psychological attitudes, (c) housing costs increased, and (d) the concentration of Blacks was increased in a predominantly Black area.

The area bisected by the freeway was 70 percent Black, had below average housing conditions but comparatively high home ownership, and a rent-to-income ratio of about 0.13. After the move, the proportion of these families living in poor housing dropped from 39 percent to 14 percent. Housing was *perceived* as better by 62 percent of the relocatees and worse by only 23 percent. Home ownership increased from 59 to 70 percent, but average rents increased 12 percent from $67 to $75 per month and monthly mortgage payments increased substantially, from $45 to $72. This latter result apparently came about because many older families who already had paid for their houses were forced to resume mortgage commitments for their new residences. In balance, as can be seen in Tables 8.13 and 8.14, the overall impact of relocation seemed to be beneficial.

The impact on certain individuals and groups was not as beneficial, however. In particular, most of the families (25 percent of the total) whose main wage earner was retired, and whose families therefore were living on generally low, fixed incomes, were forced out of houses which they owned and into less desirable rental facilities. Moreover, segregation was increased, as most of the Black displacees moved to closeby neighborhoods where, in 75 percent of the cases, Blacks already lived on the block.

Business relocation also can be quite traumatic, especially for the marginal and submarginal "family" businesses such as grocery stores and barber shops run generally by older people. Their success as an enterprise depends heavily on people in the immediate neighborhood, and if they are displaced, sales naturally decline. These losses come about primarily because of the "wet blanket" effect that relocation (which may be several years off) has on purchases and land values in the community. Losses occur also in moving and starting up anew elsewhere (if the business is not terminated). In addition, obtaining a license (e.g., for a bar) in a new location can be a difficult problem [8.50].

Interestingly, many people move themselves and do not require or make use of advisory services. This seems to result for three reasons:

1. The general mobility of families in the United States: roughly 13 million families move every year anyway [8.50].

Table 8.13 Comparisons of Residential Characteristics for Areas Occupied before and after Relocation

Characteristics	Before	After
Housing quality		
% sound	60.5	85.0
% deteriorating	34.4	12.6
% dilapidated	5.1	1.5
Overcrowding		
% dwellings with		
1.01 + persons	14.0	8.5
Housing costs		
Average house value	$9,681	$12,312
Average monthly rent	60	72
Occupancy structure		
% owner-occupied	44.0	41.0
% renter-occupied	56.0	59.0

Source: U.S. Census of Housing, 1960 [8.51, p. 76].

Table 8.14 Comparisons of Individual Housing Characteristics before and after Relocation

Characteristics	Before	After
Housing quality		
% good/fair	60.5	86.0
% poor	39.5	14.0
Overcrowding/living space		
% more space after		62.0
% no change		29.0
% less space after		9.0
Housing costs		
Average monthly mortgage		
payment	$67	$75
Average monthly rent	$45	$72
Occupancy structure		
% owner-occupied	50.0	70.0
% renter-occupied	41.0	30.0

Source: [8.51, p. 76].

2. The general desire not to live in a neighborhood that is running down.
3. The lack of any advisory relocation assistance.

The last reason is particularly disturbing. As an illustration, in one New York City urban renewal project reported by Niebanck a set of volunteer workers (local citizens and some Peace Corps people) found [8.49, p. 60]:

> ... major deficiencies in the amounts of aid being received by persons who were fully eligible for and in need of such aid. These gaps existed both in the immediate relocation program with which the project was primarily concerned and in city welfare programs generally. The deficiencies stemmed from many factors: inadequate regulations; poor administrative procedures; unnecessarily rigid policies originating at the top of the official hierarchy; poor communications between office personnel and field workers; and indifferent or incompetent people in either the office or the field.

On the other hand, one of the nicer aspects of this and other relocation projects was the friendliness of caseworkers toward the relocatees, particularly the elderly. Many older people had known only loneliness, lack of care, and insecurity until caseworkers and volunteers (sometimes church groups and neighborhood house workers) approached them concerning the expected moves. Relations were built up, particularly with other, elderly relocatees, that lasted past the movement period [8.49].

8.8 VISUAL IMPACTS

Another set of impacts that are very difficult to measure and predict are those concerned with the visual impact of a transportation facility. Relatively little has been written about appearance *of* facilities, while somewhat more has been reported about appearance *from* facilities, especially from the point of view of the highway driver. Thus, while the concern in this section is primarily for nonuser impacts, we will have to draw primarily from research related to the user.

Two points should be made initially:

1. Good appearance is not solely a result of "cosmetics," that is, changes in the visible surface of an object. It usually is an intricate part of the overall building effort.
2. Good appearance varies according to the viewer, that is, "beauty is in the eye of the beholder."

Lynch emphasizes these points heavily. In regard to the first, he says [8.47, p. 250]:

> City design is the technical core of the process of city planning, and its concerns are equally broad. Design does not focus solely on appearance, nor indeed on any single factor which is affected by form. The immediate sensuous quality of an environment—the way it looks, smells, sounds, feels—is one consequence of the way it is put together. . . .

In connection with the second point, he states further [8.47, p. 251]:

> Current campaigns for "beautification" are a reflection of middle and upper middle class taste, with its emphasis on tidiness, appropriateness, and camouflage. The junkyard, abhorrent to the garden club, is a rich mine of form for the sculptor. A lower class citizen may be attracted by visible signs of security, durability, newness, or upward mobility, and take pleasure in forms which to an upper class observer seem coarse, hard, and vulgar.

8.8.1 Visual Impact Criteria

With these points in mind, and working primarily from conscious experience rather than from detailed scientific experiments, Lynch and other researchers have developed sets of perceptual criteria that give some indication of the likely acceptability of appearance of various urban forms, including highways. Lynch divides these criteria into six major statements [8.47]:

1. Sensations should be within the range of comfort.
2. Within the range of comfort, diversity of sensation and setting are to be desired.
3. Places in the environment should have a clear perceptual identity: recognizable, memorable, vivid.
4. The identifiable parts should be arranged so that the observer can mentally relate them to one another and understand their pattern in time and space.
5. The environment should be perceived as meaningful. Identifiable parts should be related to other aspects of life.
6. The environment should play a role in fostering the intellectual, emotional, and physical development of the individual by encouraging attention and exploration.

Perhaps the greatest sources of discomfort induced by the visual environment are those scenes that portray large-scale complexity, undertones of risk and uncertainty, and culturally unacceptable sites. Most people feel uneasy, for example, when faced with a bewildering and complex array of signs, signals, and symbols. As Appleyard et al. state [8.44, p. 11]:

> Finding a way through the intricacies of a modern city is a demanding performance, and one cannot depend entirely upon such conventional aids as directional signs, at least not without some emotional insecurity.

It thus appears that when the complexity of a scene is great, the viewer is likely to be stressed emotionally and will not consider the appearance to be of the highest quality. Further, if the viewer is distressed by some scenes that obviously are not culturally acceptable, such as borrow pits, strip mines, or sanitary fills through which or by which a transportation facility must pass, he naturally will respond negatively to the scene.

Diversification appears to be a significant feature of good visual impact. Human beings seem to have an appetite for novelty and variety. Moreover, it is such change that gives them choice and helps them develop their perceptual and cognitive system. These statements have particular significance for the driver or passenger viewing the road and its surroundings as they appear and pass by. Appleyard, Lynch, and Myer allude to this when they say [8.44, p. 4]:

> The sense of spatial sequence (in views from the road) is like that of large scale architecture; the continuity and insistent temporal flow are akin to music and the cinema. The kinesthetic sensations are like those of the dance or the amusement park, although rarely so violent.

Lack of diversity in the driving experience is found in the long trips over the unchanging, featureless plains of the Midwest. The result is a monotony that in many cases leads to drowsiness and even sleep. It may be important, in fact, to make sharp changes in the otherwise smooth-flowing alignments of most roads to joust the driver out of his monotony.

Of course the amount of diversification in a view differs according to the newness of the scene to the driver and his speed. A tourist obviously sees the landscape with a fresh eye and attaches relatively few personal meanings to it. He generally is engaged in orienting himself to the large and outstanding features in it. The habitual commuter, on the other hand, probably ignores large landscape features in favor of new and unique events: construction activities, new buildings or roadway appurtenances, moving trains and airplanes nearby, and so on.

Naturally all these events cannot be seen and absorbed if the speed of the vehicle is high. In this situation the driver can only experience those views positioned almost directly ahead of him. At 25 mph his total horizontal angle of vision is about 50° to either side, and the eyes focus at a point about 600 ft ahead. At 60 mph, the focus may be nearly 2,000 ft ahead while the angle of vision has shrunk to less than 20° [8.45]. These findings indicate that it is better if the driver of high-speed roads has long, straight-ahead views of large-scale features (e.g., city skylines), while the driver of slower roads has views of more detailed and widely spaced objects.

Another way in which diversity is enhanced is through accentuated features of unique sites such as historical monuments, towers, bridges, and rock outcroppings. In addition, the masking and subsequent revelation of a prominent feature and the confinement, rotation, feinting, jogging, swerving, and sliding past of objects in the foreground all add to the diversity of the driving experience. In terms of nondriving experiences, some important variations people might seek are [8.47, p. 252]:

> The range from lonely to gregarious places, for example; or from highly defined and structured surroundings to ones which are free and loose; from calm, simple, slow worlds to rapid, complex, and stimulating ones. . . . A secluded garden opening on a busy street is one example.

While variety is important, so also is identity. Many of those who have studied the visual form of the city feel that there is a constant search by people for a "sense of place." As Lynch remarks [8.47, p. 253]:

> If the setting is vividly identifiable, the observer has a concrete basis for a sense of belonging. He can begin to make relations; he can savor the uniqueness of places and people; he can learn to *see* (or to listen or to smell).

Every street should not look like every other street.

The question now becomes that of determining what makes a setting unique. Certainly many of the features that contribute to diversity also are unique: historical monuments, bridges, cuts and other tunnel-like confinements in the roadway, and prominent natural features. Another set of situations that are dramatic and therefore create a "sense of place" are the gateway views of many large cities. Pittsburgh's freeway approach to the Golden Triangle, the view of Manhattan from the Brooklyn–Queens Expressway, or of San Francisco from the Golden Gate Bridge are examples of the spectacular possibilities. Such experiences should not be accidental.

The legible structure in space and time of these and other less imposing but still unique settings is also an important aspect of good visual impact. People seem to need the emotional security of knowing where they are and how they are going to get to where they want to go. The automobile driver should be caught up in a sequential and unfolding flow of images that must be related in some way. In this regard, the alignment of transportation guideways is important since it is through the geometric layout that the driver sets future movement patterns and predicts their success. Sharp changes in alignment thus are not encouraged unless the intent is to provide some drama to overcome monotony. Perhaps also of importance is the creation of a "beat" or "rhythm" for the traveler. If unique objects such as light poles or abutments pass by too infrequently, temporal relationships become indistinct. The same appears to be true, however, if the objects go by so quickly that their passage is blurred in the traveler's mind.

On a larger scale, the strength of the relationships between areas of the city can be increased through emphasis on contrasting features of different neighborhoods, through more and better informational signing, and, more simply, through making the outstanding objects in the city more visible from transportation facilities. In a temporal view, clues of time can be made by exposing the scars of history and the signs of future intentions, and more attention might be paid to clarifying cyclical times in the environment (day/night, winter/summer, holiday/workday).

The next step in improving visual impact is to attempt to relate visible parts meaningfully to other aspects of life: the nature site and its ecology, functional activity, social structure, economic and political patterns, and so on [8.47]. Appleyard et al. point out that the transportation guideway actually should be a kind of linear exposition [8.46, p. 17]:

> ... running by the vital centers, exposing the working parts, picking out the symbols and the historical landmarks. Signs might be used for something more than giving directions or pressing a sale. They could point out the meaning of the scene: what is produced there, who lives there, how it grew, what it stands for.

The view, in other words, should become an extension of one's self.

There are many ways in which this extension might take place. One is through more intimate contact with the environs. This could be accomplished, for example, by splitting lanes so that the driver is not quite as overwhelmed with his smallness in relation to the mass of people flowing in and out of the city. Another method is to create a succession of "goals" for the driver to achieve. These "goals" would be a series of major landmarks that could be passed after short time intervals, thus giving the driver a sense of accomplishment. It is important that all major landmarks on the horizon eventually be achieved, that is, that the road not stop short of the landmark. Otherwise, there will be a feeling of frustration on the part of the driver.

The final criterion related to the visual impact is that the visual environment should play a role in fostering intellectual, emotional, and physical development of the viewer. This criterion is not easy to demonstrate, but certainly the negative effects of a highly impoverished environment have been shown over the years. Moreover, development of the types alluded to above can come as a result of many of the other perceptual criteria previously explored: diversity, legibility, and meaning. As one example, we might think of drawing attention to the more thought-provoking elements of a scene—unique manufacturing plants, railroad yards, recreational sites (such as for rowing and sailing)—and so on. This would be done by tipping and pointing the guideway so that drivers and passengers could view the scenes with little additional effort. Or, as pointed out by the Urban Advisors to the Federal Highway Administration, these sort of developments could also be achieved by creating an appropriate foreground and enframement of the feature as viewed from the road [8.48, p. 42]. The alternatives are many and as yet hardly explored.

8.9 A CASE STUDY: BART

In the late 1970s an analysis was made of the impacts of the Bay Area Rapid Transit System (BART). Because this was the first new urban rail system built in the United States in 50 years, many people felt that it was important to investigate its impacts in depth. This provides a useful perspective by which to summarize the types and magnitudes of factors that may be significant in impact analyses. Hence a detailed review of the BART impact studies will be presented here.

A map of the BART system is displayed in Fig. 8.8 and some of the major features of the system in Table 8.15. These were as of March 1978, when most of the impact studies were near completion. Basically BART is more nearly a suburban railroad than a subway/elevated system found in most of the larger cities of the developed

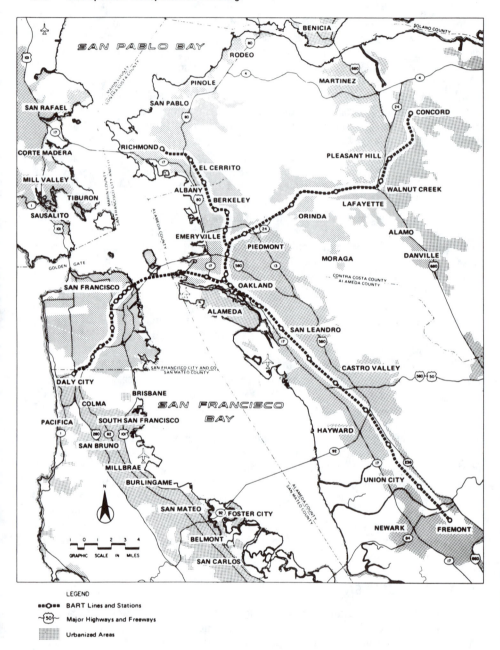

Fig. 8.8 The San Francisco Bay Area Rapid Transit (BART) system [8.55a].

world. BART was, in fact, planned and designed primarily to facilitate travel from outlying suburbs to downtown areas.

BART was approved and built during a period of vigorous growth in the San Francisco area. Suburban development was burgeoning, and major increments of highway capacity were being added. Right before BART opened, 10% of the total daily trips in the three BART counties were made on transit.

The impacts described below all are stated relative to what might have happened otherwise. Naturally "what might have been" is difficult to determine, given that it is hypothetical. In any case, a No-BART Alternative (NBA) was specified for comparison purposes. This assumed no changes to the area's freeways and bridges and a transit network and service equivalent to that right before BART was initiated. This alternate

Table 8.15   Major Features of the BART System*

Length:	The 71-mile system includes 20 miles of subway, 24 miles on elevated structures, and 27 miles at ground level. The subway sections are in San Francisco, Berkeley, downtown Oakland, the Berkeley Hills Tunnel, and the Transbay Tube.
Stations:	The 34 stations include 13 elevated, 14 subway, and 7 at ground level. They are spaced at an average distance of 2.1 miles: stations in the downtowns are less than $\frac{1}{2}$ mile apart, while those in suburban areas are 2 to 4 miles apart. Parking lots at 23 stations have a total of 20,200 spaces. There is a fee (25 cents) at only one of the parking lots. BART and local agencies provide bus service to all stations.
Trains:	Trains are from 3 to 10 cars long. Each car is 70 ft long and has 72 seats. Top speed in normal operations is 70 mph with an average speed of 38 mph, including station stops. All trains stop at all stations on the route.
Automation:	Trains are automatically controlled by the central computer at BART headquarters. A train operator on board each train can override automatic controls in an emergency.
Fares:	Fares range from 25 cents to $1.45, depending upon trip length. Discount fares are available to the physically handicapped, children 12 and under, and persons 65 and over.
Service:	BART serves the counties of Alameda, Contra Costa, and San Francisco, which have a combined population of 2.4 million. The system was opened in five stages, from September 1972 to September 1974. The last section to open was the Transbay Tube linking Oakland and the East Bay with San Francisco and the West Bay.
	Routes are identified by the terminal stations: Daly City in the West Bay, Richmond, Concord, and Fremont in the East Bay. Trains operate from 6 a.m. to midnight on weekdays, every 12 min during the daytime on three routes: Concord–Daly City, Fremont–Daly City, Richmond–Fremont. This results in 6-min train frequencies in San Francisco, downtown Oakland, and the Fremont line, where routes converge. In the evening, trains are dispatched every 20 min on only the Richmond–Fremont and Concord–Daly City routes. Service is provided on Saturdays from 9 a.m. to midnight at 15-min intervals. Future service will include a Richmond–Daly City route and Sunday service.[†] Trains will operate every 6 min on all routes during the peak periods of travel.
Patronage:	Approximately 146,000 one-way trips are made each day. About 200,000 daily one-way trips are anticipated under full service conditions.
Cost:	BART construction and equipment cost $1.6 billion, financed primarily from local funds: $942 million from bonds being repaid by the property and sales taxes in three counties, $176 million from toll revenues of transbay bridges, $315 million from Federal grants, and $186 million from interest earnings and other sources.

*As of March 1978.
[†]Sunday service began in July 1978.
Source: Preface pages in any one of the reports listed in [8.55].

thus put BART in about as favorable light as possible in the Bay Area [8.55a]. Hence, by this measure we should be greatly surprised if *any* impacts are perceptible when viewed against the totality of the area.

BART's impacts have been divided into six areas. These are summarized briefly below, with the focus being on those that may not have been as apparent beforehand.

8.9.1 Transportation and Travel [8.55b]

A major goal of BART was to improve travel times by public transit to major employment centers. Compared to the No-BART Alternative (NBA), this has been done. The average peak-period transit time has been reduced 12% from 46 to 40 min. Off-peak times have improved even more. Still, BART has not provided travel times for any trips comparable with the automobile.

Travelers perceived BART favorably with regard to its "qualitative" attributes: the comfort of the ride, safety from accident, security from crime, and (compared to autos) the opportunity to devote attention to something other than driving. Travelers were much less satisfied with parking availability at stations, BART's reliability, the time spent waiting for trains, total journey time, and (compared to buses) the cost.

BART carried 2 to 3% of all trips in its greater service area, which is about half the original estimate. Growth has been about 5% annually. The use of BART by people of different ethnic groups and income levels is roughly in proportion to their share of the service-area population (but not necessarily of the whole Bay Area).

Between 5 and 10% of BART trips probably would not have been made at all had BART not been available. In other words, BART *induced* more tripmaking. Of the remainder of BART trips, about half would have gone by bus and the other half by auto. BART had the immediate effect of reducing traffic flow by roughly 3,000 vehicle trips per day in each direction in the important and congested Bay Bridge corridor. This is about 2 years' growth in that corridor. Still, new trips by automobile have appeared to fill the road space freed by the diversion of trips to BART. In total, on the transbay corridor, it is estimated that there would have been 135,000 daily trips, on average, in the NBA. With BART, the total is 145,000. Total auto trips were down about 2%.

8.9.2 Land Use and Urban Development [8.55g]

Overall, BART has influenced land use and urban development in the Bay Area both directly (through its service) and indirectly (through effects on zoning regulations and the like). The impacts have been small, although not inconsequential. More specific results are:

1. Retailers almost completely disregard BART in their location decisions.
2. BART has affected prices and rents, but only marginally.
3. High-density residential development has not occurred in BART station areas zoned for such uses.
4. Relocation due to BART—including 3,000 households and 500 businesses—has not been disruptive.
5. BART's construction caused or contributed to declining retail sales in several areas.
6. BART has not stimulated residential relocation, nor has it been a major factor in most firms' locational decisions. Still, most developers ranked it as "somewhat important" in their decision making.
7. Residential property values within 500 to 1,000 ft of a BART station initially

rose marginally, but may have declined owing to the automotive traffic and parking nuisances.

8. Speculation occurred in 13 of the 17 station areas, but in no instance did it involve large-scale purchase or holding of land.

8.9.3 Economic and Financial [8.55h]

Several impacts in this area are interesting. First, as noted earlier, it appears that BART had almost no impact on Bay Area economic development. It did, however, increase the total goods and services purchased within the region by about $3.1 billion between 1964 and 1976. BART's construction was responsible for about 31,000 person • years of direct employment and another 44,000 of indirect. There was increased employment opportunity for minorities, but the largest share was for the laborer category, which involves little lasting skill enhancement.

Perhaps the most startling conclusion is that BART's construction and operating finance, which rely substantially on the property and sales tax, results in a heavier burden on low-income households than high.

8.9.4 Environmental [8.55f]

BART has not had much impact on the environment. The disruption of traffic and other activities during construction, traffic and parking problems at suburban stations, and train noise along aerial lines have been the most serious negative impacts. Positive ones include the landscaping, linear parks, encouragement of downtown street improvements, and generally excellent provisions for its riders. Regional air-quality impacts are insignificant because the effect on automotive traffic has not been substantial.

8.9.5 Transportation Disadvantaged [8.55c]

The principal economic benefit provided by BART for disadvantaged households was the direct employment noted above. The principal detriment was significantly lower proportional ridership of low-income minority people, who clearly are the most transportation disadvantaged. For the handicapped, while BART itself is relatively barrier-free, there remain obstacles to travel in the total environment, including boarding and use of feeder buses, absence of curb cuts, and inaccessible buildings.

8.9.6 Public Policy [8.55i]

Bay Area officials took few policy actions directly or indirectly because of BART. Most of those taken were intended to protect against, rather than take advantage of, BART. Further, public officials made policies individually, rather than as part of a coordinated strategy to provide a balanced and cost-effective total transportation system for the Bay Area.

Policy decisions generally were made using existing governmental decision-making processes. Most choices were made incrementally, and implementation of decisions did not always result.

The most significant public policy impacts seemed to be:

1. Local land use—special planning studies resulted in general plan and zoning changes. Some development areas were expanded.

2. Land-use development—for those localities that matched incentives with market demand, BART resulted in greater encouragement for development.

3. Transit finance—the Metropolitan Transportation Commission has more

authority over allocation of transit funding and monitoring transit operator performance. There has been increased involvement by the state legislature in authorizing additional *overall* transit funding.

8.9.7 Comments on the BART Impact Studies

Perhaps the most surprising results of the BART Impact studies are twofold:

1. BART led to only a slight change (an increase) in trip-making overall and in most corridors.
2. Most other impacts were relatively small.

The second actually derives from the first. If not much change in travel takes place, then it is unlikely that air pollution, noise, sales, land values, and all other factors that depend either directly or indirectly on travel will change either. The surprise is in the relatively small impact given what appears to be a large expenditure. In fact, however, the expenditure also was very small relative to total incomes in the region.

8.10 SUMMARY

This chapter dealt with transportation system impacts. Such impacts cover a broad spectrum, all the way from the destruction of historic markers to the economic development of the entire metropolitan area. Most of the models discussed were much less sophisticated than those presented in the travel forecasting section. This is not to imply that they are less important, however.

Considered in this section were the specific matters of: (1) energy consumption, (2) air pollution, (3) noise, (4) retail sales and land values, (5) relocation, and (6) visual impacts.

Energy consumption was addressed from the standpoint of the energy efficiencies of different vehicles. Forecasting of consumption was shown to be difficult given uncertainties about future efficiencies. For air-quality impacts, standards were presented and factors influencing the achievement of these standards identified. Estimation techniques were demonstrated for both microscale and mesoscale regions. Noise (sound) was shown to be measurable on dBA, L_{eq}, and L_{dn} scales. Means were presented for estimating some of these factors. The direct impacts on humans was discussed and found to be best represented by the "percent highly annoyed."

Retail sales and land values both have been affected by freeways, but most of the greater increases go to vacant land right next to new highway facilities. Relocation was shown to be an often traumatic experience, especially for older and handicapped people, yet in most cases the effects of relocation, especially on the physical quality of housing for the relocatees, has been positive.

Visual impacts were treated from the viewpoint of the user since relatively little is known about impacts on nonusers. Six criteria were explored: (1) sensual ranges of comfort, (2) diversity, (3) perceptual identity, (4) mental relationships in space and time, (5) meaningful environment, and (6) encouragement of attention and exploration.

A summary case study of the BART impact studies showed that most effects were relatively small. This is a somewhat surprising result, given the large expenditure on the system.

Impacts of the type described here plus forecasts of travel are employed in the evaluation stage in making decisions as to the best transportation system alternatives.

Previous to that, however, specific information must be collected. This is the subject of Chap. 9.

BIBLIOGRAPHY

8.1 Braun, R. R., and M. F. Rodin: *Quantifying the Benefits of Separating Pedestrians and Vehicles*, NCHRP report 109, Washington, D.C., 1970.

8.2 Gamble, H. B., and T. B. Davinroy: *Beneficial Effects Associated with Freeway Construction*, NCHRP report 193, Washington, D.C., 1970.

8.3 Environmental Protection Agency, Region X: *Guidelines for Preparation of Environmental Statements*, Seattle, Wash., April 1973.

8.4 U.S. Department of Transportation: *Secondary Impacts and Consequences of Highway Projects*, NTIS, Springfield, Va., Oct. 1976.

8.5 U.S. Department of Transportation: "Environmental Impact and Related Procedures; Final Rule and Revised Policy on Major Urban Mass Transportation Investments and Policy Toward Rail Transit," *Federal Register*, vol. 45, no. 212, Thursday, Oct. 30, 1980.

8.6 Council on Environmental Quality: "Forty Most Asked Questions Concerning CEQ's National Environmental Policy Act Regulations," *Federal Register*, vol. 46, no. 55, Monday, March 23, 1981.

8.7 Transportation Research Board: *Transportation Energy: Data, Forecasting, Policy, and Models (Several Papers)*, Transportation Research Record 764, Washington, D.C., 1980.

8.8 Claffey, P. J.: *Running Costs of Motor Vehicles as Affected by Road Design and Traffic*, NCHRP report 111, Washington, D.C., 1971.

8.9 U.S. Department of Transportation, Federal Highway Administration: *Procedures for Estimating Highway User Costs, Fuel Consumption, and Air Pollution*, Government Printing Office, Washington, D.C., March 1980.

8.10 U.S. Department of Transportation, Urban Mass Transportation Administration: *Transit and Energy*, Government Printing Office, Washington, D.C., Dec. 1970.

8.11 U.S. Department of Transportation: *Characteristics of Urban Transportation Systems— A Handbook for Transportation Planners*, Washington, D.C., July 1977.

8.12 Transportation Research Board: *Energy Effects, Efficiencies, and Prospects for Various Modes of Transportation*, NCHRP Synthesis of Highway Practice 43, Washington, D.C., 1972.

8.13 U.S. Congress, Congressional Budget Office: *Urban Transportation and Energy: The Potential Savings of Different Modes*, Government Printing Office, Washington, D.C., 1977.

8.14 U.S. Department of Transportation, Federal Highway Administration: *Energy Requirements for Transportation Systems*, Washington, D.C., June 1980.

8.15 American Association of State Highway and Transportation Officials: *A Manual on User Benefit Analysis of Highway and Bus-Transit Improvements: 1977*, Washington, D.C., 1978.

8.16 U.S. Environmental Protection Agency: *1972 National Emissions Report*, EPA-450/2-74-012, Washington, D.C., June 1974.

8.17 U.S. Environmental Protection Agency: "National Primary and Secondary Ambient Air Quality Standards," *Federal Register*, vol. 36, no. 84, April 30, 1971.

8.18 U.S. Environmental Protection Agency: "1975 Federal Test Procedure, FTP," *Federal Register*, vol. 36, no. 128, July 2, 1971.

8.19 U.S. Environmental Protection Agency: *Compilation of Air Pollutant Emission Factors*, AP-42, 2nd ed., suppl. 5, Washington, D.C., Dec. 1975.

8.20 Dabberdt, W. F.: "Experimental Studies of Near-Roadway Dispersion," 69th Annual Air Pollution Control Association Meeting, Portland, Oregon, June 27, 1976.

8.21 Cahill, T. A., and P. J. Feenay: "Contribution of Freeway Traffic to Airborne Particulate Matter," Final Report to the California Air Resources Board, Contract ARB-502, Sacramento, Calif., June 15, 1973.

8.22 Cadle, S. H., et al.: "General Motors Sulfate Dispersion Experiment: Experimental Procedures and Results," *Journal of the Air Pollution Control Association*, vol. 27, no. 39, 1977.

8.23 Sontowski, J.: "Microscale Modeling of Near Roadway Air Quality by Numerical Techniques," *Conference on the State of the Art of Assessing Transportation Related Air Quality Impacts*, Transportation Research Board Special Report 167, 1976.

8.24 Turner, D. B.: *Workbook of Atmospheric Dispersion Estimates*, U.S. Environmental Protection Agency, AP-26, Washington, D.C., Jan. 1973.

8.25 Benson, P. E.: *A Versatile Dispersion Model for Predicting Air Pollutant Levels Near Highways and Arterial Streets*, Report #FHWA/CA/TL-79/23, California Department of Transportation, Sacramento, Calif., Nov. 1979.

8.26 Benson, P. E.: *Caline 3 — A Graphical Solution Procedure for Estimating Carbon Monoxide (CO) Concentrations Near Roadways*, FHWA Technical Advisory #T 6640.6, U.S. Department of Transportation, Federal Highway Administration, Washington, D.C., March 2, 1981.

8.27 Zimmerman, J. R., and R. S. Thompson: *User's Guide for HIWAY, A Highway Air Pollution Model*, U.S. Environmental Protection Agency Report EPA-650/4-74-008, Washington, D.C., Feb. 1975.

8.28 Carpenter, W. K., G. G. Clemena, and W. R. Lunglholfer: *Introduction to AIRPOL-4: A User's Guide*, Virginia Highway and Transportation Research Council, Report no. 75-R56, Charlottesville, Va., May 1975.

8.29 U.S. Environmental Protection Agency: *Guidelines for Air Quality Maintenance Planning and Analysis*, vol. 9, EPA-450/4-75-001, Washington, D.C., Jan. 1975.

8.30 Dimitriades, B.: "Chemistry," *Conference on the State of the Art of Assessing Transportation Related Air Quality Impacts*, Transportation Research Board Special Report 167, 1976.

8.31 Kozlowski, T. P.,: "SAPOLLUT: Estimating the Air Quality Impact of Vehicular Emissions Resulting from a Traffic Assignment," *Conference on the State of the Art of Assessing Transportation Related Air Quality Impacts*, Transportation Research Board Special Report 167, 1976.

8.32 Mancuso, R. L., and F. L. Ludwig: *User's Manual for the APRAC 1-A Urban Diffusion Model Computer Program*, U.S. Environmental Protection Agency, no. EPA-650/3-73-001, Washington, D.C., Sept. 1972.

8.33 Ranzieri, A. J., and E. C. Shirley: "Examination of Regional Photochemical Models by a User," *Conference on the State of the Art of Assessing Transportation Related Air Quality Impacts*, Transportation Research Board Special Report 167, 1976.

8.34 Beaton, J. L., et al.: *Mathematical Approach to Estimating Highway Impact on Air Quality, Air Quality Manual*, vol. IV, Federal Highway Administration Report no. FHWA-RD-72-36, Washington, D.C., April 1972.

8.35 Beaton, J. L., et al., *Appendix to Volume IV, Air Quality Manual*, vol. V, Federal Highway Administration Report no. FHWA-RD-72-37, Washington, D.C., April 1972.

8.36 Beaton, J. L., et al.: "Air Quality Analysis in Transportation Planning," *Transportation Research Record* 670, 1978.

8.37 U.S. Environmental Protection Agency: *Information on Levels of Environmental Noise Requisite to Protect Public Health and Welfare with an Adequate Margin of Safety*, EPA Document 550/9-74-004, Washington, D.C., March 1974.

8.38 U.S. Department of Transportation, Federal Highway Administration: *Federal-Aid Highway Program Manual*, vol. 7, Chap. 7, Sec. 3, Washington, D.C., 1973.

8.39 Kugler, B. A., D. E. Commins, and W. J. Galloway: *Highway Noise, A Design Guide for Prediction and Control*, NCHRP Report 174, Washington, D.C., 1976.

8.40 U.S. Environmental Protection Agency: *The Urban Noise Survey*, EPA Document 550/9-77-100, Washington, D.C., August 1977.

8.41 Zettel, R. M.: "On Studying the Impact of Rapid Transit in the San Francisco Bay Area," *Highway Research Board Special Report* 111, Washington, D.C., 1970.

8.42 Horwood, E. M., et al.: *Community Consequences of Highway Improvement*, NCHRP Report 18, Washington, D.C., 1965.

8.43 Thomas, E. N., and J. L. Schofer: *Strategies for Evaluation of Alternative Transportation Plans*, NCHRP Report 96, Washington, D.C., 1970.

8.44 Appleyard, D., K. Lynch, and J. R. Nuer: *The View from the Road*, The MIT Press, Cambridge, Mass., 1964.

8.45 Cron, F. W.: "The Act of Fitting the Highway to the Landscape," in W. B. Snow (ed.), *The Highway and the Landscape*, Rutgers University Press, New Brunswick, N.J., 1959.

8.46 Lynch, K.: *The Image of the City*, The MIT Press and Harvard University Press, Cambridge, Mass., 1960.

8.47 Lynch, K.: "City Design and City Appearance," in W. I. Goodman and E. C. Freund (eds.), *Principles and Practice in Urban Planning*, International City Managers Association, Washington, D.C., 1968.

8.48   The Urban Advisors to the Federal Highway Administrator: *The Freeway in the City*, U.S. Government Printing Office, Washington, D.C., 1968.

8.49   Neibanck, P. L.: *Relocation in Urban Planning: From Obstacle to Opportunity*, University of Pennsylvania Press, Philadelphia, 1968.

8.50   U.S. Advisory Commission on Intergovernmental Relations: *Relocation: Unequal Treatment of People and Businesses Displaced by Government*, Washington, D.C., January 1965.

8.51   House, P.: "Relocation of Families Displaced by Expressway Development: Milwaukee Case Study," *Land Economics*, vol. 46, no. 1, February 1970.

8.52   Christensen, A. G., and A. N. Jackson: "Problems of Relocation in a Major City: Activities and Achievements in Baltimore, Maryland," *Highway Research Record* 277, Washington, D.C., 1969.

8.53   Levin, D. R.: "Displacement and Relocation Needs for Present and Future Highway Programs," in *Relocation: Social and Economic Aspects*, Highway Research Board Special Report 110, Washington, D.C., 1970.

8.54   U.S. Department of Transportation: *Recommendation for Northeast Corridor Transportation, Final Report*, vols. 1 and 3, National Technical Information Service, Springfield, Va., September 1971.

8.55a   U.S. Department of Transportation, Urban Mass Transportation Administration: *BART in the Bay Area: The Final Report of the BART Impact Program*, NTIS, Springfield, Va., April 1979.

Individual BART impact reports available from the same source:

8.55b   *BART's First Five Years: Transportation and Travel Impacts* (Peat, Marwick, Mitchell, and Co.).

8.55c   *Implications of BART's Impacts for the Transportation Disadvantaged* (Urban Dynamics Associates).

8.55d   *The Local Implications of BART Development* (Booz, Allen, and Hamilton, Inc.).

8.55e   *Impacts of BART on Bay Area Institutions and Life Styles* (Jefferson Associates, Inc.).

8.55f   *Environmental Impacts of BART* (Given Associates, Inc., and Deleuw, Cather & Co.).

8.55g   *Land Use and Urban Development Impacts of BART* (John Blayney Associates/David M. Dornbusch & Co., Inc.).

8.55h   *The Economic and Financial Impacts of BART* (McDonald & Grefe, Inc.).

8.55i   *The Impact of BART on Public Policy* (Booz, Allen, and Hamilton, Inc.).

8.56   Webber, M. S.: "The BART Experience: What Have We Learned?" *The Public Interest*, Fall 1976.

8.57   Knight, R. L., and L. L. Trygg: *Land Use Impacts of Rapid Transit: Implications of Recent Experience*, U.S. Department of Transportation, Washington, D.C., 1977.

EXERCISES

8.1   Pick a particular transportation project, program, policy, or strategy. Identify and describe in about 500 words the impacts that you feel are secondary or indirect, irreversible, and unavoidable. Give reasons.

8.2   For the example given in Sec. 8.3.4, assume that the volume of autos, trucks, and buses flowing *into* the downtown area is two-thirds of each outflow volume. Using the same other assumptions as in the example, compute for 1990 (*a*) the total Btu/hr for the inflow traffic, and (*b*) the average Btu/passenger · mile for the total inflow auto and bus traffic.

8.3   From the literature, find the direct energy consumption for a major activity center or people-mover system like a monorail, minirail, jetrail, skybus, elevator, or escalator. Put in terms of Btu/passenger · mile.

8.4   Assume that a "sensitive receptor" (school, nursing home, hospital, etc.) is located 246 ft from the edge of a proposed at-grade highway section that will carry a peak-hour traffic load of 20,000 vehicles/hr. For conditions of "D" stability and a cross-wind at 3 mph, calculate the expected carbon monoxide concentration at that sensitive receptor at a height of 6 ft above the surface. Assume a composite emission factor for CO of 52 g/vehicle · mile.

8.5 Assume that a new freeway system to be constructed in a metropolitan region will permit an increase in average traffic speed of 10% per year (compounded) over the next 4 years. If the average vmt increases at a compound rate of 4% per year, and no changes in vehicular emissions standards are made, show the effect of the new system on the carbon monoxide and nitrogen oxides burden for each of the four years. The composite emission factor for any speed may be estimated by multiplying the composite emission factor at a given speed by the appropriate speed correction factor for the pollutant specified. Note that nitrogen oxide emissions increase with vehicle speed. Given:

Present vmt = 1,500,000
Present average route speed, S = 32 mph
Composite emission factor for carbon monoxide @ 32 mph = 41 g/mile
Composite emission factor for nitrogen oxides @ 32 mph = 5 g/mile

Speed correction factors, V_s:
For carbon monoxide, $V_s = EXP(A + BS + CS^2)$ where:
$A = 1.241$
$B = -7.52 \times 10^{-2}$
$C = 6.09 \times 10^{-4}$
S = average traffic speed, mph
For nitrogen oxides, $V_s = A + BS$ where:
$A = 0.602$
$B = 2.027 \times 10^{-2}$
S = average traffic speed, mph

8.6 For the conditions listed in exercise 8.5, show the effect of "no-build" on the carbon monoxide and nitrogen oxides burden if it is assumed that the same vmt growth rate is used, but that, due to congestion, the average route speed *decreases* by 6% per year.

8.7 The noise levels from several different roadways, each considered separately, cause sound pressure levels of 49, 53, 47, and 52 dBA. Each of these levels is predicted for the same measuring location.
(a) What will the noise level be when all roadways are considered?
(b) Suppose the roadways causing levels of 47 and 49 dBA are eliminated. What would the total level for the other two be?
(c) Which roadways must be eliminated if you are trying to meet a 50 dBA criterion for this location?

8.8 Measurements made outdoors in a residential neighborhood have yielded the following history of A-weighted noise levels (somewhat simplified):

12 M–5 a.m.   40 dBA
5 a.m.–7 a.m.   35 dBA
7 a.m.–10 a.m. 55 dBA
10 a.m.–4 p.m. 52 dBA
4 p.m.–6 p.m.   65 dBA
6 p.m.–10 p.m. 50 dBA
10 p.m.–12 M   45 dBA

(a) Calculate L_{eq} for the day.
(b) Calculate L_{dn} for the day.

(*c*) Comment on the acceptability of this noise level and the level of annoyance it might cause.

8.9 Pick a recently completed local transportation project. Interview two or three merchants in the close vicinity of the project area. Ask them what they think the impact was on their business from the standpoint of short- and long-term property values, profits, and employment. How does this compare to your expectation previous to the interviews?

8.10 In the library, search for recent studies of relocation impacts. Summarize these in about 500 words and indicate how the results differ, if any, from those described in the text. Did the Uniform Relocation Act have any influence?

8.11 Select a particular section of street in your locality. If it could be reconstructed in the next 5 years, what changes would you make to improve its visual impact? Relate these to the criteria in Sec. 8.8.1.

9 Transportation Information Systems

The information system idea is one of the great synthesizing concepts of our time. It denotes the purposeful organization of information. It may encompass either, or both, qualitative and quantitative data. Transportation information systems may involve information collection, processing, storage, retrieval, and display for planning, construction, maintenance, and operation.

This chapter, reflecting the assumed needs of the reader, is devoted primarily to management of quantitative data. This emphasis should be put in perspective by noting the growing recognition being given to *qualitative information systems*, such as special libraries and professional communication networks. Both types of information systems are fundamental to sophisticated approaches to complex urban problems that utilize professional experience and judgment, as well as empirical fact-gathering and analysis. Throughout this book the emphasis is on a combined inductive-deductive, quantitative-qualitative approach.

In practice, quantitative information systems are in the ascendancy, and it is in this area that metropolitan transportation planning has made its unique contribution. Here we can justifiably talk about the *"spectacular breakthroughs"* of the 1960s. Thanks to efforts like the Chicago Area Transporation Study (1956–1962) and the leadership of the U.S. Bureau of Public Roads, among many other persons and organizations, the basic concepts and technology for information systems for urban transportation planning were developed and put into practice in every U.S. metropolitan area in the short span of 15 years.

These urban transportation planning studies have contributed a major prototype and reservoir of experience to the development of more advanced and inclusive urban information systems. It is this contribution that this chapter will cover, and the term "information system" will refer particularly to the quantitative type.

An *information system* is a collection of technical people, procedures, computer hardware, computer software, and a data base organized to develop the information required to support the functions of the parent organization and/or allied organizations. It is important to note that "information system" and "*data-processing system*" have quite different denotations. Data-processing system refers specifically to the computer hardware and software (e.g., computer programs). An information system includes not only a data-processing system, but also the data base and complete personnel organization for a particular information function. Our focus here will be on the particular data base needed for urban transportation planning; more general readings on data-processing and information systems can readily be found elsewhere [9.9, 9.17].

The data base plays a vital role in transportation planning. The public confidence that urban transportation planning enjoys is based in part on its demonstrated ability to simulate urban systems in the computer. Such demonstrations require vast quantities of measurements of land use, human travel behavior, and other urban characteristics. These measurements cost money—which is a second reason for the importance attached to the data base. In a typical metropolitan transportation study during the 1960s, 60–70 percent of the budget went into collecting and organizing data for subsequent analysis. In an area of 1 million people, where the overall transportation study costs about $1.50 per capita, this would mean an investment of $1,000,000 in data. Obviously, both the purposes and magnitude of the task demand careful professional attention to this phase of planning.

9.1 PROBLEMS WITH INFORMATION SYSTEMS

Generally speaking there are two distinct concepts of quantitative information systems—the "data bank" and the "management information system." Information systems for urban transportation planning have characteristics of both types. The *data bank*, or data library, offers a common data base to a number of different users. For continuous transportation planning this concept has strong appeals: (1) With data so expensive, its use by a number of agencies helps to justify its cost. (2) Use of a common data base by different agencies and for different planning purposes removes one of the causes of lack of planning coordination; put positively, it facilitates coordination.

The multiple-use idea may not have as much appeal in initial transportation studies, which are often so hard pressed to whip data into shape for their own purposes that the needs of other agencies for the data take second priority. For this and for other reasons noted below, the viable ideal of a data bank is not easily achieved.

The second type, the management information system (MIS), also has its appeal and its problems. The appeal is that of operations research—that through computer simulation of real world problems and alternate solutions the best decision can be determined. For many types of transportation decisions this has been at least partially realized. The models described in other chapters utilize basic land-use and travel data to advise decision makers which of several alternate metropolitan transportation systems would be least expensive, safest, or require the least travel time. For the freeway design engineer these models aid decisions regarding the number of lanes and the location and design of ramps. For rapid-transit planners the forecasts of ridership make possible informed decisions regarding the economic feasibility of a proposed system or line extension.

However, here too our present capabilities fall far short of the ideal, especially since the term "management information system" (MIS) connotes to many people a system that supports *all* functions of urban management. It brings to mind a command post in which managers who are fed from "real-time" sensors can make decisions that control urban processes. But in only a few narrow sectors of management—such as control of rapid-transit trains, regulation of freeway access to maintain traffic capacity, and disposition of police and fire-fighting units—are we approaching "on-line, real-time," command-and-control systems.

For both on-line and off-line, the state of the art in management information systems is still far from the ideal. "Our understanding of the urban processes is poor; the use of quantitative methods of analysis, such as mathematical modeling, is embryonic; planning in heuristic and (most) factors in the urban environment— housing, employment, education, health, welfare—are not accessible to automatic control as is, for example, the flow of traffic" [7.9]. In sum, the picture we try to paint here is one of cautious optimism.

Many management information systems are now functioning in transportation agencies—and providing powerful support for management, planning, design and operations. Nevertheless, the need for caution is underscored by many unhappy experiences. A few of the pitfalls and problems encountered, and the lessons learned, are summarized below.

Over-optimism and over-ambition. Experience shows that it is too easy to be carried away by the exciting vistas of system concepts—and to promise too much too soon. It is better to proceed incrementally, accomplishing a series of limited objectives that lead toward the goal of a more complete information system.

Error. The nature of urban planning requires data with a high degree of accuracy. But gathering the data usually depends on a small army of previously inexperienced and very human people. As a result, some studies have been near disasters; most have had some trouble. The moral is clear: Data collection demands careful recruitment, selection, training, and supervision of personnel, and quality control.

Precision. Precision problems are generally not as serious as those involving error, but over-precision is wasteful of money and under-precision affects the validity of results. Among the problems that require attention are zone size, sample size, and area measurement.

Incompatible data. When data are gathered in different surveys, at different times, incompatibilities are likely to crop up. Some of the most common problems are delineation of blocks (e.g., study blocks versus census blocks) and analysis areas (traffic zones versus census tracts), and definitions of parcel (assessors versus planners) and household. Not all compatibility problems can be avoided, but preventing them requires continuing vigilance and resolving them, statistical ingenuity.

Administrative files. Records of local agencies which would be valuable for transportation planning, even if computerized, are often formulated in ways that are unsatisfactory for planning. Common examples are assessor's records, building permits, traffic accident reports, and fire calls. Since these records are satisfactory to the agency paying for them and a change would require effort, building a common information system is often frustrated.

Random invention. A real demon! In the absence of an adequately preplanned data system with well-understood conventions and procedures, ad hoc arrangements will prevail. This kind of unwelcome innovation can occur when a member of a field crew experiences an unanticipated situation; or it can come from a top manager acting decisively but in ignorance of the operating system. There are no substitutes for (1) pretesting the system to anticipate problems and questions and (2) thorough training of all personnel.

Computer system support. Many transportation planning information systems fail or falter because of the inadequacies of computer services. They may not have their own computer and are using a remote installation. They may be dependent on a local government installation that is already overloaded with accounting and other administrative applications and does not have the systems and programming personnel competence to handle their "scientific" jobs. In some localities computer personnel competent for scientific applications are hard to find, and even the study's own installation may be inadequately staffed.

9.2 GEOCODING

A system of geographic controls is indispensible in any urban study involving the handling of data that is spatially disaggregated finer than metropolitan-wide totals. Geographic controls—or "geocoding," as the subject is more popularly called—are especially needed in metropolitan transportation studies, where data is highly disaggregated. Land use and tripmaking by their nature are usually located by parcel number and/or street address. For use in transportation models these data must be aggregated to traffic zones and/or census tracts. Similarly, specific locations such as intersections in the transportation network must be fixed in space. In addition, computer mapping, which can present the results of surveys or computer-simulation model runs almost instantly rather than taking hours or days of a draftsman's time, requires a method for positioning the computer printer or plotter so as to replicate the actual locations of urban phenomena. Consequently, transportation studies by the mid 1960s had developed a rather sophisticated system of geocoding.

The function of geographic controls in a metropolitan study is to provide a system for locating the distribution of various data in geographic space. The data can be of any type that can be converted to a metric or other computer-readable form and are of a sufficiently fine detail that they can be geographically fixed. A system of geographic controls is, of course, separate from and by and large independent of the particular data files controlled by it. The data are coded, stored, and retrieved according to the geocoding system. However, the geocoding system itself produces a sizable set of computer records, known as the geographic base file (GBF).

The antecedents of an urban geocoding system are familiar to all planners and most layman. The most common is street address, which in all cities follows some system of street names and house numbers. Another familiar concept is the *block*—the two sides of a street between intersections. The U.S. Census has popularized another meaning of the term "block"—that of an area bounded typically on all sides by streets (the definition used herein). The Census aggregates blocks into Census tracts. Planners and other urban statisticians have always used Census tracts and other subareas as handy devices for aggregating and disaggregating data.

However, none of these traditional geocoding systems were by themselves adequate for the requirements of metropolitan transportation planning. To a very large extent, the development of high-speed, high-capacity computers, about 1960, made possible the handling of the mass of geographically disaggregated data involved in metropolitan subarea analyses. Existent geocontrol systems, like street addresses and traffic zones, assumed new importance. Block and parcel data, which previously could only be handled by cumbersome manual methods, came into much wider use. The possibilities of computer mapping and graphics spurred the use of grid coordinates.

9.2.1 Geographic Subsystems

The geographic control system that emerged in the 1960s consisted of means of relating the various geocoding subsystems so that data could be disaggregated or aggregated to the appropriate scale. There is correspondingly a rough scale of size in these subsystems, rising from a point in space to larger zone systems, and to the region as a whole, as follows:

Point in urban space	Census tract
House number	Traffic zone
Assessor's parcel	Zone, ring, sector, district
Block face	Political jurisdiction
Block (census)	(municipality, county, etc.)
Grid unit	Region

A bit of elaboration may be in order for some of these geocoding subsystems. Points in urban space can be described in numeric characters through the use of an *X-Y coordinate system.* Typically, any point can be fixed precisely enough with three X digits for its "easting" location and three Y digits for its "northing" location relative to some southwest point of origin ($X = 0$, $Y = 0$). "Digitization" of a point in space is a prerequisite for modern transportation and land-use planning since computer mapping and other operations are dependent on this means of simulating urban activities in space.

The *house number or street address* is the most familiar geocoding system. It requires an address coding guide (ACG) to convert street locations to an $X-Y$-based, flexible urban information system. Other weaknesses may be in separately identifying several dwellings or businesses at one address. A frequent problem occurs in identifying locations where there are no buildings with "house" numbers: vacant or undeveloped land, parks, cemeteries, large institutions, etc.

The *assessor's parcel* is the unit of record for the local tax assessor. Typically, it is contiguous land under a single ownership. Since this generally corresponds to the way the land is used, planners have found the parcel a convenient unit for recording land use. In addition, if the assessor's records are computerized and can be accessed, valuable data for urban analysis and planning can be gained.

The *block face* (Fig. 9.3, shown later) is important as the interface between street address and the (Census) block. It is one side of a block and demarked typically by intersections.

The *block,* illustrated in Fig. 9.2, is usually bounded on all sides by streets. Occasionally a block boundary will be a river, railroad, city limit, or other feature.

The *grid unit* is a geographic subarea that since 1960 or so has come from obscurity to perform valuable services. Figure 9.1 illustrates three different sets of grid units or cells derived from three different map projections. In the illustration the Wisconsin capitol falls within the 1,000-ft square grid unit identified by its

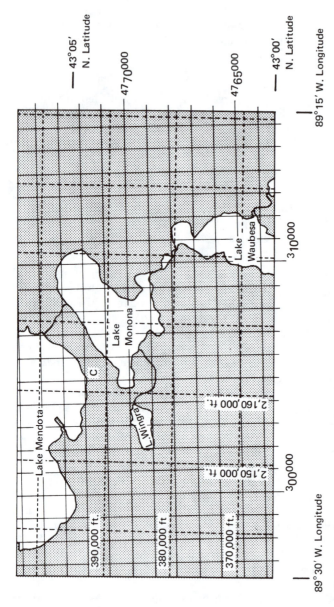

Fig. 9.1 Three systems of coordinates: The bordering ticks show longitude and latitude, the dashed lines show the Wisconsin Coordinate System, and the solid lines show the Universal Transverse Mercator (UTM) grid. [*Arthur N. Robinson and Randall D. Sale, "Elements of Cartography," New York: John Wiley & Sons, Inc., 1969, p. 29.*]

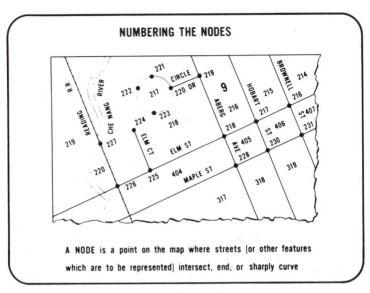

NUMBERING THE NODES

A NODE is a point on the map where streets (or other features)
which are to be represented) intersect, end, or sharply curve

Fig. 9.2 Example of sequential node numbering and block description by node (Census Tract No. 9). [*Bob Totschek, Vlad Almendinger, Ken Needham, "A Geographic Base File for Urban Data Systems," Santa Monica, Calif.: Systems Development Corporation, 1969.*]

southwest X, Y coordinates as 2164,0394, based on the state plane coordinate system. Within the same projection, grid units may be hierarchical, e.g., square mile and quarter-square mile. Grid units are very useful as a means of computer mapping, since a certain number of characters on the computer printer will form a square to represent the real world phenomena within the grid unit. Grid units are also useful when analysis subareas having equal area (or approximately equal area, as we will see below) are desired.

Census tracts (Fig. 9.2) are groups of blocks. Tract boundaries are generally stable from Census to Census and are widely used. Tracts are therefore often a useful subarea for analysis.

Traffic zones are tailored to the particular needs of transportation planning. They are formed to keep the zone size roughly equal in numbers of trips generated and somewhat homogeneous as to the kind of trip. Zone size correlates with the precision of simulations and forecasts of urban travel. In some trip-generation models zone homogeneity facilitates forecasting of tripmaking by type of trip, e.g., commercial. Often traffic zones are modifications of Census tracts, but smaller than tracts in areas of high trip generation, such as the central business district (CBD). Tract boundaries might also be modified in order to place a regional shopping center in a single zone. An alternative to modified Census tracts, usable with certain trip-distribution models, is to create traffic zones from combinations of grid units. For example, in the Niagara Frontier Transportation Study, zones varied from a quarter-square mile in the Buffalo CBD up to 6 and 8 square miles in rural areas [9.2].

Sectors, rings, and districts Traffic analysis requires that data be aggregated from census block to traffic zones in order to form meaningful transportation–land-use relationships. These zones are often further aggregated into districts for metropolitan-level studies. Traffic districts are organized into a ring-sector system by which the spatial location of districts can be readily visualized. Each district has a two digit name, the first representing its ring location and the second its sector location. The 10 sectors start at 12 o'clock and run clockwise. Thus each district has a geographic location which can be visualized without the aid of a map, a unique value of the ring-sector system.

Sometimes problems arise because the peculiar eccentricities of the $X-Y$ coordinate system are not understood. The three common map coordinate systems are shown in Fig. 9.1. The State Plane Coordinate system is the most often used. However, the Universal Transverse Mercator (UTM) projection is also growing in use, especially as more states think in terms of statewide transportation planning. Many states have two or more plane coordinate zones, which makes statewide plotting and other analyses of coordinate-based data extremely difficult when this projection is used. Both projections lend themselves to the use of Cartesian coordinates because X and Y distances (scale factors) are equal, thus simplifying many types of calculations.

In both of these equirectangular projections the X and Y lines are usually not exactly east–west and north–south due to the representation of the earth's spherical surface as a plane. This is illustrated in Fig. 9.1, which shows three systems of coordinates which generally appear on the United States Geological Survey (U.S.G.S.) quadrangles, the most popular source of grids for transportation studies. This figure is drawn from the Madison (Wisconsin) Quadrangle, at a scale of 1:62,500. The boarder ticks show the graticule, that is, the net formed by meridians of longitude and parallels of latitude, and the Wisconsin Coordinate System, South Zone. The other lines show the UTM grid. Only in latitude and longitude are the grid lines true north and south.

Locations in coordinates are given by a series of digits, the first half of which is the X location and the second half the Y location. Thus, on Fig. 9.1 the reference 0571 would locate point C (the capitol) with a 1,000-m square and 058716 would locate it within a 100-m square, these designations being the coordinates of the southwest corners of the squares. Additional digits would add greater precision in locating point C, but this would have to be done on a more accurate, larger scale map.

9.2.2 Geocoding with Census Files

The *1970 Census* instituted a new computerized procedure that incorporated many of the advances developed in the transportation studies during the 1960s, just as the studies had previously incorporated Census blocks, tracts and data into their information systems. The new Census procedures proved very helpful to continuing transportation planning, not only in simplifying geocoding preparations, but in facilitating the coordination of interactions between the transportation system and other urban systems.

The major geocoding innovation in the *1970 Census* is DIME—Dual Independent Map Encoding. DIME incorporates most of the useful geocoding components just discussed, especially the grid coordinates, the parcel/block inventory system, and the ACG. A DIME geographic base file is essentially a computerized description of block boundaries defined by its nodes (points such as intersections and other turns in boundaries). Figure 9.4 illustrates this technique. The term "Dual Independent" refers to the fact that each boundary segment is described by specifying its two end nodes and its right and left blocks (Fig. 9.3) Master coding maps with

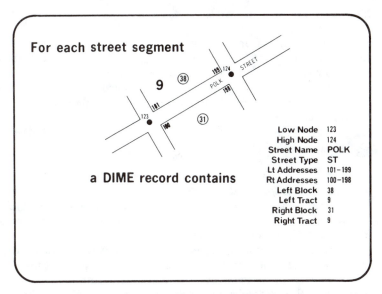

Fig. 9.3 Dual independent map encoding (DIME) system. [*Bob Totschek, Vlad Almendinger, Ken Needham, "A Geographic Base File for Urban Data Systems," Santa Monica, Calif.: Systems Development Corporation, 1969.*]

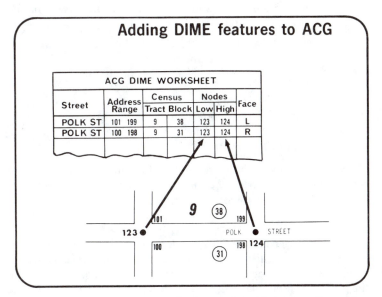

Fig. 9.4 Coding of a DIME file. [*Bob Totschek, Vlad Almendinger, Ken Needham, "A Geographic Base File for Urban Data Systems," Santa Monica, Calif.: Systems Development Corporation, 1969.*]

geographic identifiers at street intersections are available for most SMSA's. These depict most street features and are sufficiently detailed to read street address ranges and major nonstreet features. Keyed to the DIME maps, which exist in computer form, are the DIME ACG (Address Coding Guide) files; these contain street address ranges by blockface and are referenced to the DIME system through grid coordinates at street intersections.

9.3 LAND USE

A review of other chapters reveals the vast and varied data requirements of transportation models and planning. There are, however, *three basic transportation surveys* that frequently take a large amount of time and money and have specialized requirements. The next three sections are devoted to these inventories—land use, travel, and transportation facilities.

Land is a limited community resource, and it will become even less available in future years as the urban population increases and as the population places increasingly varied demands upon it. Not only is land itself valuable, but the way it is used is critical, since certain combinations of land uses tend to be compatible and reinforcing, others less so, and still others noncompatible and conflicting. For these reasons, the land-use survey, which attempts to identify and classify land uses in a systematic fashion, has been of great value to community planning.

The types of surveys that have been used in the past have varied widely in function, but most transportation-oriented land-use surveys have tried to serve four basic objectives, all of which remain valid:

1. To provide a land use base from which trip generation factors and trip generation forecasts can be derived
2. To provide necessary data for coordinating transportation facilities with other uses
3. To provide a "universe" of dwelling units from which a sample can be drawn for the home interview phase of the travel survey
4. To provide data useful for the day-to-day planning activities of city, county, and state government

With funding tighter in the 1980s, most agencies have been turning to secondary data to meet the first purpose above. This substitute approach for a land-use survey is discussed in Sec. 9.3.8.

9.3.1 Types of Land Surveys

Planning agencies conduct a wide range of land use studies to acquire basic data on land characteristics and the activities that occupy land. These data are used in evaluating current patterns of land use and in formulating the land-related aspects of the transportation plan. Chapin and Kaiser, in *Urban Land Use Planning* [9.1], describe nine basic studies of this type:

1. Land use survey
2. Vacant land survey
3. Flood damage prevention survey
4. Structural and environmental survey
5. Cost-revenue studies
6. Land value studies

7. Studies of the aesthetic features of the urban area
8. Studies of public attitudes and preferences regarding land use
9. Studies of activity systems

9.3.2 Classification of Land Uses

In the past, land-use data have been collected for one purpose and then collected again in another format for another purpose. This led to an undue duplication of effort and a lack of uniformity in results. To a very large degree these and other deficiencies have been rectified by the publication of the *Standard Land Use Coding Manual* in 1965 by the U.S. Urban Renewal Agency and the U.S. Bureau of Roads [9.11]. One of the main advantages of the *Manual's* approach is the ease with which data can be used for a variety of purposes once they are coded. Other advantages include the amenity of coded data to automated data processing techniques and the facility with which data collected at different times and in different cities can be compared. Use of this coding scheme has become increasingly common in the years since its publication, and its continued use should go far toward facilitating the analysis and interchange of statistical information and research findings.

The *Manual's* standard coding system provides *four levels of detail* on land-use activity. Further, each level is subdivided into categories: there are 9 one-digit categories, 67 two-digit categories, 294 three-digit categories, and 772 four-digit categories. The first two levels are illustrated in Table 9.1. The categories at the one-digit level identify land-use activities of a general nature, and categories at the two-, three-, and four-digit level, activities of a more specialized type. The structure of this classification system stresses, then, the activity aspect of land and permits that level of detail to be selected that is most appropriate to the analysis and presentation of data.

The coding scheme also permits a very rough correlation with the Standard Industrial Classification (SIC); insofar as possible, it uses the SIC category titles and that system's description of manufacturing activities. It does not, however, identify the four-digit land use categories by the same code numbers as the comparable categories in the SIC, nor does it use similar criteria to code activities.

9.3.3 Types of Land-Use Surveys

Once a classification system has been adopted—preferably the standard coding scheme just described or a variant thereof—the task of planning and carrying out the land-use inventory can begin. The decision as to the type of survey will be based on its purpose or purposes, and on the resources available. Land-use surveys are classified in several ways: first by whether or not dwellings and other places must be entered, and, second, by whether or not the data must be computer readable.

The first classification regarding building entry offers two types of land-use surveys, which Chapin and Kaiser call the *inspection* type and the *combined inspection–interview* type. The former is the most economical and is used whenever the survey purposes can be satisfied by inspection of the exterior of buildings. Such inventories are often termed "windshield surveys."

The combined inspection–interview survey is used when exterior inspection does not yield enough information. One such instance is when the land-use inventory is combined with another survey purpose, such as a taxable property inventory or school census, generally in cooperation with other agencies. In this way the cost to

Table 9.1   A Standard System for Identifying and Coding Land Use Activities—One- and Two-digit Levels

Code	Category	Code	Category
1	Residential	11	Household units
		12	Group quarters
		13	Residential hotels
		14	Mobile home parks or courts
		15	Transient lodgings
		19	Other residential, NEC*
2	Manufacturing	21	Food and kindred products—manufacturing
		22	Textile mill products—manufacturing
		23	Apparel and other finished products made from fabrics, leather, and similar materials—manufacturing
		24	Lumber and wood products (except furniture)—manufacturing
		25	Furniture and fixtures—manufacturing
		26	Paper and allied products—manufacturing
		27	Printing, publishing, and allied industries
		28	Chemicals and allied products—manufacturing
		29	Petroleum refining and related industries
3	Manufacturing (continued)	31	Rubber and miscellaneous plastic products—manufacturing
		32	Stone, clay, and glass products—manufacturing
		33	Primary metal industries
		34	Fabricated metal products—manufacturing
		35	Professional, scientific, and controlling instruments; photographic and optical goods; watches and clocks—manufacturing
		39	Miscellaneous manufacturing, NEC*
4	Transportation, communication, and utilities	41	Railroad, rapid rail transit, and street railway transportation
		42	Motor vehicle transportation
		43	Aircraft transportation
		44	Marine craft transportation
		45	Highway and street right-of-way
		46	Automobile parking
		47	Communication
		48	Utilities
		49	Other transportation, communication, and utilities, NEC*
5	Trade	51	Wholesale trade
		52	Retail trade—building materials, hardware, and farm equipment
		53	Retail trade—general merchandise
		54	Retail trade—food

(*See footnote on page 301.*)

**Table 9.1   A Standard System for Identifying and Coding Land Use
Activities—One- and Two-digit Levels** (*Continued*)

Code	Category	Code	Category
		55	Retail trade—automotive, marine craft, aircraft, and accessories
		56	Retail trade—apparel and accessories
		57	Retail trade—furniture, home furnishings, and equipment
		58	Retail trade—eating and drinking
		59	Other retail trade, NEC*
6	Services	61	Finance, insurance, and real estate services
		62	Personal services
		63	Business services
		64	Repair services
		65	Professional services
		66	Contract construction services
		67	Governmental services
		68	Educational services
		69	Miscellaneous services
7	Cultural, entertainment, and recreational	71	Cultural activities and nature exhibitions
		72	Public assembly
		73	Amusements
		74	Recreational activities
		75	Resorts and group camps
		76	Parks
		79	Other cultural, entertainment, and recreational, NEC*
8	Resource production and extraction	81	Agriculture
		82	Agricultural related activities
		83	Forestry activities and related services
		84	Fishing activities and related services
		85	Mining activities and related services
		89	Other resource production and extraction, NEC*
9	Undeveloped land and water areas	91	Undeveloped and unused land area (excluding noncommercial forest development)
		92	Noncommercial forest development
		93	Water areas
		94	Vacant floor area
		95	Under construction
		99	Other undeveloped land and water areas, NEC*

*NEC = not elsewhere coded.
Source: [9.11, pp. 29-31].

the planning agency is kept within acceptable limits. Other reasons for building entry is to collect data on space use by floor area or to determine the number of employment or dwelling units. Still another reason is to determine the condition of the structure. Obviously, the inspection–interview type of inventory is more elaborate and expensive than the exterior inspection type.

The second classification relates to the way the land-use data are recorded in the field. The older method is to record the data directly on maps or air photos. During the 1960s, especially in the large-scale transportation land-use studies, there was increased use of a method known as *field listing*. The use of either format is predicated on the purposes and amount of data to be acquired and the methods to be employed in storing and retrieving data. In general, the former is best suited to a map storage system, and the latter to an electronic storage system.

The map-record type of land-use inventory is the simplest. It is more frequently employed in the small town or city, or where a new planning program needs quick data, or a new planner wants to get the "feel of the area," or where the only data required are simple classifications of use. Typical blue-line prints or air photos at 1 in. = 400 ft or larger scale are used as field sheets on which the land-use information is annotated.

9.3.4 Field Listing

Field listing of land use is the near-universal method for transportation planning, and is growing in popularity for general comprehensive planning purposes. A separate field list form (Fig. 9.5) is usually reserved for each block and a line is filled out for each parcel or land-use and dwelling unit. In certain situations it has advantages over map-recorded surveys. First, it lends itself to computerized records and analyses. Computerized land use is especially suited to large metropolitan areas where statistical analyses are more used than conventional land use maps. Second, it permits automated and integrated record systems encompassing property assessment, land use, building permit and inspection, and other parcel-based records. Third, it must be used when the land-use survey provides the universe of dwelling units from which the home interview sample is drawn. Finally, it must be used when the amount of information to be included exceeds that which can be recorded on a map. Field lists can be used either for hand preparation of conventional colored land use maps, or keypunched for computer-made maps and analyses.

Figure 9.5 illustrates the basic form for "field listing" land uses. This form is designed to serve both the field form and the coding form from which computer records are created. Combining the forms saves the time of transcribing as well as eliminating the errors that would occur. The field form therefore reflects the standard 80-column record.

On the form a separate line is used for each parcel and for each separate dwelling unit or space use on the parcel. Thus every use in a multi-use building is listed separately. The lister travels in a predetermined and regular pattern so that the order of listing can be subsequently retraced on maps by office personnel. In an urban area this usually means starting at the northwest corner of a block and proceeding clockwise. Most areas can be covered by a two-man team in a car. However, where structures must be entered or traffic conditions will not allow, the field listing is done on foot.

As the first lister proceeds on his route, he lists successive parcels by entering the block number (field 7), in urban areas the street name (not shown on Fig. 9.5), the house or building number (field 12), apartment number (20) or location

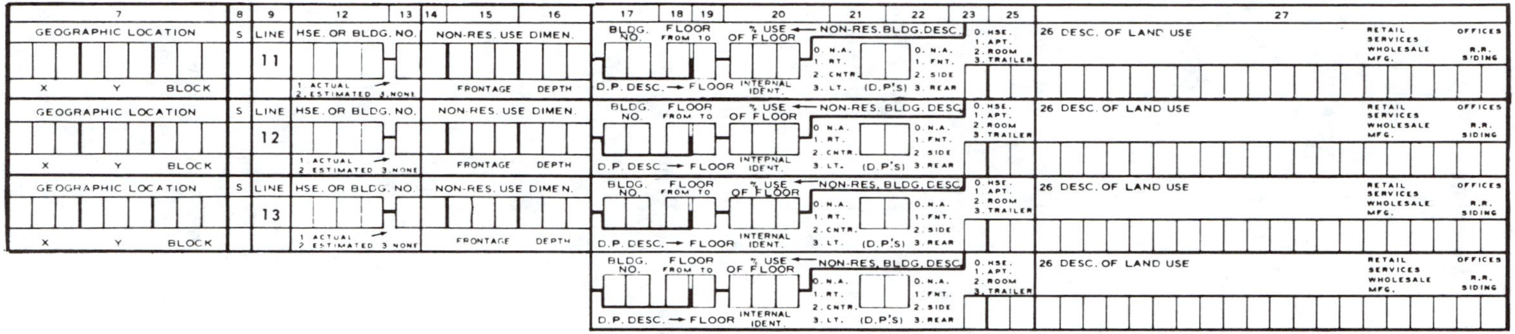

Fig. 9.5 Land use filed listing form (rural areas). [*New York State Department of Transportation, Albany, New York.*]

(21–22), and dwelling place (DP) type (23). For nonresidential uses, the field lister enters on the form the identification (17), the percentage of floor in the use (20), a written description of the use (26), and, to avoid ambiguity, a positive identification of the major use as retail, services, wholesale, manufacturing, or offices. The last entry helps eliminate a major source of error. For example, entries in field 26 like "drugs" or "home appliances" could be any one of the major uses.

Subsequently, office procedures supply parcel dimensions for nonresidential uses (15–16), the land-use classification numeric code (25), and firm name (27). Actual assignment of the land-use code is best done in the office since a look-up is required, especially if the four-digit, 772-classification system of the *Standard Land Use Coding Manual* [9.11] is used.

The field-list forms are then entered into computer files. Subsequently, contingency checks are made to screen out errors in coding, transcription, and typing. After this check is cleared, various computer runs are made to calculate land and floor areas, and to see that parcel area totals equal block areas, and that maps made by the computer printer conform to known retail, industrial, and residential uses. Lastly, summaries are prepared for the basic geographic units (block, zone, grid units, etc.) to be used in analysis, regional growth forecasts, and trip-generation calculation. It should be noted that these post-field phases of surveys are often underbudgeted in time and manpower.

9.3.5 Presentation of Land Use Data

The results of the land-use survey may be summarized in either map form or statistically. Traditionally, land-use data have been presented in the form of the land-use map. This shows land use by general category of use, that is, residential, commercial, industrial, institutional, parks and recreation, transportation and utilities, agriculture, and water. Ordinarily, these categories are shown in color in order to provide visual differentiation to the various uses, since this type of presentation is more effective in conveying information and in appealing to the public. A standard color scheme is nearly always adopted, the type varying with the diversity and types of uses and with the purpose of the survey.

In recent years, however, land-use maps have increasingly been based on printouts of data processed on a subarea basis by computer. These may require that a draftsman prepare an overlay of principal streets for orientation. Computer-made maps, described later, eliminate the time-consuming task of transcribing from the data base to the overlay.

The statistical summary of land use is typically prepared to show the total land area devoted to each category of use employed in the survey; including that in vacant or nonurban use. This information is usually broken down into subareas. These may be Census tracts, traffic zones, well-defined neighborhoods, ring-sector districts, or other areas delineated for analytical purposes. The amount of land given over to urban uses is frequently summarized in terms of percentages for the developed part of the city, the fringe areas, and the planning area of the survey. This format provides a meaningful summary for intercity comparisons or for comparisons between existing land use and proposed land use as set forth in a land development plan.

9.3.6 Secondary Data for Trip Generation

During the 1970s many agencies turned to secondary data sources as a substitute for land-use data to feed land-use and transportation models. There were several contributing reasons for this shift. First, although land-use surveys yield a vast amount of theoretically valuable data, as a practical matter the analysis of land-use activities generating trips often was reduced to numbers of persons or households "producing" trips and number of employees "attracting" trips. Second, research had demonstrated that these were as reliable indicators of trip generation as, say, acres of land by use, or square feet of floor area by use. Finally, as money became scarce, an alternative to expensive primary surveys was needed [9.13].

For the residential trip-production base, the U.S. Census has proven the best secondary source. More detailed information on household characteristics is available from the Census's Urban Transportation Planning Package (Section 9.5.2.3). Traffic zones now more often follow Census tract boundaries. In those areas where traffic zones differ, the Census, for a small charge, will tabulate household characteristics using a block-to-zone aggregation table.

Updates from the base year are sometimes necessary. Short of a survey, local records or city directories are the best source. Automobile registrations, appropriately geocoded, are commercially available.[1] A few regions are continuing to tabulate land use data regularly from local sources such as assessor's records, utility companies, and building or occupancy permits.

Creating an employment distribution file as a basis for calculating trip attractions is more difficult, and agencies use a wide variety of electronic sources, supplemented by calls or letters to larger employers. It is necessary to geocode the locations in the files. The sources include state employment security files, Dun and Bradstreet business data files, R. L. Polk directories, industrial directories, and the Census's *County Business Patterns*. None of these are adequate in themselves and some combination, manipulation, and/or primary supplementing is necessary. *County Business Patterns* has good totals but cannot be disaggregated geographically. Employment security files combine all locations for the same employee at one address, which may even be in another SMSA (Standard Metropolitan Statistical Area). Dun and Bradstreet has the establishment locations but does not cover public and other nonbusiness employment, and many small enterprises and professional offices [9.13].

9.4 TRAVEL FACILITIES

Inventories of transportation facilities—both road and mass transporation—serve three basic objectives. They (1) measure the capacity for and quality of service, (2) locate trouble spots, and (3) make possible simulation of existing and future travel. The last is especially important since it is used to devise and evaluate transportation plans. The transportation system, as noted in Chap. 5, can be divided into four components—guideways, vehicles, terminals, and control systems. The discussion here will focus on data for guideway links and networks.

9.4.1 Functional Classification of Guideways

Road component. The road component of a guideway system is subdivided into an arterial and nonarterial subsystem. In turn, the arterial subsystem consists of freeways and arterials. In transportation engineering parlance, freeways are divided roadways having no direct land access and no intersections with other streets at

---

[1] The R. L. Polk Company is the largest supplier of directories and automobile registration lists.

grade. Their prime function is to move traffic quickly, easily, and with a high measure of safety. By way of contrast, arterials have a dual function: to move traffic and to provide access to land uses, especially the highly trip-generating commercial activities.

The nonarterial subsystem is composed of local streets and collectors. The principal function of local streets is to provide access to land and primarily to residential land. While they occupy nearly 70 percent of total street mileage, they carry only 10 to 17 percent of all traffic. For this reason, local streets have no real significance at the scale of metropolitan transportation planning, and transportation network analysis and evaluation can reasonably focus on the smaller number of large facilities with little attendant loss in accuracy. The collector conducts traffic between local streets and arterials or local traffic generators such as shopping centers, schools, or parks. In commercial areas, traffic volumes often mount too quickly for the effective use of collectors; but they are used in large industrial areas. For most collectors, land access is an important function.

Transit component. There are two main types of public transit: local transportation and rapid transit. The difference between the two stems in part from the type of vehicles they use but more precisely from the type of right-of-way they operate upon.

Local transit generally operates on a public street right-of-way. As such, the mass transportation vehicle—be it bus, street car, or cable car—shares the right-of-way with automobile and truck traffic. The speed of the vehicle depends, then, on the speed of the traffic streams as well as on the number of stops per run and the distance between stops. Local transit has many subtypes: fixed route, variable route, or nonroute (e.g., Dial-a-Ride), schedule or nonschedule. Jitneys, for example, operate fixed routes, but are unscheduled.

On the other hand, rapid transit operates on a right-of-way reserved solely for its use. Because of this, it provides faster, more efficient service than most transportation facilities, especially over long distances. There are several types of rapid transit. One is the suburban railroad: it operates at relatively high speeds with intervals of .5 mile or more between stations. Another is the subway-elevated: it uses lighter, self-propelled cars with an ability to accelerate and decelerate relatively quickly, permitting more frequent stops. A third type is the light rail vehicle (LRV), successor to the street car, which may run either on street tracks or its own right-of-way. A final type is the express bus system on its own right-of-way.

9.4.2 Network Coding Concept

Until the 1950s, transportation studies were unable to represent an entire transportation system within a computer. Lacking this, they were unable to readily construct minimum time paths through the network or to simulate vehicle flows over the system. As a consequence, the quality and comprehensiveness of their findings suffered noticeably.

A procedure developed during the Chicago Area Transportation Study in 1956–1961 did much to correct these deficiencies, and thereby vastly improved the sophistication and quality of on-going transportation studies. This breakthrough was supplied by a very simple numbering system that enabled all the links of a transportation system to be identified and interconnected numerically, just as they are interconnected physically on the ground. The system is described thusly:

1. The entire metropolitan area is divided into zones, and each zone receives a three-digit number.
2. Each intersection between links in the street network is identified by a five-digit number, the first three digits being the number of the zone in which the intersection is located, and the last two digits identifying intersections within the zone.
3. Each link of the street system is identified by the number of the intersection at its beginning and end.

An illustration, Fig. 9.6, shows how this numbering system works in practice, as it did in the Chicago Transportation Study. Other studies use variants of this system. [9.2].

9.4.3 Travel-Facilities Inventory

The inventory process itself consists of locating and describing each link of the transportation system in the manner just noted. The description of each link includes a measure of both the present capacity and use of a specific link and a statement of its performance characteristics.

The first step in the facilities inventory is setting of criteria for arterials and then determining which of the many streets in a circulatory network are freeways and which are arterials, sorting these from collectors and local streets. The freeways are relatively easy to determine. But arterials are not as easily defined, since it is not always possible to determine just what streets are the major channels of movement within a city, nor what streets provide the main means of access to commercial activities. This can be and often is a subjective interpretation. It is not unusual, for example, for some streets to be classified as arterials that have a lower traffic volume than streets that are not so classified. Generally, though, roads that carry more than 2,000 vehicles per day and that are continuous in length are considered to be arterials; in any case, those are the basic criteria suggested here.

Subsequent steps in a travel-facilities inventory include preparation of a manual for training field and office personnel, and documenting the results. At the same time, the arterial network identified in the first step is digitized for computing

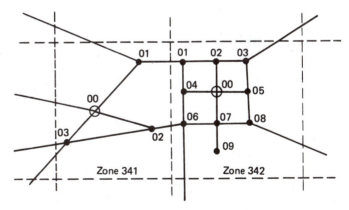

Fig. 9.6 Method of coding transportation networks. Arterials are identified by their numbers within zones and/or by their x-y coordinate locations.

purposes according to the method described in Sec. 9.4.2. Field workers then collect data needed to calculate travel time and capacity, such as link length, pavement width, right-of-way width, signals, green time, number of lanes, parking controls, speed limits, and adjacent land use. Field work is followed by checks, coding, data entry, electronic processing, reporting, and analysis.

9.4.4 Facilities Usage Survey

The function of this survey is to sample the traffic volume carried by individual links of a road network so that areas of congestion can be defined. It is also used—in an important function—to measure the accuracy of computer simulation models. This type of survey is often called the vehicles miles of travel survey (vmt).

The basic procedure for conducting the route usage inventory has been suggested by Creighton [9.2, pp. 155–156]. First the network of roads in an urban area must be mapped and classified into convenient categories such as freeways and parkways; major and secondary arterials; collector and local streets. These streets are then sampled, the sample rate depending on the volume carried by each class of street: "Limited access roads may be sampled at a rate running between 30 and 100 percent of all links. Arterials may be sampled at rates between 20 and 25 percent." Very small samples are taken of local streets; these range between 1 and 5 percent since they carry such small volumes of traffic that great reliability is required. The samples are selected on a random basis by computer.

The techniques used to measure traffic-flow vary by type of street. On freeways, counts are generally taken with portable machine counters for 24 hr or more. Arterial streets may be counted in the same way, or manually for short periods of time (usually 2 hr) with the count expanded to reflect a 24-hr period. A volume count intersection form is used to manually record traffic flow. In similar fashion, local streets may be counted manually for 2-hr periods and this amount factored to represent 24-hr traffic volumes. Volume counts can be made at intersections or midblock on both arterial and local streets, and on links of freeways.

Finally, the volume data is coded, compiled, and contingency checked. The sample volumes are then applied to other streets of similar type in the study area. The volume count for each street is multiplied by the length of the link. This produces the estimate of system vmt, which can be presented in map or statistical form.

9.5 TRAVEL ORIGIN AND DESTINATION

The origin and destination travel survey provides data on the origin and destination of all trips, the purpose, travel time, and length of trip, the mode of travel, and the land use at the points of origin and destination. This information then serves as the basis for predicting the demand levels and flow patterns of future travel. In particular, trip data are combined with land-use data to calculate trip-generation factors. In other words, tripmaking is related to land use and then forecast as a function of land use. Trip data also aid in calibration of transportation models for simulating trip distribution and assignment and modal choice for the base year. Still another use is in planning transit, both with fixed route and as special services for selected market segments.

9.5.1 Types of Trips

The origin–destination survey measures a representative sample of the trips that originate within or pass through an urban area. The vehicular and person trips that occur in an area are composed of three types: those made by persons and commercial vehicles from outside the area (external trips); those made by commercial vehicles within the area (internal commercial); and those made by persons who reside in the area (internal residential). Each of these requires a different type of sampling, interview technique, and listing.

9.5.2 Source of Travel Data

Travel origin and destination data can be obtained either through a primary survey or through secondary sources. Complete data are available only with a primary survey, generally called an "O&D"—origin and destination—survey. The complete O&D consists of three complimentary surveys—cordon, truck-taxi, and home interview.

9.5.2.1 Cordon survey The cordon survey is used to inventory external travel. As the first step, a cordon line is drawn about the study area. The line itself is located some distance from the built-up portion of the study area, where the density of roads is not too great. All roads that cross the cordon line are listed and traffic volumes on these roads counted by portable mechanical counters.

After these roads have been counted, they are arrayed in order from largest to smallest. The most heavily traveled roads are interviewed for a 24-hr period, less heavily traveled roads for 16 hr, and minor roads for some 8 to 12 hr. Generally a sufficient volume must pass through the interviewing stations to equal about 90 to 95 percent of the total average daily traffic crossing into the study area. Only a representative sample of the vehicles (perhaps 20 to 35 percent) that cross through the cordon line need be interviewed for trip data.

Vehicles are stopped at the interviewing station and the drivers are questioned as to their origin, destination, trip purpose, land use at origin and destination, and number of occupants inside the vehicle. Trucks are similarly interviewed, but buses and emergency vehicles are not stopped.

9.5.2.2 Truck-taxi survey This survey gauges the amount of commercial traffic that is based within the urban area. Trucks and taxis can be sampled from lists of registrations obtained from the state government or from municipal tax lists if they are licensed by the municipality. The sample is made on a random basis, and each selected registration is interviewed either by telephone, mail, or personal visit. The personal visit is the most expensive of the three types of interviews, but is unquestionably a more thorough and trustworthy technique. The data to be derived from these interviews include the purpose of the trip, the origin and destination of each trip, the number of passengers in the vehicle, and, in the case of trucks, the industry of ownership and truck loading. The sample rate may vary between 5 and 20 percent and is based on city size, interview technique, and needed precision.

9.5.2.3 Home interview survey The home interview survey measures the largest group of trips within an urban area, internal residential-based travel, which composes about 85 percent of all trips.

The sample in this survey is often a difficult technical problem requiring a knowledge of statistics, administrative costs, data-processing techniques, and use of data. Usually the sample is obtained through field listings, or from the Bureau of the Census, directories, or utility lists. Choice of source hinges on such factors as budget, reliability and currency of exsiting sources, and whether the source can be used to generate other types of data. Once the source has been chosen, the sample is then selected, usually so that there is an even distribution of interviews for each day of the work week over the entire metropolitan area. The sample rate may vary between 1 and 25 percent, depending on city size. Because of recent advances in sampling procedures, the sample population in these surveys has tended to decline, with a great savings in cost.

Data collected in a home interview include both characteristics of the household as well as information on trips made by members of the household. Characteristics of the household that are determined include such things as number and ages of residents, occupation, car ownership, and sometimes income. Travel data include origin and destination of trips, trip purpose, mode of travel, time of day, car loading, and land use at origin and destination. Travel data are obtained for every person 5 years of age or older. Children under 5 are considered as accompanying adults. Figure 9.7 shows portions of a typical form. The data from the three travel surveys are then expanded to reflect the entire "universe" of trips in the urban area.

9.5.3 The Urban Transportation Planning Package

An important secondary source is now available for updating home-based travel files. In 1970, and again in 1980, the Census of Population has included a sample known as Place of Work–Journey to Work. For this sample of households, Census enumerators obtained data on the place of work and its location, the mode of travel to work, household characteristics, and other related information. These data are geocoded and tabulated, on order, by Census tract and block group, and for a small additional charge, by traffic zone. This work is done by the Bureau to the specification of a local transportation planning organization. The product is called the Urban Transportation Planning Package (UTPP). Costs are estimated at about $10 per 1,000 SMSA population with traffic zone tabulations running another $2 or $3 per 1,000 [9.14]. The UTPP, of course, does not have cordon, truck–taxi, or nonwork trip data.

9.6 TRANSIT SURVEYS

Various types of surveys especially appropriate for transit are outlined in this section. An excellent source on such procedures is the *Canadian Transit Handbook* [9.31]. A few other sources are available [9.32, 9.33].

As with the automobile mode, transit planners and managers are concerned with two major types of information. The first type, "demand" data, describe characteristics of users or potential users of the system: How many riders? What are their travel-generating activities, origins and destinations, and travel needs? How are they, and potential riders, likely to react to proposed system changes? "Supply" data, the second type, concern the operating characteristics of the system and its equipment.

9.6.1 Demand Surveys

Seven types of passenger surveys are described next: counts, boarding, characteristics, attitude, nonuser and route, revenue, and special surveys.

9.6.1.1 Passenger counts Passenger counts measure the volume of riders at specific points on routes, usually the maximum load points. Counts are generally used to establish trends in ridership and forecast equipment needs. Counts may be one of four types: cordon, point, on-bus, and derived.

Cordon counts often cover all modes. For example, Toronto has a biennial program to secure person·trips by all modes for three cordons, one around the

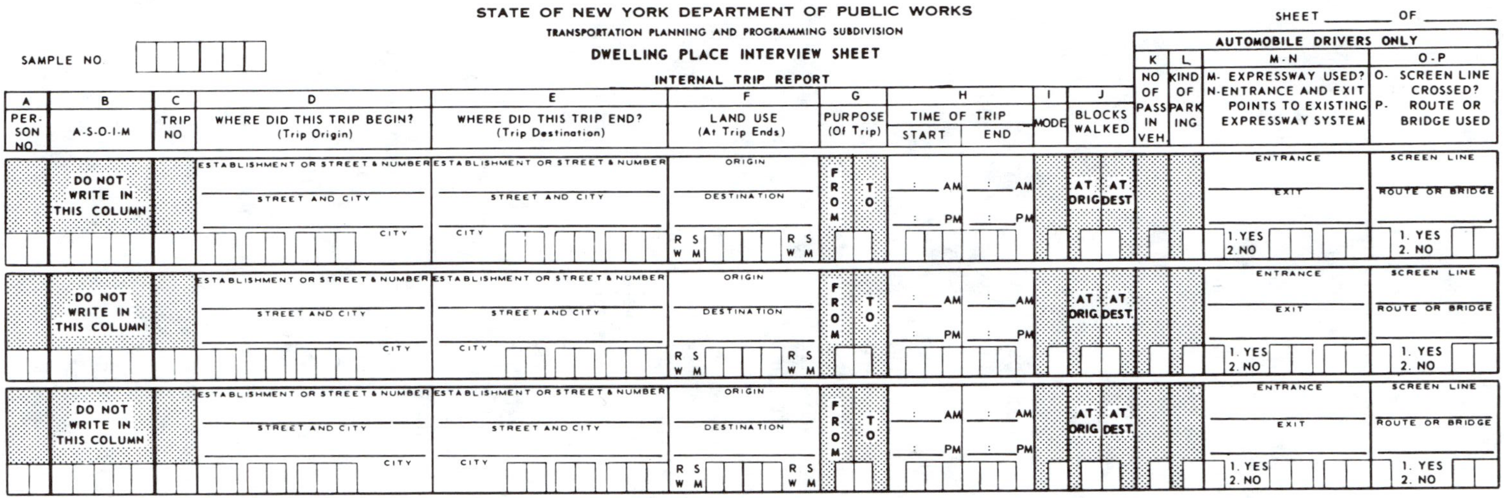

STATE OF NEW YORK DEPARTMENT OF PUBLIC WORKS
TRANSPORTATION PLANNING AND PROGRAMMING SUBDIVISION
DWELLING PLACE INTERVIEW SHEET
INTERNAL TRIP REPORT

SAMPLE NO

SHEET ______ OF ______

A	B	C	D	E	F	G	H		I	J	AUTOMOBILE DRIVERS ONLY			
											K	L	M-N	O-P
PER-SON NO.	A-S-O-I-M	TRIP NO	WHERE DID THIS TRIP BEGIN? (Trip Origin)	WHERE DID THIS TRIP END? (Trip Destination)	LAND USE (At Trip Ends)	PURPOSE (Of Trip)	TIME OF TRIP START	TIME OF TRIP END	MODE	BLOCKS WALKED	NO OF PASS IN VEH.	KIND OF PARKING	M- EXPRESSWAY USED? N-ENTRANCE AND EXIT POINTS TO EXISTING EXPRESSWAY SYSTEM	O- SCREEN LINE CROSSED? P- ROUTE OR BRIDGE USED
	DO NOT WRITE IN THIS COLUMN		ESTABLISHMENT OR STREET & NUMBER / STREET AND CITY / CITY	ESTABLISHMENT OR STREET & NUMBER / STREET AND CITY / CITY	ORIGIN / DESTINATION / R S W M / R S W M	FROM / TO	: AM / : PM	: AM / : PM		AT ORIG / AT DEST			ENTRANCE / EXIT / 1. YES 2. NO	SCREEN LINE / ROUTE OR BRIDGE / 1. YES 2. NO
	DO NOT WRITE IN THIS COLUMN		ESTABLISHMENT OR STREET & NUMBER / STREET AND CITY / CITY	ESTABLISHMENT OR STREET & NUMBER / STREET AND CITY / CITY	ORIGIN / DESTINATION / R S W M / R S W M	FROM / TO	: AM / : PM	: AM / : PM		AT ORIG / AT DEST.			ENTRANCE / EXIT / 1. YES 2. NO	SCREEN LINE / ROUTE OR BRIDGE / 1. YES 2. NO
	DO NOT WRITE IN THIS COLUMN		ESTABLISHMENT OR STREET & NUMBER / STREET AND CITY / CITY	ESTABLISHMENT OR STREET & NUMBER / STREET AND CITY / CITY	ORIGIN / DESTINATION / R S W M / R S W M	FROM / TO	: AM / : PM	: AM / : PM		AT ORIG / AT DEST.			ENTRANCE / EXIT / 1. YES 2. NO	SCREEN LINE / ROUTE OR BRIDGE / 1. YES 2. NO

Fig. 9.7 Dwelling place inventory form. [*New York State Department of Transportation, Albany, New York.*]

downtown–midtown areas, a second several miles out, and the third at the "metro boundary." Hand counters record vehicular volumes and observers estimate passengers per vehicle by mode. Computer analysis provides a number of maps and tables including person and vehicular trips by direction and by time of day for autos, light, medium, and heavy trucks, taxis, Transit Commission buses, other buses, streetcars, subway cars, and computer rail cars. Because the observer must estimate the number of people on a bus from the outside as it passes, variations of 10% from year to year can result from counting errors [9.34].

Point counts are made at bus stops to enable the observer to estimate the passengers in each bus more accurately [9.35]. On-bus counts include the following types: (1) fare-box recorders that can yield ridership for each trip, (2) automatic, electric-eye counters [9.36] that can produce continuous counts, (3) driver activated mechanical counters, and (4) on-bus counts as part of a speed and delay or rider attitude study (to be discussed further). Derived counts develop ridership data from fare box receipts.

9.6.1.2 On-board surveys Unlike the on-bus techniques just described, on-board suveys are made by an observer and can provide accurate, detailed information about riders. The observer records the number of persons boarding, leaving, and transferring at each stop, by time of stops. Analyses can then yield useful information on maximum load points, variations in loads, schedule adherence and speeds, stops volumes, and passengers·miles per bus [9.32].

9.6.1.3 Passenger characteristic surveys Passengers and/or potential riders are interviewed to obtain two types of data: (1) tripmaker characteristics—age, sex, car availability, etc., and (2) trip characteristics—time, purpose, origin, destination, walk distance to and from stops, etc. Several different techniques are available: on-board interviews, stop or station interviews, postcard mailback, home interviews, and telephone interviews. Selection of a technique depends on survey purposes and resources. The on-board and postcard surveys have sample-bias problems, but they are frequently used nevertheless since they may provide the needed information at least cost. On-board interviews can often be economically combined with on-board counts or attitude surveys. Home and telephone interviews have the advantage of development data on potential riders, but they are more expensive [9.38].

9.6.1.4 Attitude surveys These can serve three general purposes: (1) to provide a basis for system improvements most desired by riders, (2) to help design information services to the public, including advertising and public relations, to strengthen the image of the service, and (3) to help predict ridership on new routes, or as a result of fare, route, or services changes. The range of techniques parallels that for determining passenger characteristics. Design of an attitude survey is a skilled task and a poorly designed or administered survey usually produces worthless and misleading information. A number of guides on questionnaire design are available [9.39, 9.40, 9.41].

9.6.1.5 Nonusers and route marketing surveys Nonusers and potential riders can be surveyed using home or telephone interview techniques. This is most likely to be done in an area where new or improved service is contemplated. It would be based on a rigorous sample. The data would probably be used in some demand-forecasting model.

9.6.1.6 Revenue surveys These are conducted to relate types of riders and service to revenues, and may provide an important element in cost-revenue accounting for management. The revenue–passenger relationship may be developed for such rider classes [9.33] as:

1. Revenue passengers
   Adult fare: ticket, token, or cash
   Adult pass
   Contract, e.g., postal employees, college fee-paid students, etc.
   Student fares: ticket, token, or pass
   Student pass
   Senior fare
   Senior pass
   Premium fare, i.e., express, Dial-a-Bus, etc.
2. Transferring riders
3. Free-rider passengers, e.g., preschoolers, transit employees, police, etc.
4. Charter services

9.6.1.7 Special surveys In addition to the more standard surveys just described, transit planning and management will often require the design of special purpose surveys. Two examples are given here. The first is concerned with the "submodal" split of riders at transit stations. The Toronto Transit Commission (TTC) makes annual surveys of the arrival modes of riders at its subway stations. For example, at the TTC Islington Station in 1976 the distribution was feeder bus 72%, kiss-and-ride 7%, park-and-ride 8%, and walk-in 13% [9.34].

The second example is that of a pilot/demonstration project. In Ottawa, Canada, a two-step procedure was designed to estimate transit demand by handicapped persons. Step one was an interview survey of 986 potential users, a sample drawn from an estimated 16,000 handicapped persons. Interviewees were asked a wide range of questions about present modes of travel and prospective ridership given various types of special service. The second step was a pilot/demonstration project to evaluate the demand projects resulting from the interviews. In 1974 a subsample of handicapped residents were offered trips to work, medical facilities, etc., on taxis and special services. Results showed the degree of overestimating that had occurred in the interviews and provided factors for calibrating the demand forecasts [9.36].

9.6.2 Supply-Side Surveys

The second major category of transit surveys concerns the "supply" of transit services. These tend to be of two types—inventories and transit operations surveys. The former consist of periodic cataloging of the number, type, age, condition, etc., of all system equipment [9.33]. Such surveys may also include bus stops, shelter, and other ancillary items. Inventories are fairly straightforward and need not be detailed here.

Transit operations surveys typically employ an observer, a watch, and a recording from [9.32, 9.33, 9.37]. An example is the "speed and delay" study used to determine causes of delays in bus operations. A separate field sheet is used for each bus run in one direction. Sheet heading information might include bus line, direction, vehicle number, start location and time, end location and time, and weather. Column heads typically would be the following:

Control points		Stops or slows		
Location	Time	Stop/slow	Seconds	Cause

9.6.3 Intercity Comparative Data

Intercity comparisons probably play a more important role in transit management than in auto-system management. This is because more of the system operating costs are centralized, rather than dispersed among many individual auto owners. Figure 9.8 conceptualizes a model of transit performance measures. This kind of transit management information system (TMIS) serves two functions [9.44]:

1. For planning new transit systems and service, the TMIS enables forecasts to be made of key indicators such as ridership, vehicle requirement, revenue, operating cost, new cost, cost per trip, and cost per passenger·mile.
2. For evaluating the performance of existing systems, the TMIS facilitates comparison with other systems with comparable service area characteristics.

Data for operationalizing the TMIS is collected from all types of transit service areas by the Urban Mass Transportation Administration [9.45, 9.46]. Such comparative data is invaluable for the first purpose mentioned, since there is no way to measure transit travel behavior where no service exists. In contrast, auto travel behavior can be measured through an origin and destination survey.

The second purpose of the TMIS, as just noted, is also significant for the feedback it can provide to local transit officials and stakeholders. Private business operations are presumed efficient because they survive in a competitive business climate. That a similar pressure for efficiency does not exist in the public sector is

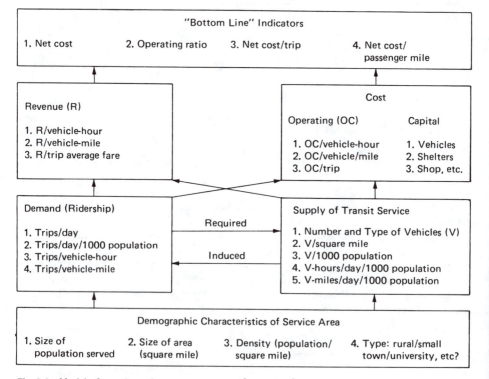

Fig. 9.8 Model of transit performance measures. [9.44, p. 9].

often lamented by theorists and taxpayers alike. However, for public transit services the TMIS provides an effective substitute. Transit managers, policy officials, and stakeholders can see how well their system is "competing" with other systems with similar service-area characteristics.

9.7 DATA-BASE MANAGEMENT SYSTEMS

Recent rapid developments in computer technology (especially in microcomputers) have led to increased use of formal data-base management systems (DBMS). These can be employed to process the various kinds of data described in the preceding sections. As of the end of 1981, there were at least 50 DBMS programs available for microcomputers [9.21].

A DBMS starts with the basic idea of a *file*, which is

a collection of information that can be read or written as a unit

A file might contain information on each trip made by an interviewed household, on each bus owned by a transit agency, and so on. Since there may be several different files, the names are kept in a *directory*.

A file consists of one or more records, which contain similar information. In the preceding example, there might be one record for each trip and one for each bus. Each record has a unique number by which it may be accessed.

A good DBMS allows the user to:

1. Define and create files
2. Maintain files by adding, scanning, inspecting, updating, and deleting records
3. Process files by sorting, selecting, and relating them[2]
4. Develop reports

The first three of these will be addressed here.

9.7.1 Defining Files

To illustrate the preceding operations, let us imagine trying to develop a plan for coordinating transportation between several human-service providers and possible recipient groups. Two files are involved, one for the providers and the other for the users. In both cases there is a need to *define* the files.

The first step is to give each file a name. Then each *field* (or characteristic or variable) must be identified and its dimensions specified. The first file, for instance, may be named PROVIDERS, and the fields may be as given in Table 9.2. Since the field name (actually a description) could be rather long, it is given a shorter identification code (ID). Also, the maximum number of characters in the field has

[2] Currently very few DBMS programs have so-called "relational data-base" capability, although their number should increase rapidly.

Table 9.2　Fields for PROVIDERS Illustration

Number	Field name	ID	Length
1	Agency providing service	AGEN	10
2	Recent annual budget for transportation	BDGT	7
3	Type of transportation provided	TYPE	25
4	Cutback budget level	CBL	7

Table 9.3 Example Set of Records for the PROVIDERS Fields

Record	AGEN	BDGT	TYPE	CBL
1	VA	30000	VAN USE	
2	TITLE XX	25000	GENERAL	
3	WIN	20000	DRIVERS	
4	TITLE 3	60000	EQUIP GRANTS	
5	MEDICAID	80000	USER GRANTS	
6	CAA	10000	VAN USE	

to be specified. In the illustration it is not anticipated that the budget figures (field 2) will have more than 7 digits (rounded to the nearest dollar). Field 4 is an example of a computed field. The cutback budget is to be calculated as a certain proportion of the corresponding figure in field 2.

9.7.2 Maintaining Files

The first step in file maintenance is to add actual records to the file.[3] In our example the result might look like Table 9.3, where information for six provider agencies has been entered (presumably in the order in which the information was obtained). The computed field is shown as empty, but it could be calculated now (or later). One possibility could be that there is an expected 10% reduction across the board in all agency budgets, so that the CBL field would be computed to contain the numbers 27000, 22500, 18000, 54000, 72000, and 9000, respectively, for each record.

Scanning involves searching the records on one or more fields for certain characteristics. Suppose, for instance, that it were desirable to find all those agencies giving grants. A scan of the file from the top would come to record 4 and then, if continued, stop at record 5. In another approach, each record could be *inspected* directly, simply by calling it by its number.

Updating is nothing more than changing a field on a particular record. It may have been found, for example, that TITLE 3 now gives user rather than equipment grants. Record 4 then would be called, and the TYPE field altered accordingly. A delete also may be performed, as for instance if the TITLE XX program (record 2) no longer supported transportation services. The file left after all these maintenance procedures would look like Table 9.4.

[3] The word "maintenance" does not seem appropriate for this first operation, but it is widely used.

Table 9.4 PROVIDERS File After Maintenance Procedures

Record	AGEN	BDGT	TYPE	CBL
1	VA	30000	VAN USE	27000
3	WIN	20000	DRIVERS	18000
4	TITLE 3	60000	USER GRANTS	54000
5	MEDICAID	80000	USER GRANTS	72000
6	CAA	10000	VAN USE	69000

Table 9.5 PROVIDERS File after Processing

Record	AGEN	BDGT	TYPE	CBL
1	MEDICAID	80000	USER GRANTS	72000
2	TITLE 3	60000	USER GRANTS	54000
3	CAA	10000	VAN USE	9000
4	VA	30000	VAN USE	27000
5	WIN	20000	DRIVERS	18000

9.7.3 File Processing

The processing of files can take several forms. First, the file can be *sorted*. To do this requires a *key*. To illustrate, the resultant file in Table 9.4 may be sorted and reordered so that the records are listed by agency alphabetically. Or they could be sorted by putting at the top all those with budgets over $35,000, then alphabetically by agency within the resultant two groups. The initial key in this second sort would be the BDGT field and the next key the AGEN field.

Compacting would involve elimination of some records and renumbering the remainder. As can be seen in Table 9.4, record 2 has been deleted. However, in most DBMSs it would not actually be eliminated, just not used. Compacting would carry out the actual elimination, and then renumbering of the records. This, along with the second sorting process described above, would lead to the file in Table 9.5.

With a file processed to the shape of that in Table 9.5, it then is possible to *relate* it to another file. These component files are known as *relational data bases*. Three main steps are involved—selection, projection, and joining. The first creates a subset of all the records in a file, the second a subset of the fields, and the third a combination of the two files.

A *selection* process on Table 9.5 might involve elimination of any agency with a budget less than $15,000. The CAA record (record 3) thus would be discarded. *Projection* might be employed to eliminate the CBL column or field, since it might not be needed any more.

Suppose now that a second file showed the need for different types of service by user groups—elderly (OLD), handicapped (HANDI), and poor (POOR)—in the example locality (Table 9.6). The two files could be *joined* on the basis of the common TYPE field (although the fields do not have to have the same name in both fields, just the same categories of items). The result, after the above three processes, is the new file, called "COMBO," in Table 9.7. Each time USER GRANTS is mentioned in the PROVIDERS file (as in record 1), the "NO NO YES" fields from the USER GRANTS record in the USERS file is added, and so on. Now the planner can start to see which agencies can be of assistance to particular groups.

DBMSs are not that complicated in theory and use. What is more complicated,

Table 9.6 Example USERS File

Record	TYPE	OLD	HANDI	POOR
1	VAN USE	YES	NO	NO
2	DRIVERS	YES	YES	NO
3	USER GRANTS	NO	NO	YES

Table 9.7 COMBO File

Record	AGEN	BDGT	TYPE	OLD	HANDI	POOR
1	MEDICAID	80000	USER GRANTS	NO	NO	YES
2	TITLE 3	60000	USER GRANTS	NO	NO	YES
4	VA	30000	VAN USE	YES	NO	NO
5	WIN	20000	DRIVERS	YES	YES	NO

however, is the identification of rules and development of programs for carrying out the types of transformations in the preceding example.

DBMSs should become common in most planning agencies in the future. A major area of application probably will be in *monitoring* [9.20]—continuing programs of survey to ensure that previously formulated standards or objectives are being met. The monitoring system set up by Hampshire County Council in England, for example [9.19], focuses mainly on housing, population, and employment, but much data is available both there and in the United States to trace the impacts of transportation systems or program changes after implementation. A DBMS would ease this process by allowing for more rapid incorporation of new data and updating of old.

9.8 CONSULTANT-BASED SYSTEMS

The kind of information described so far in this chapter does not include anything about *procedures* for planning. Put in a different way, data have been collected on the land use, transportation, and travel, but not on the logic or "rules" planners use to plan.[4]

Techniques for tracking the growth and change in rule structures can be employed to develop consultant-based systems (CBS), otherwise known as "expert systems" [9.18], such as that employed in a part of the medical field by the Stanford University Hospital [9.17]. This particular system, known as MYCIN, has been built on the past experience of doctors in their diagnoses of certain types of microbial problems. The symptoms and resultant diagnoses, together known as "rules," have been recorded (added to a data file). After a period of time, the data base has been built up to the point where it contains more "knowledge" than most individual doctors have. It then can be employed (still with some human judgments) by nonexpert (in microbial areas) doctors faced with immediate problems to diagnose.

A basic overview of a CBS as it might relate to some aspects of planning is displayed in Fig. 9.9. The heart of the system consists of the consultation and explanation sections (I and II). Initially, background data and forecasts (III) are collected and "fed" to the consultation system (I). "Knowledge" (V), as incorporated in a set of decision rules, then is applied by the consultation system to the background data (and exogenous forecasts) to produce a plan (a set of decisions). The rationale for this is brought out by the explanation system (II). Monitoring then takes place to determine whether:

1. Actual developments and the assumed changes in external factors take place as forecast under the given decisions (IV).

[4] Some of these are discussed in Chap. 12, although they are not referred to as "rules."

2. The rules by which decisions are made have been supplemented or altered (VI).

In a dynamic environment, as found in most cities, these two types of changes can be significant, so that it may be necessary to go through the whole cycle in Fig. 9.8 several times (in fact, on an almost continuous basis).

Looking more closely at the rules, we find they are posed primarily in terms of if–then statements. As a hypothetical illustration, a rule to be applied to certain subareas in a city might look like:

IF: volume to capacity ratios are below 0.8 on all arterials in the subarea,
AND: there are less than 3,000 residents per square mile in the subarea,
THEN: bus transit revenues in the subarea will be less than 50% of costs (with a confidence factor of 0.8 on a scale of 0.0 to 1.0).

Rules are "combined" through logical connections and through probabilistic inference, generally using Bayes' theorem [9.18]. The "strength" of the relationships in a rule is evidenced by a subjectively assigned confidence factor, as in the example just shown. This is presented to the user along with an outline of the rules applied.

While to our knowledge no such CBSs exist for metropolitan transportation planning, it is likely that some will be created within the next few years.

9.9 GRAPHICS DISPLAY

Transportation planning studies have always involved use of advanced graphics display techniques, particularly as a means to summarize and highlight masses of

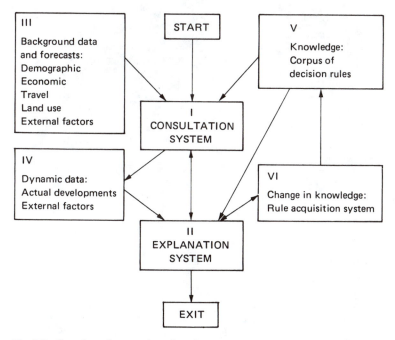

Fig. 9.9 Overview of a consultant-based system.

data. The initial Chicago Area Transportation Study [9.23], for instance, was undertaken in the 1950s and employed cathode-ray-tube-generated pictures showing the magnitude and location of trip ends of various types as well as connecting (desire) lines. These, in addition to hand-drawn multicolor maps, presumably helped portray many of the more complicated data sets while also aiding decision-makers in understanding their transportation problems.

Presently, computer graphics display techniques are undergoing very rapid change, particularly as color terminal/monitors, microcomputers, and associated software become less expensive and more sophisticated. In fact, with light pencils and similar devices, graphic displays can be utilized as a medium of input as well as output. This is the basis for CAD/CAM (computer-aided design/computer-aided manufacturing) systems [9.24] in which a designer or planner can (a) see a display of, for instance, a transit route structure, (b) change a section by erasing/drawing on the screen, and then (c) have estimates of the resultant impacts (e.g., on ridership) calculated. Technology also is making possible three-dimensional animated color graphics (for example, for microcomputers [9.25]), which in the future may be started and propagated by voice.

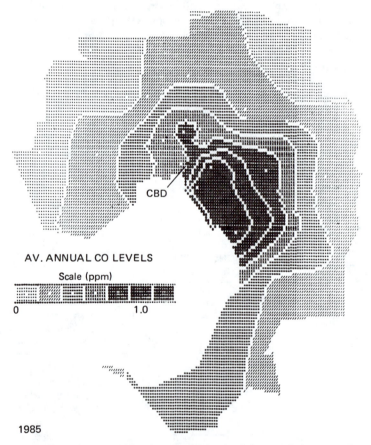

Fig. 9.10 Predicted vehicle air pollution levels without emission controls for Melbourne at 1985. [*J. F. Brotchie et al., TOPAZ, Springer-Verlag, Heidelberg, 1980, p. 77*].

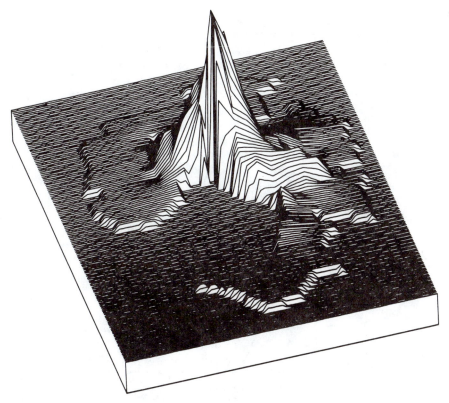

Fig. 9.11 Urban traffic energy consumption release per unit area estimated for Melbourne at 1977. [*J. F. Brotchi et al., TOPAZ, Springer-Verlag, Heidelberg, 1980, p. 146.*]

An earlier and still much-used display technique called SYMAP [9.26] had a major advantage of working with a printer rather than a plotter, since many agencies did not have the latter. A sequence of characters were printed, and the printer subsequently backed up and overprinted some of these characters. The result was a display like that in Fig. 9.10, with the darkest overprints representing the most intense annual carbon monoxide emission levels. A subsequent program, SYMVUE [9.26], required a plotter, but then enabled three-dimensional contour presentations as in Fig. 10.11.

A variety of other types of displays now can be produced using readily available graphics terminals and/or plotters. Figure 9.12 shows one for projected land-use (that is, employment) growth in various parts of the Seattle metropolitan area. This is a three-dimensional view, with blocks of different heights representing the expected increases in jobs.

Two ways of showing origin-to-destination desire lines are presented in Fig. 9.13. The first (*a*) highlights greater trip desires through the sizes of the blocks in the original zones. The second (*b*) lets the height of the rectangles connecting origins and destinations serve this purpose. A similar graphics approach is employed in Fig. 9.14, although there the emphasis is on transit ridership. That particular figure shows a variety of possibly useful data such as peak, average day, and night bus-loading as well as standees.

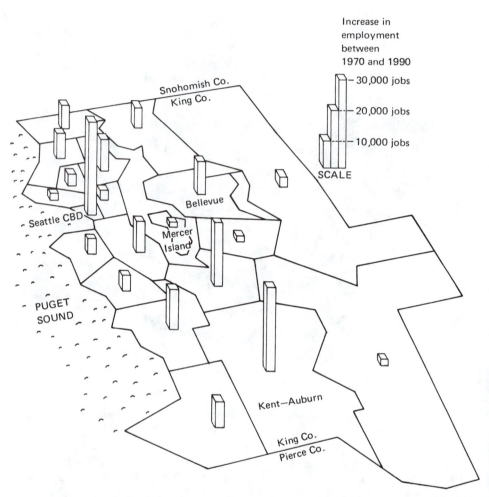

Fig. 9.12 Interim Regional Development Plan employment growth forecasts for 1970–1990 period. [9.29, p. 13].

A network map such as that in Fig. 9.15 can be utilized for several purposes:

1. As a way to check if the network has been coded as desired
2. With overstrike of different symbols or colors to point to congested links
3. For inputs to analyses of link additions or subtractions

The last of these probably will be under increased use in the future, as planners and designers use CAD systems to alter the transport network and estimate the resultant impacts.

9.10 SUMMARY

The four activities described in this chapter—geocoding, and inventories of land use, travel, and the transportation system—constitute the minimum components of a transportation study information system. Completion of these four data-collection

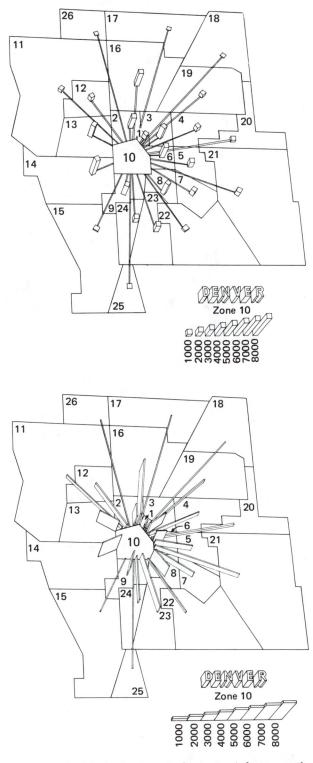

Fig. 9.13 Two displays of origin–destination pairs (desire lines). [9.28, p. 12].

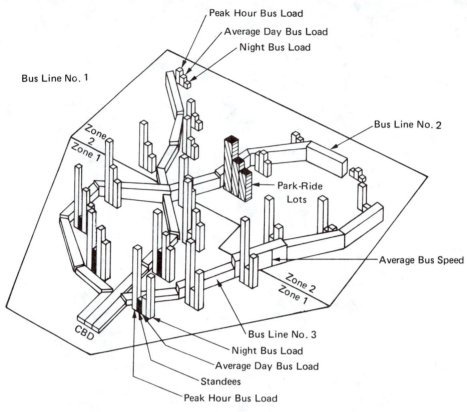

Fig. 9.14 A transit service display example. [9.29, p. 14].

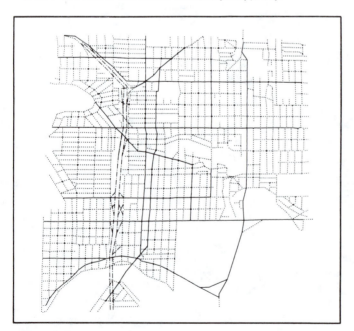

Fig. 9.15 Display of street network in northern part of Seattle, Washington. [9.27, p. 12].

activities creates the core data bank from which information can be "withdrawn" for a large variety of transportation and other purposes.

Other chapters in this book discuss uses of the information. A few examples will suffice here: Land-use data are the basis for planned or forecast future land use from which future trip generation is derived. Travel surveys provide data on current actual trip making. Land use and trip data are correlated to calculate trip-generation factors for travel forecasts. The existing transportation network is a basic input to computer simulation of current travel behavior. The present network is also usually a "given" in preparation of proposed future networks.

An elemental geocoding system is a prerequisite for handling any spatially disaggregated data. In modern transportation studies, the extensive use of the grid coordinate system has made possible simulations of complex urban travel behavior and the testing of alternate proposed programs. Decision-makers are thus given objective measures of the effectiveness of the alternatives in achieving the goals of the transportation system.

The use of coordinates also provides another requisite of an ideal management information system. Computer mapping, based on the coordinates, displays results of analyses, forecasts, simulations, and tests to decision-makers in ways that are timely and comprehendible.

Once data have been collected, they must be "managed." This involves coding, scanning, inspecting, updating, and deleting records. It also involves sorting, selecting, and relating them. Recent advances in data-base management systems should help these efforts.

The future might hold interest in consultant-based or expert systems. These could contain information or the "rules" used by planners and other urban experts and thus help to improve the planning process itself.

BIBLIOGRAPHY

9.1　Chapin, S. E., and E. Kaiser: *Urban Land Use Planning*, 3d ed., University of Illinois Press, Champaign-Urbana, Ill., 1979.

9.2　Creighton, R. L.: *Urban Transportation Planning*, University of Illinois Press, Champaign-Urbana, Ill., 1970.

9.3　Memmott, F.: "Transportation Planning," in W. Goodman and E. C. Freund (eds.), *Principles and Practice of Urban Planning*, International City Managers Association, Chicago, 1968.

9.4　Southern California Regional Information Study: *An Interim, ACG-DIME Updating System, SCRIS Report No. 4, Census Use Study*, Southern California Assoc. of Governments and U.S. Bureau of Census, Los Angeles, 1970.

9.5　Southern California Regional Information Study: *The Long Beach California Experience, ACG-DIME Updating System, SCRIS Report No. 8, Census Use Study*, Southern California Association of Governments and U.S. Bureau of Census, Los Angeles, 1970.

9.6　System Development Corporation: *A Geographic Base File for Urban Data Systems*, Systems Development Corporation, Santa Monica, Calif., 1969.

9.7　U.S. Department of Commerce, Bureau of the Census: *The DIME Geocoding System Report No. 4, Census Use Study*, Government Printing Office, Washington, D.C., 1970.

9.8　U.S. Department of Commerce, Bureau of the Census: *Use of Address Coding Guides in Geographic Coding, Proceedings to the 1970 Conference at Wichita, Kansas, Nov. 19–20*, Government Printing Office, Washington, D.C., 1971.

9.9　U.S. Department of Housing and Urban Development: *Urban and Regional Information Systems*, Superintendent of Documents, Washington, D.C., undated, c. 1968.

9.10　Urban and Regional Information Systems Association: *Geocoding-71; Papers from the Working Session on Geographic Base File Developments at the Ninth Annual Conference, Sept. 8–10, 1971*, Urban Data Processing, Inc., Cambridge, Mass., 1971.

9.11 Urban Renewal Administration and Bureau of Public Roads: *Standard Land Use Coding Manual*, Government Printing Office, Washington, D.C., 1965.

9.12 Weiss, S.: "Land Use Analysis," in W. Goodman and E. C. Freund (eds.), *Principles and Practice of Urban Planning*, International City Managers Association, Chicago, 1968.

9.13 Stuart, R. C.: *Commercial Data for Transportation Planning*, U.S. Federal Highway Administration and U.S. Urban Mass Transit Administration, 1978. Available from NTIS.

9.14 U.S. Bureau of the Census, Journey-to-Work and Migration Statistics Branch: *Urban Transportation Planning Package, 1980 Census: Specifications*, Washington, D.C.: November 3, 1981.

9.15 Sandberg, G.: "A Primer on Relational Data Base Concepts," *IBM Systems Journal*, vol. 20, no. 1, 1981.

9.16 Martin, J.: *Computer Data-Base Organization*, Prentice-Hall, Englewood Cliffs, N.J., 1977.

9.17 Shortliffe, E. H.: *Computer-Based Medical Consultations: MYCIN*, Elsevier, New York, 1977.

9.18 Michie, D. (ed.): *Expert Systems in the Microelectronic Age*, Edinburgh University Press, Edinburgh, 1979.

9.19 Hampshire County Council, County Planning Department: *Strategic Monitoring Report 1979*, The Castle, Winchester, Hants, England, Nov. 1979.

9.20 Scheurwater, J., and I. Masser: *Monitoring Spatial Planning in the Netherlands: An Outline of an Information Analysis System*, University of Utrecht, The Netherlands, 1980.

9.21 Barley, K. S., and J. R. Driscoll: "A Survey of Data-Base Management Systems for Microcomputers," *Byte*, November 1981.

9.22 Ben Herman, R., and C. G. Jameson: *CCA Data Management System (TM)*, Personal Software, Inc., Sunnyvale, Calif., April 1980.

9.23 *Chicago Area Transportation Study* (3 vols.): Chicago, 1956–1962.

9.24 Basta, N.: "You and the Computer Revolution," *Graduating Engineer*, Nov. 1981.

9.25 Budge, B.: *3-D Graphic System and Game Tool*, California Pacific Computer Co. (city not specified), 1980.

9.26 Dongenik, J. A., and D. E. Sheehan: *SYMAP User's Reference Manual*, Laboratory for Computer Graphics and Spatial Analysis, Harvard University, Cambridge, Massachusetts, 1975.

9.27 Schneider, J. B.: "Applications of Computer Graphics in the Transportation Field," *Computer Graphics*, vol. 1, no. 13, December, 1978.

9.28 Schneider, J. B., et al.: *Data Display Techniques for Transportation Planners: Experiments and Applications*, Urban Transportation Program, University of Washington, Seattle, July, 1977.

9.29 Noguchi, T., and J. B. Schneider: *Data Display Techniques for Transportation Analysis and Planning: An Investigation of Three Computer-Produced Graphics*, Urban Transportation Program, University of Washington, Seattle, June 1976.

9.30 Newman, W. M., and R. F. Sproull: *Principles of Interactive Computer Graphics*, McGraw-Hill, New York, 1979.

9.31 Shortreed, J.: "Surveys," in R. M. Soberman and H. A. Hazard (eds.), *Canadian Transit Handbook*, University of Toronto, 1980.

9.32 Box, P., and J. C. Oppenlander: *Manual of Traffic Engineering Studies*, Institute of Transportation Engineers, Arlington, Va., 1976.

9.33 Ministry of Transportation and Communications: *Municipal Transit Manual*, Toronto, Ontario, 1977.

9.34 Metro Toronto Planning Department: *Metro Cordon Count Program—1975*, Toronto, Canada, 1976.

9.35 Edmonton Transportation Planning Department: Unpublished survey description and sample computer output.

9.36 Group Five Consulting: *Ridership Data Collection and Analysis System*, Ottawa, Canada, 1976.

9.37 Institute for Urban Transportation: *Mass Transit Management: A Handbook for Small Cities*, Indiana University, Bloomington, Indiana, 1971.

9.38 Kitchener Transit: *1977 Passenger Survey*, Kitchener, Ontario, 1977.

9.39 Babbie, E. R.: *The Practice of Social Research*, Wadsworth, Belmont, Calif., 1975.

9.40 Parten, M.: *Surveys, Polls and Samples: Practical Procedures*, Cooper Square Publishers, New York, 1966.

9.41 Survey Research Center, Institute for Social Research: *Interviewer's Manual*, University of Michigan, Ann Arbor, Michigan, 1976.

9.42 Toronto Transit Commission: *Modal Split of Transit Patrons Using Finch, Islington and Warden Subway Stations,* Toronto, Ontario, 1977.

9.43 O. C. Transpo: *Transportation for the Physically Handicapped,* Ottawa, Canada, 1975.

9.44 Campbell, J., C.-T. Ho, and R. C. Stuart: "Blacksburg-Virginia Tech Transit System," Appendix 9 to *New River Valley Transit Study: Final Report,* New River Valley Planning District Commission, Radford, VA., Jan 15, 1979.

9.45 Urban Mass Transportation Administration: *Analyzing Transit Options for Small Communities,* U.S. Government Printing Office, Washington, D.C., Jan. 1978.

9.46 Urban Mass Transit Administration: *Small City Transit Characteristics,* reports MA-06-0049-76-1 through 15, U.S. Government Printing Office, Washington, D.C., 1973.

EXERCISES

9.1 Network Coding:

(a) Following the method described in Section 9.2.1, lay out a grid coordinate system for a study area 5,000 X 5000 ft at a scale of 1.25 in. = 1,000 ft (0.25-in. coordinate paper will be helpful). Delineate the 1,000-ft grids. Label the 1,000-ft intervals of both the X and Y axes.

(b) Each 1,000 X 1,000-ft grid cell constitutes a traffic zone "named" (identified) uniquely by the coordinates of its southwest corner. Delineate the boundaries of zone X, Y (4000,4000).

(c) Zone (4000,4000) has a loading node at its centroid. Plot the centroid and give its coordinates.

(d) In the study area there are four major arterials, all straight, whose X, Y terminals within the study area or at the cordon are (0000,0200) and (4500,5000); (0000,1400) and (5000,3067); (0000,0200) and (5000,2000); and (3000,2400) and (4200,0000). Plot the four arterials on the coordinate grid. Links are named by the coordinates of their intersections. Name the four links that lie within the study area.

(e) A regional shopping center is located at (3800,0800) and constitutes a loading node. Create a "dummy" link from the loading node of zone (4000,4000) to the network. Plot the path of shortest distance through the arterial network for a resident of zone (4000,4000) driving to the shopping center. List in order the start, intermediate, and end nodes of the trip.

9.2 Two sets of data have been collected on five different people-mover systems for major activity centers. One set was collected by engineers, the other by environmental planners. The results are the files TECHNICAL and ENVIRO, as shown below.

TECHNICAL File*

Record	NAME	CAP	GRADE	ROW	POS
1	TELETRIP	30200	100	5	T
2	VEYORCAR	14000	10	6	T
3	SINGLECAB	37200	10	7	H
4	SKYHOIST	9400	3.3	5	H
5	MOONCAR	37200	10	8	T

*CAP is capacity in people/hr, GRADE is in percent, ROW is single-line right-of-way required, in ft, and POS is position of vehicle, either handing (H) from the guideway, or on top of it (T).

ENVIRO File*

Record	TYPE	AP	NOISE	VIBES
1	HANG	L	H	L
2	TOP	M	M	L

*TYPE is hanging (HANG) from guideway or on top (TOP) of it. AP is air pollution level—low (L), medium (M), or high (H). NOISE is noise level—L, M, or H. VIBES is vibration level—L, M, or H.

Show the resultant files (three total) after:

(a) The capacity per 10 ft of right-of-way has been calculated as a new (last) field in the TECHNICAL file. The right-of-way of the MOONCAR is found to be 10 ft rather than 8 ft. The VEYORCAR Company has gone out of business (delete record).

(b) The file from (a) has been sorted according to, first, hanging (first) or top position; then second in descending order by capacity per 10 ft of right-of-way.

(c) A projection has been done on the file from (b) to eliminate the CAP and ROW fields. The resultant file has been related to the ENVIRO file or the TYPE (of POS) field.

9.3 Describe in about 500 words a potential example of a consultant-based system (CBS) for some aspect of transportation planning. Use the components in Fig. 9.8.

9.4 In your city a study is to be made of a radial transportation corridor with severe traffic-flow, transit, and parking problems. To understand these problems several surveys are authorized. You are in charge of designing, conducting, and analyzing the land use survey.

(a) Select a 2-square-block area of heterogeneous, intensive land use for a pretest of the land-use survey.

(b) Define specific objectives for the survey, outline the steps in gathering, processing, and analyzing land-use data to achieve the objectives, and design the survey, including a field listing form. Try to anticipate where the process could go wrong, and devise fail-safe devices to ensure the quality of data collection and coding.

(c) Conduct a pretest of the survey procedures in the 2-block area.

(d) Code the data for computer use, including use of the Standard Land Use Code (2 digits).

(e) Analyze the data according to your outline.

(f) Evaluate the quality of your results: What problems were uncovered in the pretest? How could they be corrected?

9.5 In the library, find examples of reports, articles, etc., showing the latest techniques in computer graphic display. In about 750 words describe these and discuss how they are improvements over those discussed in Sec. 9.9.

10 Evaluation I: Formal Techniques

The need to evaluate alternatives, both proposed and existing, and make decisions among them is one of the most pressing yet difficult requirements in the transportation planning process. When considering the good and bad impacts that transportation projects and programs can have, we surely must have some feelings of uneasiness about the manner in which decisions affecting these important aspects of life might be made. And we certainly would want to ensure that all possible avenues of approach had been explored so that the alternative providing the maximum benefits for the required financial outlays would be detected, selected, and maintained.

The difficulties inherent in evaluation and decision making are many and, unfortunately, of great consequence. While these will not be covered in depth at this juncture, it should be pointed out that the problems of (1) setting suitable objectives, (2) making adequate predictions of consequences, (3) determining the relative importance of these consequences, and (4) assuring that the consequences take place as desired have made the responsibility of decision making a heavy burden to public officials. Who, for example, could have foreseen years ago that the automobile would take the place of the front porch swing, that it would change building design to allow for drive-in banks and restaurants, or that it would help to create the spread of suburbia? And who can say with any assurance that a life saved from an automobile accident is worth $300,000, that a dollar spent on the appearance of subways is equally as beneficial as a dollar spent for added speed, or that private ownership of transit facilities is to be preferred by the public to governmental ownership or control? These are several of the many vexing problems that face the planner or public official.

Two main types of formal evaluation will be discussed here. The first might be considered *pre* (before) project, program, or plan evaluation. The second would be *post* (after). In the former the purpose is to assess the *predicted* levels of user service and impact levels for different proposed alternatives in order to help select the best from among them. The purpose of the latter is to assess the *actual* impacts of a previously implemented project, program, or plan to see if such should be continued as is, modified, or perhaps dropped. Postevaluation was represented in Chap. 2 by the feedback arrow in the planning process from the operations and maintenance stage to that of data collection (then to "evaluation"). In the past, preevaluation has been the most prevalent, but this is expected to change as funding becomes dearer.

This chapter is divided roughly into five main sections. First, the general theory of economic evaluation is described. This is followed by examples of methods by which benefits and costs can be compared. This presentation forms the basis for an assessment of the advantages and limitations of evaluation techniques and leads into the more general approaches of goals achievement and cost-effectiveness. The discussion then proceeds to *post* evaluation procedures and general evaluability assessment.

Having shown how evaluations theoretically *should be* made, we proceed in the next chapter to discuss some aspects of how they *actually* are made. Included in Chap. 11 is a description of various methods for citizen involvement in the evaluation/decision-making process.

10.1 A FRAMEWORK FOR EVALUATING BENEFITS AND COSTS

Economists and others concerned with the benefits and costs of various alternative policies and actions long have worked with the "willingness to pay" idea summarized by Wohl and Martin [10.5, p. 183-184]:

> For the case of public projects ... all factors or elements of concern and value to the owning public and for which value the public would willingly pay to gain, or to keep from losing, will be included. ... Generally, then, social or political factors enter the analysis only in those instances where society would be willing to forego financial or *other resources of value* in their stead. This assumption is made, first, since most tangible and so-called intangible objects of concern have a history of experience and have been valued at the marketplace (at least implicitly). ... Second, this assumption is made to point out that factors of *presumed* concern to the owning public and for which they are *not* willing to forego something else of value (which *must* be foregone to achieve the object of concern) are just that—presumed rather than real.
>
> Also, it must be emphasized that lack of willingness to pay for some social objective (or at least to forego something else of value in order to achieve that goal) suggests the lack of real value associated with the objective.

The idea of "willingness to pay" is brought into reality through the "demand curve," which shows what quantity of a given product people are willing to purchase at a given unit cost for the product.[1] In the case of highways, the product that is offered is "trips" of a given type, while the unit cost is composed of such

---

[1] Most demand curves show the relationship between quantity and *price*, not cost. But most economists view "price" as something that evolves through a market interaction. Since there is no explicit market interaction to establish a price for highway service, we prefer to use the word "cost" in this connection rather than "price."

items as vehicle operating and maintenance costs, tax payments, parking fees, and the time of the driver and passengers.

As a basis for an example of a demand curve for a highway, consider the overly simplified situation presented in Fig. 10.1. Trips are made between zones A and B

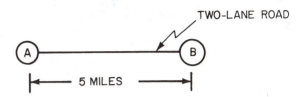

Fig. 10.1 A simple example of a roadway connecting zones A and B.

over a two-lane road that presently is 5 miles long. These trips, made during a given 1-year period, are all for the same purpose, are all done at one time of day, are all made by people of a similar socioeconomic background, and so forth. Under these conditions, and using procedures for travel prediction similar to those in Chap. 7, it is possible to construct a curve, such as that in Fig. 10.2, showing the number of trips made if the unit cost per mile of each trip were as indicated. Generally, it can be expected that as the unit cost per mile of each trip gets higher, there will be fewer yearly trips made over the road: thus the reason for the negative slope in the demand curve in Fig. 10.2.

In Fig. 10.2 it also can be seen that if the unit cost of travel were C_1, there would be v_1 yearly trips made. Further, it can be seen that some people would be willing to pay *more than* C_1, but would not be required to do so. For instance, v_2 trips would be made even if the cost were C_2 which is greater than C_1. As a consequence, there is a surplus (known as the "consumer surplus"), which accrues to the people willing to pay more: they can take the money they are willing to pay but do not have to $(C_2 - C_1)$ and use it for some other purpose. This consumer surplus thus can be thought of as a benefit arising from tripmaking, and the summation of these benefits for all trips made gives the total benefit on yearly trips made by travelers.[2]

At this juncture, it should be noted that demand curves often are difficult to establish in practice. The correlation coefficients for trip generation presented in Chap. 7 certainly verify this statement. Of particular difficulty is the establishment of the end points in Fig. 10.2. It is a rare occurrence when travel is either free or so expensive that none is made. Because of these uncommon situations, observations at the extremes have been lacking, and no firm commitment can be made of exact locations.[3] The result is that consumer surplus measurements also have been difficult to make.

Laying these problems aside temporarily, we can continue to develop the theoretical framework and at least determine what would be *desirable* insofar as an

[2] The usual definition of benefits as proposed by economists includes the entire area between the demand curve and abscissa, whereas the definition employed in the *Red Book* [10.1] follows that presented above. The distinction between the two concepts is not a crucial one for purposes of the exposition in this book, and so the *Red Book* definition will be adopted here. Economists usually denote the total consumer's surplus as the "net benefits."

[3] For an elaboration of the point see [10.21].

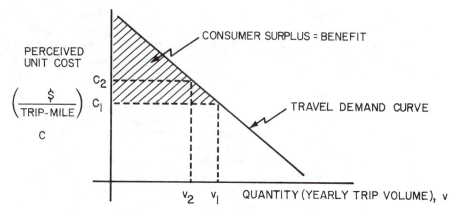

Fig. 10.2 Travel demand curve showing consumer surplus.

evaluation technique is concerned. The next step after establishing the demand curve would be to construct its counterpart, the supply surve. To do this, it is necessary to consider the short run elements of perceived cost associated with each mile driven on the example two-lane highway. The first of these costs might be for fuel (excluding taxes). This would be expected to increase with the number of yearly trips made on the highway, especially as volumes neared capacity. A curve like the lower one in Fig. 10.3 probably would be realistic.

In addition to fuel, there would be the perceived unit vehicle operating costs, engine oil, tires, maintenance, and depreciation (again excluding taxes) to consider. A reasonable assumption would be that these too would rise somewhat with increases in travel volumes since more traffic would mean more delays, more idling of engines, longer times on the road, and so forth. These prices, when added to those for fuel, would bring the total unit costs *up* to the second lowest curve in Fig. 10.3. Following a similar line of reasoning, we would anticipate that travel-time costs, the final type, might increase rather sharply with volume as congestion on the highway slows traffic and increases the time for each trip. The sum total of these unit costs,[4] calculated for each yearly tripmaking level, is represented by the topmost line in Fig. 10.3. It should be remembered that this line is indicative only of the particular highway used in the example, and depicts the perceived cost to *supply* or handle the given number of yearly trips by that existing facility.

The combination of the supply curve for the example highway in Fig. 10.3 and the demand curve in Fig. 10.2 is diagrammed in Fig. 10.4. The crossing of the two curves forms an equilibrium point (v_3, C_3), which can be interpreted as follows:

> No amount of trips greater than v_3 will be made since, after a period of time, some people will find that the cost of making the additional trips is greater than they are willing to pay (the supply curve lies above the demand curve). Similarly, no amount of trips less than v_3 will be made since, after a period of time, some people will realize that the cost of making a trip is less than that which they are willing to pay (the supply curve lies below the demand curve). Thus, additional trips will be made until the unit costs equals that which the travelers are willing to pay.

---

[4] It is not intended that these costs represent an exhaustive set, but the general feeling among transportation planners is that these are the major items which the automobile driver *perceives* as being significant. Other costs, such as car insurance, do not appear to be important to the driver in determining whether to make *additional* trips.

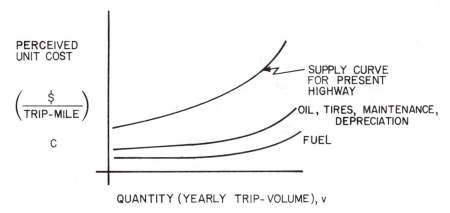

QUANTITY (YEARLY TRIP-VOLUME), v

Fig. 10.3 Supply curve for travel.

The equilibrium point (v_3, C_3) therefore indicatest the volume (v_3) of traffic that will use the example highway and the cost per trip mile (C_3) that the travelers will pay for making their trips. This cost then can be employed to calculate total benefits in a manner similar to that outlined in connection with Fig. 10.2.

The supply and demand curve concepts can be enlarged to take into account the consequences both of changes in demand and proposals for possible alternative highway improvements. By way of introduction to the first case, it generally has been true in the United States that overall income levels are rising,[5] and these increases usually lead to corresponding increases in the willingness of people to pay for certain goods or services. Thus, in referring to Fig. 10.4, it can be seen that the perceived unit cost for a given number of trips, say v_3, will tend to increase over time or, stated another way, greater number of yearly trips will be made for a given cost. This type of change is indicated in Fig. 10.5, which also incorporates the demand curve from Fig. 10.2 and the supply curve from Fig. 10.3. The "new" demand curve rises above the "old" one for the reasons cited above.

[5] For some relevant data, see Sec. 3.3.3.2 in Chap. 3.

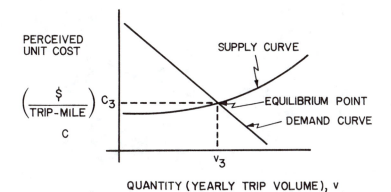

QUANTITY (YEARLY TRIP VOLUME), v

Fig. 10.4 Equilibrium of demand for and supply of travel.

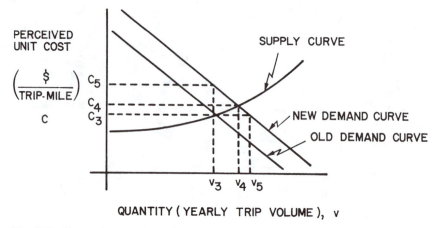

PERCEIVED
UNIT COST

$\left(\dfrac{\$}{\text{TRIP-MILE}}\right)$

SUPPLY CURVE

NEW DEMAND CURVE

OLD DEMAND CURVE

c_5
c_4
c c_3

v_3 v_4 v_5

QUANTITY (YEARLY TRIP VOLUME), v

Fig. 10.5 Changes in tripmaking and unit trip cost resulting from increased demand.

One important point to notice in Fig. 10.5 is that a new equilibrium point (v_4, c_4), results from the establishment of the new demand curve. Interestingly, both the amount of money paid for travel and the number of trips increases, the former from c_3 to c_4, the latter from v_3 to v_4. This situation implies that rising economic levels lead to increases in travel and explains to some extent why many transportation facilities are used to their capacity before expected. These increases are part of what is known as "induced traffic."

Figure 10.6 shows the effect of a proposed new highway on the unit trip cost, number of trips, and benefits as regards travel between A and B. It is assumed that the new highway will be an "improvement" over the old one (which, for purposes of this example, will be eliminated after the new one is opened) in that there will be fewer and flatter curves, slighter grades, dual lanes in each direction, and so forth. With these conditions, it then follows that the new highway most likely will lead to a reduction in both the operating and travel-time costs that help to make up the short-run supply curve in Fig. 10.3.

The former costs would be lower primarily because of decreases in motor fuel

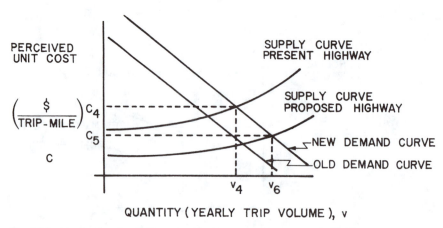

PERCEIVED
UNIT COST

$\left(\dfrac{\$}{\text{TRIP-MILE}}\right)$

SUPPLY CURVE
PRESENT HIGHWAY

SUPPLY CURVE
PROPOSED HIGHWAY

NEW DEMAND CURVE

OLD DEMAND CURVE

c_4
c_5
c

v_4 v_6

QUANTITY (YEARLY TRIP VOLUME), v

Fig. 10.6 Changes in tripmaking and unit trip cost resulting from a new highway facility.

needs brought about by the straightening of horizontal curves, the smoothing of vertical curves, and, in general, the creation of a more direct route between A and B, whereas the latter costs would be lower because of the ease of passing associated with the dual laning and, in general, the increased capacity of the new facility. The result of these effects, displayed in Fig. 10.6, would be a new short-run supply curve associated with the new highway and lying below that for the present facility.

Also resulting from these travel cost reductions would be an increase in yearly tripmaking. By building a highway with a lower unit cost of travel, we can anticipate that more people would be willing to travel, and this is the case since at the new equilibrium point (v_6, C_6), the cost has been reduced from C_4 to C_6 while the number of trips has gone up from v_4 to v_6. This increase is another major component of "induced traffic."

Another result of a reduced cost of travel usually (and in this particular example) is an increase in benefits (as defined in Fig. 10.1). Looking at Fig. 10.7, which summarizes most of the information from the previous diagrams in this chapter, we can see that the benefits, B_p, of the present facility at the present time (old demand curve)[6] are

$$B_p = \tfrac{1}{2} (C_7 - C_3)(v_3 - 0) \tag{10.1}$$

Similarly, the future benefits (new demand curve) that would result if the present highway were not replaced by the proposed one would be

$$B_0 = \tfrac{1}{2} (C_8 - C_4)(v_4 - 0) \tag{10.2}$$

Finally, if the new highway were constructed and the old one eliminated, the benefits, B_n, would be

$$B_n = \tfrac{1}{2} (C_8 - C_6)(v_6 - 0) \tag{10.3}$$

[6] For simplicity a linear demand curve is assumed in the calculations to follow.

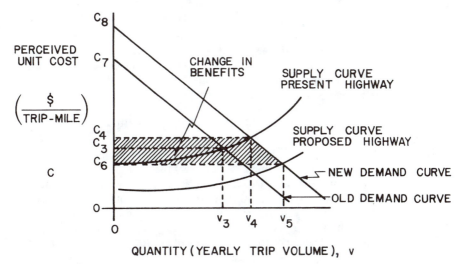

Fig. 10.7 Change in benefits from an increase in demand and from a proposed highway.

The increase in future benefits attributed to the new highway thus would be calculated via

$$B_n - B_0 = \frac{1}{2}(C_8 - C_6)(v_6 - 0) - \frac{1}{2}(C_8 - C_4)(v_4 - 0) \tag{10.4}$$

which can be reduced to

$$B_n - B_0 = \frac{1}{2}[C_8(v_6 - v_4) - C_6 v_6 + C_4 v_4] \tag{10.5}$$

Under the assumption of linearity and with a slope of $(C_4 - C_6)/(v_6 - v_4)$ and an intercept of C_8, the general equation of the demand curve is

$$C = -\frac{C_4 - C_6}{v_6 - v_4} v + C_8 \tag{10.6}$$

Since the point (v_4, C_4) falls on the line, we obtain

$$C_4 = -\frac{C_4 - C_6}{v_6 - v_4} v_4 + C_8 \tag{10.7}$$

or

$$C_8 = C_4 - \frac{C_4 - C_6}{v_6 - v_4} v_4 \tag{10.8}$$

Substituting this into Eq. (10.5) results in

$$B_n - B_0 = \frac{1}{2}\left(C_4 - \frac{C_4 - C_6}{v_6 - v_4}\right) v_4 (v_6 - v_4) - C_6 v_6 + C_4 v_4 \tag{10.9}$$

which, after algebraic manipulation, becomes

$$B_n - B_0 = \frac{1}{2}(C_4 - C_6)(v_6 + v_4) \tag{10.10}$$

which is also the formula for the area of the shaded trapezoidal section in Fig. 10.7.

A point to be stressed at this time is that benefits (or disbenefits) may accrue even if the proposed improvement is not built. In the example presented here, the change in benefits over time caused by the increase in demand is

$$B_0 = B_p = \frac{1}{2}(C_8 - C_4)(v_4 - 0) - \frac{1}{2}(C_7 - C_3)(v_3 - 0) \tag{10.11}$$

This quantity may or may not be positive, depending on the magnitude of each of the unit costs and volumes. The point is, however, that the "do-nothing" alternative is one that has to be considered in its own right: it has an impact that may be significant. In fact, the existence of this alternative is one of the main reasons why the evaluation stage has been placed before the solution-generation stage in the transportation planning process outlined in Chap. 2. It is imperative that the consequences of *not changing* from the present state be evaluated before any solutions are proposed. Otherwise, the planner has no basis with which to *compare* the benefits that may arise from various "improvement" schemes.

10.1.1 Accounting for Capital and Maintenance Costs

After the analysis of benefits is completed, it is necessary to look on the other side of the ledger—on the facility cost side. The primary component costs to be considered are those for right-of-way (land), grading and drainage, major structures, pavement, and, in a slightly different category, maintenance. The former set of costs, known as capital costs, generally are the most extensive, yet maintenance costs also can be significant. In either case, the objective at this point is to compare the benefits that will accrue from the expenditure of funds for the construction and maintenance of alternate highway facilities.

"Benefits" calculated by means of Eq. (10.10) are stated in dollar terms just as the capital and maintenance costs are. Each of the unit user costs is expressed in dollars per mile per trip, and the volume of trips is expressed in "trips per year." Therefore, since in each equation we have the unit cost multiplied by the number of trips, we get ($/trip·mile) X (trips/year) = ($/mile·year) as the units for benefits. Benefits and costs thus are commensurate, that is, measurable in the same units. Commensurability naturally is desirable in an evaluation procedure since we do not, as the expression goes, want to "mix apples and oranges."

Another consideration which must be taken into account at this stage is that the benefits and costs associated with a transportation facility vary over time. Referring to the hypothetical curves in Fig. 10.8, we can see that if a new highway facility were constructed, the benefits would not start until its completion and probably would build over time as travel increased. The costs, on the other hand, would be extremely high at the beginning when the full amount of capital had to be expended, then would decrease sharply, followed by a slight increase over time as the facility started to approach the end of its service life and subsequently required substantial reconstruction.

10.1.2 Interest of Discount Rates

A final consideration to be brought out before the comparison of benefits and costs can be made is that funds for capital costs often must be borrowed, so that interest payments must be made. Even if borrowing were not necessary, we should

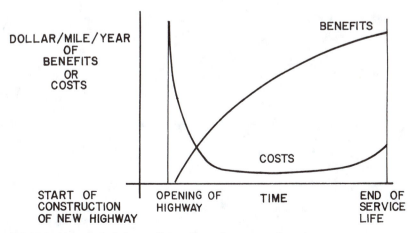

Fig. 10.8 Hypothetical changes in benefits and costs over time.

take into account the time value of money, which often can be equated to the interest rate. To understand this point one has only to remember that if an investment in a transportation facility cannot create monetary returns of at least, say, 5 percent per year (simple interest),[7] then the investor, be it a city, state, or federal government or a private firm, would be better off financially by putting its funds into a savings account at a local bank. In other words, we would expect that all funds should be allocated in such a manner that they will have a return that compares favorably with any other of their possible uses. A 5 percent return is "no return" in the sense that most other investments would give back at least this amount. The same kind of reasoning holds for benefits. Those that occur in the far distant future are worth less to "investors" (the benefactors) than those that occur right away. There thus should be an interest rate, otherwise known as a *discount rate*, applied to future benefits to equate these to present ones. The discounted benefits from a proposed transportation facility then should exceed the discounted costs in order for it to be a worthwhile investment.

What discount rate to use in a given situation is a matter of some debate. Of course, if funds are being borrowed, the actual interest rate can be employed. If not, then some other means must be found for making such a determination. Wohl and Martin [10.5] and Winfrey [10.26] go into a considerable discussion on this issue, and their books should be consulted for further information. Recent high interest rates and changes in interest-bearing instruments have tended to outdate their discussions, however, leaving little in their place.

10.1.3 Calculating Discounted Costs and Benefits

With the adoption of a discount rate, it then is possible to calculate the equivalent current values for costs and benefits. This is accomplished through the use of a multiplier to determine the present worth (PW) of a single sum. For example, if the interest or discount rate is i and we are looking for benefits, B_n, in the nth year from the start of the project or program, then the equation for the present worth of these benefits, PWB (i, n), is

$$\text{PWB}(i, n) = \frac{B_n}{(1 + i)^n} \tag{10.12}$$

These would be summed for the benefits in each year over the service life (N) of the project or program to give the net present benefits, NPB:

$$\text{NPB} = \sum_{n=1}^{N} \frac{B_n}{(1 + i)^n} \tag{10.13}$$

As an example of the use of these two equations, let us take the numbers for the first project alternative listed in Table 10.1. Assuming that the costs and benefits are discounted at a 15% rate from the *beginning* of the first year,[8] we find net present capital plus maintenance costs, NPC, of

---

[7] This figure has been chosen to represent a probable lower bound on the interest rate.

[8] Another alternative is to start the discounting at the end of the first year. The two approaches yield almost the same results if the number of years is large (as it would be in most real world cases) [10.1].

Table 10.1 Costs and Benefits for Four Hypothetical Alternatives ($1,000)

Project alternative	Year	Change in cost Capital	Change in cost Maintenance	Change in cost Total	Change in benefits (user savings)
I	1	21,000	3,000	24,000	14,000
	2		3,000	3,000	14,000
	3				
	Total	21,000	6,000	27,000	28,000
II	1	21,000	3,000	24,000	14,000
	2		3,000	3,000	14,000
	3		3,000	3,000	4,000
	Total	21,000	9,000	30,000	32,000
III	1	21,000	3,000	24,000	10,000
	2		3,000	3,000	14,000
	3		3,000	3,000	10,000
	Total	21,000	9,000	30,000	34,000
IV	1	21,000	3,000	24,000	14,000
	2		3,000	3,000	13,000
	3		3,000	3,000	7,000
	Total	21,000	9,000	30,000	34,000

$$NPC = \frac{24,000}{(1 + 0.15)} 1 + \frac{3,000}{(1 + 0.15)} 2 = \$23,137$$

and for benefits,

$$NPB = \frac{14,000}{(1 + 0.15)} 1 + \frac{14,000}{(1 + 0.15)} 2 = \$22,759$$

10.2 COMPARISON OF BENEFITS AND COSTS

There are several ways of comparing benefits and costs. Let us assume that the simplified figures in Table 10.1 are estimates for the four best alternatives to the do-nothing or null possibility—that is, to making no changes in the road shown in Fig. 10.1. The benefits would represent differences in user costs, or user cost savings, over continued utilization of the current road. These would be derived by considering fuel consumption, time, tire wear, motor oil, and other user costs described in Chap. 6. Similarly, the "costs" considered would be for the difference in capital and maintenance expenditures for a given alternative versus that for the present road.

To simplify the calculations, we will assume that the alternatives can be built in very little time, so that user savings will result immediately. Also, the roads will have a service life of only 2 or 3 years.

10.2.1 Payback Period

The first formal method of comparison is the payback period. It is the length of time from the beginning of the project before the net benefits (undiscounted) return the cost of the capital investment (undiscounted). The payback period is a common, rough means of choosing between investments in business enterprises, especially where there is a high degree of risk. It has not been a common measure in past urban transportation projects because there has not been much risk of technological obsolescence. The world is changing quickly, however.

Looking at the four alternatives, we can rank them according to payback period. This is accomplished using Fig. 10.9. The cumulative costs and benefits are plotted versus time, and the payback point is where the associated lines cross. This leads to:

Alternative	I	II	III	IV
Payback period (yr)	1.91	1.91	2.40	2.00
Ranking	1	1	4	3

In this situation, projects I and II both have the same payback period, but we know by inspection that the latter will continue to return benefits in the third year. Hence the payback period has its deficiencies as a sole criterion to choose between these two alternatives.

If we were to modify projects III and IV so that the capital plus maintenance cost of both were $34,000, they would each have equal payback periods of three years. Yet we can see that project IV is better because the time sequence of the cash flow is such that that project will have more benefits earlier in the sequence, obviously desirable since the earlier the benefits are received, the earlier they can be reinvested (or consumed), and hence the more valuable they are. The payback-period method thus has two major weaknesses as a measure of investment worth:

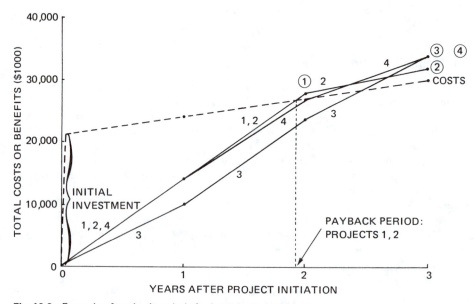

Fig. 10.9 Example of payback periods for hypothetical projects.

(1) it fails to consider benefits after the payback period, and (2) it fails to take into consideration differences in the timing for the same amount of benefits.

10.2.2 Benefits per Dollar of Outlay

Sometimes investments can be ranked by the total benefits divided by the total amount of the investment:

Alternative	I	II	III	IV
Total costs ($1,000)	27,000	30,000	30,000	30,000
Benefits ($1,000)	28,000	32,000	34,000	34,000
Benefits per dollar of outlay	1.04	1.07	1.13	1.13
Ranking	4	3	1	1

Again, no discounting is taken into account.

10.2.3 Net Present Value (or Worth)

The net present value (NPV) technique attempts to overcome many of the difficulties in the preceding two approaches by employing the discount factor to show the value of a cost or benefit now versus at some time in the future. For the first of the hypothetical alternatives in Table 10.1, assuming a 15% discount rate, the NPV would be the difference between the net present benefits (NPB) and costs (NPC):

$$NPV = 22,759 - 23,137 = - \$378 \quad (\$1,000)$$

For all four possibilities,

Alternative	I	II	III	IV
NPV ($1,000)	−378	279	746	1,495
Ranking	4	3	2	1

These show that (1) alternative IV will have the greatest excess of discounted benefits over costs, (2) project II has a greater excess than I, since the benefits last longer, and (3) the excess of project IV is more than that of III because the benefits come earlier in time.

If the NPV of a project turns out negative, as for alternative I, the discounted costs exceed discounted benefits, and the project should not be undertaken unless there are some other, noneconomic benefits of great value.

10.2.4 Benefit/Cost Ratio

Some analysts prefer to see a ratio of benefits to costs rather than a difference (as in the NPV). In some senses a ratio is an indicator of "efficiency," showing the dollars of benefits achievable for a given outlay of (discounted) costs. For alternative I we thus would have a benefit/cost ratio (BCR) of

$$BCR = \frac{NPB}{NPC} = \frac{22,759}{23,137} = 0.98$$

For the four projects,

Alternative	I	II	III	IV
BCR ($/$)	0.98	1.01	1.03	1.06
Ranking	4	3	2	1

This leads to the same ranking of projects as the NPV, and thus the same conclusions. If the BCR were less than 1.00, this would mean the costs were greater than benefits, and the project should be undertaken only if other, noneconomic benefits were substantial.

It should be noted that one project can have a higher NPV than another but a lower BCR, and vice versa. If project A, for example, had an NPC of 10,000 and an NPB of 20,000, the NPV would be 10,000 and the BCR 2.00. Yet project B may have an NPC of 5,000 and an NPB of 12,000, giving an NPV of only 7,000 but a BCR of 2.40. The two criteria thus are not always correlated.

10.2.5 Internal Rate of Return

A difficult question in both the NPV and BCR methods is the value of the discount rate. As noted earlier, there is no clear answer to this question. One way to avoid the issue is to compute the NPV for several discount rates to find the one at which the net present benefits just equal the net present costs (or, in other words, NPV = 0). This point is known as the rate of return or internal rate of return, IRR.

Calculation of the IRR is complicated because it cannot be done directly. Trials have to be set at different levels of the discount rate to the point where a negative NPV is found. The IRR then is established by interpolation. One way of doing this is shown in Fig. 10.10 for project II. IRR values were computed for discount rates

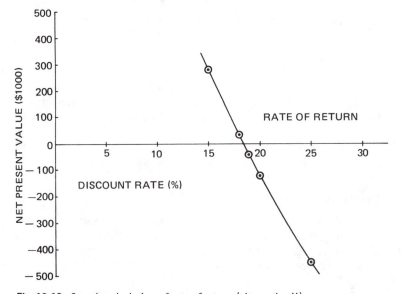

Fig. 10.10 Sample calculation of rate of return (alternative II).

from 15% to 25%. The NPV values were found to be negative for greater than 19%. Graphical interpolation indicated an IRR of about 18.5%. Another way of doing this is to extrapolate from one side or the other. About the same result would be found by extending the line for the top two or bottom three points until it crossed the x axis.

The ranking of the four alternatives by IRR gives

Alternative	I	II	III	IV
IRR (%)	10%	19%	20%	31%
Ranking	4	3	2	1

The order is the same as in the previous two situations. It is possible, however, to get different rankings coming from the last three techniques.

10.3 COMMENTS ON THE DIFFERENT ECONOMIC APPROACHES TO EVALUATION

The preceding approaches to evaluation have been used extensively in the past by many different agencies and companies. The techniques offer the distinct advantage of neutrality: the numbers used as inputs lead to an exact determination of the alternative which is best, and there can be no inference made that the evaluation process has been interfered with for personal reasons.

In this light, the approaches are extremely valuable in that, if the theory behind them is agreeable to everyone, the outputs also must be accepted. In general they give the appearance of a certain mathematical purity—a purity that cannot be tampered with and is instrumental in providing the decision maker with results unbiased by emotional factors.

The different approaches do have many common difficulties nevertheless. Some are technical in nature, while others stem directly from failure to take into account many "human" factors. Each of these types of difficulties will be discussed in turn in succeeding paragraphs.

10.3.1 System Effects

One of the foremost difficulties that arise in the utilization of the above approaches in any real-world situation is that proposed alternatives generally are part of a system and do not stand by themselves as in the previous example problem. A change in a route alignment between two points or a decrease in travel time between two points often affects travel not only on that route but on many other nearby routes.

After construction of a new facility from D to C, people traveling from A to C and presently using the section of highway from B to C may rearrange their route, going to A to D instead of from A to B to C. The result is that the number of trips on the sections from A to B and B to C is reduced *because* of the change in the route from D to C. The effects of improvements on one route thus permeate over other routes and present the evaluator with a difficult problem in accounting for *all* the benefits.

10.3.2 Unequal Alternatives

Alternatives, by definition, are different ways for accomplishing the same objective or solving the same problem. Quite obviously, we would not treat proposals for rapid transit either in Los Angeles or in Atlanta as alternates for the present mass transit situation in Minneapolis. The former two do not serve the same populations or the same travelers, nor would they necessarily result in the same type of physical system as in Minneapolis. But do any transportation alternatives ever serve the same purpose?

Referring to the examples in Table 10.1, we see that the four alternatives (including the existing one) do not serve the "same" population of travelers, the major difference probably being in the number served. The construction of any of the four new facilities thus changes the travel situation and, in effect, produces a new problem to be solved. The main point to be made here, however, is that summary figures developed for evaluating alternatives (e.g., a benefit-cost ratio) rarely compare "equal" situations because the problem under study usually is modified by the alternatives proposed.[9]

10.3.3 Risk and Uncertainty

Inherent in all evaluation techniques, including the above, are problems of risk and uncertainty. Wohl and Martin [10.5, p. 223] make the distinction between these two entities as follows:

1. Problems of *risk* are those whereby the future outcomes or consequences have a known probability of occurrence; thus while the chances of a particular outcome may be known, no assurance can be given about which particular outcome will take place.
2. Problems of *uncertainty* are those whereby even the probabilities of the future outcomes or consequences are unknown and whereby the probabilities can be determined only subjectively.

The determination of whether an outcome is subject to risk or uncertainty is a difficult matter. What is important here is that almost all of the entities used in the calculations—costs, travel volumes, interest rates, unit user service lives, and capital and maintenance costs—have to be predicted for the future and therefore fall prey to inaccuracy.

Capital costs, for example, usually are thought of as being easily predictable, but factors such as inflation and unanticipated expenses resulting, perhaps, from the discovery of rock to be excavated or, in another situation, from the need for funds for drawn out legal cases, make even cost predictions hazardous. As a consequence, if the uncertainty (or risk) is anticipated to be great, the evaluator should take this feature into account, either by weighing each outcome by its probability of occurrence, or, as is done in many cases, by increasing the discount rate so that the investor gets a larger and quicker return to make up for the riskier situation. Such techniques can be incorporated as part of some of the procedures described, but are rather cumbersome and data-consuming.[10]

10.3.4 Inclusion of Various Benefits and Costs

The discussed approaches demand that the set of benefits and costs to be included be identified explicitly. This is usually desirable, yet the problem arises as to how to account for *all* benefits and costs. How about disbenefits or diseconomies

---

[9] The creation of unequal alternatives is not a difficulty related to these techniques per se, but is common to almost every evaluation procedure.

[10] For more information on handling risk and uncertainty, see [10.16] and [10.24].

resulting from increases in noise and air pollution levels? How about engineering, planning, and administration costs? The first question is more difficult to answer, mainly because we often are not sure of the extent of the effect of transportation on the two entities.

It may be, for instance, that air pollution in a certain sector of a city is created mostly by an industrial plant there and not by automobiles, trucks, buses, and so forth, so that air pollution should not be counted as a disbenefit of transportation. Even a prominent item like travel is not caused by the transportation system alone, but is a function of land use and other factors.[11] From these examples we can conclude that an accounting of benefits should include only those attributable to the particular alternative under study but that such an approach would require the identification of cause and effect relationships about which, in many cases, little is known.

Another aspect of the benefit and cost identification problem is that of the inclusion or noninclusion of benefits or costs passed from one level of government to another. Wohl and Martin [10.5, p. 181] pose an interesting example of such a situation:

> Should a state highway agency, in deciding among various highway projects (including the null alternative), consider only the consequences to the state highway users or those to the entire state populace or should it adopt a broader national point of view? Also, should the state highway agency consider the economic feasibility of only the *state* expenditures on construction, maintenance, and administration, or should it be concerned with the feasibility of total outlays, whether federal, state, or local?

Most people would argue for the national point of view, but the vote would be far from unanimous.

10.3.5 Measurement of Benefit Factors

If, for purposes of evaluation, an attempt were made to list all of the factors affected by a given transportation alternative, a major difficulty to be faced would be that of measuring (or actually defining) the factors. Of course, travel time and number of trips are two entities which are fairly easy to measure, but, as the time-worn argument goes, beauty is not easy to gauge.

We might try to utilize such individual measures as color (wavelength), hue, contrast, brightness, and so forth in combination, but to date no one has originated a single, mutually satisfactory measure of beauty. The problem that this situation creates for the evaluator is that, without an acceptable definition of beauty, a suitable prediction cannot be made of whether an alternative transport project or program will add to or detract from the appearance of the setting into which it is thrust. As a consequence, some of the systems benefits (or disbenefits) cannot be estimated. Lack of a measurement device thereby implies a possible miscalculation of benefits (and sometimes also costs).

10.3.6 Commensuration

Earlier in this chapter the problem of "mixing apples and oranges" was presented in the context of putting benefits in dollar terms. Assigning a "value" to travel time is one example of this problem. In Table 6.2 in Chap. 6 it was suggested that about $2.40 per traveller·hour be utilized for work trips in which median time savings (5 to 15 min) resulted. How realistic is this?

---

[11] See Chaps. 6 and 7.

By far and away the most cogent example of an attempt to put benefits in dollar terms is that found in the AASHTO *"Redbook"* where, in one case, the cost per fatal accident was set at $307,210 [10.1, p. 64].

Obviously, any attempt to put all benefits, such as reductions in fatalities, on a common monetary basis will attract the label of "mercenary," and, to some extent, this charge rings true. Nevertheless, if the evaluator does not endeavor to make all benefits commensurate, the risk is run of *implicitly* assigning a value out of proportion to actual worth. If, for example, a highway costing $10 million and resulting in five deaths during its lifetime is chosen over a mass-transit facility costing $2,000,000 and resulting in one death in the same period, then, all other factors being equivalent, the price of four deaths (five minus one) has been implicitly set at $8 million ($10 million minus $2 million), or at $2 million per life. This value, many people would agree, is too high. But the point is that many transportation planners and decision-makers are faced directly with the unenviable task of deciding on the relative worth of the life of each citizen in the population.

Another interesting aspect related to attempts at commensuration is that values associated with a given item often vary according to the quantity of the item and the kind and amount of substitutes available. As an example of the first case, if 1 hr of travel time were saved from use of a new transportation system, it may be worth only $1 (per hour) to the traveler. Yet, if 2 hr were saved, they may be worth $4 total, or $2 per hour. As an example of the second case, suppose that in the previous illustration a second transportation system were built which also saved the traveler 2 hr. Because there now can be a choice in route of travel and because of the corresponding increase in dependability (that is, if one route is closed, the other can be taken), travelers may devalue the importance attached to travel time to, say, $1.50 per hour. These, then, are some of the considerations that make commensuration a difficult task.

10.3.7 Inflation

It was presumed in the cost–benefit comparisons that neither user costs (and thus benefits) nor those for construction and maintenance changed over time.[12] This obviously is not true, as demonstrated by the rises in the consumer price indices shown in Table 6.14. To make matters worse, different cost components have inflated at drastically different rates, with gasoline representing the one extreme with which almost everyone is familiar.

It obviously seems desirable to take inflation into account. To do so, however, complicates the evaluation in at least three ways:

1. It requires another set of data.
2. It requires making predictions of future prices, which most analysts think is particularly difficult.
3. It requires inclusion of the predicted prices in each cost component, which means many more calculations.

None of these difficulties is insurmountable, but they do make evaluation a much more time-consuming, expensive, and uncertain process.

---

[12] Or, alternatively, that all inflated at the same rate so that, relative to each other, there was no change in constant dollar values.

10.3.8 Inappropriate Prices

Hidden in all of the costing procedures in the five foregoing techniques is the assumption that the costs or prices employed represent the "real" value of the resources needed to produce the particular good or service. That this is not always the case is demonstrated aptly by transit fares. Rarely do these cover even a major portion of the expense of providing that service. Subsidies take care of the rest. Hence it would be inappropriate to take just reductions in fares paid as a user benefit. This usually would not indicate the "real" reduction in the value of resources to produce the service.

The "real" costs often are referred to as shadow or accounting prices. And an evaluation done with such prices is called an economic evaluation, as opposed to a financial one in which actual prices are utilized [10.8]. The difficulty, of course, is in determining the shadow prices. Some subsidies (or overcharges) may be very subtle, as they may be included in the tax or legal structure. Oil companies, for instance, have received certain special tax deductions and credits. These, in effect, help to lower the cost of gasoline so that the price at the pump is not really indicative of its "real" value. Similarly, minimum wage laws tend to raise the cost of some laborers, which makes price of construction and maintenance "artificially" higher. How much difference exists between "real" and actual costs at most times is almost impossible to determine.

10.3.9 Perceived versus Actual Benefits and Costs

A perplexing decision in most evaluation procedures is that of whether to use actual benefits and costs that accrue as a result of transportation systems or the ones *perceived* by the people affected by the system. Travel time again provides an interesting example. Suppose that, through verifiable calculations or empirical studies, it is determined that 50 min are saved on a given journey over a new transportation system. The user, however, feels that time savings are less,[13] say 40 min, and judges the worth of the system using this figure. Which figure should the evaluator use?

On the cost side of the picture, there also might be significant differences between actual and perceived costs. In fact, these differences are used to advantage by many business operations through the use of the charge account—it is much less agonizing to charge a $10 item than to pay for it in cash. The perceived cost of a cash payment is much higher.

It is important at this point to note the *significance* of the differences between perceived and actual costs and benefits. If a comparison were made between highway and mass-transit facilities, for example, we probably would find that vehicle purchase costs generally are *not* considered as part of travel costs (prices) by the highway user[14] but that vehicle costs for transit would be important since they must be included in the fare, which is all too prominent. Because automobile travel *appears* less expensive, more trip-making is done by that mode and less by transit, a situation that naturally affects the stability of transit service.

---

[13] In most cases, those affected by a transportation system do not have the opportunity to determine *exactly* how they are being affected.

[14] The mode choice model presented in Chap. 7 indicates this clearly. Automobile and insurance costs are *not* shown to have an effect on choice of mode of travel.

10.3.10 Discounting of Benefits and Costs

Earlier we described briefly the role of the discount rate. It is the *unevenness* over time of the streams of benefits and costs that causes most practical discounting difficulties. The amoritization of costs over time may be fairly uniform and end after a period of 40 to 50 years. Amortization costs, of course, would continue as long as the facility existed. But, on the other side of the ledger, benefits may continue to accrue deep into the future, perhaps even at an increasing rate (Fig. 10.8).

An interesting example of this type of situation is the famous Appian Way (Appia Antica). Opened by Claudius Appius in 312 B.C. and running outside the ancient walls of Rome, this facility still is providing service benefits to travelers after some 2,300 years of use.[15] The prolonged nature of this service brings to the fore the question of how to compare in a correct manner the different time-sequenced and widely divergent streams of benefits and costs.

If the period for benefit analysis is chosen to be relatively short, say 20 years, then the probability exists that a large amount of benefits will be ignored. Suppose, for example, that people value a dollar's worth of benefits today at 94¢ a year from now. Keeping this rate, it turns out the benefits 20 years from now are still worth 31¢. Thus, despite difficulties in estimating benefits at such a future date, it appears to be important not to overlook them since they are significant from an absolute standpoint (because of the probable increase in benefits over time) as well as from a percentage standpoint.

10.3.11 Double Counting

Another perplexing problem facing the evaluator is that of the possible double counting of benefits and costs. It would not be correct, for instance, to include both the service-station charged price for fuel *and* the tax on gasoline as components of the unit cost of operating an automobile on the highway. This would be an obvious case of double counting, since the tax already is incorporated in the service-station price.

Taxes in themselves present an interesting dilemma for analysts. Since fuel taxes are part of the pump price, they obviously affect the auto or truck driver's behavior. All other things being equal, if there is a rise in the "cost" of taxes, the trip-maker will travel less. User benefits thus would be reduced. On the other hand, taxes really are a transfer from the user to government, which in turn utilizes the revenues to construct roads, help pay for transit systems, and the like. Because some of these expenditures already are taken into account in the "cost" component of, say, benefit–cost analysis, it would be double counting to consider them also with respect to the user. Current thinking is tilted toward *not* including taxes in user cost savings [10.10].

Other opportunities for double-counting stem basically from the transfer of benefits[16] from one person or group of persons to another. Mohring and Harwitz [10.28, p. 12] give the following example of such a transfer:

> ... completion of an expressway which reduces the time and dollar costs of travel to the center of an urban area may enable a suburban apartment house owner to charge higher

---

[15] Information taken from A. Storti, *Rome: A Practical Guide*, E. A. Storti, Venice, Italy, 1965.

[16] To be discussed in more detail in Sec. 10.3.14.

rents to his commuting tenants than would otherwise have been possible. To the extent that he is able to do this, he has, in effect, extracted some of the highway benefits initially received by these tenants. He has, that is to say, forced them to transfer some or all of their highway benefits to him.

The conclusion to be drawn from this example is that it would be improper from an accounting standpoint to include both decreases in travel costs and increases in apartment rents (and thus in land values) as benefits from the expressway. They are "two sides of the same coin." Likewise, it would be improper to count any benefit or disbenefit until it has been shown to be separate and distinct from any others under consideration. To make the distinction, however, is extremely difficult.

10.3.12 Determining Who Benefits, Pays

Perhaps the most critical comment that can be made of benefit-cost and similar approaches to evaluation is that they do not indicate who is receiving the calculated benefits or who is paying the costs. Both are totaled but nothing is said about their distribution among the poor or rich; young or old; user or nonuser; trucker, or railroad, or airline; white or black; those who live in one part of the city or region or those who live in another; and so on. Needless to say, the question of who benefits is an important one. As John A. Volpe, President Nixon's original Secretary of Transportation, has stated:

> I would submit to this group (the Greater Dallas Planning Council) tonight—as I have done before the President and before my Cabinet colleagues—that all the job training centers, employment opportunities, health facilities, educational institutions, recreational areas and housing projects—all things that are needed in virtually all of our cities—will never be fully utilized if the people cannot get to them inexpensively, safely, and efficiently.
> We must have a new mobility in this nation if we are to fulfill our pledges to the disadvantaged, the young, the poor, the elderly, and the physically handicapped.[17]

By *not* looking at the types of groups affected by a given change in a transportation system, the evaluator may *inadvertently* (and disproportionately) benefit one group at the expense of another. If family members do not have access to an automobile (and many of the poor would fall in this category), they would have difficulty benefiting from any highway improvement, at least as compared to the gains they would receive from a mass transit improvement. Interestingly, conventional economic theory assumes that poor people value (have a greater utility for) money than the rich [10.29]; that is, the poor use a dollar more sparingly. Hence benefits to the poor should be counted more heavily, although they rarely are.

10.3.13 Criterion Form

The form of the criterion or objective function used in evaluation is of extreme importance. Five forms were demonstrated earlier, but the choice as to which one to use appears to have been extremely subjective in the past. The point is that this *subjective* decision between criterion forms has a significant effect upon the evaluative decisions that come out of the *objective* technique (such as benefit-cost). The payback period criterion, for example, gave preference to alternative I, while it was least preferred using the other forms. These discrepancies cast some doubt on the supposed objectivity of the overall procedures.

[17] Remarks prepared for delivery by Secretary of Transportation John A. Volpe before the Greater Dallas Planning Council, Tuesday, September 9, 1969, Dallas, Texas.

10.3.14 Transfer of Benefits and Costs

Assume we have identified those people we would like to see benefit from or pay for a new transportation system. How can we insure that they will, in fact, be the ones who do? In actuality they may be forced (or are able) to transfer them to some other group.

Suppose, for instance, that in the Mohring and Harwitz example presented earlier, it had been decided to attempt to direct as many of the benefits as possible to the expressway user. This attempt would have resulted in failure. The users are the *first* recipients of the benefits of reduced operating and travel time costs, but are forced to pay equivalently higher rents in order to live close enough to the expressway to get the travel benefits. In the end, the user has no actual gain in capital, services, or land. Instead, the landlord has made the gain (assuming he or she also is not forced to pass it on).

Quite obviously, this transference of benefits and costs creates a perplexing situation for the evaluator. To make matters worse, there really has not been enough research to provide a basis for predicting the ultimate recipients and their shares of such transfers.

10.3.15 Multiplier Effects

While perplexing, transfers can also be valuable in that they create multiplication of benefits (and costs). Many studies on investments in transportation facilities, especially highways, have shown substantial multiplier effects [10.31]. Decreased travel costs associated with a new transportation system allow the user to take the money saved and invest it elsewhere, perhaps at a profit greater than the total of the reduced travel costs. This profit then is invested in another, more profitable venture, and so the cycle goes. This creation of new benefits (or possibly disbenefits) is one of the major reasons why many people are interested in having new transportation systems in their region—their benefits generally permeate the whole area and grow rather significantly at the same time. Unfortunately, the five approaches to evaluation described previously do not, in their present form, take into account multiplier effects.

10.3.16 Conformance with Goals

The final problem with previously discussed techniques for evaluation is representative of the underlying nature of most of the problems discussed in this section. It has to do with conformity to goals. In Chap. 4, goals for transportation were developed not only for direct service factors but also for other factors affected by them. The impacts were seen to be both broad and pervasive, playing a role in changing such diverse factors as ecology, business sales, and even church attendance. As a consequence, goals had to be set up for these and many other factors to ensure that the impact of transportation "improvement" was guided in the most advantageous directions. Viewed in the light of this wide scope of intent for transportation, the preceding approaches seem to have an extremely short range of concern.

Table 10.2 shows 10 goals of probable general importance to transportation. Alongside each is a subjective assessment of whether or not the goal is recognized in, say, the benefit-cost assessment. Several goals are not covered. For example, the rather significant goal of reducing air pollution (goal 10) is not considered at all in the benefit-cost procedure, nor is the goal of dependability (goal 3). Two goals are

**Table 10.2   Extent of Consideration in Benefit–Cost Procedure
of Goals of Probable General Importance to Transportation**

Goals	Take into account in the benefit–cost procedure?
Goals for direct transportation service factors	
Provide a transportation system that will:	
1. Offer low door-to-door travel time (with emphasis on low waiting and transfer time).	Yes
2. Have a low door-to-door travel fare and/or cost of operation (if user owned).	Yes
3. Offer adaptability and flexibility in routes, schedules, types of goals hauled, etc., to meet variations in demand of different sorts.	No
4. Be dependable in all weather, traffic conditions, etc.	No
5. Enable the greatest returns on investments.	Yes
Goals for factors affected by transportation	
Provide a transportation system that will:	
6. Better the economic position of each and every individual.	Partially
7. Cause the development of more and better activities and facilities.	No
8. Offer a reduced need for land of various types.	Partially
9. Offer a high level of safety to those in contact with the system.	Yes
10. Not add to air pollution or give off toxic gases externally.	No

given only partial recognition. Moreover, as has been emphasized throughout the discussion in previous sections, we must be concerned not only with the *extent* of the benefits but also the time at which they accrue, the amount they are multiplied in passing from one person to another, and finally, and perhaps most important, the nature of the ultimate recipients. The amount of effort involved in the type of evaluation implied above should be recognized. It would require gathering data and making predictions for an extremely wide range of factors, much wider than has usually been the case. In fact, one of the main reasons why techniques such as benefit–cost analysis have been utilized so much in the past has been the relative ease of data collection and prediction.

10.4   THE GOALS–ACHIEVEMENT TECHNIQUE

In reviewing the preceding comments and criticisms on benefit–cost and similar evaluation procedures, one has to be somewhat dismayed with the seemingly overwhelming complexities facing the decision maker. One also gains some appreciation for the position of the politician or manager who must react to and give solutions for these types of problems every day. The question, then, is what, if anything, can be done to improve decision-making procedures.

One technique for broadening the evaluation and decision-making process is known as the goals-achievement method [10.11]. In this approach each potential project alternative is assessed in terms of its impacts in comparison to the objectives proposed for it. Quantitative measures are employed in this process, although some may be subjective and even probabilistic. This is exemplified in Table 10.3 for the four hypothetical projects discussed in the preceding section and for the goals in Table 10.2. One impact is the floor area of new activities (land uses) expected within a quarter mile of the new facility. Since it is difficult to forecast this impact, the resulting numbers are little more than "guesstimates." Similarly, it is difficult sometimes to determine fatal-accident rates resulting from a project alternative, so the corresponding likelihood is reported as a probability.

After having established measures and estimates for each goal, our next step in the goals-achievement procedure is to assign the best level of each factor a value of 100, and then show the scores on the other projects as a proportion of that maximum. For instance, the IRR is highest for project IV (31%), so it is assigned a value of 100. The IRR of project I, by contrast, is only 10%, so that it receives a normalized effectiveness measure on that dimension of $(10/31) \times 100 = 32$.

If the "best" level of a factor is the lowest, as is the case for fatality-rate probability (criterion 9), the procedure again is to assign the best a score of 100, but then for each other alternative divide the raw score of the best times 100 by the raw score of the particular alternate. Project IV, as an example, has the lowest probability, so it is given a normalized effectiveness measure of 100. The measure for project I then is $(1/0.53) \times 0.45 \times 100 = 86$.

The third and final step in the goals-achievement approach is to multiply the normalized effectiveness measure of each factor by the weighting (or rating) it has received. A common procedure for arriving at such weightings is to take a total of 100 points and divide it among the impacts according to their importance. This is shown hypothetically in Table 10.4, where the peak period travel time (criterion 1)

Table 10.3   Hypothetical Impacts of Four Project Alternatives

Criterion	Alternative			
	I	II	III	IV
Direct transportation service				
1. Peak period person travel time between selected points (min)	25	23	21	22
2. Peak period travel cost between selected points ($)	3.25	3.00	2.50	2.70
3. Peak period person travel time on second fastest route (min)	27	30	31	30
4. Days per year when arrivals and destinations are at least one hour later than expected	5	5	5	6
5. Internal rate of return (%)	10	19	20	31
Factors affected by transportation				
6. Transportation benefits ($)	28,000	32,000	34,000	34,000
7. Floor area of new activities within $\frac{1}{4}$ mile of new facilities (million sq ft)	0.5	1.0	0.4	0.6
8. Total land taken for new facilities (1,000 acres)	5	15	20	18
9. Probability of having a fatality rate of 2.0 or more deaths per 10 m	0.53	0.93	0.50	0.45
10. Total hydrocarbon emissions (million tons/yr)	6	9	8	7

Table 10.4   Normalized Goals-Achievement Effectiveness Measures

Criterion*	I	II	III	IV	Weight

Criterion*	Alternative I	II	III	IV	Weight
1	84	91	100	95	18
2	77	83	100	93	14
3	100	90	87	90	4
4	100	100	100	83	4
5	32	61	65	100	15
6	82	94	100	100	11
7	50	100	40	60	6
8	100	33	25	28	12
9	86	49	90	100	9
10	100	67	75	86	7
Total	811	768	782	835	

*See Table 10.3 for a description.

is given the highest rating (18 points out of 100) and criteria 3 and 4 the lowest (each 4 points). Multiplication of the effectiveness measures by the weightings thus gives a weighted final score, as for project I:

$$84(18) + 77(14) + \cdots + 100(7) = 7746$$

The project with the highest weighted final score is deemed most acceptable.

In our example the rankings of project alternatives by weighted score turn out to be:

Alternative	I	II	III	IV
Weighted score	7,746	7,415	7,898	8,502
Ranking	3	4	2	1

So project IV would seem most desirable. It should be noted, however, that the scores differ by no more than 10%, so that one alternative is not substantially better than another.

The goals-achievement technique has the main advantages of explicitness and comprehensiveness. It also opens the possibility for citizen and/or politician input in helping to determine the impacts to be assessed and ratings to be applied to each factor. On the other hand, the quantification still might be subject to some question, especially if the forecasts are nothing more than the crudest, subjective guesses.

10.5   THE COST-EFFECTIVENESS TECHNIQUE

An alternate approach to the goals achievement procedure is the cost-effectiveness technique. It actually is a much less sophisticated procedure than one might at first suppose. It works on the basic premise that better decisions will arise if clearer and more relevant data are supplied to the decision-maker. No specific attempt is made to put all benefits and costs in common units such as the dollar. As Thomas and Schofer [10.4, p. 52] remark on the cost-effectiveness approach:

> Because many of the consequences and outputs from the transportation system are intangible and otherwise difficult to value in some common metric, the decisions regarding the conversion to a single dimension—and hence the plan selection decisions—are necessarily subjective in nature, at least at the present time. . . .
>
> What might be more useful at this time is a technique for providing the kind of informational support for the selection of alternative plans which recognizes the complex nature of these transportation decisions. Such a decision supporting framework would not attempt to *make* decisions, but instead *would structure the information required for making a subjective but systematically enlightened choice* (our underlining). At the same time, however, the framework . . . must be sufficiently flexible to permit the adoption of more sophisticated techniques, such as analytic methods for realistically implementing benefit-cost analysis on ranking schemes, when such techniques are appropriate.

Thomas and Schofer specify three criteria that any framework for evaluation should satisfy:

1. It should be capable of assimilating benefit-cost and similar methodological results *in addition* to other informational requirements.
2. It should have a strong orientation toward a system of values, goals, and objectives.
3. It should allow for the clear comparison of *tradeoffs* or compromises between objectives by making explicit the relative gains and losses from various alternatives.

These criteria can be inferred from the criticisms in the preceding section of this chapter and tend to reinforce the needs brought out there. The cost-effectiveness approach, as a later example will show, seems to satisfy all three of the criteria.

10.5.1 Description of Cost-Effectiveness Framework

In the application of the cost-effectiveness analysis, the attributes of the alternative relevant to the decision are separated into two classes—costs and indicators of effectiveness. Costs are defined as the monetary outlays necessary to procure all of the resources for the construction or purchase, operation, and maintenance of the facility during its useful life cycle. Of course this assumes that the pricing mechanism operates so that all items expended on the project can be valued in terms of dollar prices. Where this is not possible, it may be necessary (and entirely realistic) to consider costs in other units, such as hours of labor and tons of steel, as well. This approach to costing is contrary to that in most present evaluation schemes and allows for a certain flexibility in cost analysis.

Effectiveness is defined as the degree to which an alternate achieves its objectives. The definition, by itself, helps to overcome one of the major objectives to the benefit–cost approach in that goals are specified explicitly and are not covered by an all encompassing "benefit" term. In benefit–cost analysis, for example, "benefits" are related to reductions in user operation costs, user time, and so forth, but in a particular situation these factors may be of less concern. The objectives to be met may be associated with an entirely different set of factors.

Information regarding the costs and effectiveness of the alternatives is presented to the decision-maker who, in turn, makes the subjective choice of the one which seems best. While planners may provide all the supporting data and estimates from these data, and may even suggest which alternative appears best to them, the ultimate choice is left to the duly appointed decision-maker(s). No hard and fast decision rules, such as those inherent in the benefit–cost approach, are permitted to make the selection "automatically." It is quite permissible, of course, to provide the decision-maker with information concerning the benefit/cost ratio and the like; however, these are and should be kept from being the sole determinants of choices among alternatives.

The value of the cost-effectiveness approach lies in several areas:

1. It stimulates, to some extent, the process by which actual decisions are made.
2. It allows for the clearer delegation of responsibilities between analyst and decision-maker.
3. It makes it easier to provide relevant information, structured in an understandable form, so that the choice process is simplified.

10.5.2 Cost–Effectiveness: An Example

The example of the cost-effectiveness approach that follows is based on an article by Millar and Dean [10.14] describing one part of the *Manchester (England) Rapid Transit Study.* While the article itself did not deal directly with cost-effectiveness as an evaluative technique, it did seem to fit very neatly into the framework described in the preceding section.

The government agencies concerned with the transit problem in Manchester made several recommendations to the study group before a detailed investigation was initiated. As quoted from Millar and Dean [10.14, p. 155]:

> The Ministry of Transport recognized that this study would not only be of value in the context of providing a *well-balanced* and *economical* overall transport system for Manchester, but would also yield information of wider application and interest.

Moreover:

> It was stipulated that this evaluation should investigate the characteristics of any system which could be *built by 1972;* that the *quality of service* which each could give should be assessed; that the likely *environmental effects* be explored; and that reliable estimates should be provided of the *capital and operating costs.* (emphasis ours)

The underlined statements can be thought of as general goals toward which the decision regarding a rapid transit system should be directed. Thus, in a general sense, the evaluation was goal-oriented, as is desired in a cost-effectiveness approach.

After preliminary elimination of some candidate transit systems (mainly because they could not be built by 1972), the study group settled on four possible alternatives: (1) Safege monorail, (2) Electric railway (duorail), (3) Westinghouse skybus, and (4) Alweg monorail. These systems are pictured collectively in Fig. 10.11. We will not detail their technical characteristics here.

Data were gathered on each system. The first items collected were responses to direct questions. Could the system be built by 1972? Did the system have a route capacity of 7,500 persons per hour (pph)? Would the system fit in with the present British Rail system? As seen in Table 10.5, the answers to the first two questions were "yes" for all alternatives. However, only the duorail system would be compatible with British Rail.

The second set of data dealt with performance characteristics and structural dimensions. The main differences between systems appeared to be that

1. The skybus had a slightly lower maximum speed and mean acceleration rate.
2. The Safege monorail, hanging below the guideway, would require a taller structure and would also need a major structure on ground level.
3. Switching would be easiest for the duorail.
4. The duorail and the skybus required the least diameter tunnel.
5. The Safege monorail would have the longest elevated beam span but also the widest.

Other characteristics did not seem to differ significantly.

Envrionmental considerations were reduced to two factors: noise levels and visual intrusion. The duorail was found to be somewhat louder than the others. No information was available at the time on noise levels from the skybus. (Note that evaluative decisions still must be made even in cases where some relevant information cannot be obtained.) Visual intrusion, being a fairly subjective matter, was

(a)

(b)

Fig. 10.11 Four transit system alternatives: (a) Safege monorail, (b) Electric railway or duorail. (*Continued*)

(c)

(d)

Fig. 10.11 (*Continued*) Four transit system alternatives: (c) Westinghouse skybus. (d) Alweg monorail.

Table 10.5 **Effectiveness and Cost Characteristics for Four Possible Rapid Transit Systems for Manchester**

Effectiveness measures	Safege monorail	Duorail	Westinghouse skybus	Alweg monorail
Could be built by 1972?	yes	yes	yes	yes
Route capacity of at least 7,500 pph	yes	yes	yes	yes
Compatible with existing British rail system	no	yes	no	no
Maximum speed (mph)	50	60	40	50
Mean acceleration rate (mph/sec)	3.3	3.0	2.3–3.0	2.7
Car capacity (person)	173	279	120	360
Height of guideway above ground (ft)	over 16.5	16.5	16.5	16.5
Beam span (ft)	104	60	60	65
Width of elevated span (ft)	30.3	27.5	19.8	15.5
Use at ground level	Suspended	On ground	On ground	On ground
Tunnel diameter (ft)	17.0	15.6	14.0	18.3
Switching	Slow	Fast	Undeveloped (?)	Slow
Noise level (internal) dB(A) over drive unit	68	71	*	81
Noise level (external) dB(A) 25 ft away	81	88	*	80
Total car requirements for				
10,000 pph	72	44	110	70
20,000 pph	144	88	220	135
30,000 pph	216	132	330	204
Train headway (min) at				
10,000 pph	2.65	2.84	2.87	3.64
20,000 pph	2.65	2.84	2.87	2.76
30,000 pph	2.65	2.84	2.87	2.44
Cos.				
Total capital costs (£) (at 30,000 design hour cap.)	81,110,000	61,090,000	66,240,000	66,920,000
Annual operation and maintenance costs (£) (at 30,000 design hour cap.)	2,040,000	1,410,000	1,800,000	1,760,000
Tota annual cost (£)	7,350,000	5,330,000	6,130,000	6,090,000

*Information not available.

judged on the basis of reaction to a set of photomontages (Figs. 10.12 and 10.13) where mockups of the guideways of the four systems were superimposed over pictures of buildings and streets along the proposed route. The planners and designers felt there was no significant visual difference between systems based on these photomontages.[18]

The final set of data was the capital and operating costs for the four systems. On a total annual cost basis, the duorail system was estimated to be least expensive, about 15 percent lower than that for the next lower system—the Alweg monorail. Capital costs for the duorail would be approximately £20 million (about $50 million) less than for the Safege monorail—the most expensive system.

[18] This visual elevation was only of the structures, not the vehicles.

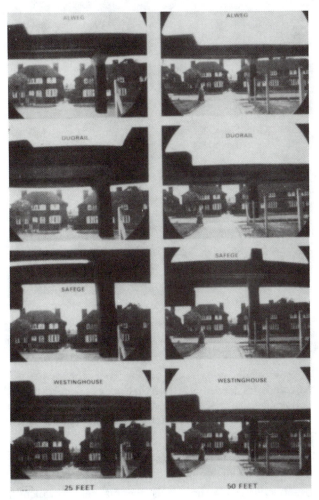

Fig. 10.12 Mockups of the guideways of the four systems superimposed over pictures of residential buildings and streets, at 25 and 50 feet.

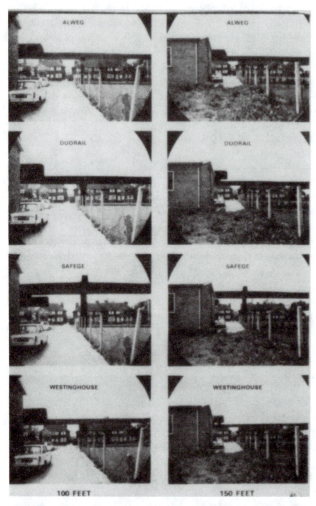

Fig. 10.13 Mockups of the guideways of the four systems superimposed over pictures of residential buildings and streets at 100 and 150 feet.

The study group, in looking over the tradeoffs between different system characteristics, apparently felt that, except for costs and adaptability to the British Rail system, all alternatives were essentially equal. The duorail, because it dominated the other systems on the two exceptional characteristics, thus was recommended for adoption.

10.5.3   Comments on the Cost-Effectiveness Technique

The preceding example has illustrated that an important characteristic of the cost-effectiveness technique is the manner in which information is presented to clarify relationships between alternatives and to outline tradeoffs or compromises that must be made to choose one alternative over the others. The cost-effectiveness framework does not indicate which system to select. It illustrates tradeoffs between alternatives, and it identifies dominated systems. It clearly lays out the expected accomplishment of each system and the related costs.

Whether the cost of additional effectiveness makes one alternative more worthwhile than another is a subjective matter and is therefore left to the decison-makers. It can be argued that leaving this decision on a subjective level is not helpful and that what is needed is an approach that will determine the worth of additional effectiveness. However, it is just such information that is difficult to provide, particularly in the evaluation of more complex transportation plans. To obtain such worth measurements involves great capabilities in working with large sets of interrelated objectives. In the cost-effectiveness approach, the decision-makers just need to be sensitive to these issues and can then secure information indicating the cost and consequences of meeting a particular goal.

The choice itself, if there are no other factors to consider, is made when the decision makers determine which of the alternatives results in a relation of cost to effectiveness acceptable to them. It is the decision makers, through the choice itself, who establish the relationship between cost and effectiveness, thus placing a boundary value on the measure of effectiveness.

In other words, if the duorail alternative were eventually chosen, the given level of effectiveness would be worth *at least* that particular cost. The duorail system is thus termed "cost-effective," for it provides decision makers with a level of effectiveness they deem satisfactory at what they consider to be a fair price. The value of the level of effectiveness is merely bounded in that the decision-makers might be willing to pay more than the cost of the duorail system for the same level of effectiveness. This amounts to a subjective consideration of the criterion of efficiency. The duorail results in sufficient returns on its resource costs to the decision-makers at one point in time.

In many cases there will be other factors to consider, of course. For example, there may be a *required* level of effectiveness that must be achieved. When such a requirement is set in advance, it remains only to select the alternative which meets it at the lowest cost. Similarly, there may be a budget constraint that cannot be violated. In this case, the decision-makers might try to achieve the highest level of effectiveness while staying within the budget constraint.

There are several aspects of the cost-effectiveness technique that are extremely valuable and should be presented briefly at this time. First, all forms of information, regardless of degree of sophistication of description, are admissible in the framework. Pictures, diagrams, and even sound tapes can be entered in the tally. Second, estimates of the extent of effectiveness factors and costs at various points of time in the future can be presented in a series of charts, thereby giving a much-needed time orientation to benefits and costs and allowing for the all-important weighting of effectiveness and cost according to time of occurrence. Third, separate impact calculations for individuals or various groups of interest can be developed.

Perhaps the greatest advantage of the cost-effectiveness technique is that it is complimentary to the "values-goals-objectives-standards" format presented in Chap. 4. Looking back at Table 10.5, we see that the measures (or criteria) presented for evaluation of the Manchester rapid transit system all flowed directly from the preliminary goals set up by the Ministry of Transport and others. Thus, the cost-effectiveness technique is a natural extension of attempts to plan using goals and objectives.

The major difficulty with the cost-effectiveness approach, and one that affects almost any evaluation technique, is that of considerable data requirements. This problem has been discussed earlier, but its importance cannot be overstressed. Many current transportation studies have found themselves inundated with data and subsequently unable to perform even some of the simple analyses required of them. This type of situation naturally is not desirable and requires extra effort on the part of analysts and decision-makers to identify and extract only those factors of relevance to the evaluation.

10.6 POSTEVALUATION

Evaluation of projects, programs, or plans after they have been implemented can take several forms. Under the best conditions these evaluations would be part of an experimental design in which relevant comparative data are collected both before and after the alternative has been implemented. In so doing, it then is possible to look at the difference between before and after conditions and attribute that difference to the alternative being assessed.

In the most vigorous experimental design there is both a control and an experimental group. The members of both are selected at random (unless sufficient funds are available to include everyone in the population) to insure an equally likely spread across characteristics. If the groups are people, for instance, we want to insure that any before/after differences noted are due to the alternative (or experimental treatment) rather than to variations in age, sex, race, income, and the like. So, through randomization, these are dispersed with equal likelihood between the two groups. Only the experimental group, however, is subjected to the treatment.

The pattern for this type of experimental design looks like:

	Before	After
Experimental group	Y_{BE}	Y_{AE}
Control group	Y_{BC}	Y_{AC}

where the Y terms are the average values for the particular variable(s) of concern in the experiment. The impact, I, of the alternative or treatment then can be calculated as

$$I = (Y_{AE} - Y_{BE}) - (Y_{AC} - Y_{BC}) \qquad (10.14)$$

In other words, we look at the after/before difference for the experimental group and compare (subtract from) that the associated difference for the control group.

Notice what might happen if we did not have the control group. Suppose, for illustration, we were undertaking a demonstration project in which free fares were offered for a particular type of transit service in one area of a city. If the price of auto fuel happened to drop at the same time the demonstration (experiment) were going on, we might actually find a decrease in transit ridership of, say, 1,000 passengers, contrary to our expectations with the lower fare. Yet, if there were a control group, not offered the free fare, the decline in their ridership may be even greater, say 1,500. The relative impact of the free fare by Eq. (10.14) thus still would be positive:

$$I = -1000 - (-1500) = 500$$

Rarely is it possible to set up the desirable type of experiment suggested above. The simple reason is that government would find it very difficult politically to offer a potentially beneficial service to one selected group and not to another. In addition, it may take many years to implement the treatment (e.g., a major rail line or highway), in which case commitment of funding for before/after data collection[19] may be extremely tenuous. Another type of experimental design thus may be necessitated.

The three most likely candidates, none involving control groups, are:

1. "Before" data collection with forecasted "after" impacts
2. "After" data collection, with recall to obtain (some) "before" information
3. Coordinated "before/after" data collection.

The first is nothing more than the procedures described earlier in this chapter. By implication, however, no substantial attempt is made to gather any "after" data to see if the forecasted impacts occurred as predicted. The second option occurs most frequently when a decision is made, after implementation, that a program should be evaluated. Some governmental agencies have their own evaluation offices (such as the General Accounting Office at the Federal level) to do exactly this type of assessment. The third type occurs most frequently when the project, program, or plan is considered from the beginning to be a demonstration, particularly of a unique, new alternative, and there is a desire to see if it is "successful."

An example of the second approach was a study conducted by the General Accounting Office (GAO) of hindrances to coordinating transportation of people participating in Federally funded grant programs. The GAO sent a team to each of 12 localities (mostly rural) to assess efforts to coordinate human-services transportation programs. Such efforts already were underway in all of these localities, so the evaluation was primarily one of noting the experiences and on-going problems of the systems.

One system investigated was the Delaware Authority for Specialized Transportation (DAST). This was operated in New Castle County (which includes the City of Wilmington) as well as two other, more rural counties. During 1975 DAST provided transportation for 39 social-service agencies, who acquired such through purchase-of-service agreements. The GAO investigators found that DAST operated on a hand-to-mouth existence, particularly because the funding capabilities of the serviced agencies as well as supporting counties always were somewhat uncertain. DAST's precarious position occurred despite the fact that [10.19, p. 46]:

> Several social service agency officials commented that (1) a coordinated system is the most effective means of providing specialized transit, (2) transit should be left to transportation specialists who can do a better and more efficient job, (3) local social service agencies spend hours operating their own transit systems which are not most effective due to fragmented delivery, and (4) there are inherent efficiencies in developing one system.

Despite these overall advantages, there were some agencies not interested in service by DAST. Some reasons advanced for this were:

- Need for immediate service that could not be met by DAST's 24-hr-in-advance scheduling.
- DAST was too expensive or could be more expensive than present transportation costs.

---

[19] Of course, there almost always are some data, such as traffic counts and population, collected continually and available for a variety of uses, such as before/after studies.

- Staff members of the social service agency accompanied clients on trips and also drove the vehicle. An emergency shelter for neglected and abandoned children required that staff members who drove vehicles be present with the children at all times.
- DAST vehicles did not have the capacity to carry a large number of children.

These types of findings, although they cannot be compared to a "before" (or baseline) situation, still help lead to an understanding of the current situation and to potential areas for improvement.

An illustration of a detailed before and after evaluation is the studies on BART—the Bay Area Rapid Transit System—reported in Chap. 8. Another example comes from ASAP—the Alcohol Safety Action Projects. These experimental efforts were carried out in 35 localities around the country for periods from 2 to 5 years. The objective was to "test the capability of the traditional community institutions to control the drinking driving problem" [10.13, p. 1]. To this end, a two-part approach was undertaken:

1. To deter drunk driving among social drinkers, both police departments and the courts were provided with resources to handle a greatly increased number of DWI (Driving While Intoxicated) offenders.
2. To remove and treat problem drinkers, a system was developed to identify such among the DWI offenders and refer them to special treatment centers [10.13, p. 2].

The primary criterion used for evaluation was the decrease (or increase) in nighttime fatal crashes before versus after the project (in some cases in comparison to daytime fatal crashes, which generally involve far fewer drunk drivers).

The results from the 35 ASAP sites demonstrate in part some of the confounding factors that can enter any before/after study. First, the population in some of the areas grew substantially. All other things being equal, this should lead to increased numbers of crashes. Second, the 55-mph speed limit was introduced during the operational phase of many of the projects. This, of course, led to fatality reductions across the nation as a whole. Third, the second part of some of the projects, involving rehabilitation, itself took up almost the whole time span of the projects, so that the results, in terms of fewer nighttime fatalities, would not show up in the statistics for the period of operation.

While it was difficult to take into account these confounding conditions, it still was determined that 12 sites had statistically significant reductions in nighttime fatal crashes while only two sites showed any increase in such. It was computed that 563 fatalities were "forestalled." Because the cost of the overall program was about $88 million, this came out to be $156,306 per fatality forestalled. By way of perspective, this figure was about midway in value among 37 countermeasures ranging from mandatory safety-belt usage ($505) to roadway alignment and gradient ($12,100,000) [10.13, p. 31].

10.7 EVALUABILITY ASSESSMENT

As more and more projects, programs, and plans are subjected to before/after evaluation, it will become clear that there are certain times when evaluations can and cannot be done. Some road and rail projects, for instance, can take up to 10 years before completion, much less before benefits start to arise. In fact, it is a rare project that can be implemented in less than a year. To evaluate these before completion has some obvious disadvantages.

On the other hand, whatever the time required, external conditions can change rapidly, so that an assessment may be needed before completion. It is better, we assume, to stop a large, expensive project midway through than to let it go to completion and produce poor results for the cost involved. And it must be appreciated that political leaders often are pressed by the distrusting public to show immediate, "tangible" results of their decisions.

The question, then, is when should an evaluation be undertaken. This leads to the concept of an evaluability assessment, which is defined as:

> a descriptive and analytic process intended to produce a reasoned basis for proceeding with an evaluation of use to both management and policymakers [10.18, p. 2].

An evaluability assessment (EA) begins by obtaining management's description of the program. The description then is systematically analyzed to determine whether it meets the following requirements:

1. It is complete.
2. It is acceptable to policymakers.
3. It is a valid representation of the program as it actually exists.
4. The expectations of the program are plausible.
5. The evidence required by management can be reliably produced.
6. The evidence required by management is feasible to collect.
7. Management's intended use of the information can realistically be expected to affect performance [10.18, pp. 2, 3].

Each of these is a question to be answered by the evaluator, working with the manager. If this analysis reveals gaps or problems in management's description, the evaluator would be involved in devising alternative descriptions which would help to fill these gaps or resolve the problems.

The most desirable conclusion from an EA is that the program description is evaluable. In other words, the description gives measures such that there is reasonable assurance that a more detailed evaluation can be done and the predetermined expectations realized. If this is the case, the full product of an EA would be expanded to include:

1. Potential changes in expectations of policymakers
2. Potential changes in the design of the program
3. The resultant evaluable program description
4. The evaluation information to be purchased
5. Organization and staff needed to implement the decisions

In contrast, there are three types of "unevaluable" conclusions that can be reached [10.18, p. 3]:

1. Comparison of program management's description with the list of prescribed elements shows it is incomplete. For example, management may not have agreed on a definition of objectives.
2. Comparison of the evaluator's and operator's description with program management's description indicates the latter is implausible, invalid, not cost-feasible to evaluate, or not useful.
3. Comparison of policymaker's and program management's description shows that the latter is unacceptable. This could occur, for instance, when

management has left out an objective or set of activities felt important to policymakers.

To our knowledge there has been no application of EA directly in the transportation field. An indirect illustration has been given by Wholey [10.23]. This involves health planning, with some attention given to greater accessibility of health services. More specifically, Public Law 93-641, the Health Planning and Resources Development Act, authorized and provided resources for federal, state, and local health planning and regulatory activities which were intended to (1) improve the availability, accessibility, acceptability, continuity, and quality of the health care system; (2) restrain rising health care costs and prevent costly duplication of services; and (3) improve the health of the American people.

The evaluators explored the objectives and expectations of managers and policymakers, explored the reality of program operations, and identified evaluation/management options for more effective program performance. They first read documents indicating Congressional intent, program regulations, reports and journal articles, memoranda, letters, documents describing the relevant organizations, and requests for evaluation contracts. They then made visits to national through local planning agencies. These activities led them to identify the program interrelationships indicated in Fig. 10.14.

In reviewing those interrelationships, they found that measures of success fell far short of those intended by managers. Furthermore, bureau managers had not agreed on outcome-oriented measures for any of the programs. The EA team then identified 14 evaluation/management options which appeared to have the potential for improving the performance of the program. The managers selected a combination of three of these options which involved, among other aspects,

1. Monitoring planned versus actual schedules for production of regulations and guidelines
2. Sampling of opinions of local staffs about the utility of regulations and guidelines
3. Monitoring the plans produced by states and localities to see if realistic, measurable objectives were developed

Presumably the above efforts would make it much easier to determine the extent to which states and localities were progressing in meeting the defined objectives. This, in turn, would make more precise before/after evaluations possible.

10.8 SUMMARY

The broad aim of this chapter has been to present formal techniques by which metropolitan transportation alternatives can be evaluated. In the first section a detailed presentation was made of the concepts underlying economic evaluation. This was followed by a description of five procedures for comparing benefits and costs. Attention then turned to the major issues connected with evaluation of this nature.

Two alternate techniques then were described. The first, the goals-achievement procedure, emphasizes the utilization within evaluation of the objectives proposed for various alternatives. The second, the cost-effectiveness approach, softens the quantitative aspects of the preceding techniques. Even pictures can be employed in the procedure, and the weighting of importance of impact factors, whether implicit or explicit, is turned over to decision-makers.

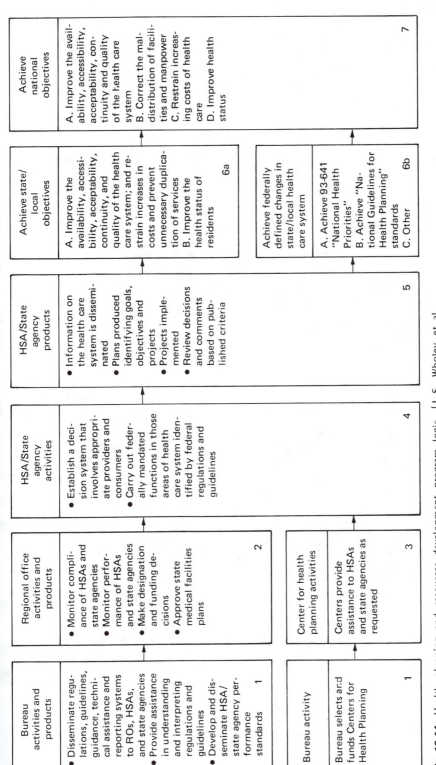

Fig. 10.14 Health planning and resources development program logic. [J. S. Wholey et al., *Evaluability Assessment for the Bureau of Health Planning and Resources Development: Bureau Program II: Administering the Law Nationwide,* Urban Institute Contact Report, November 1977, p. 5].

Postimplementation evaluation approaches then were described, with particular emphasis on before/after experimental designs. Assumed in all of the preceding techniques, however, was that the subject project or program was in a position to be evaluated. This is not always so, in which case an evaluability assessment may be needed to determine how to create a worthwhile assessment.

BIBLIOGRAPHY

10.1 American Association of State Highway and Transportation Officials: *A Manual on User Benefit Analysis of Highway and Bus-Transit Improvements 1977*, Washington, D.C., 1978.

10.2 U.S. Department of Transportation: *Evaluating Urban Transportation Systems Alternatives*, Washington, D.C., Nov. 1978.

10.3 Suchman, E. A.: *Evaluative Research*, Russell Sage Foundation, New York, 1967.

10.4 Thomas, E. N., and J. L. Schofer: *Strategies for Evaluation of Alternate Transportation Plans*, N.C.H.R.P. Report 96, Washington, D.C., 1970.

10.5 Wohl, M., and B. V. Martin: *Traffic Systems Analysis for Engineers and Planners*, McGraw-Hill, New York, 1967.

10.6 Hatry, H. P., Winnie, R. E., and D. M. Fisk: *Practical Program Evaluation for State and Local Government Officials*, 2nd ed., The Urban Institute, Washington, D.C., 1980.

10.7 Winnie, R. E., and H. P. Hatry: *Measuring the Effectiveness of Local Government Services: Transportation*, The Urban Institute, Washington, D.C., 1972.

10.8 United Nations Industrial Development Organization: *Guidelines for Project Evaluation*, United Nations, New York, 1972.

10.9 Boots, A. J. III, et al.: *Inequality in Local Government Services: A Case Study of Neighborhood Roads*, The Urban Institute, Washington, D.C., 1972.

10.10 International Labour Office, World Employment Programme: *Roads and Redistribution*, Geneva, 1973.

10.11 Steger, C., and R. C. Stuart: "Goals Achievement as an Integrating Device for Technical and Citizens Inputs into Urban Freeway Location and Design," *Man-Environment Systems*, vol. 6, 1976.

10.12 U.S. Department of Transportation, Federal Highway Administration: *Community Improvement in Highway Planning and Design*, Washington, D.C., May 1975.

10.13 U.S. Department of Transportation, National Highway Traffic Safety Administration: *Alcohol Safety Action Projects: Evaluation Methodology and Overall Program Impact*, vol. 3, Washington, D.C., April 1979.

10.14 Millar, J., and J. Dean: "Practical Considerations of Rapid Transit—A Summary of the Manchester Study," *Journal of the Town Planning Institute*, vol. 54, no. 4, April 1968.

10.15 U.S. Congress, Office of Technology Assessment: *Technology Assessment of Changes in the Use and Characteristics of the Automobile*, Washington, D.C., Feb. 1978.

10.16 U.S. Department of Transportation, Federal Highway Administration: *Evaluation of Traffic Operations, Safety, and Positive Guidance Projects*, Washington, D.C., Oct. 1980.

10.17 U.S. General Accounting Office: *Evaluation and Analysis to Support Decisionmaking*, Washington, D.C., Sept. 1976.

10.18 Schmidt, R. E., Scanlon, J. W., and J. B. Bell: *Evaluability Assessment: Making Public Programs Work Better*, Human Services Monograph Series No. 14, Project SHARE, U.S. Department of Health and Human Services, Washington, D.C., Nov. 1979.

10.19 U.S. General Accounting Office: *Hindrances to Coordinating Transportation of People Participating in Federally Funded Grant Programs, Vol. II—Case Studies*, Washington, D.C., Oct. 17, 1977.

10.20 Baltimore Regional Planning Council: *Transportation Control Plan, Vol. 1: Summary*, Baltimore, Sept. 29, 1978.

10.21 Little, I. M. D.: *A Critique of Welfare Economics*, Oxford University Press, London, 1958.

10.22 Baltimore Regional Air Quality Task Force: *1980 Status Report: Transportation Control Plan—Two Years Later*, Baltimore, 1981.

10.23 Wholey, J. S.: *Evaluation: Promise and Performance*, The Urban Institute, Washington, D.C., 1979.

10.24 The Rand Corporation: *Measurement and Evaluation of Transportation System Effectiveness*, National Technical Information Service, PB 185 772, Springfield, Va., 1969.

10.25 Dodson, E. N.: "Cost-Effectiveness in Urban Transportation," *Operations Research*, vol. 17, no. 3, May–June 1969.

10.26 Winfrey, R.: *Economic Analysis for Highways*, International, Scranton, Pa., 1969.

10.27 English, J. M. (ed.): *Cost-Effectiveness, the Economic Evaluation of Engineered Systems*, Wiley, New York, 1969.

10.28 Mohring, H., and M. Harwitz: *Highway Benefits: An Analytical Framework*, Northwestern University Press, Evanston, Ill., 1962.

10.29 Squire, L., and H. G. van der Tak: *Economic Analysis of Projects*, Johns Hopkins University Press, Baltimore, 1975.

10.30 Winch, D. M.: *The Economics of Highway Planning*, University of Toronto Press, Toronto, 1963.

10.31 Harris, C. C. Jr.: *Regional Economic Effects of Alternative Highway Systems*, Ballinger, Cambridge, Mass., 1974.

EXERCISES

10.1 The current street in a CBD is six lanes wide with parking in both curb lanes. Average one-way traffic volumes are 150 vehicles per hour on a 6% plus grade. Traffic stops frequently (7 times per mile). A leveling and traffic-signaling project is proposed in which the grade will be reduced to +2% and the number of stops is expected to drop to 2 per mile. Over a 4-mile section, this will result in a 10-min time saving per trip and a 10% increase in traffic. Assume that (1) all vehicles are autos, (2) all trips are for work, with 1.2 persons per vehicle, (3) the value of time is $3.00 per passenger·hour, (4) fuel costs $1.50/gallon, (5) fuel efficiencies and other conditions are the same as in Tables 6.3 and 6.4, and (6) all trips cover the 4-mile section. Determine the hourly dollar benefits for the one-way traffic for the proposed project (versus doing nothing).

10.2 A newly developed safety plan for a community involves a program of more stringent auto inspections. Over a 3-year demonstration period the program is expected to save the number of lives, and cost the amounts listed below. It has been agreed that the value of a life in each succeeding year will be $200,000; $250,000; and $300,000. A discount rate of 15% is assumed. What is the (*a*) payback period; (*b*) benefits per dollar of outlay; (*c*) net present value; (*d*) benefit/cost ratio; (*e*) rate of return?

Year	1	2	3
Estimated added lives saved	2	4	2
Cost of program	500,000	600,000	700,000

10.3 Using the references shown in the text and others found in the library, write about 500 words on considerations in determining a discount rate.

10.4 Two policy plans for transporation of the elderly and handicapped have been developed by a locality. The impacts of the two relative to the proposed goals (criteria) and weights of importance are shown below. Which plan is more desirable if an evaluation is done using the goals-achievement procedure? Give the final scores.

Criterion	Alternative		Weight
	I	II	
1. Average access time (min)	25	35	30
2. Percent target group served	10	15	25
3. Average waiting time (min)	30	60	5
4. Cost ($m)	2	1.5	30
5. IRR (%)	20	18	10

10.5 Pick a particular proposed transportation project or program in your community. Identify 8 to 10 major impact factors, including cost(s). Estimate as best as possible the likely impacts 1 and 5 years after implementation of two alternatives to the project or program and set up a cost-effectiveness matrix to display these.

10.6 An investigation was made of the percentage of eligible household members holding jobs before and after a special van service program was provided to a low-income neighborhood. The results were compared to a similar neighborhood that did not receive the service. Unfortunately, a depression hit in the middle of the program, so overall employment dropped in both areas. Was the impact of the program still positive?

	Program neighborhood	Comparison neighborhood
Before	50	60
After	35	40

10.7 Find a project, program, or policy. Do an evaluability assessment to see if an evaluation would be worthwhile. If not, make recommendations on how to make it so. (Note: Do not take a big project. To keep this exercise manageable, talk briefly with only two or three people involved and then make your determination.)

11 Evaluation II: Decision Making

The benefit–cost evaluation technique detailed in Chap. 10 typified the transportation planner's role in the decision-making process for several decades. In the first section below, we note the dramatically changed environment for transportation decision making and the resulting expansion of the planner's role. In the second section, the new role expectations are made more concrete with the help of selected case studies. The third section outlines a new "decision-support" role model with a stress on community involvement. The final section discusses the most frequently used involvement techniques in the decision-making process.

11.1 THE CHANGING CLIMATE OF TRANSPORTATION DECISION MAKING

The discussion of evaluation in the preceding chapter might lead a person to a straightforward model of the decision process (Fig. 11.1). This model was in fact held by many planners and engineers prior to the 1960s, and it may have been appropriate to a simpler world than we have today. In times gone by, "everything was not connected to everything else," as it is today. The professional thus was the expert in the field. If his or her recommendation was not adopted, it was probably due to "politics."

As noted in Chap. 1, the goals of the early transportation studies in the 1940s and 1950s focused almost exclusively on traditional engineering factors. Using operations research approaches, a single goal to be optimized might be "to select the transportation plan that will carry 20-year travel demand at the least cost,

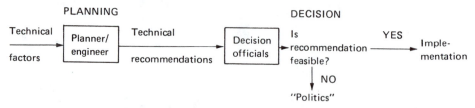

Fig. 11.1 Model of decision making based solely on technical role.

within the constraint of public policy." Given this single objective, the transportation models could be called upon to indicate the best transportation plan. The engineer–planner woud likely have based a recommendation on benefit–cost or some other form of economic analysis. The need for citizen participation probably seldom touched the minds of these pioneer planners. Some years ago, Melvin Webber described the role that was expected of the highway planning professional [11.1, p. 47]:

> A persisting difficulty derives from the way we organize to produce and distribute public services and from the ways we do our bookkeeping. Highway engineers, charged with installing a road between two points and with doing so efficiently, are thereby compelled to find a short route. If parkland should happen to lie along the way, so much the better; it is probably cheaper to build there than along a path occupied by houses and other buildings. If the "best" path should happen to require removal of a venerated building of some sort, well it is probably cheaper to remove the building than suffer the greater land-acquisition costs of a longer route. On the account sheets of the highway department, the least cost solution defines the correct alignment.

In this traditional style of decision making (Fig. 11.1), there are only two actors and their roles are sharply differentiated. The planner's function is technical, professional, and expert; the decision maker is lay and political. The planner is constrained by professional training, which is specialized, with a narrow set of goals; the decision maker is bounded by the goals of his or her constituency, which may be quite broad and conflicting.

The model of technical evaluation in Fig. 11.1 provides a baseline, circa 1960, for understanding the revolutionary changes in the roles played by transportation planners in decision making. Such role changes occur in a societal context. Thus a look at four major trends affecting transportation decision making is pertinent. The four are (1) increasing complexity of problems, (2) increasing activism of citizens in public issues, (3) legislation legitimizing citizen involvement, and (4) an anti-government, promarket mode of thought and action. These trends are profoundly influencing not only transportation decision making, but the entire transportation approval process (discussed in Chap. 13).

11.1.1 Increasing Complexity

Urban transportation planning got its start as an extension of engineering. The engineer is results-oriented—so if we have to have a metropolitan transportation plan, make one—and get on to building it. History, starting in the 1960s, moved to frustrate this instinct for construction. Metropolitan planners were confronted with issues of new complexity. It became impossible to focus transportation decisions on a single, optimizable goal. The tradeoffs between goals became publically and politically sensitive. Urban transportation could no longer be considered as an isolated, discrete sphere of expertise.

Several illustrations of the increasing complexity are pertinent. One is the Interstate Highway Act of 1956, which required a freeway system that would link all of the major traffic generators of the nation. State highway agencies, often with little urban experience, were planning and building freeways through urban areas to central business districts. There were many unanticipated, or at least unprovided for, consequences. Local officials and civic leaders often were not prepared for the large-scale displacement of homes and businesses, the loss of neighborhood activity patterns, the decentralization of employment centers (and property tax base), often to suburban jurisdictions, and other socially, economically and politically sensitive impacts of the highway program. Public protests—outrage is the more accurate word—were strong enough to stop or prevent Interstate construction in some localities. The most publicized early instance—the Embarcadero Freeway in downtown San Francisco—had subsequent counterparts in scores of American Cities.

Another facet of growing complexity—the accelerated sprawl, with the flight of jobs to outlying locations inaccessible to central-area residents—further aggravated the problems of the 50% of Americans who, by reason of age, poverty, handicap, or high-density residence, are unable to afford or drive a car [11.58]. "The major transportation problem of our time . . . is toward ways of increasing mobility for those who are comparatively nonmobile" [11.1, p. 50]. A concern for equity, along with the desire to minimize the social, economic and political costs of urban freeways, led to public policy attention to various forms of mass transit.

Still another example of emerging complexity is that of energy and environmental concerns: automobile emissions were found to cause approximately two-thirds of the unhealthy air in our cities. Finally, increasing petroleum prices, and a dependency on foreign countries, created an "energy crisis." On all of these (and other) issues, automobile-oriented transportation planners were caught without basic solutions. The problems called for policy research, not merely technical production competence.

The transportation planner's response to these more complex problems was often a natural extension of the technical role model (Fig. 11.1): broaden the base of expertise—use interdisciplinary systems analysis. The operations research team approach was successfully used in 1955–1965 to develop the land-use and transportation models that stimulated metropolitan growth and future travel. Science and technology, which had triggered a green revolution and put a man in space, could, if we assembled the right team of experts, also get us to the airport. The methods of science, merged with those of engineering, would prove effective in solving transportation planning problems involving neighborhood well-being, urban form and services, transportation for the disadvantaged, air pollution, and energy efficiency.

Interdisciplinary expertise, too, proved to have limitations, In the 1960s, design concept teams, with interprofessional competence, were assembled to address the interrelated problems of transportation corridors in large cities. The teams failed, largely because the residents of the corridors felt that their needs and values had been overlooked. They had not been educated to the facts and analyses, or consulted on the alternatives and tradeoffs that would affect them. Just as represented in Fig. 11.1, when the teams made their recommendations, the citizens would not permit their lives to be drastically altered by experts, and the officials rejected the plan. Technical professionals produced their own preferred "solution," leaving the citizens the option, not of participation, but of protest. As Melvin Webber summarizes [11.1, p. 43]:

After having been through an extensive and rich learning experience during several decades and having finally accumulated a high level of technical sophistication in transportation planning, responsible governmental officials are not only saying there can be no technically correct solutions to transport problems. More than that, they are saying that the acceptable answers are only those that have been derived politically, only those that result from open bargaining among contesting publics. That must be one of the more notable commentaries of our time.

11.1.2 Citizen Activism

Not surprisingly, the increased technical complexity addressed by transportation officials is paralleled by increasing activism by citizen groups. Upper-income economic interests have always had access to the ears of decision-makers (see the BART and MARTA case studies in Sec. 11.2). A major development of the 1950s and 1960s, with continuing momentum into the 1970s and 1980s, is activism of citizen groups who feel that they do not have adequate influence over decisions affecting them. First manifested in the lunchroom sit-ins, bus boycotts and Freedom Marches led by Martin Luther King and other Southern Blacks, successful activism can also be traced (1960) to Saul Alinski, The Woodlawn Organization (TWO) of the predominately Black South Chicago area. Activism spread with the Vietnam protests, the women's movement, environmental protests, and other causes. Such broad-based citizen movements, then as now, were generally motivated by strongly held values that were not part of the political values of established institutions.

Transportation projects became targets of citizen activism from the start of the 1960s. Scores of cities had "freeway revolts," resulting in abandonment of transportation projects, often after millions of dollars had been spent on engineering, and even in construction.

Unfortunate as the controversies, litigation, and wasted monies were, they were recognized as part of a widespread change in American and Western society. To quote the proceedings of a Transportation Research Board (TRB) conference [11.1, pp. 47–48]:

> In a society as pluralistic as this one, it is virtually impossible to find any design, any plan, that would suit all groups and individuals. Because Mr. A hates what Mr. B loves and because there is no way to say who is right, there can be only persisting difference and latent conflict.
>
> That may be the most important observation we can make in this setting, and yet most systems analysts and systems engineers seem not to know about it. Probably because they were trained to think in the contexts of bounded and tamed problems in such fields as mathematics, physics, and operations research, where there are findable solutions, systems engineers have come to believe that there are findable solutions to social problems, too. More, they believe that there is one best answer that, once found, is indisputable. But with problems that touch on society, and thus on pluralities of publics holding to pluralities of value systems, there can be only a plurality of answers, sometimes one for each participant in the affair. There is no one best answer to socially related problems. There are no set solutions. There is no way to find what is right. Indeed, there is no one right to be found.

11.1.3 Governmental Action

Citizen protests proved remarkably effective, as they could be expected to be in a democratic government. Legislatures and agencies responded with all manner of programs aimed at the disfavored conditions. Metropolitan transportation decision making has been profoundly altered by Congressional acts, administrative regulations, and major court decisions with respect to environmental impacts, open space protection, civil rights, metropolitan area coordination, citizen participation, and other aspects of the planning process [11.3]. These are detailed in Chap. 13, but for emphasis the key national actions should be repeated here.

The 3-C provision of the 1962 Highway Act, which made local governments partners in transportation planning, was the first of a number of Congressional mandates during the 1960s that successively strengthened community involvement in planning. This was followed by Section 4f of the Federal Highway Act (1966), the Metropolitan Development Act ("A95," 1966) and the National Environmental Policy Act (1969). The general climate for citizen participation was reinforced by Model Cities (1966), community action, the Freedom of Information Act (1966), and participation requirements for most capital grants to urban areas. The Congressional intent to encourage and require community involvement was reinforced by administrative regulations, training programs, professional association activities, and other local experimentation and experience with participation in planning.

These governmental actions have generated and guided the work of transportation planners, and at the same time have compounded their problems. A top state transportation official said of the Boston Transportation Planning Review (BTPR), [11.23, p. 107]:

> Our process was designed to deal with controversy, and we found it very useful to take the federal law seriously because that law, too, is an attempt to deal with controversy.

Nevertheless, many local transportation officials felt that Federal and state regulations were reaching an intolerable level of complexity, rigidity, and dysfunction. This culminated in the early 1980s in an extensive effort within the transportation planning community to streamline Federal requirements. Whatever the outcome, metropolitan transportation decision making will have been drastically and permanently altered by governmental policy changes during the 1960s and 1970s.

11.1.4 Antigovernment, Promarket Solutions

The reaction against overregulation by government, in other sectors as well as in transportation, could be seen as part of a larger antigovernment, pro-free-market movement with ideological overtones. One aspect was the taxpayer revolts, typified by passage by the voters of California of the Proposition 13 referendum in 1978. In 1980 airlines were largely deregulated. In 1981, the new Reagan administration moved to free up private enterprise and stimulate the national economy through reduction of social programs, tax cuts, and removal or curtailing of regulations originally designed to protect health, safety, and the environment.

The full impact of this movement on transportation decision making, while already significant, has yet to be realized. In 1981 the regulation for mandatory automatic seat belts or air bags was deferred. Federal pressures for air quality and energy conservation were eased with the partial dismantling of the Environmental Protection Agency and the Department of Energy. Federal aid for transit was curtailed. Highway construction in many states remained unfunded as a result of more fuel-efficient automobiles and consequent reduction of gasoline tax revenues, coupled with inflation and rising highway maintenance costs. In most states taxpayers successfully resisted adding to the gas tax.

One interesting result of the movement is that it has stimulated some observers to suggest that the market system serves analogously as a model to replace the technical evaluation model of Fig. 11.1. Juri Pill, Manager of Corporate Planning for the Toronto Transit Commission, says [11.4, p. 13]:

The bureaucratic culture—and one can include consultants here as an appendage—has conservatism to spare. For better or worse, bureaucrats are repsonsible for keeping the system running, and therefore have an admirable pragmatism, a respect for experience and "the way things are done," and a classic reluctance to rock the boat.

The result is an incremental style of decision making, a lack of innovation in planning, and disinterest in community goal setting. In contrast, in a market economy [11.4, p. 6]:

> In a great many cases, the optimism is justified and rewarded, as the incentive structure and the diversity of perspectives lead to the eventual solution of most problems. Major breakthroughs occur without much thought about the consequences, and in this sense a democracy with a market economy can be characterized as playful and almost childish in its view of the future. This playfulness of course leads to the sorts of innovations that rarely occur in the much more serious atmosphere of a planned economy and a totalitarian regime, and it is important that this attitude be maintained.

This view is derived from both classical economic and political thought and from general systems theory. It fosters greater reliance on social systems, like the market, that are general self-organizing, self-regulating and self-correcting. This market-like model for transportation decision making was also put forth at the TRB Conference mentioned earlier [11.1, p. 44]:

> In direct contrast with the systems engineers, who see themselves as the potential designers of these systems, people of this (market type) persuasion seek to minimize the roles of central decision-makers. They aim to disperse decision-making among the millions of individuals who constitute the society and thereby retain the autonomous processes that initially created social organization.

Because standardized government services are sure to displease some consumers, promoters of market-style decision making [11.1, p. 45]

> would treat services like education and highways in a manner rather like that accorded housing and investment credit. Rather than permitting professionals or systems engineers in government to decide how much of what should be produced, they would seek to permit individual consumers to make those decisions.

In the market, only the consumer can give the final answer. Whether the transportation innovation is compact cars, the Concorde SST, airfare deregulation, demand-responsive transit, free-entry taxis, road pricing, or whatever [11.1, p. 51],

> . . . only by field tests under work-a-day conditions can we be confident that any of these proposals is acceptable, and thus right.
>   In this sense, the market strategy becomes an effective medium for concrete citizen participation, for here the citizen participates where it matters and in ways that do not rely on forensic, social, political, or cognitive skills. In such marketlike settings, the work of the systems engineers merges with that of the individual citizen to provide the only concrete answers we can expect to find in these fields. But the answers will be provided, it should be clear, only if a differentiated array of services is offered at a range of prices such that citizen consumers has a spectrum of choice. Anything less would bring us back to where we have been, with the technical expert producing his own preferred "solution," leaving citizens the option, not of participation, but of protest.

11.1.5 The Technical Role Model in Retrospect

At the beginning of the chapter we described a traditional transportation planning model in which the planning role is primarily technical (Fig. 11.1). Some assumptions of the model still hold: (1) decision officials are the legally elected representatives of the citizenry, and their role in a decision is to (2) weigh the planner's recommendation against the needs and desires of their constituents, and (3) either implement or reject the recommendation. But, with radically changed times, most of the assumptions implicit in the technical role model are no longer valid. The increased complexity is only partly caused by the interface of transportation with other problem domains. If this were the case, interdisciplinary, interprofessional design concept teams, might have been able to cut through the complexity. There proved to be no "technical fix."

Much of the complexity of transportation planning is due to the impacts that transportation decisions have on the lives of diverse interest groups in the community. The assumption that technical planning could be isolated from these groups' values, aspirations, and ambitions has not stood up. In the face of politicized citizen activism there is often no "right answer." Often several groups or agencies hold a veto power over a project.

In an attempt to prevent irresolvable, deadlocked decisions from occurring, Congress and state and local legislators have enacted legislation that requires a participatory transportation planning and decision-making process. Thus, for complex planning problems, the technical role model is no longer effective: it is literally illegal, as many court decisions in the 1970s attest.

Finally, the technical role model grew out of the turn-of-the-century movement to replace graft-ridden political machines with professional managers and technicians. During the 20th century there was a public faith that professionalized government could solve many of society's problems. In the last third of the century this faith has been eroded somewhat. The demand is for more participatory decision making, more innovation and policy leadership from professionals, and more attention to making institutions (private as well as public) more responsive to people's needs and pocketbooks. The technical role model does not provide much guidance for meeting these demands.

11.2 ACTUAL TRANSPORTATION DECISION MAKING: THREE CASE STUDIES

Transportation and decision making, profoundly influenced by social trends of the magnitude just described, has become a very complex art and science. While certain identifiable parameters and principles apply, each decision has a unique setting, a "cast of actors and factors." Case studies have the capacity to illuminate the interactions of the actors and factors, and the following have been selected for that purpose.

11.2.1 BART: A Metropolitan Strategy

The San Francisco Bay Area Rapid Transit System (BART) became operational in 1972, but preliminary discussion on similar ideas began as far back as the Second World War. A unique and certainly provocative outlook on decision making relative to BART is presented by Beagle, Haber, and Wellman [8.32], who contend that its real purpose is to serve downtown banking and insurance interests:

> The second element in the [corporate metropolitan] strategy is the creation of a rapid transit network which will connect the central city to the outlying consumer markets and labor pools.
> The push for a rapid transit system in the Bay Area began in the early fifties with Carl Wente (chairman of the board of the Bank of America), Kendric Morrish (a Wells Fargo director) and Mortimer Fleishhacker (a Crocker Citizens Bank director connected with both BAC and the Blyth-Zellerbach Committee, a corporate group supporting urban

renewal). These men initiated feasibility studies for what was to become the Bay Area Rapid Transit District (BART). In 1962 voters approved an initial bond issue for the construction of a high speed transit system embracing San Francisco, Contra Costa and Alameda counties and, ultimately, San Mateo and Marin.

The first chairman of the BART board of directors was Adrian Falk, a retired vice-president of S&W Fine Foods and past president of the California Chamber of Commerce. According to Falk, BART's basic function was to make possible the centralization of certain executive functions in downtown San Francisco. "It's the only practical way," he told a local newspaperman. "Certain financial, banking, and industrial companies want to be centralized, want to have everyone near each other. They don't want to have to go one day to Oakland, the next day to San Jose, the next day to San Francisco."

The major contributors to the public relations fund during the 1962 bond election were the three downtown banks plus a large number of companies which stood to benefit directly from construction contracts: Westinghouse, Kaiser, Bethlehem, Bechtel and the Downtown Property Owners and Builders. Bank of America's Carl Wente was head of the finance committee. BART was sold to the electorate as a crusade against the auto lobby. In fact, it ran into little trouble from this direction. The construction of thirty-two additional freeway lanes is projected for this area in the next ten years (there are forty-eight now). From the outset, BART was conceived of more as a commuter railroad than a true public transit system. It makes no pretense at carrying the great bulk of local traffic. Traffic on the Oakland–San Francisco Bay Bridge is still expected to reach the point of absolute capacity by 1975.

BART will have many consequences: first, it will greatly encourage downtown congestion and density. It has already stimulated a substantial building boom. Almost immediately after construction began, the three major banks put up high-rise headquarters buildings downtown, and increasingly the downtown San Francisco landscape is spotted with new BART-oriented construction sites. According to the Chamber of Commerce, a "direct dividend" of BART's construction will be the new "Embarcadero Center," a Rockefeller venture of great ugliness. The Embarcadero Center will involve three high-rise buildings on the waterfront, and gradually plans are being announced for redevelopment of the entire waterfront area.

More important, though, BART will guarantee the growth and renewed prosperity of downtown business. Essentially, it expands many times over the labor market area and the marketing area for goods and services. The "best workers" can be recruited for downtown jobs, choosing from the whole three-county area. And likewise, the richest, most discriminating consumers are given easy access to the prestige retailers of the downtown complex and the professional services in which it specializes.

Also, property values all along the transit route will soar. In Toronto, they increased up to tenfold adjacent to the new subway line. And BART officials expect a comparable rise in their domain. Millions of dollars will be made by the public-spirited businessmen who pushed the plan and then made their services available to construct it. And the taxpayers will be stuck with paying off the bond issues and debts of $2 billion or more. That BART will actually be profitable, that it will contribute significantly to the retirement of its debt, is highly unlikely. BART has already run into financial troubles, spending far more than its initial capitalization. The public is about to pay for these profits, inefficiencies and costs of inflation out of a special hike in the sales tax.

But the problem is not that business will make money off the construction and financing of public services; nor even that business will do a bad job and end up providing uncomfortable, ugly, and congested services. The problem is that it serves the rich and is paid for by the poor. By increasing the public debt and tax burden and by raising property values along the route, BART insures an increased squeeze on those least able to pay. Its effect on housing is obvious. Rents will be forced up as tax costs are passed on, and homeowners will be deprived of their property as the costs of ownership increase.

BART doesn't even have the saving grace of helping workers from the black and brown ghettos get to industrial jobs outside the city. The trains do run both ways. But the routes link the central city with the rich suburbs, not the industrial hinterland. And the trains will pass through ghettos only incidentally: Hunters Point is not on the route, and there are no stations in the Oakland ghettos. BART will make little contribution to an anti-poverty policy of connecting poorer workers with jobs and a wider employment area.

"The end result of BART is that San Francisco will be just like Manhattan," according to an influential insurance broker. "It's not a question of whether it's desirable," he continued, "but what's the practical matter. As a practical matter you can't have eighteen different banking and insurance centers. You have to concentrate them with all the various services around them. The people who run these centers want all their services—the people they work with—advertisers, attorneys, accountants—around them. It's a complete part of the way we do business in this country."

While we do not necessarily agree in total with the above statement, it certainly does indicate that decisions relative to transportation can be made at levels substantially different from those assumed, say, in the benefit-cost approach. In particular, the BART example illustrates that technical answers to complex problems usually have complex social, economic, and ecological (SEE) impacts. Such answers may generate new problems as serious as those "solved." High-impact solutions seldom can be implemented without the acquiescence, if not the support, of the affected group. As Webber pointed out earlier, there are "no technical fixes." Closely related is the point made in the last chapter (Sec. 10.2.10) that the benefit–cost technique obscures the differential distribution of benefits and costs on different groups in the population: BART "serves the rich and is paid for by the poor."

Another important point is that high-impact metropolitan planning strategies, not susceptible to conventional benefit–cost analysis, can be formulated by self-serving actors with access to the requisite planning and implementation resources. The case study indicates that private economic interests centered on the downtown area sufficiently dominated government decision making and the media to conceive and execute the BART plan. It is also noteworthy that they were using transportation to achieve nontransportation goals, and that their real goals were not their publicized goals.

11.2.2 MARTA: Metropolitan Strategy Revised

The BART example implies that lower income voters either did not know or did not care what was happening to them. This was not so in Atlanta. The Metropolitan Atlanta Rapid Transit Authority (MARTA) case has some similarities to BART. Its inception can be traced to a downtown economic elite. In the mid-1950s, transportation policy studies [11.56, 11.57] by the Metropolitan Planning Commission laid groundwork for a more general supporting public opinion, which also was influenced by extensive freeway construction or plans affecting neighborhoods, displacement, housing supply and costs, etc. Despite a widespread assumption that low-income people would benefit by transit, the inner-city Blacks perceived MARTA as primarily designed to bring suburban employees and shoppers into Central Atlanta. In the first referendum, to many people's surprise, the Black vote was instrumental in defeating the issuance of transit bonds.

The rapid-transit proposal was then revised, with active participation from Blacks and other groups, to provide better connections via buses between inlying neighborhoods and outlying jobs. In addition, a 15-cent fare was underwritten by an areawide 1% retail sales tax. In 1971 the revised proposal was approved in a second referendum [11.2, p. 97].

Adding to the BART study, the MARTA history illustrates the point that even in a large metropolitan area with a distinct "power structure," significant power may still be diffused among many publics. These groups then must be brought into the decision process if strategic objectives are to be accomplished. In Atlanta the coalition for the transit objective finally included both inner-city poor and suburban wealthy. Metropolitan planning and the media, as collectors, analyzers, and disseminators of information, played vital roles in catalyzing the coalition.

11.2.3 Blacksburg: Major Street Planning

Another case example is that of major street planning in a small town. Blacksburg is a community of 33,000 persons, including 20,000 students at Virginia Tech. In 1974, the Virginia Department of Highways and Transportation (VDHT) initiated an update of the Blacksburg Area Comprehensive Transportation Plan (systems plan), collected, data, and completed the plan in 1976. Local officials were well impressed by the technical craftsmanship of the VDHT team working on the study. At the request of the town, a trip-distribution model was calibrated and used. Origin-and-destination and transit surveys were made. Attention was given to downtown circulation; special work was done on a proposal for a one-way street system [11.24].

VDHT in every respect met minimum citizen participation legal regulations. There was a small citizens advisory committee (CAC). A Delphi panel[1] was used to forecast population. A sizeable study technical team was present at each of the several meetings. Strong feelings of good will and respect developed between local officials and the VDHT team. The plan was adopted unanimously by the citizens advisory committee and the planning commission.

Was the plan worth the money and time put into it? Certainly one important criterion for judging a systems plan is its effectiveness in guiding subsequent transportation decisions. By this measure the Blacksburg Transportation Plan failed its first two tests. In both cases, Town Council action inconsistent with the plan was due to citizen opposition.

The first test was the proposed Patrick Henry Drive Extension, which would have provided a much-needed, safe, circumferential road across the north side of town to relieve serious downtown congestion and serve a developing area. The route would have been through vacant land; the taking of only one house would have been required. The route was not for immediate construction. The Council was being asked only to put the route on the offical map[2] so as to protect the right-of-way (ROW) under Virginia law. Most of the ROW would have been acquired at the time of subdivision. VDHT had agreed to provide financial aid for advance acquisition, if necessary.

At a joint Planning Commission–Council meeting, the latter body appeared unanimously in favor of the action. Yet, at the Council meeting when three property owners spoke in opposition, four members of Council voted against the plan. Only the president of the League of Women Voters appeared in support of the plan.

The second test was the proposed widening to four lanes of Tom's Creek Road. This proposal was not only in the plan, but in the town's capital improvement program[3] for several years with no apparent opposition. Yet at the location and design hearing, when the project was $40,000 into preliminary design, it generated instant, effective neighborhood organization and opposition. A crowd of 100 well organized, well prepared, vocal citizens attended the Council hearings and meetings.

---

[1] A "panel" in which each person remains anonymous to the others and there are several rounds of interactions, through an intermediary, to obtain forecasts (see Sec. 11.5.16).

[2] See Chap. 13 for a definition and description.

[3] See Chap. 14 for a definition and description.

And despite positive support for the plan from the Town Manager and the Town Planner, it was unanimously rejected by Council. No citizens appeared in support.

In retrospect, it can be noted that in this Tom's Creek situation there were alternative vacant land locations for a route that would have siphoned off the excess future traffic from Tom's Creek Road, if it were not widened to four lanes. Why were these alternatives not given more consideration in the formulation of the systems plan?

A speculative answer, supported by at least some evidence, is that the busy citizens and local and state professionals working on the Advisory Committee and Planning Commission felt that they just did not have the time to go into the social, economic, and ecological impacts; to secure broader participation; or to handle controversy if it arose. One can suspect that they yielded to publicizing the plan with only generalized lines on a map, and to not publicizing the more detailed "functional plans." Thus, controversy could be avoided and the systems-planning job could be completed on schedule.

Such exclusive concentration at the systems stage on technical transportation considerations, ignoring broader social, economic, ecological, and participatory considerations, has unfortunate consequences. One is the externality of not producing a community-supportable plan, and of postponing controversy until the project stage. By then the controversy is more intense, polarized, and expensive in its disruption of engineering studies and capital improvement schedules.

11.2.4 Implications

Several conclusions can be drawn from the preceding case studies:

1. The systems-planning stage must be more than good technical studies. It must uncover issues and build community support—prior to the project design stage. This requires:
   a. Presentation at public meetings, and in the media, of explicit functional diagrams of road-improvement proposals.
   b. Exploration and presentation of alternatives.
   c. Identification of social, economic, and ecological impacts of alternatives (much more general than in project environmental impact statements).
   d. Budgeting for social, economic, and ecological assessment and citizen participation.
   e. Redefinition of the role and composition of the Citizen's Advisory Committee (CAC), and the use of a range of participatory techniques.
   f. Adoption of the plan by the governing body, on recommendation of the CAC.

2. The purpose and manner of functioning of the CAC needs to be redefined as follows:
   a. The purpose of the CAC should be twofold: to determine the best transportation-systems plan (technical and political feasibility), and also to speak out in support of the plan whenever projects come up in the capital improvement program or become issues. CAC members should be briefed on this latter purpose both before and after their selection to serve.
   b. Committee composition should include people from potentially impacted neighborhoods as well as those not directly affected; civic leaders known for their public interest stands; and members of the community power

structure. In each of these categories an effort should be made to get people who have a stake in an efficient, effective community.

c. The CAC needs continuity for the timespan from systems planning to project design. Citizen interest in such a continuing group might better be sustained if the responsibilities of the CAC covered all broad policy aspects of community development, rather than just transportation.

d. The plan should not be considered an immutable blueprint for transportation development, but rather one that would be reviewed annually by the CAC and amended as changing conditions warrant. Also, all plan elements should require current decisions to secure informed public support, protect ROW, etc.

e. Annual reviews should be integrated or coordinated with reviews of the local comprehensive development plan and capital improvement program, where these exist.

f. The CAC should be provided with techniques for, and assistance in, evaluating the tradeoffs between various consequences of the transportation alternatives.

3. Adoption by local governing bodies would assure a higher degree of local responsibility toward the plan.

   a. Adoption would be on recommendation of the CAC.

   b. The Council could exempt from approval, and request further study by the CAC and professional staffs, of alternatives that it considered not technically or politically satisfactory. An unadopted plan, or element, would clearly signal unresolved issues and indicate politically unbuildable projects.

4. Continuity and comprehensiveness in citizen support groups (2c above) requires Federal interagency coordination of policy, including funding, and the possibility of Congressional action.

5. The development of options and alternatives (1b above) suggests the need for better understanding between local and state officials, and citizens, of the current steps that can be taken to preserve ROW for future road improvements. Often this approach will provide the best alternative for protecting community values [11.25].

11.3 PARTICIPATORY DECISION MAKING

In many situations, perhaps for all complex planning problems, the technical evaluation model of Fig. 11.1 is not viable by itself. The case studies revealed that public decision-makers on any complex issue are confronted with a wide group of actors, each with its own goals, power base, and internal rationality. Each group has high stakes in the decision, standing to gain or lose. The result is that the decision-maker seems inevitably plunged into controversy, needs more technical information, and requires assistance in reconciling technical and political demands. The resulting decisions often tend to be reactive and incremental, designed to remove political "heat" and achieve short-range "solutions."

11.3.1 Participatory Role Model

Under these pressures the transportation planners' roles are changing. They are expected to develop implementable solutions—politically as well as technically feasible. This means that they must not only be aware of the social, economic, and ecological impacts of transportation proposals, but must be sensitive to how these proposals will be viewed by affected groups. Action proposals must address a wider range of goals, involving more actors, and consideration of more complex and innovative alternatives. Decision officials and the public, better educated and more sophisticated than before, require more information regarding the problem, the alternatives, and the impacts.

Some generalizations are possible regarding the emerging roles of the transportation planner under these new circumstances. Figure 11.2 represents the new circumstances and roles. Unlike the technical evaluation model, which often operated well under a homogeneous, middle-class value system and a stable socioeconomic elite, the current decision climate often has a heterogeneous set of actors with conflicting goals. The power of the "new actors" is often strengthened by "iron triangles" of lobbying–bureaucratic–legislative mutual interests, and by intergroup coalitions. Implications for the planner's role include:

1. Involvement of affected groups in the formulation of recommendations
2. Competence in communication and involvement skills
3. Sensitivity to group goals, local needs, and political perspectives
4. New geographic forums for the interplay of actors—metropolitan and subarea (e.g., corridor), as well as city and state
5. Responsiveness to current issues, including policy analysis roles and short- and long-range consequences of the decision options
6. Greater flexibility in long-range transportation plans because of the necessity to trade off against other interrelated public goals.

11.3.2 Roles of Planners and Citizens

Compared with the technical model (Figure 11.1), the planners' role today is likely to be more circumscribed in one way, but expanded in others. No longer are planners the "sole rulers of a piece of technical turf." Rather, planners work much more closely with decision officials to develop and execute policy affecting a larger domain. More of their success in this expanded role will depend on skills and techniques for collaborating with a wide range of actors—decision officials, other local, state, and federal agencies, private organizations, economic interests, and citizens groups. In brief, *community involvement* is critical in modern transportation planning.

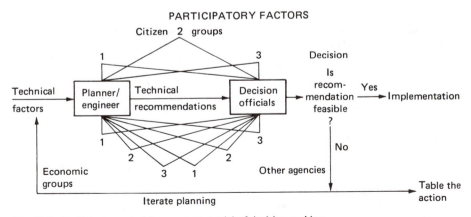

Fig. 11.2 Participatory, decision-support model of decision-making.

Community involvement—the participation of other agencies, groups, and individual citizens—touches many phases of the planning process from identification to implementation (see Table 11.1). It is covered here in a single section in order to treat this important subject in a coherent, integrated fashion. We have located the section in the evaluation chapter to stress the point that giving participants an appropriate measure of influence in actual transportation decisions generally enhances the quality of the decisions and the probability of successful implementation.

In Sec. 11.1 we noted that until recent decades, transportation planners did not always view community involvement as part of their role. They generally regarded themselves as working within a field of technical competence within which they, as the experts, could perform their legally mandated mission with relatively little outside "advice." They used economic analysis techniques such as cost–benefit (Chap. 10) and mathematical models (Chaps. 5–7) to provide the right, professional answer.

We noted that events of the 1960s and since have changed this self-image. Urban transportation decisions are complexly interrelated with factors requiring technical expertise far beyond those of the planner. Moreover, the decisions impact economic interests, neighborhood residents, and other groups in ways that stimulate their political activism. As a result the transportation planning process is now constrained by legal requirements for participation and environmental impact analysis. Further, all of these trends are occurring within a flux of public opinion concerning the role of government.

One unhappy effect of these events, in many instances, was to increase conflict between planners and "outside groups." The planning staffs and citizen groups often saw themselves as adversaries competing for the ear of decision-making officials. Unfortunately, many transportation planners perceived the citizen activism as a challenge to their professional competence.

Gradually, however, a redefinition of the proper roles of the citizen, the planner, and the public decision-maker has emerged. It has become essential, sound

Table 11.1 Participation in the Planning Process

Transportation planning process steps	Roles of participants*			
	Citizens	Other agencies	Staff planners	Decision officials
1. Defining the problem	X	X	O	
2. Determining goals and objectives	X	X	O	X
3. Model and study design			X	
4. Data collection		x	X	
5. Model calibration and use (analyses, forecasts, impacts)			X	
6. Developing alternatives	X	X	X	
7. Evaluation and decision making	X	X	O	X
8. Specification			X	
9. Implementation	X	x	X	X
10. Monitoring and feedback	X	X	X	X

Roles: X = major, x = minor or possible, O = facilitating or supporting.

practice to include at the model (or study) design stage (Fig. 11.1) an explicit participation process that clearly defines when, at what stage, and in what role citizens and other agencies will be involved [11.7]. In current practice staff planners, other agencies, citizens, and officials are all participants in a collaborative planning process. This partnership concept has eliminated much of the adversarial atmosphere of earlier transportation planning.

Each participant is seen as having a legitimate stake, a gain or loss that is at risk, in transportation planning. What then are the functions of each of these participants at each stage? What particular insights, skills, and responsibilities does each bring to the transportation planning process? Table 11.1 suggests the complementary roles that each plays in the successive steps of the planning process. The following discussion addresses these questions in terms of the interests, potential contributions, and appropriate roles of each of the participants at each stage.

11.3.3 Citizens

Citizens come to the planning process as taxpayers, voters, transportation users, and adversely or beneficially impacted residents. It is their city, their neighborhood, their quality of life that is affected. Whether or not the ensuing plan is actually implemented will depend greatly on their support. They should be considered the experts in the identification and shaping of values, goals, and objectives. They then play major roles in weighing these factors in study design (steps 1-3) and in developing and evaluating alternatives (steps 6, 7). Citizens' values are often critical in determining support or acceptance of plan implementation (step 9). Finally, citizen satisfaction with postimplementation system performance and impacts will influence their attitudes toward further transportation improvements (step 10).

11.3.4 Other Agencies

Many agencies, at all levels of government, have stakes in transportation decisions. Land-use agencies are concerned with the growth patterns that may be stimulated by transportation improvements. They are also concerned about the level of transportation service to traffic generators, and the adverse effects on neighborhoods and community amenities. Environmental agencies have responsibilities in areas of air and water quality. Social agencies look for transportation aid for the elderly, handicapped, poor, and other transportation–disadvantaged groups. Agencies, like citizens, are concerned that particular values be given due consideration in the design of transportation studies and in the formulation and evaluation of alternatives. They may also be a valuable source of secondary data.

11.3.5 Staff Planners

Transportation planners need to bring two sets of expertise to the process. Their role is primarily concerned with technical knowledge and skills. In addition, because they must work closely with citizens, other agencies, and officials, transportation planners also need "process skills"—listening, communicating, and facilitating understanding and agreement.

Applying both types of professional knowledge and skills, planners play important roles in all steps of the process. Their technical expertise is dominant in study design (step 3) and management, in data collection (step 4), model calibration and use (step 5), development of alternatives (step 6), and in specifying, implementing and monitoring the chosen alternative (steps 8-10). In all steps, planners use their process know-how to facilitate collaboration on the nontechnical aspects of the process.

In addition, the planners' analytical competence can aid in the value-laden steps of the process. For example, planners may help citizens to better define goals by clarifying linkages between land-use objectives and transportation, or the tradeoffs between economic development, environmental protection, and transportation. Better integrated and understood goal sets fashioned early in the process will facilitate planning, evaluation, and implementation in the later stages. Planners usually play a supporting role in helping citizens state values and goals. They may conduct attitude surveys. Planners are also likely to play a lead in presentation of the planning results and recommendations to decision officials, although this role is more effectively filled by a combination of planner and citizen. The staff planner often has responsibilities to help organize citizen participation and meetings, and to employ suitable means of informing citizens and securing their contributions to the process.

11.3.6 Decision Officials

As the legal elected or appointed representatives of the public, decision officials have an ultimate responsibility for the proper establishment and conduct of transportation planning, and specific responsibilities for setting of goals and policies, adoption of plans and improvement programs, adoption of ordinances and laws, and approval of budgets. The nature of their elected, directly responsive role, and the due-process requirements such as public hearings, often help ensure that the planning process does not become dominated by narrow interest groups or bureaucracies. Decision officials have key formal (and often informal) roles in four steps of the process—goal setting (step 2), evaluation and decision making (step 7), implementation (step 9), and monitoring (step 10).

11.4 DECISION-AIDING TECHNIQUES

In the study design the planner should tailor the participation aspects of the planning process to the situation. The "situation" may include the community's usual channels and style of decision making, previous events, the seriousness of the anticipated impacts, the need to involve all impacted groups, and available funding and skills. The work plan can then identify the roles anticipated for each participant at each stage. Then the most appropriate form or technique of participation can be selected. A list of potential participation techniques, such as that in Table 11.2, can provide a helpful start. The techniques are described more fully in Table 11.3 and the latter part of this section after selection of techniques; the final design of the work program adapts them to the situation and integrates them into the technical phases.

In the past, transportation planners have often found it difficult to elicit citizen involvement in system-wide, comprehensive planning. Citizens saw such planning as long-range, too general, subject to change, and without immediate effect on their lives. Transportation planners sometimes contributed to this lethargy by underfunding involvement, by inadequately publicizing impacts, and by not providing suitable channels (e.g., overreliance on public hearings). For example, in a small city the state highway agency updated the transportation plan. Funding was provided for new surveys and transportation models, but not for participation. The advisory committee was made up largely of local officials. Newspaper maps and reports of

Table 11.2 Participatory Techniques for Transportation Planning

Technique	Appropriate process steps*	Function of technique†
1. Advocacy planning	1-2, 6-7, 10	I, c, d, r
2. Arbitration	7	DM
3. Attitude survey	1-4, 7	C
4. Brainstorming	1-4, 6-7, 10	C
5. Brochures, pamphlets, or newsletters	1-10	D
6. Charette	1-2, 6-7, 9-10	I
7. Citizen advisory committee	1-2, 6-7	R
8. Citizen employment	1-10	C, D, S, i, r
9. Citizen honoraria	1-3, 6-7, 10	S
10. Citizen review board	7	DM
11. Citizen training	1-2, 6-7, 9-10	S
12. Community-sponsored meeting	6-7, 10	C, r
13. Community technical assistance	1-2, 6-7	S, i
14. Coordinator or coordinator-catalyst	1-2, 6-7	DM, S
15. Decision process explicit	3	DM
16. Delphi	5	C
17. Demonstration project	7-9	R
18. Design-in	6-7	I
19. Display models	7-9	R
20. Drop-in center	1, 9-10	D, C
21. Electronic balloting	1-2, 7	DM
22. Environmental impact statement	7	R, DM
23. Field office	1, 9-10	D, C
24. Fishbowl planning	1-2, 6-7	
25. Focused group discussion	1-4, 6-7, 10	C, i, r
26. Games and simulations	2, 7	S
27. Goals achievement	7	DM, S
28. Group dynamics	1-10	S
29. Hotline	1-10	D, C
30. Information meeting	1-2, 6, 9-10	D
31. Interview key people	1-2, 6-7, 10	C, R, d
32. Mailing list	1-10	S
33. Mail or telephone ballot	1, 7	R, DM
34. Media ballot	7	R, DM
35. Mediation	7	DM
36. Monitor mass media	1-10	C
37. Neighborhood meetings	1-2, 6, 9-10	R
38. Neighborhood planning council	1-2, 6-7, 9-10	R
39. News release	1-10	D
40. Nominal group technique	1-2, 6-7, 10	C
41. Ombudsman	1-3, 5-11	C, DM
42. Open files	1-10	D
43. Participatory radio, TV	1-2, 7, 9	D, C
44. Policy Delphi	1-2	I, r, i, c
45. Program design participation	3	S
46. Public hearing	7, 9	C
47. Public-information programs	1-2, 5-7, 9-10	D

(See footnotes p. 388).

Table 11.2   Participatory Techniques for Transportation Planning (*Continued*)

Technique	Appropriate process steps*	Function of technique[†]
48. Reaction to models or exhibits	1–2, 4–7	C, D, R, C
49. Referendum	7	DM
50. Retreat	1–2, 6–7, 10	I, R
51. Role playing	2, 7	S
52. Scenario	1–2, 5–7	O, i, r
53. Seminar	1–10	S
54. Small-group meeting	1–2, 6–7, 10	S
55. Task force	1–2, 6–7, 9	S
56. Workshop	1–2, 6–7, 9	S

*Planning process steps from Table 11.1: 1. Problem, 2. Goals, 3. Model design, 4. Data collection, 5. Model use, 6. Alternatives, 7. Evaluation, 8. Specifications, 9. Implementation, 10. Monitoring.

[†]Function of techniques: C = information collection, D = information dissemination, I = initiative planning, R = reactive planning, DM = decision making (and conflict resolution), S = participatory process support. Major function is given in capitals and lesser functions in lower case.

the plan proposals did not contain the detail that the projected "street improvements" along residential streets was intended to be four lanes. There was little public reaction, and the City Council adopted the plan.

Only after one of the street improvements was well into project design did property owners discover that they would lose much of their front yards and landscaping, have difficulty getting out of driveways, and experience the noise of heavy traffic. The then polarized controversy continued for several years and resulted in a veto of the project. Clearly the issue should have surfaced during the systems stage: the planners and the community as a whole could have considered alternative route locations, then the tradeoffs between travel time and neighborhood protection, and made a decision that would have greatly facilitated project design and construction. The means for achieving this would have been broader participation, more detailed published maps, and channels for citizen interaction. The purpose of this careful attention to creating a collaborative partnership approach to transportation planning is the establishment of transportation plans that are both technically and publicly desirable.

Attitude survey (technique 3). Obtain a cross section of community opinions and/or values based on scientific sampling and careful construction of questions. Surveys may be used early in a planning study to determine values, needs and wants, concerns, and goals of individuals in the population. Attitude surveys identify and measure "inductive goals," in contrast to the "deductive goals" that may be developed by the planning study. Use of an attitude survey in this way should be tempered by an awareness that citizens' basic values are often weighted very differently when the citizens are confronted with specific project impacts. A survey may also be used in the evaluation stage to ascertain public reaction to a proposal. For example, a random-sample survey was used by a university administration to determine student support for an increase in fees to finance a transit system. A caution here: the stability of survey results depends on how knowledgeable the surveyed public is about the project [11.6, 11.3].

Table 11.3   Profile of Participatory Techniques for Transportation Planning*

---

1. *Advocacy planning:* Interest group utilizes independent professional assistance to advance and protect its interests. Also used for the form of pluralistic planning or community technical assistance where the planner is on the public, central-agency payroll, but endeavors to develop the confidence of the community or neighborhood and advocate its views to the agency. [11.3; 11.6, p. 163].

2. *Arbitration:* Resolving conflict between different interest groups who consent to binding negotiation by using a mutually agreed-upon arbitrator as a third party to synthesize positive aspects of various viewpoints and, if necessary make the final decision. [11.3, 11.8].

3. *Attitude survey**

4. *Brainstorming**

5. *Brochures, pamphlets, or newsletters:* Publications keep publics informed about meetings, events, progress, and proposals. [11.8, 11.26].

6. *Charette:* Develop solutions to complex problems within a strict deadline through involving all interest groups in one or more intensive and highly interactive meetings. [11.3, 11.8, 11.25, 11.27, 11.28].

7. *Citizens advisory committee (CAC)**

8. *Citizen employment:* Direct employment of client community representatives which can result in continuous input of client values, interests, attitudes, and levels of knowledge. [11.3, 11.8, 11.25, 11.27, 11.35].

9. *Citizen honoraria:* Payments used as an incentive for participation of low-income citizens by dignifying the status of the citizen by placing a value on his or her participation. [11.3, 11.8, 11.17, 11.33].

10. *Citizen review board:* Delegates decision-making authority to the members of an appointed or elected citizen board. [11.3, 11.8, 11.27, 11.34].

11. *Citizen training:* Various educational techniques for instruction issues, planning, or leadership, which employ workshops, seminars, short courses, and more informal educational vehicles to improve the competence of participants. [11.3].

12. *Community-sponsored meeting:* An assembly organized by a community group generally to focus on a proposal with the objective of providing a forum for the discussion of various interest-group perspectives. [11.3, 11.6].

13. *Community technical assistance:* Provision of professional staff to neighborhood or interest groups so they may develop alternative proposals, articulate objections, or otherwise develop positions on the issues at hand. [11.3, 11.6, 11.18, 11.31, 11.36].

14. *Coordinator and coordinator–catalyst**

15. *Decision-making process explicit:* To inform participants how the agency, in general and/or on the issue at hand, will proceed with planning and decision making. [11.3, 11.8].

16. *Delphi**

17. *Demonstration project:* Test and/or demonstrate the feasibility of a proposal so that the agency and other affected parties can assess its performance and impacts. [11.8, 11.25, 11.37].

18. *Design-in:* A variety of planning methods in which citizens work with maps, scale representations, and photographs to provide a better idea of the effect on their community of proposed plans and projects [11.3; 11.5, p. 4; 11.25–11.27; 11.38].

19. *Display models:* Replications and exhibits to illustrate the completed project.

20. *Drop-in center/field office:* Manned information distribution points where a citizen can stop in to ask questions, review literature, or look at displays concerning a project effecting the area in which the center is located. [11.3; 11.5, p. 4; 11.8].

21. *Electronic balloting:* Permit votes, rankings, or ratings by participants in a meeting or workshop to be instantaneously recorded and displayed by means of electronic equipment. [11.8].

(*See footnotes p. 391*).

Table 11.3 Profile of Participatory Techniques for Transportation Planning* (Continued)

22. *Environmental impact statements (EIS):* Legally required for all Federally funded projects, and other projects in some states, draft EISs detail all of the possible alternatives and their anticipated impacts. The draft EIS must be circulated to all interested or affected persons, at least 60 days prior to a final decision. Channels for comment or objections must be provided. The final EIS must contain a summary of the objections and the decision agency's response to them.

23. *Field office:* See Drop-in Center [11.8].

24. *Fishbowl planning**

25. *Focused group discussion**

26. *Games and simulations:* Experimentation in a risk-free setting with various alternatives (policies, programs, plans) to determine their impacts in a simulated, competitive environment where there is no actual capital investment and no real consequences at stake [11.3; 11.5, p. 5; 11.8; 11.40–11.43].

27. *Goals achievement**

28. *Group dynamics:* A generic term referring to either interpersonal techniques and exercises to facilitate group interaction or problem-solving techniques designed to highlight substantive issues. [11.3].

29. *Hotline:* Any publicized telephone-answering service connected with a planning process and used to answer citizens directly, to record questions to be answered by a return call, or to provide recorded information on the process. [11.5, p. 4].

30. *Information meeting:* An open public assembly designed by the agency to present problems or proposal information and details to interested publics and secure reactions. [11.6, p. 177; 11.8; 11.17; 11.18; 11.47; 11.48].

31. *Interview key people:* Through talks with selected *key* people, a quick definition of the situation, identification of issues, and "pulse" of the community is obtained. Key people are opinion leaders, elected officials, and leaders and doers in clubs, agencies, and organizations. [11.8].

32. *Mailing list:* A list, or set of labels, of names, addresses, and/or phone numbers as a basis for information dissemination, surveys, or personal contacts. [11.8, 11.26].

33. *Mail or telephone ballot:* Once a well constructed mailing list is developed, it can form the basis for a poll or opinion survey. [11.8, 11.26].

34. *Media ballot:* Forms printed in a newspaper and returned by readers, which can stimulate community thought on an issue and encourage a self-selected sample to make their views known. [11.3, 11.8, 11.25, 11.49].

35. *Mediation**

36. *Monitor mass media:* Systematic covering of newspaper and other key media is a basic technique to (a) develop awareness of local issues, values, and impacting developments, (b) learn the scope and reliability of the information the public is receiving, and (c) evaluate the effectiveness of agency news releases and media interviews. [11.8, 11.51].

37. *Neighborhood meetings:* Meetings held for residents of a specific neighborhood that has been, or will be, affected by a project or plan. [11.3; 11.5, p. 5; 11.6].

38. *Neighborhood planning council (NPC):* A reactive structure for obtaining participation on issues that affect a specific geographic area; the council serves as an advisory body to the public agency in identifying neighborhood problems, formulating goals and priorities, and evaluating and reacting to the agency's proposed plans. [11.3; 11.5, p. 5; 11.6; 11.52–11.54].

39. *News release:* The most common technique for securing attention uses the media. Preparation of release requires knowledge of media deadlines and rules, and attention to appropriate style. [11.8].

40. *Nominal group technique (NGT)**

41. *Ombudsman:* An independent, impartial official who serves as a mediator between citizen and government to seek redress for complaints, to further understanding of each other's position, or to expedite requests [11.3, 11.5].

(See footnote p. 391).

Table 11.3   Profile of Participatory Techniques for Transportation Planning* (Continued)

42. *Open files:* Goes beyond the requirements of the Freedom of Information Act to make key project information readily accessible to the public at the main agency office, a field office, drop-in center, or public library. [11.8].
43. *Participatory radio or TV:* Use of electronic media to involve large numbers of people. Typically in this technique the top officials responsible for a project or decision are available on television or radio to explain the project or issue and answer questions telephoned in on dedicated lines. [11.8].
44. *Policy Delphi**
45. *Program design participation**
46. *Public hearing**
47. *Public-information program:* Coordinated use of a variety of tools for information dissemination such as press releases, feature stories, direct mailings of reports, newsletters, or brochures, radio or TV public service programs, spot announcements, speeches or debates; displays, slide presentations, documentary films, videotapes, and other audiovisual tools; exhibits; legal notices; speakers' bureau; and other responses to inquiries from the public. [11.3; 11.6, p. 182; 11.8].
48. *Reactions to display models or exhibits:* Use of models or exhibits to elicit opinions, values, suggestions, and other responses. [11.8].
49. *Referendum:* Voting by citizens for or against a proposal, or between alternatives. May be official and statutory, or informal and advisory. [11.8, 11.25, 11.27].
50. *Retreat:* An intensive working session, generally on a weekend and away from customary distractions, that can aid consensus formation on large, complex issues in the intensive, interactive, intergroup environment that can be created at a retreat. [11.8, 11.55].
51. *Role playing**
52. *Scenario**
53. *Seminar:* A forum structured so that various concerned groups or agencies present their positions regarding a proposal or issue. [11.8, 11.25].
54. *Small group meeting**
55. *Task force**
56. *Workshop**

---

*References are shown in brackets. Asterisked (*) items are discussed more fully in the text.

Brainstorming (technique 4). A widely used technique to obtain quickly a large number of uncritical ideas that then can be evaluated critically, brainstorming is appropriate for in-house, interagency, or citizen groups. It attempts to create a supportive, uncritical atmosphere where the brain's right hemisphere can function freely. The procedure is simple:

1. A question is carefully framed in advance. It may concern identification of needs and concerns, components of a problem, or possible solutions.
2. Large groups are divided into small groups of not more than 8-10 persons. Newsprint, visible to all, is provided along with a person to record.
3. Participants are briefed on the "Rules for Brainstorming":
   • Get out as many ideas as possible, as quickly as possible
   • No discussion, comments or criticism permitted
   • Redundancy is OK
4. Ideas are recorded.
5. Ideas are put to use. Ask for comments or prioritization. The ideas may now be discussed critically.

For its purpose of idea generation, brainstorming is a well tested, widely used, inexpensive technique [11.8; 11.25; 11.26].

Citizens advisory committee (CAC) (technique 7). A form of reactive planning in which a panel of citizens are called together by the agency to represent their groups and/or communities. The CAC is widely used with varying results. On the positive side, given good leadership and staffing it can offer ideas, evaluate current and proposed programs, interpret the agency to the public, and develop a panel of knowledgeable citizens. Disadvantages include a potential for time-consuming and frustrating experience for citizens. Members may be unequal in status, knowledge, and experience, with some therefore reluctant to participate actively. A similar impasse may develop between members and staff. Members frequently feel they are coopted and can only "rubber-stamp" the agency's proposals. CACs are often not representative, have no accountability to a larger community, and can be a barrier to using other means to secure citizen participation [11.3, 11.17, 11.29, 11.30, 11.31, 11.32].

Coordinator and coordinator-catalyst (technique 14). This is an individual who has the responsibility for providing a focal point for participation, being in contact with all parties and channeling their views into the process. The catalyst has the additional role of encouraging constructive interaction among the parties. The coordinator-catalyst is a visible, identifiable, and responsible official of the particular planning or program structure who can speak with authority about the planning process and make programmatic decisions either alone or in concert with other high-level staff.

> Once the coordinator is identified, citizens know who to contact and can look to the coordinator to acquaint relevant critical actors. The Coordinator can also communicate directly with appropriate officials, handle grievances, and otherwise insure that the process operates up to its expectations. The assumption of such a role on the part of a program manager or someone in authority close to the planning process is usually a sign that there is a strong commitment to citizen participation on the part of the agencies involved [11.3].

If the coordinator-catalyst can successfully secure the confidence of the various parties, this approach has a high potential for conflict resolution and development of an acceptable proposal. A disadvantage is that:

> much rests on the capability and integrity of one person in this kind of role. Particularly in a highly controversial program, it may be difficult to find an individual with the diplomatic skills, the technical knowledge, the credibility, and the personality to relate to a wide variety of different groups and interests, and to cope with the constant pressure such a role involves. The Coordinator places himself or herself squarely in the cross-fire of conflict, and if he or she cannot speak with authority, he becomes useless for resolving issues [11.3].

Delphi (technique 16).

> A method for systematically developing and expressing the views of a panel of individuals on a particular subject. Initiated with the solicitation of written views on a subject, successive rounds present the arguments and counterarguments from the preceding round for panelists to respond to as they work toward a consensus of opinion or clearly established positions and supporting arguments.
> Delphi was originated as a forecasting technique by the RAND Corporation and has been widely used in industry and business. In the latter field, it has been hailed as one of the most satisfactory forecasting devices for use when trends are discontinuous [11.5, p. 4].

Used as a forecasting tool (as contrasted with policy Delphi), a panel is named representing expertise in various aspects of the forecast—such as population shifts, economics, or possible technological innovations. Members of the panel are not known to each other, and all communications are in writing. Answers to first-round questions form the basis to the second round, etc. For example, a state highway agency felt that the local population forecast to be used as a basis for estimating need for future transportation facilities was too high. It convened a Delphi panel composed of local bankers, realtors, and Chamber of Commerce interests as well as the planners. After three rounds, there was considerable convergence among panelists' judgments on a forecast about 20% lower than the original.

Advantages include the ability to select the panel members for their competence or perspective, the anonymity of panelists thus preventing status bias, and the iterative interaction between panelists. Disadvantages include the time demands on panelists, which may be a problem unless the panelists are paid. In the latter case, Delphi may prove expensive. Considerable staff and calendar time may be required. A variant, where only numerical responses are required, is to run Delphi at a meeting or conference where electronic equipment will permit quick recording and calculating of responses [11.3, 11.6, 11.8].

Fishbowl planning (technique 24).

> A process involving citizens in restructuring a proposed plan before adoption: Fishbowl planning uses public meetings, public brochures, workshops, and a citizens' committee; brochures provide continuity between successive public meetings [11.5, p. 5].

Fishbowl planning is an open reactive planning process, in which all affected groups can express their support or opposition to an agency proposal, and have the opportunity to restructure it or affect its adoption. It is not to be confused with the group dynamics technique with the same name. Fishbowl planning, originally developed for use within the Department of Defense, has subsequently been used widely by the Corps of Engineers (COE) in water-resources planning. In this form, it makes use of a combination of other techniques, particularly public meetings, brochures, workshops, and citizen committees. These are woven into a sequential process.

The agency initially presents its proposals or alternatives (possibly through an EIS) in a brochure. These brochures are widely distributed before and discussed at a public meeting, at which the issues are identified and citizen committees (CIT-COMS) initiated. CITCOMS are self-selected and represent interest groups and governmental agencies. They suggest additional alternatives and considerations to restructure the proposal(s). Workshops composed of both CITCOM representatives and agency planners to provide technical assistance (see community technical assistance) develop the restructured proposal(s) more fully in a new brochure, which is then circulated widely in preparation for another public meeting. This process may be repeated through several cycles.

Advantages stem from the openness of the process, which permits early involvement, identification of parties-at-interest, and full opportunity for parties to inform themselves and understand each other and work toward consensus. While COE use of the technique has been criticized, postevaluation indicates that with improvements it has a high potential for resolving issues and generating an acceptable, supported plan. In any case, it demonstrates the way in which several techniques must often be combined to form an effective program [11.6, 11.3, 11.20, 11.17, 11.39, 11.40].

Focused group discussion (technique 25). Small meetings (8–10 people) guided by a trained moderator using a prepared outline, and based on the synergistic idea that the group collectively has more insight than the individual members. This technique can be used in combination with the CAC, citizen training, or other techniques. It has great potential for eliciting values, concerns, and proposals, for developing understanding of the various facets of a problem, and for aiding development of agreement [11.3].

Goals achievement (technique 27). Uses a matrix to separate and display value factors and technical elements of a decision. It permits different interest groups to independently weigh the goals and thus test the sensitivity of the decision to the value factors. It thus provides a framework for further discussion and negotiation.

Goal achievement was discussed and illustrated in the previous chapter as a technical evaluation approach. It also has a number of advantages as a participatory technique. As a matrix, it displays the essential elements of a decision: alternatives, goals that the alternatives serve, criteria for measuring goal achievement, scores for each alternative's goal achievement, and relative weights assigned each goal. It can be used, in the absence of data, as a framework for judgment, yet it can also accommodate data-based scores. It enables parties-at-interest to display their particular value set, and indicates the effect of their values on the possible decision. By separating value and technical factors, it clarifies the roles of citizens, planners, and officials. It clearly indicates points of agreement, points of insignificant differences, and points that need to be resolved. In successive iterations, goals that do not affect the choice and alternatives that are scored low can be eliminated so that attention can be concentrated on the key goals and promising options. Goals achievement is readily combined with workshops, task forces, charettes, and other techniques.

Goals achievement becomes unwieldy for both decision-makers and citizens when there are too many goals and/or alternatives. Some degree of understanding and skill in the use of goals, criteria, and other aspects is required of participants. The planner, or other actor, usually must also be trainer and facilitator. If used in a data-based, rather than a judgmental, mode, it may be expensive and time-consuming [11.8, 11.44, 11.45, 11.46].

Mediation (technique 35). Use of a third party to act as facilitator between conflicting interest groups *who consent to negotiation.* The procedure is more direct and less formal than arbitration. While basic principles are simple, it can be time-consuming and can require an experienced mediator. Advantages include demonstration of agency concern for an equitable solution, and the development and expression of group values. Disadvantages may be the bypassing of officials and other groups not included, the difficulty and expense of finding an acceptable mediator, the intensive effort required by participants, and the potential for long-term bitterness (as well as improved understanding). The use of agency (or other governmental) personnel with mediation skills may reduce costs and make mediation less pressured [11.3, 11.8, 11.25, 11.27, 11.50].

Nominal group technique (NGT) (technique 40). A small-group process in which ideas or proposals are generated nonvocally, then pooled, discussed, and ordered by priority. The technique is generally appropriate to (1) identify elements of a problem or (2) a solution, and to (3) establish priorities. NGT is good in that more vocal individuals cannot dominate the group, more reticent persons can get their ideas out, and the planners have the benefit of both individual and group deliberations in writing. It requires a group that has a familiarity with and interest in the problem, and 60–120 min. NGT is not appropriate for information exchange, coordination, or negotiation.

In a typical NGT process, 7–10 people might take four steps: (1) each individual silently generating ideas in writing, (2) a round-robin feedback from individuals with a recording of each idea in a terse phrase on a flip chart, (3) group discussion of each idea to clarify and exchange opinions, and (4) individual voting on priorities using 3 X 5 cards to retain anonymity. Weighted voting is usually used with individuals assigning the highest priority 5, the second 4, etc. The weights are then added up for each idea to provide a group priority rating. Ideas and scores from any number of small groups can be consolidated [11.10].

Policy Delphi (technique 44). Application of the Delphi techniques to policy or project development (see 11.5.16). Policy Delphi must be carefully tailored to the particular situation. Three key early questions to be resolved are how to (1) set the objectives of the exercise, (2) secure informed, motivated respondents to represent the various viewpoints of parties at interest, and (3) structure the problem so that the respondents can interact constructively and yet not preclude their unique insights and innovations. For example, if the issue were that of the mix of financing sources to be used for transit, the first round might focus on identifying and defining the elements of the problem, including the goals and values, and suggesting alternatives. In the second round the respondents could then rate the alternatives based on the impacts. The third round permits the respondents to rerate the impacts based on the comments made in the second round. Additional rounds would determine the degree of opinion convergence that can be achieved. After the last round, a meeting of the panel, face-to-face, may be appropriate.

Advantages are that policy Delphi provides a test of whether agreement on a policy or proposal can be achieved; Since panelists remain anonymous and participate in writing, no one or two individuals can dominate the group; replies are more considered; and the planners can interact with the panel. Disadvantages are that it takes committed individuals since it may be time consuming; it requires considerable staff and calendar time; and participation is limited to a small panel, with the larger community uninvolved [11.6, 11.11–11.15].

Program design participation (technique 45). Involvement of citizens in the design of the participatory process itself, including if, when, and how participation should take place. Such involvement is required by Federal regulation (23 CFR 795) in the preparation and revision of each state's Environmental Action Plan. When participatory programs are designed solely by agency staff, there is a tendency for the programs to reflect the needs and convenience of the staff. This reduces the level of participation. The reverse is also true: participation in program design by citizens and other agencies better reflects their needs and leads to greater motivation for subsequent participation [11.5, p. 9; 11.8].

Public hearing (technique 46). A method usually required by law when some major governmental program is about to be implemented or legislation considered. It is characterized by procedural formalities, an official transcript or record of the meeting, and open participation by an individual or representative of a group to present views for the official record [11.5]. The public hearing fills a unique, but limited, niche in the repertoire of participatory techniques. As a legal requirement, public hearings usually provide a capstone to the participatory process, measuring the degree of support for the proposal. Citizens, well prepared for the hearing through the process, can make forceful and persuasive presentations. The hearing then provides an official record that is useful in the actual decision. Limitations are that hearings offer only one-way communication, usually just before the final decision—consequently, more heat than light may be generated. Hearings are most likely to attract hostile, emotional views, which then are discounted by decisionmakers as irrational and confused [11.6, p. 143; 11.3].

Role playing (technique 51). This is a form of training in which participants prepare to handle possible future events constructively by playing their own or other "actors' " roles. Participants aid the professional trainer in discussing the effectiveness of alternative ways of handling the situation (see Sec. 11.5.26). The best-known version of role playing is psychodrama. It has become a recognized training method to enable corporate managers to develop human relations and leadership skills, FBI agents to negotiate with hostage takers, and other similar preparation for nonroutine events [11.9]. Related to a transportation problem, psychodrama can be varied depending on the number of "real-world" actors available to participate. Usually available actors play themselves, unless the resulting "solution" is not satisfactory, in which case role switching may generate new understanding and a more satisfactory solution. Design of the role playing involves identification of likely evaluation, decision, or confrontational points in the future course of the project process; staging (or even scripting) the initial (problem) part of the skit; and permitting the actors to play out, and replay under different conditions, the rest of the scenario. With imagination, role playing provides a powerful tool for understanding the perspectives and motives of other actors and modifying negotiating strategies accordingly. Limitations include the unavailability of some of the actors and their not being bound by the insights gained in the role playing [11.8, 11.3].

Scenario (technique 52). A scenario may be (1) a "possible future," communicated concisely and credibly, or (2) one of several techniques that uses the participation of small teams to generate a view of the future. In the first meaning "scenario" connotes, in contrast to "projections" or "forecasts," an image of the future that is multivariant, holistic, but only one of several alternative futures. The development of scenarios can serve several participatory purposes: (1) make the image of the future visible and explicit; (2) examine the premises, the assumptions regarding cause and effect; (3) bracket the range of potential events; (4) provide a more comprehensive, more considered, forecast.

There are a number of scenario-generating techniques, and combinations of techniques, available. None provide cookbook procedures. Three discussed here are (1) consensus techniques, (2) iteration through synopses, and (3) cross-impact matrices. Delphi (see Sec. 11.5.16) is an example of the first—the systematically developed consensus of "wise old men." A variant is "iteration through synopses," in which independent scenarios are developed within a discipline and then made compatible through interdisciplinary interaction [11.16]. A third, more sophisticated, technique is use of a cross-impact matrix to identify key variables and linkages, and the type, strength, and timing of the linkage expected. A combination technique might include the following steps to be conducted by a task force, Workshop, Delphi, NGT panel, or staff for subsequent wider review:

1. Determine the domain of concern (preferably the goal or objective).
2. Brainstorm those factors that will affect the future, the attainment of the goal.
3. Select or cluster those factors, controllable or uncontrollable, most likely to influence future goal achievement.
4. Identify "themes"—the fundamentally different futures that could occur from major change in an impacting variable.
5. Outline alternative scenarios based on the themes—that is, what is likely to happen to dependent (impacted) variables if the major changes take place. A cross-impact matrix is a useful technique for this step. Flesh out the outline (or matrix) into a concise word-picture.

The foregoing five steps constitute the basic scenario development process. It is often desirable to go further to develop a contingency plan. The five following steps are a guide.

1. Identify and select the best policy response for each scenario. To identify possible policies, reverse the usual "if . . . then" game: if this scenario is taken seriously, what can we do to maximize opportunities and minimize threats? Then test alternative policies by forecasting their consequences: If we do this, what will happen? Choose the best policy—that one that best achieves the goals identified in step 1—and reject those alternatives that are less desirable or feasible.

2. Identify a decisionmaking sequence for each plan. What objectives (steps) need be taken in what order to implement the policy?

3. Cross-coordinate each policy. The scenario technique addresses the problem of uncertainty—the essential unknowability of the future. Until we know that a particular scenario is unlikely, it is desirable to keep its policy response option open. This is done by cross-coordinating the plans: that is, we study whether by implementing one plan, we will have precluded the others.

4. Develop an operation plan with the goal of preserving as many of the desirable options as possible. This will include identifying the "trigger points"— conditions, trends, or events that might indicate a shift from one policy to another.

5. Specify assumptions made at all steps in the scenario and policy development process so that later planners and decision-makers can see whether the scenario and policy response are still on target.

Scenarios and contingency planning offer a series of participatory steps that address the persistent problem of uncertainty about the future [11.6, p. 92].

Small group meeting (technique 54). The familiar informed tête-à-tête of selected representatives—a "summit conference" that encourages frank, direct interaction and close working relations between leaders. Informality does not make careful preparation less essential. The purpose of the meeting and the agenda should be determined in advance, confirmed at the start of the meeting, and amended as needed. Selection of group representatives requires care (see key persons). Skilled facilitation is often necessary to bring out views, secure interaction, relate to the agenda, and maintain interest. Unless techniques affording broader participation are also used, there is danger of the small group becoming elitist. Advantages are greater freedom to communicate openly, the briefer preparation and lead time, and the opportunity for closer personal relations [11.3, pp. 130–135; 11.8].

Task force (technique 55). An ad hoc citizen or interagency committee in which the parties-at-interest are actively engaged, under a strong chairman, in a well defined problem-solving or other specific task effort. Created by the sponsoring agency, the task force may be part of a larger, on-going participatory structure, such as an advisory committee or planning council, or it may be a single independent effort. The purpose of the task force is limited to a specific problem or task objective in the planning process. Therefore, it is a temporary structure whose function and existence generally terminates with the accomplishment of the task or solution of the problem. The directed purpose of the task force necessitates a membership large enough to represent all groups, views, or disciplines, but sufficiently limited in number to allow all members to interact actively and effectively. A membership of from 8 to 20 participants is characteristic. Task force members may be selected and appointed by the planning agency or self-selected by interest groups identified by the agency and the formally appointed by that agency. Generally, task force members tend to be fairly equal in status and responsibility, except for the chairperson who, having ultimate responsibility for the results, occupies a key and pivotal role in promoting interaction among members between and during meetings, directing the progress of the task force, coordinating its efforts, and preparing (or directing preparation of) the final report. The task force generally relies on agency staff for technical assistance and support [11.6, p. 126].

Advantages include the task force's specific, well defined objective, small size, agency staffing, and diverse viewpoints, which motivate participation. This maximizes probability of usable results. The limited size and diverse membership of the task force encourage intense discussion and interaction, creating opportunities for tradeoffs, negotiation, change in perspective, and compromise. It is an excellent technique for issue resolution and consensus building if the agency staff is skilled in managing groups and is receptive to citizen inputs. Potential disadvantages are lack of responsiveness to the larger community, and manipulation by agency technical personnel or by dominant members [11.6, 11.8, 11.27].

Workshops (technique 56). This is a structured, product-oriented working session. Workshops provide a framework that enables the parties-at-interest to engage in mutual education by thoroughly discussing an issue or idea and by trying to reach some basic understanding about its nature, relevance, or role in the particular study or project. Like task forces, workshops are most suitable to provide for participation around a specific issue that has a specific focus and a specific set of potentially interested parties. Workshops can be a useful adjunct and supplement to public information meetings to follow up on specific issues raised but not resolved at these larger meetings [11.6, p. 70].

Unlike in the task force, in a workshop the responsibility for results rests on all members. Effective workshops generally focus on a specific topic, and data are presented in language readily grasped by laymen. They usually involve a limited number of direct participants—from 12 to 25 is a manageable number, although workshops of from 40 to 100 people can be operated successfully if the group is subdivided into small work groups.

Workshops can be open to all or by invitation only, but they usually work best when limited to the active parties-at-interest who are at the same general level of knowledge or understanding of the particular issue at hand. While there is usually a workshop chairman or discussion leader, the meeting is characterized by less agency presentation and more mutual discussion than larger, open meetings can permit. Workshops usually result in the production of a written document describing the issues and positions developed, the suggestions made, and the consensus, if any, reached by the participants. In a long planning process, it is advisable to have a series of workshops that coincide with important milestones of the process—e.g., alternative futures, goal development, alternative options, impact analysis, etc.

Workshops offer one of the best ways to introduce value-laden views of both agencies and citizen groups, and to debate to negotiate these views with an aim of reaching some conclusion. Properly designed and facilitated, workshops are a proven mechanism for defining problems and issues, devising and testing alternatives, reducing differences, and building consensus. However, workshop effectiveness may dictate selectively in the designation of participants, resulting in alienation of uninvited persons. This can usually be avoided by sequentially interrelating workshops and other, more open, participatory techniques (see fishbowl planning). As in other techniques, workshops have the unfortunate, but preventable, potential for manipulation by either the sponsoring staff or other interests [11.6, 11.17–11.22].

11.5 SUMMARY

Actual decision making usually involves a much wider range of factors than can be encompassed by the economic evaluation models, such as benefit–cost or investment–return, discussed in Chap. 10. Furthermore, many of these factors are value judgments that can best be taken into account through a participatory planning process that involves other agencies and public groups. Some stages of the planning process—such as modeling, data collection, and program management—are largely technical. Other phases—such as problem and goal identification, evaluation, and monitoring—often also involve nontechnical participants. Thus, the transportation planning professional will often need to draw on his or her participatory, as well as technical, skills.

Chapter 11 examined case studies of actual decision making and described the trends that have changed the transportation planner's role from a primarily technical one to a combined technical–participatory one. The current planning model is one that provides broad support for decision officials. The cost-effectiveness and goals-achievement techniques are often more appropriate for evaluation in the new decision-making climate. In describing the decision-support model, the chapter provided the roles of major participants and outlined a number of techniques useful in conducting participatory planning.

BIBLIOGRAPHY

11.1   Webber, M.: "On the Technics and the Politics of Transport Planning," *Citizen Participation in Transportation Planning*, special report 142, Highway Research Board, Washington, D.C., 1972.

11.2   Bradley, P.: "Citizen Participation and the Citizen Viewpoint," *Citizen Participation in Transportation Planning*, special report 142, Division of Engineering National Research Council, Washington, D.C., 1973.

11.3   U.S. Department of Transportation: *Effective Citizen Participation in Transportation Planning: Volume I Community Involvement Processes*, Federal Highway Administration, Washington, D.C., 1976.

11.4   Pill, J.: "Corporate Planning and Urban Policy Analysis: The Process is the Product," in *Systems Analysis in Urban Policy Making and Planning*, NATO Advanced Research Institute, Oxford, England, September 1980.

11.5   Torrey, W., and F. W. Mills: *Selecting Effective Citizen Participation Techniques*, U.S. Department of Transportation, Federal Highway Administration, Washington, D.C., 1977.

11.6   Stuart, R. C.: *Techniques for Strategic Planning: A Technical Supplement to Module Number One of Policy/Program Analysis and Evaluation Techniques*, Center for Urban and Regional Studies, Virginia Polytechnic Institute and State University, under contract to The Urban Management Curriculum Project, The National Management and Development Service, Washington, D.C., 1976.

11.7   Gil, E., and E. Lucchesi: "Chapter 19: Citizen Participation in Planning," in F. S. So, I. Stollman, F. Beal, and D. Arnold (eds.), *The Practice of Local Government Planning*, International City Management Association, 1979.

11.8   Smith, D. C., R. C. Stuart, and R. Hansen: *Manual for Community Involvement*, Center for Urban and Regional Studies, Virginia Polytechnic Institute and State University, under contract to Federal Highway Administration, U.S. Department of Transportation, May 1975.

11.9   "Executives, Cops, G-Men Prepare for Crises Through Psychodrama," *Wall Street Journal*, November 25, 1981 (CXCVIII: 104), p. 1.

11.10 Delbecq, A. L., A. H. Van de Ven, and D. H. Gustafson: *Group Techniques for Program Planning*, Scott, Foresman, and Co., Glenview, Ill., 1975.

11.11 Brown, B. B.: *The Delphi Process*, The RAND Corporation, Santa Monica, Calif., P-3925, 1968.

11.12 Dalkey, N. C.: *The Delphi Method: An Experimental Study of Group Opinion,* The RAND Corporation, Santa Monica, Calif., RM-5888-PR, 1969.

11.13 Reisman, A., S. J. Mantel Jr., B. V. Dean, and N. Eisenberg: *Evaluation and Budgeting Model for a System of Social Agencies,* Technical Memorandum No. 167, Department of Operation Research, Case Western Reserve University, Cleveland, Ohio, 1969.

11.14 Turoff, M.: *The Design of a Policy Delphi,* National Resource Analysis Center, Systems Evaluation Division, Executive Office of the President, Office of Emergency Preparedness, Technical Memorandum T.M.-123, 170, Washington, D.C., 1968.

11.15 Turoff, M.: "Delphi Conferencing: Computer-Based Conference with Anonymity," *Technological Forecasting and Social Change,* vol. 3, pp. 159–204, 1972.

11.16 Brech, R.: "Britain 1984," in *Planning Prosperity: A Synoptic Model for Growth,* Danton, Longman and Todd, London, 1964.

11.17 Arthur D. Little, Inc.: *Effective Citizen Participation in Transportation Planning,* U.S. Dept. of Transportation, Federal Highway Administration, Washington, D.C., 1976.

11.18 Boston Transportation Planning Review: *Report of Community Liaison and Technical Assistance Staff,* Boston Transportation Planning Review, Boston, Mass., Jan. 1973.

11.19 Creighton, J.: *Citizen Participation/Public Involvement Skills Workbook,* Synergy, Los Gatos, Calif., Jan. 1973.

11.20 Sargent, H. L.: "Fishbowl Planning Immerses Pacific Northwest in Corps Project," *Civil Engineering,* vol. 42, Sept. 1972.

11.21 Sloan, A. K.: *Citizen Participation in Transportation Planning: The Boston Experience,* Ballinger, New York, 1974, pp. 125–142.

11.22 U.S. Department of Agriculture: "Analysis of Public Inputs to the Salmon River Study and the Idaho and Salmon River Breaks Primitive Areas Study," U.S. Department of Agriculture, Forest Service, Hamilton, Mont., June 1973.

11.23 Wolford, J.: "Public Participation in Balanced Transportation Planning," *Citizen Participation in Transportation Planning,* Special Report 142, Highway Research Board, Washington, D.C., 1972.

11.24 Stuart, R. C., and W. B. Issel: "Citizen Participation in Systems Planning: Comments to the U.S. Department of Transportation in Response to Notice 79-15," in *Federal Register,* August 9, 1975, Blacksburg, Va.

11.25 Yukubonsky, R.: *Community Interaction in Transportation Systems and Project Development,* New York State Department of Transportation, Sept. 1973.

11.26 Pennsylvania Department of Transportation: *A Manual for Achieving Effective Community Participation in Transportation Planning,* prepared by Veland and Junker, and Portfolio Assoc., April, 1974.

11.27 Arthur D. Little, Inc.: *Effective Citizen Participation in Highway Planning,* draft report, March 1975.

11.28 Kahn, S. D.: *Experiment in Planning an Urban High School: The Baltimore Charette,* report from Educational Laboratories, New York, November 1969.

11.29 Bleiker, H., J. S. Suhrbrier, and M. L. Manheim: "Community Interaction as an Integral Part of the Highway Decisionmaking Process," *Highway Research Record No. 356,* Highway Research Board, Washington, D.C., 1971.

11.30 U.S. Department of Housing and Urban Development: *The Model Cities Program: A History and Analysis of the Planning Process in Three Cities,* U.S. Government Printing Office, Washington, D.C., May 1969.

11.31 Warner, K. P.: *Public Participation in Water Resource Planning,* National Water Commission, Arlington, Va., July 1971.

11.32 U.S. Department of the Army, Corps of Engineers, Institute for Water Resources: *Public Participation in Water Resources Planning,* Springfield, Va., December 1970.

11.33 Stenburg, C.: "Citizens and the Administrative State: From Participation to Power," *Public Administration Review,* May–June 1972.

11.34 Arnstein, S.: "Maximum Feasible Manipulation," *Public Administrative Review,* September 1972.

11.35 Taran, F. B.: *The Utilization of Non-professional Personnel in Social Work: A Review of Relevant Studies,* Mobilization for Youth, New York, 1963.

11.36 U.S. Department of Housing and Urban Development: *Citizen Participation in Model Cities,* technical assistance bulletin no. 3, Washington, D.C., 1968.

11.37 California Division of Highways: *Century Freeway Relocation Housing Project Progress Report No. 1,* Right of Way Dept., Sacramento, Calif., 1968.

11.38 Dixon, J. M.: "Planning Workbook for the Community," *Architectural Forum*, December 1969.

11.39 (a) Aggerholm, D. A.: "Evaluation of Seattle Districts Fishbowl Planning," July 1973 (unpublished).

11.39 (b) Ragan, J.: "Seattle District Public Participation Program: An Evaluation," March 7, 1973 (draft report).

11.40 Cunningham, M., J. Carter, C. Reese, and B. Webb: "Toward a Perceptual Tool in Urban Design: A Street Simulation Study," *4th Annual Proceedings*, Environmental Design Research Association, 1973.

11.41 Fry, F. F.: "Route Location Game," *Traffic Quarterly*, January 1970.

11.42 Long, N. E.: "Ecology of Games," *American Journal of Sociology*, vol. 64, November 1958, pp. 251–261.

11.43 Raser, J. R.: *Simulation and Society*, Allyn and Bacon, Boston, 1969.

11.44 Steger, C., and R. C. Stuart: "Goals Achievement as an Integrating Device for Technical and Citizen Inputs into Urban Freeway Location and Design," *American Society of Civil Engineers Transportation Conference*, Montreal, July 17, 1974.

11.45 Stuart, R. C., and J. W. Dickey: "Modelling the Group Goals Component of the Plan Evaluation Process," *American Institute of Planners, Confer-In-West*, San Francisco, October 24–28, 1971.

11.46 Thomas, E., and J. Schoefer: *Strategies for the Evaluation of Alternative Transportation Plans*, National Cooperative Highway Research Program, Transportation Research Board, 1971.

11.47 National Association of Regional Councils: *Regional Council Communication, A Guide to Issues and Techniques*, National Association of Regional Councils, Washington, D.C., January 1973.

11.48 Northeast Illinois Planning Commission: *Public Participation in the Regional Planning Process*, Chicago, Ill., January 1973.

11.49 MIT Operations Research Center: *New Technology for Citizen Involvement*, Preliminary Report, July 1971.

11.50 Douglas, A.: *Industrial Peace Making*, Columbia University Press, New York, 1962.

11.51 Berlo, D. D.: *The Process of Communication*, Holt, Rinehart, and Winston, New York, 1960.

11.52 New York City Planning Commission: *Community Planning Handbook*, New York, January 1974.

11.53 Sparer, G.: "Consumer Participation in OEO-Assisted Neighborhood Health Centers," *American Journal of Public Health*, June 1970.

11.54 U.S. Department of Housing and Urban Development: *What Is a PAC Citizen Participation in Urban Renewal*, HUD-176-R, Washington, D.C., 1971.

11.55 Pill, J.: "The Delphi Method: Substance, Context, a Critique and Annotated Bibliography," *Socio-Economic Planning Sciences*, vol. 5, 1969.

11.56 Metropolitan Planning Commission, Atlanta, Georgia: Now—For Tomorrow, 1955.

11.57 Metropolitan Planning Commission, Atlanta, Georgia: "Transportation Policy: CBD Cordon Capacity," 1956.

11.58 Stuart, R. C.: *Transit Technical Study*, New River Valley Planning District Commission, Radford, Va., 1979.

11.59 Greenberger, M., M. A. Grenson, and B. L. Crissey: *Models in the Policy Process*, Russel Sage Foundation, New York, 1976.

11.60 Altschuler, A.: *The Urban Transportation System, Politics and Policy Innovation*, The MIT Press, Cambridge, Mass., 1979.

EXERCISE

11.1 For each of the following situations, assume (imagine) the detailed setting and concisely (1) state a final objective, (2) PERT, CPM, or outline the major technical and participatory activities necessary to achieve the objectives, (3) indicate the techniques you would use for each of the activities. Now (4) consider the above as a hypothesis, that is, "if (2) and (3) are done, (1) will result." Does the hypothesis have any weaknesses? Are they remediable?

How? Obviously, much of what you decide, especially in (4), will depend on the detailed setting that you assume. Specify your assumptions wherever they might not be evident to the reader of your answers.

(*a*) You and the County Attorney have been asked by the County Board to talk to all interested parties in a hot controversy over the transportation of hazardous materials through the county. After discussion, you must draft an appropriate ordinance.

(*b*) You are Director of Transportation Planning for a large city. The state transportation agency has just advised you that it wishes to undertake a corridor study leading to a possible new radial freeway that, while much needed in terms of present and future travel demand, would affect a number of residential areas.

(*c*) As Director of Planning in a small community, you have just completed a feasibility study that shows that, given Federal and state aid, a local transit system would meet the transportation needs of many people at a reasonable cost to local taxpayers. Nevertheless, the idea is very controversial. The Manager and Mayor have indicated that it is your job to "educate the public" prior to a vote by Council.

12 Generation of Alternate Solutions

In using the phrase *generation of alternate solutions*, we are attempting to portray a situation in which components such as vehicles, networks, terminals, and controls are combined to form a systematic, integrated alternative. This might take the form of a project, program, policy, or strategy, depending on the location in the policy chain (see Chap. 2). In any case, a plan usually is created to describe the solution and the means to bring it into existence. It is hoped that when the resultant plan is evaluated it will be worthwhile or, in the most promising case, optimal in some sense. The objective of this chapter is to present some examples of the ways in which transportation plans have been formulated or "synthesized."

The first section contains a display of a sample of new transportation technologies. The second deals with transportation systems management, in which relatively low cost, unobtrusive alternatives are sought as part of near-term "management" of the existing system. In the next section a variety of types of plans are presented. A discussion then follows on ways in which such plans can be formulated. These are illustrated by a description of the development of a pedestrian-mover system for a central area and a metropolitan-wide system. In the last two sections, alternate approaches to the "using transportation to solve transportation problems" syndrome are discussed. In one approach, land-use arrangements are altered to reduce the need for travel (and corresponding transportation systems). In the second, transportation is employed to create beneficial impacts on other types of development, in this case the mixing of families of difference income types throughout a region.

No attempt will be made in this chapter to discuss all of the many ways in which transportation plans can be developed, simply because the variety and breadth of transportation problems would make such a discussion lengthy and arduous. Instead, we will try to display an assortment of both techniques and application situations through a set of representative examples.

12.1 TRANSPORTATION TECHNOLOGIES

An important part of any potential solution to transportation problems involves "technology," which in this context refers primarily to a "hardware" system. Naturally there are many types of transportation technologies, from the more mundane—motor buses, highways, traffic signals and signs, parking lots, subways, trains, and the like—to the more esoteric—air cushion vehicles, linear induction motors, pneumatic tube trains, and the like. In recent years the number of technologies potentially applicable in urban transportation situations has increased markedly, especially as attempts have been made to alleviate such annoying problems as air and noise pollution, lack of comfort and privacy in mass transit, and high operating and travel costs.

It is difficult in a text of this nature to describe all of the technologies that may be available and applicable in various transportation situations. Consequently, we first will outline the main elements of a transportation system so that different technologies can be classified and put in their proper role, and then we will describe some representative and unique technologies. These have been chosen to display a wide spectrum of possibilities and to indicate the types of functional characteristics that should be of some consequence in the assessment of technologies for potential application in a particular situation.

12.1.1 Taxonomy of a Transportation System

Transportation is a complex set of objects and events that are difficult to summarize in a simple fashion, yet the need to plan requires an articulation of the elements that form the entity known as transportation. Many classifications or taxonomies have been proposed, and the one in Table 12.1 is a condensation of those ideas.

Table 12.1 A Transportation System Taxonomy

1. Network
   a. Links
   b. Nodes (switching or loading points)
   c. Physical traveled way (if any)
2. Vehicle
   a. Passenger and cargo space
   b. Suspension system
   c. Propulsion system
   d. Braking system
   e. Communication system
3. Terminals
   a. Loading and unloading area and system
   b. Storage area
   c. Maintenance area and system
4. Control system
   a. Physical
   b. Legal and governmental
   c. Managerial—policy and budget

Perhaps the easiest way to understand the makeup of this taxonomy is to envision a situation in which a consultant is called in by the developers of a potential new town site presently having little or no transportation system. Since they are interested in an almost totally new system, the consultant is asked to describe all of the elements that might go into a complete transportation "package" that could be used in the new town.

Inevitably this "package" would contain, in some form, a network composed of individual links, nodes, and perhaps tangible surface (traveled way) on which trips could be made. Moreover, the consultant might specify a vehicle or set of different vehicles having certain propulsion, suspension, braking, and communications systems and, of course, a compartment of some sort in which passengers could ride. There also would have to be terminals for the concentrated loading, unloading, and storage and maintenance of these vehicles, as well as a control mechanism for overseeing and directing the operation of the whole system. This, briefly, would be the contents of the "package" presented to the new town developers.

As in most classification schemes, there are items that do not fit neatly into one category or the other: what is the "vehicle" in the cases of the moving sidewalk or pipeline? What are the "links" for a craft that hovers on a cushion of air, allowing it to travel almost anywhere? What is the traveled way or the control system for a vehicle running in a tube under the force of gravity? Obviously, these questions cannot be answered simply, but the circumstances to which they pertain are exceptions rather than general situations. The taxonomy displayed in Table 12.1 thus is a useful one under most circumstances.

12.1.2 Examples of Transportation Technologies

As mentioned previously, different technologies will be described here.[1] The discussion will focus on their functional characteristics. Pictures of some of the described technologies are found on pages 407–412. All prices are assumed to be in the year of the associated reference. We start with technologies that exist, and in some cases are in use now, then proceed to those on the horizon. The latter have been selected either because they show promise of being employed extensively or, alternately, because the impacts they might generate could be both unique and consequential. Descriptions of other technologies can be found in the books by Richards [12.23], McGean [12.70], and Anderson [12.71].

Transit bus As part of a large research and development effort to create an advanced bus (the TRANSBUS), several manufacturers came up with protoype designs. The bus that, in 1981, has been produced on an interim basis and has captured a better portion of the new market is the Grumman Flxible 870.

The largest version of the 870 is 40 ft long and $8\frac{1}{2}$ ft wide. It seats 48. Some of the advantages of the 870 over older buses are that it (1) is built mostly of aluminum, for easier maintenance, lighter weight, and fewer parts (particularly fasteners); (2) has a "kneeling" action, which can reduce the first step height from $13\frac{1}{2}$ in. to 8 in., a great convenience for semiambulatory, nonwheelchair riders; (3) has a wider doorway with a 34-in. platform for easy boarding; and (4) has larger, tinted side windows to permit 50% more visibility for passengers. In addition, the bus can be fitted with a front-door electro-hydro lift to accommodate wheelchair riders [12.69].

---

[1] Some technologies have been discussed elsewhere. Monorails, such as the Alweg and Safege, for instance, were described and pictured in Sec. 10.5.2, along with the Westinghouse Skybus and a typical duorail system.

Specialized paratransit vehicles Over the years a need has been shown for specializing vehicles designed to serve the handicapped. Two experimental versions of such vehicles have been built by the Advanced Systems Lab (previously a division of AMF) and Steam Power Industries (Dutcher Industries). The vehicles feature a capacity of four to six passengers as well as convertibility to accommodate one or two wheelchair riders. The floor is low and flat, for improved accessibility via a transportable ramp. Compared to larger minibuses, the vehicles have higher maneuverability in traffic, higher fuel economy, and lower cost of acquisition and maintenance [12.20, p. 37].

An interesting aspect of these two experimental vehicles is that they were built initially with steam engines. This was done to meet very strict air-pollutant emission design standards. Later these engines were replaced with commercially available gasoline versions and comparative tests undertaken [12.20, p. 4].

The Morgantown People Mover (automated guideway transit) This system is an automated self-service transit system operating with a fleet of electrically powered, rubber-tired vehicles on a dedicated guideway at 15-sec separations either scheduled or on demand. The system was designed and constructed by the Jet Propulsion Laboratory and the Boeing Co. It is capable of transporting 1,100 passengers in 20 min between two stations 1.5 miles apart. It can operate 24 hr per day and provide nonstop origin-to-destination service through the use of off-line stations.

A vehicle carries up to 21 passengers, with 13 standing. It is 15.5 ft long and 6 ft wide, weights 8,600 pounds empty, and can have speeds of up to 30 mph. Rubber tires and an air-bag suspension provide a quiet and comfortable ride. Unique features include a heated guideway for operation during icing conditions, on-board steering, and a synchronous point-follower control system to manage all system operations via computers [12.20, p. 46].

Initial costs were quite high. As of 1977, Federal contracts have totaled $134 million [12.20, p. 49] for what is now a 3-mile system. These costs, of course, include a substantial research and development component, but costs for other, subsequent systems probably still will not be low.

Vector bike In 1980 the "bike" shown in the photo set a world record for a one-person-powered vehicle of 56.66 mph. The vehicle is 116 in. long, 25 in. wide, and 32 in. high. It weighs only 51 pounds, including the protective fiberglass and Lexan shell. Steering, braking, and gear-shift functions all are on a single joystick [12.61, p. 65].

It is doubtful that many people would be anxious to lie in the awkward position required in order to pedal the bike. Nor will the current (1981) price tag of about $10,000 encourage many to buy it. Still, more practical models of this type may be developed in the future, possibly making "bike" riding to work and shopping a more common experience. Toward this end, a particular advantage of the Vector bike vis-à-vis current models is the cover for protection from inclement weather.

Propane-fueled automobiles Alternative fuels and assorted engines of several different types have been under experimentation for many years. In Brazil, for instance, alcohol has become a relatively common fuel, used as a means to help replace gasoline, which is relatively scarce and expensive there.

One alternative that may have great potential in the United States is propane-fueled automobiles. Some camping vehicles have run on propane for years, so it should not be too difficult for an auto manufacturer such as Ford to develop a practical propane car. This should mean fuel cost savings, since in 1981 propane was

The Grumman Flxible 870® bus.

Specialized Paratransit Vehicle. (*Courtesy: James R. Dumke, Transportation Systems Center, U.S. Department of Transportation*)

A station on the Morgantown People Mover System. (*Courtesy: James R. Dumke, Transportation Systems Center, U.S. Department of Transportation*)

The Morgantown People Mover System—vehicle and guideway. (*Courtesy: James R. Dumke, Transportation Systems Center, U.S. Department of Transportation*)

Interior of an experimental vehicle for an automated highway. Side-mounted control stick located on seat. (*Department of Electrical Engineering, Ohio State University*)

Light rail transit vehicle. (*Courtesy: James R. Dumke, Transportation Systems Control, U.S. Department of Transportation*)

Montreal (duo-rail) subway vehicle. (*Montreal Urban Community Transit Commission*)

Unimobil/Habegger monorail (minirail) system. (*Universal Mobility, Inc.*)

Multi-modal Capsules. (*General Motors Research Laboratories*)

SR.N4 Hovercraft. (*British Hovercraft Corporation*)

just over half the price of gasoline. In addition, propane burns more cleanly, which means fewer changes in spark plugs, longer oil life, etc. But fuel tanks and pressure valves have to be sturdier. The propane auto should have about the same startup, acceleration, and speed characteristics as the gasoline-powered auto [12.60, p. 18].

Automated highway The automated highway involves a specially designed roadway with control cables buried in the pavement. Usually there would be two unidirectional lanes and a third reversible lane under control, with additional outside lanes for transition. Each vehicle using the automated highway would have to be equipped with a special electromechanical package for control of speed, spacing, and lateral placement. These controls would be activated by a computer located at some central point. Information signals would be transmitted via the buried cables and then induced into the vehicle.

The special "add-on" control package for the vehicle would cost about $140, while the guideway cost is anticipated to be about $3,843,000 per lane mile. Operating costs for the vehicle and system would run about $0.105/passenger·mile [12.23, 12.24].

The automated highway would appear to be much safer than a regular highway, and the passenger, relieved of the driving task, would be able to move at an average trip speed of 58 mph. With maximum vehicular speeds of 70 mph, the automated highway would be capable of accommodating 9,000 vehicles/hr·lane. Access points or terminals would be at minimum intervals of 2 miles [12.23, 12.24].

Light rail transit vehicle Light rail transit (LRT) involves rail vehicles operating on predominantly reserved, but not necessarily grade-separated, rights-of-way. Electrically propelled rail vehicles operate singly or in trains. LRT provides a wide range of passenger capacities and performance levels at moderate cost. In some European cities, light rail is introduced as "pre-metro" for future upgrading to standard rapid transit. Light rail transit may be considered as an outgrowth of street railway technology. The current standard light rail vehicle (SLRV) design provides for both high-level platform and street-level boarding with both fixed and movable steps. In order to assist elderly or handicapped persons who are unable or have difficulty climbing steps, Boeing Vertol is designing and fabricating a lift device to carry passengers from a low street-level loading position into the SLRV and vice versa. The Massachusetts Bay Transportation Authority (MBTA), working with the San Francisco Municipal Railway, the Southeastern Pennsylvania Transportation Authority, and other U.S. transit authorities, developed a standard specification for new light rail vehicles under UMTA funding. As a result, some 275 new SLRVs are now in production for Boston MBTA and San Francisco MUNI [12.20, p. 29].

Duo-rail subway system: Le Metro Le Metro is the subway system in Montreal. It was planned to give efficient service along principal arteries of the city, following closely the main streams of traffic. Use has been made of streets adjacent to the major arteries in order to prevent traffic congestion and interference with local business. Work began on the network in May, 1962, with the system put into service in October, 1966. The guideway cost $250/linear ft in rock, $900/linear ft in earth, and $600 to $1,000/linear ft in open cut areas [12.44].

Average trip speed on the system is 22 mph, with headways between trains in peak periods of 130 sec. The 15.5-mile network is serviced by 26 stations. A nine-car train has a capacity of 1,500 passengers, giving a capacity of 60,000 passengers/hr·lane in each direction. The use of pneumatic tires on the vehicles, and concrete rails, as compared to steel wheels and rails on most subways, results in a smoother ride, a substantial decrease in noise [12.44], and greater acceleration and deceleration.

Cars will be guided, controlled and spaced by an electronic control system embedded in the highway

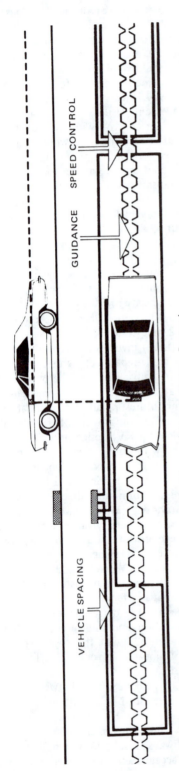

VEHICLE SPACING

GUIDANCE SPEED CONTROL

Electronic guideway configuration for an automated highway. (*General Motors Research Laboratories*)

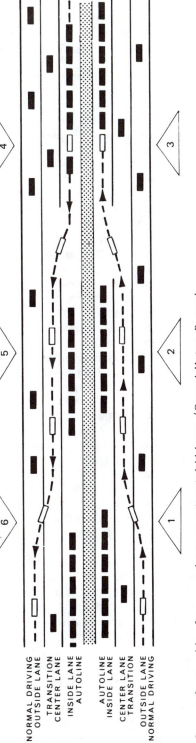

1 A motorist traveling in a normal lane but wanting to enter the Autoline lane would move into the transition lane and signal his desire to enter the Autoline.

2 By putting his car on automatic control, his speed and position would be monitored and adjusted.

3 The car would be automatically guided into position at the end of the first available group on the Autoline lane.

4 To leave the Autoline lane, the motorist would first signal his intention to the system.

5 His car would move automatically into the transition lane at the first safe opportunity.

6 He would return his car to manual control and then move into a normal lane.

NORMAL DRIVING
OUTSIDE LANE

TRANSITION
CENTER LANE

INSIDE LANE
AUTOLINE

AUTOLINE
INSIDE LANE

CENTER LANE
TRANSITION

OUTSIDE LANE
NORMAL DRIVING

Concept for transition from a regular to an automated highway. (*General Motors Research Laboratories*)

Minirail The minirail, developed by Habegger Maschinen Fabrik in Switzerland and Universal Mobility, Inc., of Salt Lake City, Utah, is comprised of a series of small cars joined into trains that run on top of a small rail. It has seen service as a transportation system in the Swiss Exhibition at Lausanne in 1964, at Expo in Montreal in 1967, in Charlotte, North Carolina, and elsewhere.

The minirail has a top speed of 28 mph when stations are spaced at approximately 0.3-mile intervals. Each car accommodates up to 12 seated passengers, with a train usually being made up of 9 cars. Under these conditions, the guideway would have a capacity of 6,000 passengers/hr in each direction [12.41, 12.32].

The minirail is suitable as a distribution system in central business districts, at airports, and other areas of intense activity, accommodating many trips normally made on foot. The system has been designed to run through buildings.

Major advantages of this technology are the relatively low costs of construction and operation (an automatic version would require no on-board operating personnel), unobtrusive appearance, and relatively clean, quiet operation. One disadvantage it shares with other pneumatically tired vehicle systems (e.g., automobiles) is its particular susceptibility to snow and ice.

Multi-modal capsule systems The Multi-Modal Capsule System, developed by General Motors, is comprised of a group of small passenger capsules transported by special highway vehicles similar to flat bed trucks [12.42].

The guideway is estimated to cost about $630,000/mile and create an associated operating cost of about $0.032/passenger·mile. The top speed for the Multi-Modal Capsule is intended to be 65 mph, with an average trip speed (including waiting time) of 22.1 mph. Following the design employed in test applications, the actual guideway probably will have access points spaced at approximately 5-mile intervals. Each capsule would have a capacity of two passengers, permitting unusual privacy for a mass transit system [12.24].

Hovercraft Developed in Britain and manufactured there by organizations such as the British Hovercraft Corporation Ltd., Hovermarine Transport Ltd., and Vosper Thornycroft Ltd., the hovercraft is a vehicle gaining support over land or water by means of an air cushion. Various designs of craft exist: amphibious, semi-amphibious, and sidewall. Only the amphibious craft can travel over a variety of surfaces apart from water and retain no contact with the surface other than the skirt. Hovercraft may operate at a "hover height" of up to 8 ft above water level for the largest craft, SR.N4, for example, although for most craft it is usually 2 or 3 ft.

The "Mountbatten" class SR.N4 hovercraft is the largest in the world and weighs almost 200 tons. It measures 130 ft long, 78 ft wide, and when hovering is 45 ft high. Several craft have been constructed and are engaged in ferry operations between England and France across the Dover Straits. Each craft can carry up to 280 passengers and 37 cars and operates at speeds of up to 70 knots. Normally crossings take 30 to 40 min. Almost 1.5 million passengers and 250,000 vehicles are carried annually by these hovercraft services, which began on August 1, 1968.

This type of transport is particularly useful for traversing water in areas where bridges or other fixed links are not possible, and where shallow water prevents ships from taking shorter, more direct routes between landmasses or across rivers. It is also possible to take advantage of the craft's amphibious qualities to site new terminals outside existing port or city areas, and usually only simple slipways need be used. The hovercraft's faster speed and ability to take shorter routes may mean that service capacity can be better tailored to traffic demand as well as offering more frequent timings [12.48, 12.49].

Quickie 02 Canard This type of plane, known as a "Canard" because of its forward wing, seats two people. There also is a 4-cu-ft baggage space for 40 lb of goods. The plane is about 20 ft long, with a wingspan of about 17 ft. The takeoff distance with two people aboard is around 650 ft. The Canard can be taken apart easily in the middle of the fuselage and rolled onto an 8 X 17-ft trailer.

With the 64-hp engine, the plane can reach a speed of 180 mph. At a cruise speed of 170 mph it can get 44 mpg; at 130 mph, it can obtain 60 mpg. In 1981 a kit could be obtained for about $10,000. This did not include fabrication, which was estimated to take 500 hours of time [12.52, pp. 58–60].

The implications of such a comparatively low-cost, high-speed transport vehicle are significant. It may be that many more people can live at great distances (say, 90–150 miles) from work places and still commute there in less than an hour's time.

12.2 TRANSPORTATION SYSTEMS MANAGEMENT

A special group of transportation alternatives has become known as transportation systems management, or TSM. These alternatives are designed to address the short-term transportation system needs through more efficient use of existing transportation facilities. TSM encompasses a range of improvement strategies that are nonfacility and low-capital oriented and that use both demand management and supply optimization to capitalize on existing highway and transit-related facilities to achieve transportation-related goals [12.62]. The TSM planning process is now the focus for the short-range transportation element in transportation improvement plans (TIP) (see Chap. 14) that have been prepared for U.S. urban areas.

12.2.1 Elements of TSM

TSM projects selected by urbanized areas for inclusion in their transportation improvement programs (TIP) are envisioned to involve a wide range of actions with low-capital investment requirements that can improve transportation service in the short term. There are a number of systems for categorizing TSM action elements. The one in this section is based on material presented in *Transportation System Management* [12.66], which includes a list of TSM actions on the basis of seven major categories (see Table 12.2). Two points should be made clear with regard to this list. First, it is representative of the available TSM actions but it is not necessarily exhaustive; other TSM actions may exist or evolve in certain categories. Second, the development of a TSM plan can result from a "top-down" approach in which one first examines local transportation goals and objectives and works toward achieving these by identifying supportive TSM actions; or one can look at local transportation problems, and associated TSM actions aimed at alleviating them, in developing a local TSM plan—a "bottom-up" approach.

12.2.1.1 Improved vehicular flow The actions taken to improve vehicular flow are generally traffic-engineering measures related to an alteration in the traffic-control strategy and/or the utilization of available road space. These actions, if successful, have the desired effects of decreasing congestion and delay, increasing average travel speeds, and reducing travel times. They can therefore lead to reduced energy consumption, noise pollution, and engine emissions. These actions can help to postpone, or even eliminate the need for capital-intensive measures aimed at increasing street capacities. The benefits may be lost, however, if automobile usage

Table 12.1   Spectrum of TSM Actions

1. Improved vehicular flow:
   Improvements in signalized intersections
   Freeway ramp metering
   One-way streets
   Removal of on-street parking
   Reversible lanes
   Traffic channelization
   Off-street loading
   Transit-stop relocation
2. Preferential treatment of high-occupancy vehicles:
   Freeway bus and carpool lanes and access ramps
   Bus and carpool lanes on city streets and urban arterials
   Bus preemption of traffic signals
   Toll policies
3. Reduced peak-period travel:
   Work rescheduling
   Congestion pricing
   Peak-period truck restrictions
4. Parking management:
   Parking regulations
   Park-and-ride facilities
5. Promotion of nonauto or high-occupancy auto use:
   Ridesharing
   Human-powered travel modes
   Auto-restricted zones
6. Transit and paratransit service improvements:
   Transit marketing
   Security measures
   Transit shelters
   Transit terminals
   Transit-fare policies and fare-collection techniques
   Extension of transit with paratransit services
   Integration of transportation services
7. Transit management efficiency measures:
   Route evaluation
   Vehicle communication and monitoring techniques
   Maintenance policies
   Evaluation of system performance

Source: [12.66].

is encouraged because of a perceived increase in capacity. A balance must be struck between those measures that improve vehicular flow and discourage the use of low-occupancy vehicles, and those that encourage transit usage. Actions such as one-way operation, reversible flow, and left-turn prohibitions can have dramatic impacts upon capacity, accident reductions, and travel-time improvements, as reported in Table 12.3.

12.2.1.2 Preferential treatment of high-occupancy vehicles The preferential treatment of high-occupancy vehicles can be accomplished through the reservation of street space, during certain periods of time, for high-occupancy auto and transit vehicles and/or through special traffic-signal systems that permit buses to preempt normal signal functioning to their advantage. This action promotes more efficient use of the existing road space by increasing the people-carrying capacity per lane. The designation of bus-only lanes has become relatively commonplace, particularly on major urban arterials and urban streets during peak travel periods. Perhaps one of the most publicized examples of a bus-only lane is the Shirley Highway in suburban Washington, D.C. In this application a special roadway was constructed in the median of an expressway and is now used by buses and high-occupancy commuter vehicles.

In nearly every application a major component is the physical separation of automobile and transit vehicles. Results have shown reductions in travel time and time variations incurred by transit vehicles. It has been realized that high-occupancy commuter vehicles can also benefit from the improved traffic flow of a low-volume traffic lane, and that use of the bus lane further encourages car-pooling and van-pooling without a major deterioration of the benefits that buses gain with the specially designed facility. Transit-service reliability has improved for Shirley Highway buses from 33% of the buses arriving early or on time to 92% as a result of the busway installation. Use of the Shirley Highway busway for carpools of four or more passengers has increased from approximately 700 high-occupancy carpools in 1973 to 4,000 as of March, 1979, a nearly sixfold increase in 6 years.

12.2.1.3 Reduced peak-period travel This group of TSM actions has the effect of improving peak-period traffic flow by reducing the vehicle·miles traveled (vmt) during peak periods. Work rescheduling and/or the implementation of flexible work schedules, where possible, have the effect of spreading travel demand and smoothing peak-period flows. They may also discourage car-pooling because of the variations in schedules, but that problem must be looked at separately. Figure 12.1 illustrates the smoothing effect upon employee arrivals of three flexible work schedule programs in San Francisco.

The imposition of peak-period truck restrictions is one of a number of actions that can reduce traffic demand during the peak period. However, it can also be quite detrimental to urban goods movement. Such restrictions on truck movements must be imposed very carefully because of pick-up and delivery scheduling limitations, especially where peak-period deliveries are common, and where off-street loading space is not readily available.

Congestion pricing in the form of higher road and bridge tolls and transit fares during peak hours may have the effect of spreading demand within the peak period, or it may simply penalize those who have no choice but to travel during this period. Parking surcharges may discourage auto users during peak periods and lead to more pooling arrangements. The concept of congestion road pricing is theoretically

Table 12.3 Reported Ranges of Change—Large-Capacity Increase Alternatives

Alternative	Range of capacity increase (%)	Range of accident reduction (%)	Range of travel-time reduction (%)	Average travel-time reduction (%)
One-way operation	20–50	0–60	20–40	30
Reversible flow	20–50	0–30	20–50	30
Left-turn prohibition	14–40	40–70*	35–100	50

*Assumes no separate turn lane.
Source: [12.53].

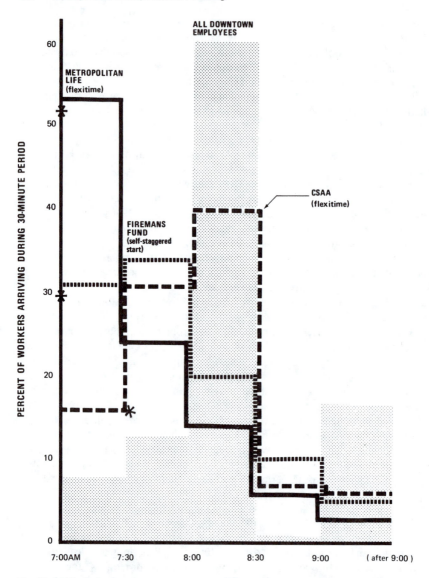

Fig. 12.1 Workplace arrival time distributions, San Francisco. *Indicates earliest sanctioned start time [12.8, p. 77].

attractive in that it requires those who cause congestion to pay for its consequences. Yet the application of congestion road pricing with a truly continual readjustment of user charges awaits the solution of a number of theoretical and practical problems. It is unlikely that a pure version of such a system will be imposed anywhere in the free world in the foreseeable future.

12.2.1.4   Parking management   No other operational control measures can have as dramatic an effect on traffic flow as an extensive program of management of an area's parking supply. Auto drivers have proven to be very resilient to many flow-control measures and persistent in the face of measures aimed at discouraging auto use. The ultimate test of this persistence, however, is met when they have reached their destination and cannot dispose of their vehicles without a large penalty.

Parking management involves the establishment of a plan and an operational strategy involving (1) the location of parking, (2) the amount of off- or on-street parking to be made available, (3) the parking prices, and (4) time restrictions. Three factors act to limit the effectiveness of a parking control effort: private parking supply, through traffic, and parking/traffic enforcement. Moreover, for a parking-management program to be effective, the majority of spaces in an urban area must be subject to controls. Yet it may be difficult to establish the legal basis for controlling spaces located on private property.

If parking management and controls have the effect of reducing congestion, additional traffic might be encouraged to take advantage of the improved flow conditions. There may be a need to include some disincentives to through traffic to avoid negating the benefits of the parking control programs. Restrictions on parking may, however, have the effect of increasing vehicular movement; systems are being developed that indicate to drivers the routes to parking areas that have spaces available. In Aachen, Germany, 40 variable-message signs direct motorists to available spaces in up to 12 parking areas [12.28]. To be effective, a parking-management plan must also include accommodation of a thorough enforcement effort, possibly supported by parking revenues, and provision of parking spaces for people who live in the affected areas through special permits or reduced parking fees.

A parking-management program need not be restricted to limiting and/or controlling supplies in congested areas. It may be useful to include park-and-ride facilities at strategic locations on major arterials that may be integrated with bus transit routes. Characteristics and usage of fringe parking facilities in eight locations are exhibited in Table 12.4; note that not all lots are utilized to capacity.

12.2.1.5 Promotion of high-occupancy and nonvehicular travel modes These actions represent a mixture of possibilities with a common objective—that of minimizing the total vehicular miles of travel, or vmt. These actions include ride-sharing, encouragement of walking and bicycling, and the promotion of auto-restricted zones.

Ride-sharing represents the wide range of car-pooling and van-pooling plans, as well as the programs and services that can assist those planning and promoting the services throughout the community.

The formal and, to some extent, the informal ride-sharing programs that have evolved have been well documented, and the potential benefits to individuals and to the community have been found to be significant. The savings to the individual increase as the travel distance and the number of participants increase, and manifest themselves primarily in cost decreases, reduced vehicle wear and tear, and reduced strain due to driving. Community benefits include drops in vmt, energy demand, and pollution, as well as, possibly, a reduction in the total number of vehicles needed within the community. Additionally, the supply of parking spaces provided by employers might be decreased.

The encouragement of walking and bicycling also can result in lowered vmt, congestion, pollution, and energy demand. Certain short-distance trips, if replaced by walking or bicycling, actually yield a very high energy and pollution savings per unit of travel because of the inefficiencies of motor vehicles for this type of travel. Bicycling can be encouraged through the implementation of bike lanes and paths,

Table 12.4 Characteristics of Bus Fringe Parking Facilities and Service

	Milwaukee	Seattle	Vancouver	Miami	Shirley Highway
Lot type	Shopping center	Park/ride only	Exhibition park	Park/ride only	Park/ride only
Bus service	Freeway express	Freeway express	Express	Arterial express	Freeway express
Bus headway (min)	12	15	5–10	10	15
Mid-day/evening service	Local bus	Local bus	NA‡	NA	NA
Distance to CBD (miles)	10	9	5	10	16
Priority facilities	None	Exclusive ramp	None	Exclusive lane	Exclusive lane
Highway congestion	Moderate	Moderate	Moderate	Light	Severe
Access to/from highway*	Good	Good	Good	Good	Poor
Amenities	Lighting/shelter	Lighting/shelter	Lighting/shelter	Lighting/shelter	Lighting/shelter
Tolls, CBD Parking cost	$1.25	$1.00	$.82	NA	$1.45
Park/ride daily cost	$1.00	$.70	$.50	$1.20	$1.45
Usage capacity	150/300	475/475	600/-	400/950	250/400

	Hartford	Washington, D.C.	Atlanta	Santa Monica	San Francisco
Lot type	Shopping center	Shopping center	Shopping center	Park/ride only	Park/ride only
Bus service	Freeway express	Arterial express	NA	Freeway express	Freeway express
Bus headway (min)	10	18	15	15	10
Mid-day/evening service	NA	NA	Mid-day only	None	NA
Distance to CBD (miles)	7	10	5	13	9
Priority facilities	None	None	None	Exclusive lane	Exclusive bridge lane
Highway congestion	Moderate	NA	NA	Severe	Severe
Access to/from highway*	Good	Good	NA	Poor	Good
Amenities	Lighting/shelter	Lighting	Lighting/shelter	Attendant/shelter	Lighting/shelter
Tolls, CBD parking cost	NA	NA	$.50–$.60	$1.50	NA†
Park/ride daily cost	$.90	$1.60	$.50	$1.00	NA†
Usage/capacity	200/250	80/150	40/200	30/300	60/165

*"Good" access is within approximately $\frac{1}{2}$ mile of the major highway.
†The cost difference is believed to be significant due to high San Francisco parking costs and toll on the Bay Bridge.
‡NA = information not available.
Source: [12.8, p. 97].

bicycle parking facilities, and transit/bicycle "piggyback" arrangements. Locations with consistently favorable weather conditions have a decided advantage, but nearly any city in the United States can point to substantial cost (and other savings) that can be accrued through the promotion and acceptance of walking and bicycling.

Restricting vehicles from selected areas is not a new idea but is being considered more frequently in highly congested cities in the United States. Auto-restricted zones have been widely implemented in Europe, particularly in areas where the physical dimensions of streets and the continued demand of pedestrian and vehicle traffic required some resolution that discouraged vehicular traffic. In addition, in historic central areas, the reduction of pollutants has retarded the deterioration of building facades. It is interesting to note that restrictions to auto movement within residential areas have been successfully applied in a number of locations in the United States.

12.2.1.6 Transit and paratransit service improvements Existing public transit services may be improved in a number of ways, including increases in the frequency of service, improvements in service regularity, reliability, and comfort, and cost reductions. Many of these actions have been identified as being related to transit marketing, which refers to not only selling the transit system to prospective riders but also continually monitoring and evaluating it in terms of its responsiveness to the patrons' service needs [12.66]. It is essential that any attempt at increasing usage of available transit services be based on a realistic assessment of existing and potential market segments and the provision of services to these segments. Moreover, the installation of bus-only lanes, park-and-ride facilities, and signal preemption, as well as other measures designed to reduce congestion and travel time, must be integrated within the overall transit operating strategy. The continuing process of identifying the transit riders' needs must be carefully coordinated within the overall TSM planning process. Information regarding transit improvement response has often been gathered in after-the-fact analyses; for example, it has been reported that observed responses to bus-service frequency improvements appear to be greatest when the route involved serves middle- and upper-income areas, when the prior frequency was less than 3 buses or so per hour, and when the travel market involved is predominantly comprised of short trips.

Paratransit operations, which closely approximate the personalized service of the private automobile, are an essential feature of complete transit service. Paratransit is characterized by small minibus or van-type vehicles, flexible routing and/or scheduling, and service to low-demand, low-density areas. The well-known dial-a-ride or subscription services are the most prominent examples, although shared-ride taxis offer another potential alternative. Paratransit offers an economically feasible alternative for feeder service to line-haul transit, for coverage in low-demand areas beyond current fixed routes, for off-peak, low-demand service, and for meeting the needs of low-mobility groups.

12.2.1.7 Transit management efficiency measures Transit operations represent a major area for potential system improvement through a range of actions involving transit route planning and evaluation, vehicle communication and monitoring techniques, maintenance policies, and system performance evaluation. Adjustments to transit routes involving route location, transit-stop location, scheduling, and route layout may lead to significant service improvements and increased transit usage. The middle range of estimated ridership growth in response to overall expansions of bus transit is a 0.6 to 0.9 percent increase per 1 percent increase in regional bus miles of service. Advances in computer packages that provide the transit planner with assistance in route planning and scheduling have led to more efficient and more effective planning by allowing the computer to deal with the complexity of the factors involved, such as the multiplicity of choices, limitations of time, driver selection preferences, and labor restrictions [12.26].

The results of a number of recent studies indicate a significant benefit to transit management if there exists real-time information exchange between buses and the operations base. Primarily this capability allows supervisory personnel to respond with minimal delay to emergency situations and breakdowns. In addition, modern electronic equipment can be used to transmit data and information to a central computer. This data—on vehicle location, schedule adherence, passenger loading, and mechanical equipment—all can be used to provide more reliable and secure service to the transit user.

Maintenance program planning is a necessary adjunct to the successful operation of a large fleet of mechanical equipment. The main element of a coordinated and comprehensive program is improved scheduling. Since vehicle performance is affected by a range of factors including route length, road and weather conditions, stop frequency, loading, traffic density, driving styles, and the age and type of vehicle, records of vehicle operations and maintenance as well as costs form the basis for maintenance planning schedules. There is a growing body of computerized management tools that help in the automation and systematization of the tasks of collecting vehicle service information and monitoring inventory and repair costs. As an additional benefit, because of the frequency of reporting and the orderly review procedures inherent in the system, oversights are less likely to occur.

Transit-system performance analysis has not received much attention in the past for a number of reasons, including lack of sufficient data to allow for comparisons and assessments across the range of operating systems that make up the industry. Newly mandated Federal requirements will make available a range of data that will enable annual reviews of system efficiency and effectiveness. There are a number of current efforts underway aimed at developing suitable performance methodologies for transit systems of various sizes that will provide several benefits for transit-system management through internal and external review and analysis of system operations.

TSM has come to the forefront of transportation planning and engineering activity and has proven to be a very opportune approach to solving local transportation problems, particularly relative to peak-period congestion. The next step in the evolutionary development of transportation planning and operation will no doubt involve the integration of short-term and long-term planning at the local and regional levels, so that the impacts of TSM projects and regional development activities can be coordinated more rationally. For more information on TSM, particularly on traveller response, see [12.8].

12.3 TYPES OF PLANS

There are many different types of plans. One way to classify these is according to the level in the policy chain—strategy, policy, program (or system), or project. Another way is by topical area. Examples would be elderly and handicapped, fuel conservation, and air-pollution control. Still another type of plan focuses on unlikely but possibly highly damaging occurrences. Such an approach, useful in helping to ameliorate the bad effects of, say, natural disasters or energy shortfalls, comes under the heading of a contingency planning. It will not be possible to describe and give examples here of each of these types of plans. Some of the more interesting and useful types have been selected for discussion, however.

12.3.1 Policy Plans

On the top level of the policy chain is the policy itself. While "policy plans" currently are not well known in the United States, they are becoming more common. Policy plans contain fairly general proposals, but they usually come after an analysis and review of experiences and thus do show some certainty of intent to implement without further investigation and discussion.

The so-called structure plans developed for many of the county councils in Britain are good examples of policy plans. Table 12.5 presents 11 transport policies from the structure plan for Oxfordshire (which includes the famous city of Oxford). These policies had been worked out through a series of administrative and legal

Table 12.5 Transport Policies from the Oxfordshire Structure Plan

Policy	Description
T.1	To give priority for highway expenditure to new roads and road improvements to serve towns which are proposed for substantial expansion.
T.2	To make provision for the by-passing of all towns and villages on designated major through-traffic routes. At the present time, other than trunk roads, the only designated major through route is Swindon to Oxford (A420).
T.3	To secure more car sharing and the giving of lifts.
T.4	To give priority to the provision of an intertown network of public transport services, preferably unsubsidized, with subsidized feeder and other services.
T.5	To allocate subsidy for public transport services so as to secure the largest possible network consistent with the provision of convenient, reliable and efficient services for the maximum number of people.
T.6	To avoid and prevent any action that would prejudice the retention of all existing rail lines that would be capable of reopening to passenger and freight use at minimum cost to the County Council.
T.7	To minimize the danger of transport accidents, particularly by the segregation of cyclists and pedestrians from other road users.
T.8	Within Oxford, to allow maximum freedom of movement commensurate with: (1) An acceptable quality of environment. (2) Convenient conditions for those without cars. (3) The viability of the commercial center of the city.
T.9	To encourage those traveling in Oxford to use the bus by making public transport an attractive alternative means of travel by: (1) Continuing and developing bus lanes. (2) Enabling buses to put down and pick up passengers as near to the shopping center and railway station as possible, ultimately under cover. (3) Allowing buses to travel on certain roads in the center not available to private cars. (4) Exploring the possibility of developing means of public transport, to serve the center of the City, of a type that would not cause pollution to the atmosphere.
T.10	To monitor indicators of transport needs and effects continuously, to ensure that action achieves the stated objectives to maximum effect.
T.11	To develop a system or lorry (truck) routes.

Source: [12.39, pp. 58-59].

actions. They also had been developed in conjunction with policies for other functional areas dealing with employment and workplaces, population and housing, the built environment, countryside and villages, shopping, and recreation [12.39]. One of the main purposes of a policy plan thus is to lay out the policies in a broad variety of areas. Checks then can be made for complementarity and consistency so that, for example, land-use patterns do not evolve that create unexpected demands on the transport system. The policy plan also provides a vehicle for discussion of proposed actions before expensive detailed plans and designs are undertaken.

12.3.2 Program or Systems Plans

One version of a program plan is the systems plan. This has been the most common output of planning efforts in urban transportation studies done throughout the world. It is particularly useful when large, capital-intensive projects like expressways and rail systems are anticipated. These can be identified and tested using travel and impact models like those described in Chaps. 7 and 8. The important point is that the influence on the whole network of reductions in travel times and costs on one link can be estimated. The plan then truly can be said to concern the whole system.

The transit section of the long-range transportation systems plan for the Washington, D.C., Metropolitan Area is displayed in Fig. 12.2. Emphasis is on the Metrorail network emanating from the center of Washington, D.C., itself. Part of this had been constructed by 1980, and active efforts were underway to find sufficient funding to finish the approved 101 mile network. The latest estimates for the overall capital costs were over $7 billion.

Also shown on the map are proposed transit corridors (primarily for scheduled bus service), commuter rail lines, and a short automated guideway transit line. Not shown are any bus priority lanes or suggested changes in bus routes and schedules. These generally are considered too detailed at this level of the plan.

A chart depicting a sample of proposed improvements to controlled principal arterials in the Washington Metropolitan Area can be found in Table 12.6. Options included constructing, widening, upgrading, and reconstructing these highway segments. Past practice had been to stage or program the improvements over time—usually four 5-year periods—according to likely funding levels available. Recently, however, such funding has become uncertain (see Chap. 14), so that long-term, sequenced programs of this type are less common.

12.3.3 Project Plans

Plans at the project level usually are referred to as "designs" and focus on individual items such as those listed in Table 12.6. Still, there often is an intermediate level between the system and the project. This is called the "corridor" level, with associated corridor plans. As an example, these might specify some combination of a rail line along the length of the corridor, with bus and "kiss-and-ride" access at the origin end and bus and walk distribution at the destination. The transit corridors indicated in Fig. 12.2 would be candidates for detailed corridor plans.

12.3.4 Traffic Safety Plans

In recent years planning requirements have been developed for a variety of special purposes. These include, for example, program plans for traffic safety,

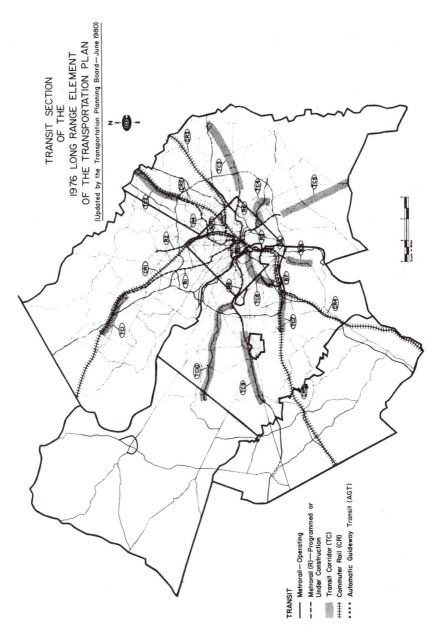

Fig. 12.2 Transit section of the 1976 long-range element of the transportation plan (Washington, D.C. metropolitan area). (Bus priority lanes not depicted.) [12.34, after Table 2].

Table 12.6 Proposed Improvements to Controlled Principal Arterials, Washington Metropolitan Transportation Plan

Reference number	Improvement	Name of facility	Location/segment		1976 Plan reference
			From	To	
			Maryland		
CP1	Widen	Md. 4 (Pennsylvania Ave.)	I-95 (Capital Beltway)	D.C. line	7 CM
CP2	Upgrade	Md. 5 (Branch Ave.)	Md. 223 (Woodyard Road)	D.C. line	2 CM
CP3	Upgrade	U.S. 29 (Columbia Pike)	Howard County line	Md. 650 (New Hampshire Ave.)	9 CM
CP4	Construct	U.S. 29 Spur	U.S. 29 (Colesville Road)	Md. 193 (University Blvd.)	31 CM
CP5	Construct/upgrade	Md. 115 (Eastern arterial)	Montgomery Village Avenue	Md. 609 (Norbeck Road)	28 CM
CP6	Construct	Great Seneca Highway	Md. 118 (Darnestown–Germantown Road)	Md. 28 (Darnestown Road)	29 CM
			Virginia		
CP7	Widen	Va. 7 (Leesburg Pike)	Loudon County line	I-495 (Capital Beltway)	32 CM
CP8	Widen	Va. 28 (Sully Road)	Va. 7 (Leesburg Pike)	Prince William County line	6 CM
CP9	Construct	Va. 28 Bypass	Rte. 28, South of Manassas	Rte. 28, South of Centreville	24 CM
CP10	Upgrade	U.S. 50 (Arlington Blvd.)	Courthouse Road	I-495 (Capital Beltway)	5 CM

Source: [12.34, Table 3, p. 27].

428

mobility for the elderly and handicapped, control of air pollution, and fuel conservation.

An illustration of the first category was the National Highway and Traffic Safety Agency (NTSA) Alcohol Safety Program. This resulted in 35 Alcohol Safety Action Projects (ASAPs) in communities throughout the Nation.

As explained briefly in Chap. 10, ASAPs utilized the traditional community agencies that always have dealt with the alcohol safety problem—the courts, police, schools, etc. But there were two novel features [12.38, p. 2]:

> First, a special management office was to be established that would ensure an integrated, systems approach to the drunk driving hazard. This would ensure that both the police departments and the courts were provided with the resources to handle a greatly increased number of Driving While Intoxicated (DWI) offenders, and that a public information effort would be mounted to support this activity. In this way it was hoped to deter social drinkers from drunk driving. A second novel portion of the ASAP effort was directed at developing systems for identifying problem drinkers among the DWI offenders and referring them to appropriate treatment agencies. Through this effort it was hoped that the problem drinkers, who could not be deterrred from drunk driving, might be treated for their alcohol problem and ultimately removed from the drinking driver population.

Specific project plans then had to be developed to deal with four main components:

1. Enforcement—processes for streamlining of arrest procedures and using breath testing devices
2. Courts—procedures for much more rapid and efficient processing of DWI cases while preserving traditional judicial safeguards for the defendants
3. Treatment—education or direct treatment procedures, as compared to fining only
4. Public education—mass media programs to enlist public support for ASAP operations

These plans focused on the interrelated development of the four components and their acceptance and coordination among the community agencies.

12.3.5 Plans for the Elderly and Handicapped

Another type of specialized plan that has been developed in many localities focuses on the mobility of the elderly and handicapped. The breadth of such plans usually is somewhat wider than has been the case traditionally, since there is a concern on the one hand for walking and on the other for specialized transportation, particularly that provided by human-service agencies. These concerns are in addition to transit and private-auto usage.

An example of some of the elements proposed for a plan for the elderly and handicapped in Lancaster County, Pennsylvania (including the City of Lancaster), is displayed in Table 12.7. The pedestrian-system improvement section emphasized the creation of curb cuts, changes in sidewalks standards (e.g., for width and curb cuts in new subdivisions), and more rapid snow and ice removal. Automobile accessibility was to be aided primarily through development of more on-street handicapped parking spaces, of standard dimensions.

The transit element is interesting in that a recommendation is made against retrofitting half the existing bus fleet with lifts. This recommendation comes despite the feeling that retrofitting would help about 2,700 people in the County. The main reasons given for not taking this approach were [12.46]:

Table 12.7 Sample Elements from the Proposed Lancaster County Transportation Plan for the Elderly and Handicapped

A. Pedestrian System
  1. Improvements should be made in local ordinances with respect to:
    a. curb cuts
    b. sidewalk standards
    c. snow and ice removal requirement
  2. Curb cuts should be included as part of Transportation Systems Management program.
  3. Accessibility information should be more widely distributed to the handicapped and to those who can have an effect upon the quality of the street environment.

B. Automobiles
  1. The availability of on-street parking for the handicapped should be considered in the context of a City parking study to be completed.
  2. Reserved handicapped parking spaces should meet established dimension standards.
  3. Eligibility for use of reserved handicapped parking spaces should be designated through use of a parking permit card.

C. Transit
  1. About 50% of the chronic transportation-handicapped persons in the County who can use transit can benefit from the following transit improvements:
    a. pulling bus to curb to stop
    b. more courteous bus drivers
    c. shelters with seats at important bus stops
    d. not starting movement of vehicle until all elderly and handicapped are seated
    f. announcement of bus routes
    g. review of schedule changes
  2. Retrofitting half of the existing fleet is not recommended.

D. Specialized Transportation
  1. The RRTA* specialized transportation program for the semiambulatory and nonambulatory handicapped in the urbanized area will be limited virtually to medical, work, and food shopping trips. It must be considered adequate until operational experience indicates otherwise. Without operational experience, it is impossible to accurately forecast program costs.
  2. Specialized transportation should be purchased through LISTS† to realize the benefits of coordination and reduced costs.
  3. Private nonprofit agencies should be encouraged to use Federal funds to purchase wheelchair-accessible vehicles and coordinate their services through LISTS.
  4. LISTS should not purchase any wheelchair-accessible vehicles unless anticipated purchases by private carriers do not materialize and the demand for more of this vehicle type is adequately demonstrated.

*RRTA is the Red Rose Transit Authority.
†LISTS is the Lancaster Integrated Specialized Transportation System.
Source: Adapted from [12.46, pp. 46–48].

1. Too many barriers in the street environment make it impossible to make maximum use of the accessible bus.
2. Route deviation service or increased time delays of operating a lift could have a significant negative impact on revenue producing ridership.
3. Specialized transportation provides a better and necessary quality service.

The last element in the plan deals with specialized transportation. An ongoing problem that is addressed is the great number of uncoordinated transportation services being provided by human-service agencies [12.35]. Emphasis is given to integrating RRTA and LISTS services along with those offered by private nonprofit agencies. This highlights the often difficult task of planning for coordinated efforts when some of these are being made by nongovernmental groups or agencies.

12.3.6 Air-Pollution Control Plans

Another type of special purpose plan was that for air-pollution control. These were otherwise known as transportation-control plans. Many of them, developed in adherence to the National Clean Air Act and resultant regulations, were admittedly excessive. They would have required, for instance, that over 80 percent of the auto travel in Los Angeles be eliminated and that all of the gasoline pumps in the Washington, D.C., area be covered (at a cost of about $10,000 each) to reduce gasoline vapor emissions. Some of these plans showed clearly why implementation can be extremely difficult.

One of the more reasonable transportation control plans was that developed by the Regional Planning Council of Baltimore [12.30]. As seen in Table 12.8, the plan was divided into two major groups. First were those suggested for implementation by 1982 (the plan was published in 1978). Second were those for which more study was proposed. Aside from construction of a major rail system and some standard TSM actions, the plan had some unique alternatives. One of these was a motor-vehicle inspection and maintenance program in which vehicles weighing 8,500 pounds or less were to be inspected periodically to ensure they were properly tuned, had operational antipollution devices, and were producing acceptable levels

Table 12.8 Proposed Transportation Control Plan for the Baltimore Region

Measures for implementation	Measures for study
The following measures are to be implemented by 1982:	The following measures are to be studied before the end of 1980:
Incentives for high-occupancy vehicles • park-and-ride lots • improved rail transit • car-pooling • van-pooling • preferential parking for ride-sharing vehicles • bus-service improvements	Incentives for high-occupancy vehicles • exclusive lanes for buses and ride-sharing vehicles • reduced transit fares
Measures for cleaner vehicles • inspection and maintenance of motor vehicles	Measures for cleaner vehicles • cleaner fleet vehicles • retrofit of emission-control devices • cold-start emission reduction programs
Measures to reduce congestion-related emissions • traffic flow improvements • on-street parking restrictions • reduction of extended restrictions • reduction of extended idling • encouragement of bicycling	Measures to reduce congestion-related emissions • staggered or flexible work hours • encouragement of moped use
Measures to reduce automobile use • residential permit parking	Measures to reduce automobile use • increased parking fees • increased fuel taxes • auto-free zones • road tolls
Other measures • land-use management • public-education campaign	Other measures • episodic vehicle controls

Source: [12.30, p. 1].

of emissions. Another was a residential permit parking system which would discourage commuting by automobile. This would be done by allowing only vehicles with valid parking stickers (neighborhood residents and their guests) to park during the day in those neighborhoods close to commercial, employment, and similar sites.

The latter approach obviously would be a controversial one. But the plan notes that [12.30, p. 1]:

> Even after all measures now specified for implementation are fully applied, the region still will not achieve EPA's (Environmental Protection Agency) air quality standards by 1987.

This highlights the tradeoff in plan development between suggesting aggressive plan actions and finding ways to get them accepted.

12.3.7 Energy Contingency Plans

Another type of special-purpose plan created in several localities across the country focused on energy conservation in times of greatly reduced supplies. Interestingly, these plans were made in the hopes that they never would need to be implemented. This highlights the contingency concept, where the proposed alternative only would be employed in an extreme emergency.

The contingency plan developed for the Southeastern Virginia region (Norfolk area) had three stages, with the second and third representing successively greater restrictions in response to increasingly lower fuel availability. The first stage, as seen in Table 12.9, would involve some relatively simple TSM measures along with public appeals for reductions in nonwork travel. In the second stage, much more preference would be given to mass transit and private ride-pooling. The excess transit demand would be met by using the reserve fleet owned by the transit operators. In the third stage, the emphasis would be on discouraging and actually

Table 12.9 Proposed Elements in Southeastern Virginia Transportation Energy Contingency Plan

STAGE I
Expand formal ride-sharing activities
Establish ride-sharing staging areas and additional fringe parking facilities
Encourage combining home-based trips for non-work travel

STAGE II
Provide for increased use of mass transit by putting reserve fleet in state of readiness
Increase supply and promote use of private/community bus service
Give preference to car/van-pool vehicles at staging areas, fringe and employee parking lots
Expand computer ride-sharing services

STAGE III
Odd–even gas-station use
Invoke commercial parking rates on non-van/car-pool drivers
Improve public sector energy efficiency (develop public awareness of reduction)
Federal gasoline rationing
Establish information services to inform the public
Utilize state gasoline "set-aside" to maintain emergency services

Source: [12.33, pp. 49–51].

restricting auto usage, primarily through rationing schemes (particularly in conjunction with proposed Federal efforts). At the same time, state gasoline supplies would be called upon for emergency conditions.

An interesting aspect of this type of plan, and one that probably will become more common, is the adjustment or adaptation to external forces as they change. Different alternatives are proposed depending on the severity of the problem as it evolves. Since forecasting of most problems (particularly fuel shortages) is difficult, more plans should start to focus on responses that would be undertaken only under certain circumstances. A heavy emphasis then would be given to monitoring to determine the level of the problem and the time required to respond to is [12.47].[2]

12.4 CASE STUDY—FORMULATION OF A PEDESTRIAN-AID TRANSIT SYSTEM

A good example of a fairly creative attempt to plan a new transportation system was that by the Working Party on the Introduction of a New Mode of Transport in Central London. In their publication *An Aid to Pedestrian Movement* [12.13], they go into detail on the stages in the synthesis process, the factors and constraints considered in each stage, and the guidelines developed as a result. In the interest of displaying these components of the synthesis process, we will summarize the study here.

12.4.1 The Working Party Study

The investigation was divided into twelve major sections:

1. The "transport gap"
2. Route principles
3. Effect on developments
4. Possible routes
5. Potential demand
6. System concepts and journey times
7. Flexibility and reliability
8. Capacity
9. Stations and depots
10. Comfort
11. Environmental impact
12. Costs and revenues

This classification itself might prove useful in creating a new system since the planner must identify at the start the kinds of elements for which he must have a concern (see Chap. 4 for goals, objectives, and constraints and this chapter for transportation system components). Also, as part of the planning endeavor, it usually is necessary to state some generally accepted objectives. In this study they seemed to be:

1. A maximum walking distance of 3 min (690 ft) should be observed.
2. The system should serve as many high activity zones as possible.
3. There should be minimum initial provision of track.
4. Room for extension should be allowed.
5. Connections should be made to present transport stations (suburban rail, subway, car parks).
6. Visitors to London should be considered.

[2] Similar reasoning also holds in responding to potential beneficial, rather than problematic, situations.

The final and perhaps most important objective was to fill the so-called "transport gap." This relates to travel ranging in length from about 0.25 to 1.5 miles.[3] Shorter trips can be done by foot, longer ones by automobile (or taxi, bus, subway, and so on). In between lies a gap where it is inconvenient to go by any of these modes.

Based on these objectives (not stated specifically as such in the report), eight "route principles" were developed. These are quoted in full here since they provide useful guidelines for the synthesis endeavor.

1. In any attempt to install a new transport system in the existing fabric of central London serious limitations on space would apply at street level because of the dense vehicular and pedestrian traffic which is already carried and the network of utility services which exists just below ground level. To install a segregated system at ground level would mean that conventional traffic would have to be re-routed at intersections and frequent pedestrian ways constructed from one side of the street to the other, either by bridge or subway. On the other hand an elevated system would need to provide the minimum statutory headroom of 5.03 m (16 ft 6 in.) over the highway so as to give clearance to buses and other large vehicles, and the supporting columns would need to occupy positions in which they did not obstruct either vehicular or pedestrian traffic significantly. Provision would also need to be made for stations.

2. A further consideration which would have to be borne in mind is the existence of conservation areas in the central area. Although these areas are not necessarily "hard" in planning terms, their penetration by any new transport system would have to be considered in relation to the character of the area concerned and the physical characteristics of the system under consideration.

3. Although the likelihood of installing a new system within the existing fabric need not be completely ruled out, particularly now that much lighter transport systems can be envisaged, in the light of the foregoing any new system should be regarded basically as one which could be integrated successfully into the framework of major redevelopments, that is, having regard to the location of such areas, to the necessity to avoid routing through the existing fabric, and to the anticipated time sequence of the redevelopment of central London.

4. Within these areas of redevelopment it might be possible for the buildings and shopping malls to be designed around a system so that it played an important part in their functioning. Stations could be situated either within or behind the new buildings but the route could emerge into the main pedestrian areas between stations so that passengers would be able to see and enjoy the surrounding environment, and be brought into contact with the shops, restaurants, theatres, hotels, and other amenities situated along the route. It is clear in these circumstances that the possibility of suspending a system below the pedestrian deck could be disregarded, and that in order to provide this flexibility of routing, and also so that it would be able to skirt existing "hard" development where necessary, a comparatively tight turning circle might be required.

5. With a new transport system designed basically to integrate satisfactorily within redevelopments, the question would also arise as to whether it should be at pedestrian level or placed overhead. The prime advantage of a system at pedestrian level would be its ease of accessibility for intending passengers. On the other hand, it would have the disadvantage of dividing the pedestrian areas through which it passed. An overhead system might overcome this difficulty and avoid any vehicle/pedestrian conflict while at the same time taking up less space at pedestrian level. It might also be possible to move the structures so that alternations could be made to short lengths of the route because of such things as redevelopments with the minimum of interference to pedestrian movement. On balance, therefore, the indications are that an overhead system might be preferable to one at pedestrian level.

6. Use of an overhead system would necessitate careful investigation being made into the siting of stations. Should a system be brought down to pedestrian deck level at stations or should intending passengers be taken up? The provision of stations at pedestrian level could give rise to considerable technical and operational problems, particularly if standing passengers were carried. For example, additional stress would be placed on

---

[3] Generally 50 percent of all urban travel is less than 3.0 miles in length [12.14, p. 31].

braking systems and power units in having to overcome gradients when entering and leaving stations. Journey times could be longer, and the continual rising and falling at stations could provide an unpleasant journey for passengers. Furthermore, where stations are situated within redevelopments, the cost of the extra space required to accommodate the track gradients into and out of stations could prove prohibitive. Elevated stations might overcome these difficulties and although it could be argued that ideally stations should be sited at pedestrian level, in all probability the great majority of stations might have to be situated at mode level. In some circumstances, however, as for example where there were natural changes of level, it might prove possible to site stations at pedestrian level.

7. On the question of stations generally, in view of the probably high capital cost involved, including land costs, station sizes would have to be kept to the minimum practicable.

8. If it can be accepted that an opportunity appears to exist for the introduction of a new mode of transport capable of complimenting existing modes in circumstances where they might be inadequate or unable to penetrate, the next step would be to select in broad principles the areas through which it might operate. Rights-of-way would need to be protected, and it would be important that firm guidance was given as early as possible to architects involved in redevelopments through which a system might pass so that these rights-of-way could be protected and the financial implications evaluated. It would be essential that no major redevelopments were carried out in such a way that they blocked the best routes, unless a system and the buildings were so designed that the system could pass through the building. In such circumstances, limitations on the size of the envelope of the system immediately become apparent.

A significant point to be noted about these principles is that attention given initially (and somewhat predominantly) to land use and environmental concerns. The Working Party apparently recognized from the beginning that synthesis could not be accomplished by first determining where routes ought to go to save the *users* the most inconvenience. The solution resulting from this approach probably would have required much relocation of buildings, significant visual intrusion, and, in general, a level of inconvenience to the nonuser that probably would not have been tolerated. Instead, an immediate effort was made to locate routes in redevelopment areas *in the general vicinity* of the optimum user routes. In these areas, the impacts would not be as crucial and in some cases would have a positive effect.

The Working Party also appeared to anticipate that with such a great emphasis on the use of redevelopment areas, there would be a heavy dependency on developers, especially insofar as their reactions to potential plans and timetables for improvements were concerned. The next part of the synthesis process thus dealt with potential difficulties that could arise in these areas. First, there were land costs. An effort had to be made to encourage developers to absorb most of these costs based on the incentives of increased land values and commercial sales and improved appearance.

Timetables were especially difficult to judge. Often developments are not built on schedule (if at all), and a transportation system tied into such potential developments obviously would be subject to the same uncertainties. To allow for these possibilities, the Working Party looked at several alternate locations for routes.

The next two steps in the synthesis process were those of specifying route locations in more detail, then estimating the potential demand for the services provided. It was thought that a pilot study would be needed initially because of the uncertainty about demands, costs, impacts, and the progress of future redevelopment plans. If successful, the pilot project could be expanded in coordination with redevelopment efforts. Several full-scale route plans were tested, however. The primary factors considered in the location of routes were:

a. Connection with rail and subway stations
b. Closeness to a major arts complex

c. Avoidance of areas with architectural character
d. Possible installation problems
e. Possible areas of redevelopment
f. Availability of alternate crossing of the Thames River
g. Access to theaters, movies, major shopping and tourist areas
h. Possibility of later extensions of routes

Of course, potential ridership also was a major factor in route location. Estimating this ridership proved to be a difficult task. It was necessary to predict the number of passengers on the proposed system who previously would have been pedestrian or bus, taxi, auto, subway, or rail riders. Moreover, there was the possibility of people (especially tourists) being *induced* to make more trips because of the increased transport service. Unfortunately no well-tested and reliable models existed to predict mode choice and induced traffic for the situation being studied, so most predictions were made under somewhat arbitrary assumptions. For example, it was assumed that all passengers on bus trips wholly within the limits of the network would transfer to the new system.

After having specified possible route locations and corresponding traffic volumes, the Working Group evaluated several possible technologies to be used on the route. These included low-speed belts, continuous trains, and independent vehicular systems. The first two were eliminated from consideration. The moving belt proved to be slow (2 mph) and had serious limitations on curvature (500 ft minimum when superelevated) while the continuous train was felt to require acceleration/deceleration structures that were too large. The operation of the independent vehicle system then was simulated under various conditions of network switching capabilities, and corresponding journey times were calculated.

The next step was to consider various vehicle types and sizes and the resulting capacities under different headway and station stopping time conditions. It was felt that, at least during rush hours, some people had to stand, otherwise the system could not possibly support itself financially. With standing, the turning radii would have to be 25 ft at a minimum (for passenger comfort). Actual capacities were taken to be 66 percent of predicted, theoretical capacities because of daily variations in demand, station stopping times, weather conditions, accidents, and so on. Three car trains were anticipated, although it had yet to be determined if they would cause a visual intrusion or would fit into the stations.

Stations were preliminarily designed so as to fit within new buildings at mode level. Consideration was given to ways of loading and unloading trains, safety zones and capacity, passenger information systems, revenue collection, and need for attendants to repair breakdowns and reduce vandalism. Thought was also given to a process by which potential passengers could be monitored, then metered, so that stations (and the system as a whole) would not get overcrowded.

User comfort was judged on the basis of four major factors:

1. Acceleration, deceleration, and jerk rates both laterally and longitudinally
2. Degree of body support
3. Heating
4. Ease of entry and exit from the vehicle

One result of this part of the study was that the lateral acceleration rate was limited to 8 ft/sec^2, a not uncomfortable level for most people. Heating was felt to be a difficult problem in vehicles with large doors that were opening constantly.

The final two sections of the study dealt with the environmental impact of the system and its expected costs and revenues. In the environmental impact section, such items as noise, visual intrusion, vibration (external), and fumes and dirt were taken into account. A vehicle with electrical propulsion and air or pneumatic tire suspension was assumed, so that most of the potential environmental problems would be eliminated. Visual intrusion (and overshadowing) were checked by taking pictures at critical points on the.proposed system routes and making sketches from the pictures with the system overlayed.

The cost and revenue balance appeared to be somewhat unhealthy. Using two proposed pricing schemes, the Working Group found that in both cases traffic receipts would not even cover operating costs, much less capital costs. In addition, there was an unfavorable impact on London Transport services, so that the overall effect on public transport finance was sizably negative. Of course the estimates of ridership could be quite conservative, but it is doubtful that the new transit system would ever pay for itself entirely.

12.4.2 Comments on the Working Group Study

The preceding discussion has been an attempt to summarize the main points of the Working Group Study for a "people-mover" system in Central London. It is difficult in an actual study to separate the setting-of-objectives stage from the evaluation stage or from the transportation synthesis stage, which is of concern here. Subsequently, there may be some confusion as to what constitutes the solution generation or *synthesis* stage. In the study outlined above, synthesis was shown primarily through the process by which the various transportation system elements were brought together and through the identification of factors of primary concern in each stage of the process.

Looking first at the factors, we find that consideration was given to such diverse items as structural weight of the guideway, scheduling of redevelopment plans, land value increases, and effect on other transport systems. Quite obviously, a useful synthesis of transportation elements can be much enhanced by taking into account a wide variety of factors that would appear to be important to different citizens and citizen groups. From these considerations then comes a set of general principles (e.g., the route principles) that can be used as guides in the planning endeavor.

The nature and sequencing of stages in the synthesis process also is of importance. The stages correspond roughly to the general elements of the transportation system (vehicles, guideways, terminals, controls—Sec. 12.1) and to the general types of objectives (transportation service and impact). The sequencing of these stages in the study proceeds something as follows:

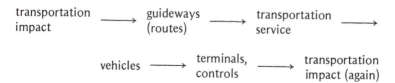

As was brought out earlier, the first stage in the process was not directed primarily to the servicing of traffic. Instead thought was given to possible acceptable locations (e.g., redevelopmental areas) of routes somewhere in the vicinity of the best user service locations. In the next stage, attention was given to the possible reactions of renewal area developers. Then (and only then) was consideration given to more precise locations of routes for the users and to the travel produced as a result. Also of interest is the fact that the routes and general guideway configurations were chosen before the vehicle type was specified. Subsequently, the vehicle type was used to dictate in large measure the type of stations and controls needed. Finally, environmental impacts and revenues and costs were estimated (note also that pricing schemes were considered almost last in the process).

In conclusion, we have found in the Working Group's study one type of process for combining transportation elements to form a system having certain service characteristics and impacts. This process may serve as a guide for future endeavors of a similar nature.

12.5 GENERATION OF SOLUTIONS AT THE METROPOLITAN LEVEL

An approach to system generation at the metropolitan level must respond to the demands placed on the transportation planner, especially at this juncture of the planning process. The goal of the generation phase can be stated as the creation of alternative plans (or courses of action) that satisfy as best as possible the goals of the project.

The phrase *satisfy as best as possible the goals* needs to be fully understood as it applies to public planning. Transportation planning, like other public planning, often must be conducted in a "goldfish bowl" where it is observed intently by many groups—public agencies, elected officials, special interests, civic associations, businesses and families—affected actually or potentially by the planning effort. Each group has its sets of goals and attitudes—often conflicting with those of other groups—which it tries to impose on the planners. The complexity of transportation planning problems is greatly compounded by the range of goals to be served, the vitalness of the goals to the groups holding them, and the conflicts between these goals. While it is the function of the political system to resolve conflicts between groups, the planner cannot escape pressures to create solutions that will bridge the conflicts, or at least to maintain credibility with the different groups. Every reasonable effort must be made to serve their goals.

12.5.1 Special Requirements for Solution Generation

The demands of complex transportation problems place certain special requirements for creativity, relevance, political credibility and economy, and effectiveness of effort on the planner or engineer in the synthesis phase. Seven such requirements are identified here:

1. Clear definition of project goals, including some resolution of goal conflicts (Chap. 4).
2. A *strategy* to balance the conflicting demands for (*a*) consideration of broad range of relevant factors with (*b*) the constraints of time, staff, and money.
3. An *information system* of all relevant and available information, including the necessary analyses and forecasts (Chaps. 5 to 9).
4. Organization and presentation of this information in a form meaningful and "instantly" available for the plan-making task.
5. Design of alternatives relative to the goals.
6. Documentation of design decisions.

7. An effective feedback relationship between the plan-synthesis phase and the plan-testing and evaluation phases.

As noted previously several of these requirements are discussed in some depth elsewhere in this text. In this section they are related specifically to the metropolitan synthesis stage, with examples drawn from several transportation systems planning endeavors.

The first three requirements, relating to goals, strategy, and information, would largely be fulfilled in preceding stages of the transportation planning or problem solving process. The requirement of goal definition and resolution corresponds to the situation noted above that most complex transportation problems involve multiple, conflicting goals. A solution that satisfies one goal may adversely affect another. This dilemma is not always easily resolved when faced by a single individual and is compounded when multiple public groups are also involved.

Technical devices for resolving goal conflicts through combining, ranking, or weighing goals are discussed in Chap. 10 and [12.14 and 12.54]. Sometimes these devices are ignored and goals are deliberately blurred and ambiguous. Often, and we hope increasingly, however, both real and political solutions are served by precise, operational definitions of goals (i.e., criteria that permit measurement of the effectiveness of each alternative in achieving the goals). The sharpening of tensions on the planner often enhances the possibility of creative solutions that bridge conflicts.

Without goals so sharpened and confirmed with the decision-making body, the planner cannot effectively and efficiently structure the data-collection, modeling, plan-making, testing, or evaluation phases. Among the devices used to do this in metropolitan transportation systems planning are the following:

1. Combine all goals to the extent they can be measured in dollars into a single cost/benefit ratio (Chap. 10). This device was extensively used until the early 1960s but generally is no longer acceptable by itself [12.55].
2. Use one goal as a decision rule to select the best plan from among those that meet minimum standards with respect to other goals [12.14].
3. Develop effectiveness measures for all goals and set weights as to the importance to be attached to each goal [12.54].
4. Develop effectiveness measures for all goals and rank the goals (see Chap. 10).

The second requirement relates to the strategy needed in part to cope with the major dilemma of complex planning problems. The horns of the dilemma are relevance and feasibility. On the one hand, the numerous relevant goals seem to require use of an overwhelming amount of information. On the other hand, resources of calendar time, staff time and skills, and money are too scarce to permit consideration of all relevant factors. To be effective, the strategy must be set during an initial policy and study design stage to make best use of staff in the data collection stage (where often 70 percent of the budget goes in metropolitan transportation studies) and in the modeling stage so as to provide the best preparation possible for the generation model use, and evaluation stages.

In several typical studies [12.56, 12.57, 12.58], the dilemma between relevance, complexity, and incommensurate calendar time was resolved basically by this strategy:

1. Assume continuation of existing land use trends and policies of local government (i.e., no exploration of greenbelts, new town, etc.).

2. Use least cost as a decision rule and minimum standards in respect to other goals (i.e., device 2 above).
3. Concentrate first on developing and evaluating a system of *expressway* corridors. Then do subsequent work on the transit system, the arterial system, and construction scheduling. Leave route location and specification to be accomplished incrementally as scheduled.
4. Investigate spacing of links in relation to future trip density to zero in on the scale of network needed.
5. Test all system and route proposals economically and rapidly by combining them into a "composite network" of corridors.

Requirement 3, the need for a data base for systems synthesis, is probably the most obvious. In metropolitan-wide synthesis this means, at a minimum, completion of data collection and usually also of modeling. The latter is desirable for greatest efficiency, as it is important to have as precise a picture as possible of future travel demand before starting preparation of alternate plans. In the Chicago (CATS) and Niagara Frontier Studies this was accomplished by analysis of spacing of urban freeways in relation to anticipated future trip density [12.14].

To save calendar time, much can be done toward developing alternate plans prior to and subject to completion of the models, which are the testing mechanism. In Fig. 12.3, for example, much of steps 2B, C, and D—the Transportation and Barrier Maps and the Principles Statements—as well as much of the measurement of Corridor Conditions (3C), could be done in advance of completion of Optimum Spacing (2D) or Testing of Alternates (4A). The resulting gain of calendar time at the expense of efficiency is a question of strategy for a particular study.

As noted previously, the bulk of the work required to satisfy the first three requirements must be substantially completed before the systems synthesis stage can progress far. This is represented in Fig. 12.3 by step 1. However, readiness for the phase should be reviewed, undone work expedited, and any last-minute changes in strategy or tactics made.

Requirement four, organization of data, is represented in Fig. 12.3 by step 2. The purpose of this step is to organize information needed for plan design so as to be (1) most meaningful to the planner or engineer and (2) "instantly" available. "Meaningful" organization of information recognizes that the manner in which data are displayed can stimulate ideas for solutions—and facilitate developing and checking them. "Instant availability" is required to reduce the time spent by the planner in securing the particular data needed. If the data are not conveniently accessible, the planner may choose to do without them rather than take the time, with a resulting loss in creativity and realism in planning. What is needed is a "planning-oriented information system" in which the relevant information is displayed accessibly, coherently, and comprehensively for the plan-designer.

A plan generation information system is illustrated in Fig. 12.3 in step 2. Each block represents a set of information presented so as to convey to the planner or engineer the parameters of the problem being faced.

A. *Transportation factors map*
1. Regional influences: Factors beyond the immediate urban area that must be taken into account include existing and potential transportation routes, cities, recreation areas, and other traffic generators, their size and probable growth; new town plans, natural factors shaping growth such as bodies of water and mountains. This information is displayed on a small-scale map so as to encompass an area at least 100 miles in each direction.

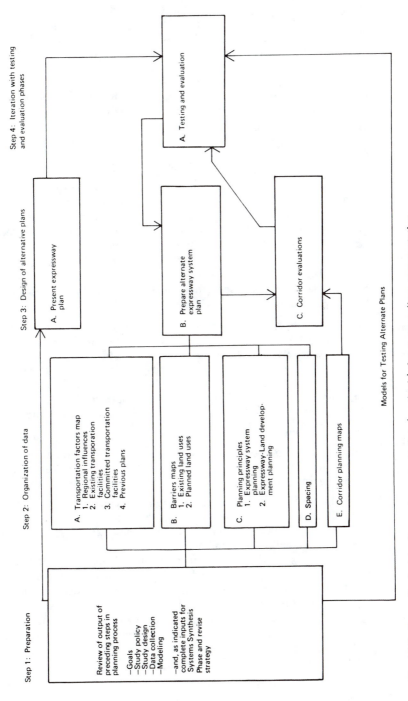

Fig. 12.3 Generalized steps for systems generation (synthesis) for metropolitan transportation planning.

2. Existing transportation facilities: Arterials, expressways, bus lines, rail transit, railroads, airports, ports, and terminal areas. What were the urban transportation facilities in existence at the time of data collection? What facilities have or are being added? What interfaces between modes?

3. Committed transportation facilities: What proposed facilities must be taken as "givens" in planning new systems? Not always an easy question to answer. The safest indicator is that right-of-way acquisition or construction that has actually started. This map should also show the proposed facilities that are not so firmly committed—those in official plans or improvement programs, or for which engineering plans have been prepared.

In addition to providing the "given" base for the planning of new systems, the committed system performs two other functions: (1) to test the comparative value of alternate immediate improvements, and (2) to provide one alternative for the future, the "do-nothing plan," or null hypothesis against which the benefits and costs of other plans can be measured.

4. Previous plans: This map or series of maps depicts historical transportation system or route plans, both those built and those not. In retrospect how realistic were the plans? What are the lessons?

B. *Barrier maps.* This map or series of overlays shows the location of existing or planned land uses that should be avoided by, or sometimes adapted to, or weighed against, new transportation facilities.

1. Existing land uses:

Cemeteries	Forest preserves
Railroad lines and yards	Scenic conservation areas
Public and private institutions	Historic sites
Water and wetlands	Neighborhood conservation sites
Poor soil condition areas	Shopping centers
Airports	Major industries
Parks	Ethnic and religious neighborhoods

2. Planned land uses:

Residential neighborhoods	Institutional expansion areas
Redevelopment areas	Industry and business expansion
School "enrollment" areas	areas

If all of these barriers are taken as impenetrable, construction of new transportation facilities often would be impossible, as may well be the case in some instances. Desirably, therefore, the map indicates the relative impenetrability or modifiability of each area.

C. *Planning principles.* The goals of the particular planning project should be translated into explicit principles and standards, and made available to all concerned. A considerable body of professional experience has been accumulated in recent years. Principles are well established for any one system, like the road system; multimodal principles are more general and still in the process of development governing the interface between the transportation system and land development.[4] By way of illustration, Creighton provides six planning principles in *Urban Transportation Planning* [12.14]:

---

[4] See [12.55, pp. 22–29]. The entire report, *The Freeway in the City*, is a summary of urb transportation-land development principles.

1. Continuity: Transportation systems should provide for direct, continuous motion and should not contain jogs, stops, or enforced changes in direction.
2. Lane balance: The number of lanes of expressway entering an interchange should be the same as those leaving it.
3. Even distribution of investment: Expressways should not be crowded together in some places and spaced too far apart elsewhere.
4. Dispersion: Concentration of traffic on segments of road systems (e.g., at the core of the city) is to be avoided.
5. Sufficient arterials: There should be a fine enough spacing of arterials so that they are not overloaded by traffic, forcing some travel to dangerous neighborhood streets.
6. Enclosed spaces: Spaces which exist within the mesh formed by major roadways should be areas which can readily contain efficient and pleasant groupings of activities or land uses.

Creighton also outlines some more specific principles, which can be found in his book.

D. *Spacing.* How far apart should expressways and arterials be in each part of the region to carry the anticipated future travel volumes? Should the future system be rich or lean? Adequacy of transportation is the major goal of transportation planning. Since spacing of expressways is a function of trip density, speed and length, and construction and right-of-way costs, it is possible to calculate it once the regional growth model stage of the process is complete. The result provides a "ballpark" estimate of richness or leanness of the future system as a guide to the planner [12.14].

E. *Corridor planning maps.* Air photos on detailed land-use maps, at a scale of 400 ft to 1 in or better and keyed to the information on the barriers maps, are required as the basis for measuring corridor conditions and costs affecting route desirability and feasibility.

Requirement 5, actual development of alternative plans, is represented in Fig. 12.3 by step 3.

A. *Present expressway plan.* Every major urban area in the United States now has a transportation plan with some degree of commitment by the responsible agency and other officials. Because of the rapid changes in local conditions, U.S. policy requires that these plans be regularly reviewed and modified as necessary. For reasons of continuity, courtesy, and strategy, it is customary that the first alternative tested is the existing plan.

B. *Prepare alternate expressway system plan.* This is the system synthesis activity for which previous steps have been prepared and with which the model use and evaluation stages are iteratively linked. This linkage tends to balance creativity and realism. The number of alternatives to be developed must be determined by the strategy. In most metropolitan studies the number has varied from 2 up to the 28 in the Chicago Study [12.56]. These systems plans lead to two other steps in the synthesis process. The first of these is to abstract the proposed systems from their real world setting, represent them mathematically, and test their ability to carry future traffic (step 4).

C. *Corridor evaluations.* The other step, conducted in parallel, is to examine each of the proposed transportation systems in its real world setting. This is done through assessing the conditions in each corridor relevant to possible route locations and the goals of the project. The various conditions indicated on the Barriers map (2B) are evaluated in greater detail, and the impact of each alternative is measured.

The conditions revealed by this last step are of great importance in metropolitan planning and may determine whether a particular corridor—and therefore a particular system—is acceptable.

The sixth requirement, documentation of decisions in the development of alternatives, is important from several viewpoints. First, in the synthesis work that follows each iteration with model use and evaluation, good documentation facilitates building on the work previously done. Secondly, it facilitates preparing the defense of the recommended system in the face of anticipated public questions and criticism, a regular feature in metropolitan transportation planning these days. Finally, since plans must be reviewed and updated over the years, usually by different persons, documentation facilitates replication and verification of the original decision.

The final requirement, the cyclical or iterative relationship with the model use and evaluation stages, balances creativity with relevance of proposals to goals. Evaluation makes possible a decision on an alternative in a way that directly relates the selected alternative to the goals of the project.

12.6 NONTRANSPORTATION SOLUTIONS FOR TRANSPORTATION PROBLEMS

The discussions in the previous two sections have emphasized the process by which transportation elements can be synthesized to form a modified system. It should not be assumed from these discussions, however, that congestion can be relieved and access improved only by altering the transportation system. In fact, there are many nontransportation means by which these objectives might be achieved.

One set of means already has been discussed under TSM procedures. This included various types of taxation or pricing so as to encourage or discourage certain kinds of travel. Also included were staggering of work hours, reducing the number of work days (e.g., a 4-day, 10-hr-per-day work week), and simply restricting various kinds of travel (e.g., by trucks on certain routes and in specified areas).

12.6.1 User-Side Assistance[5]

Since the amount of travel which a household undertakes is highly correlated with its income, one way to increase mobility and accessibility is through income-transfer programs.[6] This, in fact, is one of the (indirect) objectives of general support programs such as Social Security, Unemployment Compensation, Aid to Families with Dependent Children, and the like. These kinds of approaches appear to be particularly useful to the elderly and handicapped, since they are disproportionately poor in the U.S. population.

---

[5] Some of the discussion here is taken from J. Dickey, R. Kirby, and U. Ernst, "A Transportation Credit Program," in [12.22, pp. 13–14]. See also [12.25].

[6] See Chap. 14 for estimates of the expenditure impacts of such programs.

An illustration of a more direct approach of this type is called user-side assistance (USA). This has been employed to some degree in public transportation, although few applications have been monitored carefully enough to permit a comprehensive evaluation of the administrative costs or the quality of services obtained by client groups from the providers. The Medicaid program (medical care for the poor), has been subsidizing taxicab rides for its clients for some time, and several communities have used discretionary monies such as revenue-sharing funds to institute user assistance schemes for groups who have limited mobility.

An example of USA is in Los Gatos, a small city of some 23,700 people, where elderly and disabled residents may purchase a maximum of 10 taxicab tickets a month at a cost of 50¢ per ticket. They can use one ticket per trip anywhere within the city limits. For each ticket used, the city reimburses the taxi operator $2.10 out of revenue-sharing funds. To prevent cash-flow problems for the taxicab operator, the city pays the operator a monthly advance based on average ticket use.

In December 1974, the City of Oak Ridge, Tennessee, started selling tickets at 25¢ each to people 60 years of age and over. Each ticket can be used in lieu of up to $1.00, for a taxi fare, with the user paying the balance. The city pays 90¢ for each ticket turned in by the taxicab operator. On those rides with fares less than 90¢, the taxi operator makes a small profit, while on those over 90¢ the taxi operator sustains a small loss.

For another example, automobiles have been provided to one special group, disabled veterans. Those who are able to drive using various kinds of adaptive equipment can apply to the Veteran's Administration for help in purchasing an auto or other vehicle and the required adaptive equipment. The one-time grant had a limit of $3,800 in 1979.

A similar program, which includes all travel modes, but likewise is restricted to the severely handicapped, is the Mobility Allowance Program in Great Britain. About 100,000 people get up to £10 (about $20) per week to help them purchase any form of transportation, including autos. In addition, a new voluntary organization, Motability, has been set up to help recipients to apply a portion of their allowance toward the lease or purchase of an auto or similar vehicle. This is being done in conjunction with the provisions of £100 million in loans from banks on favorable terms.

This approach has several advantages over traditional public transportation policies for the disadvantaged. However, most current USA programs prevent recipients from taking advantage of the private automobile or truck. The possibility of subsidizing auto use has been recognized. For instance, various Congressional ideas for oil deregulation and gasoline rationing have called for broad user-side mechanisms to help reduce the impact of increased fuel prices on the poor [12.10].

12.6.2 Substitution by Communications

Another nontransportation scheme is through substitution of communication for transportation. This has been an oft-proposed approach that, until the mid-1970s, really was not a realistic way of reducing travel. Particularly because of rising real incomes, people seemed to be able to afford to do more of both. The picture has changed since then, however. Increases in the costs of travel and comparatively substantial drops in the costs of communications have made the latter much more appealing. In addition, the technology of communication, especially through computers, is changing rapidly, making available a much wider range of options.

As one example, the visiphone spurred interest at its introduction in the early 1960s, but never was used very much. Recently, however, it has gained a resurgence of interest, and it may yet become a common household (or at least business) device. By 1983, AT&T is expected to have set up public and private picture-phone meeting services involving two-way simultaneous audiovisual teleconferencing in 39 cities [12.51, p. 34]:

> Public meeting room facilities will include a conference table with the main control panel, seven color television cameras, and three monitors. Such rooms will accommodate six conferees, while private ones will allow a minimum of two and a maximum of ten.
> A graphics display unit equipped with a camera will be utilized to transmit images of three-dimensional objects as well as graphs, charts, transparencies, and similar material. A hard copy machine will be available to produce images displayed on the incoming monitor. A video tape recorder also can be employed in the public rooms. This would be used to record either the incoming or outgoing signal and the audio signal from both locations to provide documentation of a meeting.

With technologies like these evolving rapidly, it will become less necessary to travel, particularly for person-to-person meetings in which facial expressions often can be as important as the words said.

12.6.3 Land-Use Organization

Still another nontransport solution is through the organization of land use, either to decrease amounts and lengths of trip-making or to induce changes of mode. The latter approach is demonstrated most aptly on many college campuses

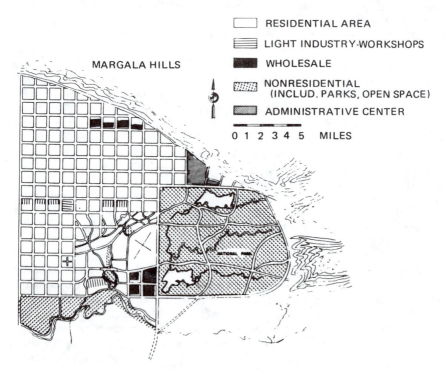

RESIDENTIAL AREA

LIGHT INDUSTRY-WORKSHOPS

WHOLESALE

NONRESIDENTIAL (INCLUD. PARKS, OPEN SPACE)

ADMINISTRATIVE CENTER

MARGALA HILLS

0 1 2 3 4 5   MILES

Fig. 12.4 Master plan of Islamabad. (*Adapted from Capital Development Authority, "Islamabad—The Project and Perspective," Government of Pakistan, Islamabad, 1970.*)

Fig. 12.5 Layout of Shalimar—6, one of the communities within a superblock in Islamabad. (*Adapted from Capital Development Authority, "Islamabad—The Project and Perspective," Government of Pakistan, Islamabad, 1970.*)

where buildings and open spaces are arranged to encourage pedestrian travel. On a regional scale, the same kind of idea has been attempted around Stockholm, Sweden, where satellite new towns have been organized in radial corridors emanating from the city [12.2]. Rapid transit lines have been constructed in the corridors so as to induce people to use the higher-capacity transit systems and forsake the purchase and/or operation of the automobile.

The arrangement of land use to reduce the amounts and length of travel has been attempted primarily in new town developments, found mostly in Europe but increasingly in the United States and other parts of the world. Both Islamabad, the capital of Pakistan, and Chandiagarh, capital of the Punjab State in India, have been built on the "superblock" concept where major roads are spaced at approximately 1.25 mile intervals in both directions (see Fig. 12.4 and refs. 12.3 and 12.4). A kind of self-contained community (Fig. 12.5) then is organized within each superblock so that few people need to travel outside the immediate area. The layout of Belconnen, a satellite new town to Canberra, the capital of Australia, follows somewhat similar principles (see Fig. 12.6 and ref. 12.17]). Neighborhoods, consisting of 3,500 to 4,000 people, were organized so that almost everyone would be within 0.5 mile walking distance of local stores, primary schools, recreation areas, and so on.

Land use also can be organized for similar purposes around existing cities, although this has not been tried in many cases yet. The Stockholm plan is one exception. Some research has been done, however, to indicate the extent of travel

Fig. 12.6   Neighborhood of Aranda, Belconnen: 1. Shops; 2. Primary school; 3. Parish center and school; 4. Joint church; 5. Preschool, Mothercraft center, 6. Group housing; 7. Pedestrian underpass; 8. Recreation.

savings that could be obtained through proper land use arrangement. One such endeavor [12.6] involves an application of TOPAZ (Technique for the Optimum Placement of Activities in Zones), a mathematical programming procedure. This will be discussed now to demonstrate both the types of results that can be obtained and the technique itself.

The objective function of TOPAZ is the minimization of establishment plus travel costs in an urban area. Establishment costs include those for building units, water and sewer, local streets, and electricity. Land-value increments are perceived as negative establishment costs or benefits. Travel costs are predicted using a simple gravity model, and a constant cost per mile of travel. With these unit costs, TOPAZ is employed to allocate land acreages of various types (residential, commercial, etc.) predicted to be needed by 1990. The acreages are allocated to various zones where vacant land is available and, again, where total costs would be a minimum.

In applying TOPAZ, an initial feasible solution (plan) is needed as input. Such a plan for Blacksburg, Virginia, a town of roughly 10,000 population (in 1970), consisted primarily of allocations of land to the northwest side of town. The total cost of this solution was estimated to be $90 million (1970), of which $70.5 million was for establishment costs and $19.5 million for travel. We thus see that travel is a major expense in urban areas, accounting in this solution for about 22 percent of all costs [12.6].

In the TOPAZ-generated minimum-cost solution, total costs were reduced to $84.1 million with about half of this reduction being in travel ($16.6 million). This result shows that it is possible to decrease travel costs about 14 percent through proper land-use arrangement in this small town.

Interestingly, the minimum-cost solution does not indicate a need for expansion to the zones utilized in the initial solution. Instead, growth is spread to those substantially developed zones fairly close to the town center.

TOPAZ also was employed in studies strictly to minimize and to maximize travel costs. The minimum was found to be $15.4 million, the maximum $20.7 million. If the latter value were used as a base, we would see that travel costs could be as much as 25 percent higher in the most expensive arrangement (future development scattered to the far periphery). Improper land-use arrangement thus can be seen as a significant contributor to traffic congestion and unneeded travel. However, proper land organization, especially if employed in connection with other nontransport schemes like these presented earlier in this section, could be of considerable help in alleviating these problems.

12.7 TRANSPORTATION SOLUTIONS FOR NONTRANSPORTATION PROBLEMS

While nontransportation solutions to transportation problems may be of great benefit, the opposite may also be useful: utilizing transportation to help solve other problems. Much of the literature dealing with developing countries gives heavy emphasis to the role of transportation in the development process [12.18]. Even in the United States, over $1 billion has been invested in highways in the Appalachian region primarily to induce economic development [12.19]. Transportation, then, is obviously thought to be a significant part of the solution to "development" problems, whether they be economic, social, or environmental.

We will not attempt to show how various transportation systems have been evolved to solve nontransportation problems. The possibilities and actualities are both so broad and pervasive that even an attempt simply to *list* applications would fall far short of being exhaustive. Instead, we will present one semihypothetical example that indicates the *potential* of utilizing transportation to achieve other goals.

12.7.1 Economic Segregation and the Transit System

The study to be described here [12.7] involved the use of the EMPIRIC land-use model (see Chap. 6) in conjunction with goal programming, a particular mathematical programming technique. The EMPIRIC model can be employed to predict the number of families in each of four income classes that will reside in a given zone of an urban area in the future. Factors affecting the growth of residences in a zone include the future *regional* population and employment, the present number of families and employees of different types in the zone, land areas, water and sewer system types and availability, and, most important to this study, interzonal highway and mass-transit travel times. The mass-transit travel times were chosen for more intensive analysis in this study.

The objective of the goal program was to minimize the "extent of segregation" of families by income throughout all zones in the region. In other words, an attempt was made to create a more homogeneous distribution of poor and rich families throughout the metropolitan area. In an operational sense, this means that the proportion of families in each income level in each zone should be as close as possible to the corresponding proportion for the region as a whole. There could be a surplus of a given type of family in a given zone as compared to the regional proportion, or a dearth (slack), or no difference. The objective of the study thus became to determine how interzonal mass-transit travel times could be changed to minimize the slack or surplus of families of all types in all zones.

Six zones out of the Boston metropolitan area were chosen for investigation. The first was Boston itself, while the five other zones were surrounding suburbs. Three situations were investigated, the first in which segregation by income was to be minimized, the second in which there was no change made in future transit travel times, and the third in which segregation was to be maximized. These latter two situations were studied to provide bases by which to judge the relative impact of transit on segregation. The last case is particularly relevant since it shows the absolute worst condition of segregation by income levels and thus can be used as a yardstick for measuring the success or failure of the other two situations.

The outcomes of the three investigations are displayed in Table 12.10. A plus value indicates a surplus of families of the given type over that which would occur if there were no segregation. A minus shows a slack. The total amount of slack and surplus for the minimum segregation case was 29,842. This can be compared to 37,317 for the "no change" and 49,021 for the worst segregation cases. The magnitude of these three totals leads to some interesting implications. First, and probably of foremost interest, is the fact that transit changes can have an effect on segregation. The difference between the best and worst cases is 19,189 families, which represents about 7.5 percent of the families in the region. This difference is not insignificant, especially since it is almost equivalent to the total amount of what might be called "inherent" segregation (the 29,842 families) that would be left if transit travel times were improved to their most desirable levels. In other words, proper transit improvements could reduce the amount of segregation by income in the region by approximately 40 percent as compared to what it would be if the transit changes with the worst impacts were implemented.

The actual effect of transit probably is not quite as significant, however. Comparison with the "no change" case, for instance, shows that the maintenance of the *status quo* insofar as transit is concerned would lead to a segregation level roughly in the middle between the best and worst situations (37,817 verus 29,842 and 49,021). The overall effect that transit might have in this situation would thus amount to a level of 7,975 families, which would be approximately 20 percent of the segregation by income that would exist if the "no change" alternative were followed.

One other aspect of interest is that, even if all necessary transit changes were made, there still would be a discrepancy of 29,842 families from the most desirable levels. This amount possibly could be reduced by changes in other municipal services (highways, water systems, and sewer systems) shown to be relevant in the EMPIRIC model. Such changes probably would not cause a significant reduction, however. To reduce segregation to its absolute minimum probably would require the

Table 12.10   Slacks (—) and Surpluses (+) in Families in Each Income Class in Each Zone Resulting from Minimum Segregation, No Change, and Maximum Segregation Solutions

Zone and income class	Minimum segregation	No change	Maximum segregation
Boston (1)			
$0–$4,999	+2,320	+1,047	+8,515
$5,000–$9,999	0	−4,631	0
$10,000–$14,999	−373	−1,306	+107
$15,000 +	+1,900	+1,636	+2,206
Lexington (2)			
$0–$4,999	−2,515	−2,236	0
$5,000–$9,999	0	+1,014	+5,419
$10,000–$14,999	+1,449	+1,258	+2,232
$15,000 +	+1,667	+1,725	+2,030
Natick (3)			
$0–$4,999	−3,468	−3,212	−4,661
$5,000–$9,999	0	+933	0
$10,000–$14,999	+1,421	+1,609	+1,328
$15,000 +	+882	+936	+823
Weymouth (4)			
$0–$4,999	−4,106	−3,683	−5,991
$5,000–$9,999	0	+1,541	0
$10,000–$14,999	+902	+1,213	+756
$15,000 +	+54	+442	+262
Peabody (5)			
$0–$4,999	−4,196	−4,067	0
$5,000–$9,999	0	+459	+7,219
$10,000–$14,999	+862	+957	+2,491
$15,000 +	+644	+670	+1,168
Stoneham (6)			
$0–$4,999	−1,760	−1,575	−2,502
$5,000–$9,999	0	+675	0
$10,000–$14,999	+544	+681	+486
$15,000 +	+272	+311	+235

Source: [12.7].

action of laws and politics not currently operative in urban areas (and thus generally not incorporated in the EMPIRIC model).

While the total slack and surplus for each case is important, the distribution of these totals also is relevant. It should be noted first that in all three cases the amount of slack exceeded the surplus, with the maximum segregation case having the biggest difference of 23,713 families. From this result comes the not too surprising inference that the process by which segregation is heightened involves the restriction of movement of families into a zone rather than an overexcessive migration of families. Yet this conclusion is not upheld in all zones and for all types of families. As should be expected, there are many variations which prevent generalities from being accepted unconditionally. Referring to Table 12.10, we see that the impacts on various zones can be quite different, with both large and small slacks and surpluses. The effect, though, on a per population basis is somewhat more noticeable. Boston, the zone having the largest number of families, is left with about the same amount of slack and surplus to overcome as the suburban zones, which means that, no matter what transit changes are made, the suburban zones will still be faced with a rather difficult task if they desire to overcome segregation by income.

Another interesting aspect of the distribution of slacks and surpluses is that they vary according to the particular case under study—minimum segregation, "no change," or maximum segregation. Although large differences between the three are not common, one example does bring out an extraordinary result: if the "no change" strategy were chosen, there would be a large slack in the number of families with incomes between $5,000 and $9,999 in Boston, whereas, if either of the other two extreme strategies were chosen, there would be no slack or surplus. Similar results hold for zones 3, 4, and 6, except that there would be large surpluses instead of slacks. These results mean that in certain situations conservative policies, such as the "no change" one, may produce *much more exaggerated* conditions than if a policy involving some changes were followed.

The patterns of transit travel-time reductions needed to bring about the impacts discussed above are very distinctive. There are only four reductions needed to obtain a minimum segregation, and these all relate to Boston. In contrast, maximum segregation is obtained by reducing most of the travel times within and between the suburban zones and by leaving Boston alone. These results are reasonable. To get minimum segregation, one must get the low-income families out of Boston, and this is done by providing better transportation in these directions. On the other hand, to increase segregation, one simply isolates Boston travel-time-wise and spends available funds on the intersuburban transit system.

As a final remark, it should be mentioned that the improvements needed to achieve minimum segregation really are not as extensive as was first imagined. Only four changes are needed, and only one of these is to the lower travel-time limit, although all four changes would be in Boston where improvement costs would be highest. What might be a significant finding at this point, though, is that attempts by low-income groups to hinder or even stop construction and operation of transit facilities in the inner cities would only do harm to the cause of integration, because the needed travel-time improvements in and from the city would not be realized. We might also expect similar results insofar as urban highways are concerned.

12.8 SUMMARY

In this chapter an effort has been made to display a variety of plans and of "principles" involved in developing them for transportation and related problems. The first section highlighted a variety of transport technologies that could be employed as part of a potential solution. In the next section the low-cost, non-capital-intensive alternatives represented by transportation system management (TSM) were presented. Subsequently, a cross section of plans was illustrated, including those at different levels of the policy chain (strategy or policy, program or system, and project) and those for special purposes (mobility for the elderly and handicapped, alcohol safety, air-pollution control, and contingent energy-consumption reduction).

BIBLIOGRAPHY

12.1   Owen, W.: *The Accessible City,* The Brookings Institution, Washington, D.C., 1972.

12.2   Genteli, G.: *The Satellite Towns of Stockholm,* Department of Planning and Building Control, Stockholm, 1960.

12.3   Doxiadis, C.: "Islamabad, the Creation of a New Capital," *Town Planning Review,* vol. 36, no. 1, April 1965.

12.4   Evenson, N.: *Chandiagarh,* University of California Press, Berkeley, 1966.

12.5   Hugh, W. L.: *Cumbernaud New Town Traffic Analysis Report,* Cumbernaud Development Corporation, Glasgow, 1958.

12.6   Dickey, J. W., P. A. Leone, and A. R. Schwarte: "Use of TOPAZ for Generating Alternate Land Use Schemes," *Highway Research Record* 422, 1973.

12.7   Dickey, J. W.: "Minimizing Economic Segregation Through Transit System Changes: A Goal Programming Approach," in G. Newell (ed.), *Proceedings of the Fifth International Symposium on Traffic Flow Theory and Transportation,* Elsevier, New York, 1972.

12.8   U.S. Department of Transportation, Federal Highway Administration: *Traveler Response to Transportation System Changes* (2d ed.), Washington, D.C., July 1981.

12.9   Wohl, M., and B. Martin: *Traffic Systems Analysis for Engineers and Planners,* McGraw-Hill, New York, 1967.

12.10 Congress of the United States, Congressional Budget Office: *Urban Transportation for Handicapped Persons: Alternative Federal Approaches,* Government Printing Office, Washington, D.C., Nov. 1979.

12.11 Watson, P. L., and E. P. Holland: *Relieving Traffic Congestion: The Singapore Area License Scheme,* World Bank Staff Working Paper No. 281, Washington, D.C., 1978.

12.12 U.S. Department of Transportation, Urban Mass Transportation Administration: *Transit Actions: Techniques for Improving Productivity and Performance,* Washington, D.C., Oct. 1979.

12.13 Working Party on the Introduction of a New Mode of Transportation in Central London: *An Aid to Pedestrian Movement,* Westminster City Council, London, 1971.

12.14 Creighton, R. L.: *Urban Transportation Planning,* University of Illinois Press, Urbana, Ill., 1970.

12.15 U.S. Environmental Protection Agency, Federal Energy Administration, and Department of Transportation: *Joint Strategies for Urban Transportation Air Quality and Energy Conservation,* vol. 1, Washington, D.C., Jan. 1975.

12.16 U.S. Department of Transportation, Urban Mass Transportation Administration: *Analyzing Transit Options for Small Urban Communities: Transit Service Objectives and Options,* Government Printing Office, Washington, D.C., Jan. 1978.

12.17 National Capital Development Commission: *Tomorrow's Canberra,* Australian National University Press, Canberra, 1970.

12.18 Owen, W.: *Distance and Development,* The Brookings Institution, Washington, D.C., 1968.

12.19 Friedmann, J.: "Poor Regions and Poor Neighbors: Perspectives on the Problem of Appalachia," *Appalachia,* vol. 1, no. 8, April, 1968.

12.20 U.S. Department of Transportation, Urban Mass Transportation Administration: *Innovation in Public Transportation,* Washington, D.C., 1978.

12.21 U.S. Department of Transportation, Office of the Secretary: *Coordinating Transportation for the Elderly and Handicapped* (Executive Summary), Washington, D.C., May 1979.

12.22 U.S. Department of Transportation, Office of the Secretary: *Transportation for the Elderly and Handicapped, Programs and Problems 2,* Government Printing Office, Washington, D.C., Oct. 1980.

12.23 Richards, B.: *New Movement in Cities,* Reinhold Publishing Corporation, New York, 1966.

12.24 Canty, E. T., and A. J. Sobey: *Case Studies of Seven New Systems of Urban Transportation,* General Motors Corporation, Warren, Mich., January 13–17, 1969.

12.25 Kirby, R. F., and F. L. Tolson: "Improving the Mobility of the Elderly and Handicapped Through User-Side Subsidies," Working Paper 5050-4-4, The Urban Institute, Washington, D.C., May 1977.

12.26 Hasselstrom, D.: *Public Transportation Planning—A Mathematical Programming Approach,* Department of Business Administration, University of Gothenburg, Gothenburg, Sweden, 1981.

12.27 *AMF-Safege Monorail: The Transportation of Tomorrow-Today*, American Institute of Planners 1964 Convention, Robert Treat Hotel, Newark, N.J.

12.28 Organization for Economic Cooperation and Development, Road Research Group: *Integrated Urban Traffic Management*, Paris, Dec., 1977.

12.29 Chilton, E. G.: *Future Urban Transportation Systems*, vol. 2, Stanford Research Institute, Menlo Park, Calif., May, 1967.

12.30 Baltimore Regional Planning Council, *Transportation Control Plan, Vol. 1: Summary*, Baltimore, Sept. 29, 1978.

12.31 *Popular Science*, vol. 199, no. 5, Nov., 1971.

12.32 Barton-Aschman Associates, Inc.: *Study in New Systems of Urban Transportation: Guidelines for New Systems of Urban Transportation. Urban Needs and Potentials*, vol. 1, Chicago, May 1968.

12.33 Southeastern Virginia Planning District Commission: *Southeastern Virginia Energy Contingency Study*, Norfolk, Va., June 1981.

12.34 Metropolitan Washington Council of Governments: *The Transportation Plan for the National Capital Region*, Washington, D.C., May 21, 1980.

12.35 U.S. Department of Health and Human Services: *Planning Guidelines for Coordinating Social Service Agency Transportation*, Washington, D.C., 1981.

12.36 U.S. Department of Transportation, Urban Mass Transportation Administration: *Transportation Energy Contingency Planning: Local Experiences*, Washington, D.C., June 1979.

12.37 U.S. Department of Transportation, Urban Mass Transportation Administration: *TSM Prototype Planning Study: Portland*, Washington, D.C., Nov. 1979.

12.38 U.S. Department of Transportation, National Highway Traffic Safety Administration: *Summary of National Alcohol Safety Action Projects*, Government Printing Office, Washington, D.C., Aug. 1979.

12.39 Oxfordshire County Council: *First Structure Plan for Oxfordshire: Consultative Document*, Oxford, England, Nov. 1975.

12.40 Steiner, G. A.: *Strategic Planning*, The Free Press, New York, 1979.

12.41 Richards, B.: "Urban Minisystem," *Architectural Forum*, vol. 128, no. 1, Jan.–Feb. 1968.

12.42 General Motors Research Laboratory: *New Systems Implementation Study, Vol. 1, Summary and Conclusions*, Research Publication GMR-710A, Warren, Mich., Feb. 1968.

12.43 Transit Research Foundation of Los Angeles, Inc.: *City and Suburban Travel*, issue 3, June, 1970.

12.44 Operations Research, Inc.: *Requirements for Transit Car Specifications*, report 24, Federal Clearinghouse, Springfield, Va., PB 169 564, April 1964.

12.45 Virginia Department of Welfare: *Commonwealth of Virginia Comprehensive Annual Plan for Social Services Under Title XX of the National Social Security Act*, Richmond, Va., 1981.

12.46 Lancaster County Planning Commission: *Elderly and Handicapped Transportation Study: Lancaster County, Pennsylvania*, Lancaster, Pa., Nov. 1978.

12.47 Ministry of Housing and Physical Planning, National Physical Planning Agency: *Process Monitoring on Behalf of Physical Planning*, The Hague, Netherlands, circa 1980.

12.48 Maxtone-Graham, J. A.: "Flying the English Channel: Altitude 7 Feet," *Popular Mechanics*, Jan. 1969.

12.49 "Hovercraft Proved," *Engineering*, vol. 208, no. 5393, Sept. 5, 1969.

12.50 U.S. Department of Commerce, National Bureau of Standards: "The Shirley Highway Express Bus-on-Freeway Demonstration Project—Project Description," Washington, D.C., Aug. 1971.

12.51 "A Conference at Two Places at Once," *Modern Office Procedures*, Aug. 1981.

12.52 "180-MPH Kit Canard," *Popular Science*, Aug. 1981.

12.53 U.S. Department of Transportation, Federal Highway Administration: *Alternatives for Improving Urban Transportation, A Management Overview*, Technology Sharing Report 77-15, Washington, D.C., Oct. 1977.

12.54 Stuart, R. C., and J. W. Dickey: "Modeling the Group Goal Component of the Plan Evaluation Process," Confer-In West, American Institute of Planners, San Francisco, Oct. 1971.

12.55 The Urban Advisers to the Federal Highway Administration: *The Freeway in the City*, Government Printing Office, Washington, D.C., 1968.

12.56 *The Chicago Area Transportation Study:* vols. 1–3, Chicago, 1959, 1960, and 1962.

12.57 *The Cleveland Seven-County Transportation Study*, Cleveland, 1969.

12.58 *The Niagara Frontier Transportation Study*, vols. 1 and 2, Albany, N.Y., 1963 and 1964.

12.59 Miki, T.: "Kama-kura's Shonan Monorail," *Permanent Way*, no. 45, 1971.

12.60 Dunne, J.: "Ford's Propane Cars," *Popular Science*, Oct. 1981.

12.61 "Now You Can Buy a Record Holding 56.66-MPH Bullet Bike," *Popular Science*, Sept. 1981.

12.62 Orski, C. K.: "Transportation System Management in Perspective," *Traffic Engineering*, Nov. 1976.

12.63 Patricelli, R. E.: "Definition, Objectives, and Implications of TSM," in *Transportation System Management*, Transportation Research Board Special Report 172, Washington, D.C., 1977.

12.64 Organization for Economic Corporation and Development, Road Research Group: *Integrated Urban Traffic Management*, Paris, 1977.

12.65 Levinson, H.S.: "Urban Transportation Programs: Myths and Realities," *ITE Journal*, Nov. 1979.

12.66 U.S. Department of Transportation, Urban Mass Transportation Administration: *Transportation System Management: State of the Art*, Washington, D.C., Feb. 1977.

12.67 U.S. Department of Transportation, Urban Mass Transportation: *Transportation System Management: An Assessment of Impacts*, Washington, D.C., Nov. 1978.

12.68 DiRenzo, J.: "Measures of Effectiveness for TSM Planning and Evaluation," presentation at Transportation Research Board, Conference Session on TSM Evaluation Methodology, Washington, D.C., Jan. 1980.

12.69 Holman, G. P.: "Bus Manufacturer Uses Extrusions and Innovations with Success," *Precision Metal*, March 1979.

12.70 McGean, T.: *Urban Transportation Technology*, Lexington Books, Lexington, Mass., 1976.

12.71 Anderson, J. E.: *Transit Systems Theory*, Lexington Books, Lexington, Mass., 1976.

12.72 Stuart, D. G.: "Generic Applicability of Automated Guideway Transit," *Journal of Advanced Transit*, vol. 15, no. 2, Summer 1981.

12.73 Nilles, J. M., et al: *Telecommunication–Transportation Tradeoffs: Options for Tomorrow*, Wiley Interscience, New York, 1976.

12.74 National Transportation Policy Study Commission: *The Impact of Telecommunications on Transportation Demand Through the Year 2000*, NTSPC Special Report no. 3, Washington, D.C., Nov. 1978.

EXERCISES

12.1 Go to the library. Select a new transportation technology similar to the ones discussed in Sec. 12.1. In about 500 words, describe the characteristics of the chosen technology and the circumstances under which it would/would not be advantageous.

12.2 Pick a particular study (a small city or part of a large one). Indicate what TSM improvements might be made, and why. What might be the potential level of impact on vmt given these improvements?

12.3 Locate a special purpose transportation plan. Indicate how it could be made into a contingency plan by considering different alternatives given various external events.

12.4 Develop and describe briefly a set of principles that might be emphasized at the policy planning level.

12.5 Investigate a nontransportation means for solving a transportation problem. Summarize briefly the results of a literature search, then indicate what might be the range of levels of influence of the selected approach on the problem itself and on other relevant impact factors.

12.6 Busing has been employed with at least two objectives of reducing racial segregation and improving the education of minorities in primary and secondary schools. According to the literature, what seems to have been the effect relative to these objectives?

12.7 For urban areas of 100,000, 500,000, and 1,000,000 population, prepare example TSM solutions that would be appropriate to areas of these sizes but not necessarily appropriate for all sizes. Explain why these differences should exist for different-size urban areas.

12.8 Explain how TSM solutions can be evaluated to assess their effectiveness; utilize the concepts developed for evaluating transportation system alternatives in Chap. 10.

12.9 In the case of an urban area with severe peaking characteristics in the morning and evening periods along three to four major corridors, what alternatives are available to reduce the peaks with particular attention to work trips?

13 Implementation I: Legislation and Organization

While it is important to develop solutions that are technically optimal, or at least feasible, an equally important test of a good solution is its implementability. Far too many plans and designs have been developed only to gather dust in some remote corner of an agency office. Planners sometimes feel that technical analysis is the most complex task and that implementation is "just a detail." The fact is that implementation is an equally complex and difficult challenge—and one that can be impossible if the groundwork is not properly laid during the planning stage. The purpose of this chapter is to discuss implementation as it bears on planning.

13.1 PROBLEMS OF IMPLEMENTATION

There are many reasons for difficulties in achieving prompt and effective implementation. Among them are [13.1, pp. 88–89]:

1. Poor metropolitan management and coordination of planning, programming, and implementation processes:
   a. No single, responsible person, office, or agency
   b. Roles of actors not agreed upon
   c. Turnover among professional staff
2. Lack of support from decision-makers
   a. Failure of planning to address the needs and goals of local decision-makers
   b. Lack of an adequate continuing information process for decision-makers (brought about in part by use of planning jargon)

    c. Treatment of transportation as an end in itself rather than a means to other community goals

    d. Too short-range political orientation of decision-makers

    e. Lack of public understanding

    f. Dearth of a protransportation constituency

    g. Lack of involvement of implementers in planning

3. Inadequate understanding and use of local regulatory powers, for example,

    a. Local decisions on land use, zoning, subdivision regulations, and official maps can either support or subvert transportation plans

    b. Lack of awareness that the integrity of transportation plans is essential to achieve nontransportation goals such as neighborhood quality, convenience of shops and services, and economic development.

    c. Lack of state–local coordination on right-of-way (ROW) protection

    d. Inappropriate role of private land developers in transportation implementation

4. Transportation plans and programs that are too far off-target

    a. Ignorant of current, local needs

    b. Overly ideal and too expensive with respect to available financial resources

    c. Inflexible and not adaptable to changing conditions

    d. No regular, thorough reviews and updating

    e. Uncertainty of future conditions regarding financing, energy, lifestyles, etc.

    f. Inflexible categorized Federal aid

5. Excessive adjudication and public opposition

    a. Highly litigatable impact-analysis requirements

    b. Inadequate citizen participation and information

Many of these problems are discussed in this and the next chapter. The sections to follow in this chapter focus first on Federal legislation, then on local/state legal tools for implementation. Attention then turns to transportation organization and coordination as a way to help overcome many of the problems. More formal implementation approaches revolve around the approval process, described in the succeeding section. This process is somewhat limited, however, in that it does not bring involvement from a broad cross section of local implementors and interest groups. Hence this is addressed in the next section, with the following showing how transportation-plan implementation really comes as part of analogous comprehensive urban development planning efforts. Finance and budgeting (addressing such problems as 4b, 4e, and 4f in the list just given) is the subject of Chap. 14.

13.2 LEGISLATION

As just noted, many of the implementation problems have received considerable attention from Congress, the state legislatures, and the transportation agencies at the two levels. The resulting Federal and state legislative acts and administrative regulations have had the effect of structuring the transportation planning and development process. The full impact of these acts and the directions in which they are aimed can be appreciated best in historical perspective. We thus have divided urban transportation legislation into three historical stages:

1781-1944   Exclusive local responsibility; Federal laissez-faire attitude toward cities

1944-1964   Federal and state aid to cities for highway purposes only

1964-   Expansion and restructuring of transportation planning and development processes to:

a. Provide aid for transit
b. Aid accessibility by the elderly and handicapped
c. Attend to social, economic, and natural environmental impacts
d. Create a metropolitan intergovernmental and interagency comprehensive planning and implementation process
e. Expand the Federal approval process

Subsections next profile each of these periods and trends.

13.2.1 Local Responsibility, 1781-1944

During this period almost no attention was paid by the Federal government to urban transportation. And, since most state legislators were dominated by rural interests, they too gave little attention to the topic. It generally was considered to be an exclusive local concern.

The Transportation Act of 1920 was the first really comprehensive transportation regulatory act passed in the United States. It was known primarily as a statement of national purpose, which took into account a new coalition of interests supporting Federal aid to highways. The coalition included the farmers, other agricultural interests, auto owners and manufacturers, bicycle enthusiasts, mail-order businesses, postal interests, and the railroads themselves.

Like the Act of 1920, the Transportation Act of 1940 was more a policy statement than an implementation vehicle. Until 1940, governmental regulation was restrictive and punitive in nature and without a formal policy statement of the needs, directions, and purposes of the transportation industry. The 1940 Act represented a realization on the part of the Federal government that the transportation needs of the country were multimodal, not simply a problem of highway construction. (Highway travel had become a more and more dominant mode with the decline of the railroads.) The 1940 Act was intended to be an overall guide to the direction of policy making agencies and organizations (especially Congressional) and an attempt to enable them to work toward a common goal of maximum efficiency in the transportation industry. The Act was to serve as a rough indicator of intended allocation of natural resources between the various transportation modes and other industries.

Another forerunner policy trend, Federal aid for roads, can be dated from the Federal-Aid Road Acts of 1916 to 1920. The Act of 1916 was the first authorizing Federal funds for intercity highways, giving great impetus to state road-building projects. It initiated a 30-year period in which Federal road aid was almost exclusively for rural and intercity roads. In fact, the 1916 Act specifically prohibited aid to urban areas of 2,500 or more, except routes where the houses averaged more than 200 ft apart [13.2, p. 553].

The 1916 Act established a basic framework of Federal-state cooperation in road construction. Indeed, for three decades all Federal-aid highway acts were amendments to the 1916 Act. It set a Constitutional basis for Federal-state cooperation by requiring specific assent by state legislatures to obtain and use funds in compliance with Federal law.

During the Depression of 1930 to 1940, Federal funds did go into highway construction in cities, not in recognition of their now serious transportation problems, but to provide work [13.4, p. 124]:

It was one of the "radical" schemes born out of the desperate national plight. When Franklin D. Roosevelt began the vast New Deal program to relieve unemployment, highway construction was among the first items on the list—especially those highways going in and out of our great cities.

The legislative basis for urban highway construction in 1933–1936 was emergency work relief. The Highway Act of 1936 regularized the basis by authorizing extension of Federal-aid intercity highways into and through urban areas. However, there was no requirement that funds be spent where most needed (such as in cities). Nor was there any specific allotment of funds (until 1944) that must be spent in cities. Nevertheless, during this period the urban problem was recognized and the groundwork for Federal aid was laid.

13.2.2 Highways-Only Aid, 1944–1964

The Federal-Aid Highway Act of 1944 is the milestone legislation marking the start of the period of Federal aid to urban road systems. That Act created a new appropriation category, the Federal-aid urban system, with monies apportioned among the states according to their urban populations. Authorized annual sums for the several Federal-aid systems are given in Table 13.1. Each state had the option of shifting up to 20 percent of its funds from one system to another. The percentage portions among the three systems were constant for the next quarter century. Federal aid was limited to 50 percent of construction and $33\frac{1}{3}$ percent of right-of-way costs.

The Act of 1944 is also noteworthy for providing for the creation of a 40,000-mile "National System of Interstate Highways" to connect "the principal metropolitan areas, cities and industrial centers, to serve the national defense . . ." [13.4 and 13.5]. Following the 1944 Act, the state highway agencies prepared, often for the first time, comprehensive highway plans for urban areas, indicating the preliminary locations of the proposed Interstates.

Implementation of the Interstate system was not specifically financed until the Federal Highway Act of 1956. It provided an extraordinary 90–10 split of capital costs. For most of a 13-year period, appropriations of $2,200 million were authorized to cover the Federal 90 percent. It expanded the system by 1,000 miles to 41,000 miles to accommodate urban bypasses and connectors. An important innovation in the 1956 Highway Revenue Act was the creation of the Highway Trust Fund, which for the first time linked Federal tax income from user charges, such as the gasoline tax, with highway expenditures to make the program self-financing [13.4, pp. 131–133].

Table 13.1 Annual Sums for Federal-Aid Systems

$225 million	45%	Primary system (intercity) including urban extensions
$150 million	30%	Secondary and feeder system (rural farm-to-market roads, mail and school bus routes)
$125 million	25%	Urban system
$500 million	100%	Total annual authorization

13.2.3   Transportation Planning and Development, 1964-Present

The last two decades have witnessed five major developments in urban transportation legislation—funding of transit, special provisions for the elderly and handicapped, environmental protection, and organization for coordinated discussion making.

13.2.3.1 Transit   By the 1950s, burgeoning human and auto populations, and deteriorating transit service forced the larger cities to undertake planning for transit. Several cities were outstanding in terms of intensity of effort and actual results. The San Francisco-Oakland Bay area counties in 1951, through the California Legislature, set up a transit commission with a $750,000 state grant. The resulting studies led to creation of the Bay Area Rapid Transit District (BART) and construction of the first new rapid transit system in the United States in 50 years. It is noteworthy that the project was initiated through a local bond issue and surplus Bay Bridge auto tolls, with no initial Federal aid.

Philadelphia created the Urban Traffic and Transportation Board in 1953 and, after rebuff from the state legislature, in 1958 started leasing improved commuter service from the railroads through a technique that was later (1960) institutionalized as the Passenger Service Improvement Commission [13.5, pp. 142-145]. Cleveland in 1954 began rail rapid transit service financed by revenue bonds repaid completely from the fare box.

Chicago has a history of land-use and transportation planning that can be traced back to the Burnham Plan of 1906. In the 1950s the city launched two projects of note here. The first was the inclusion of rail rapid transit in the median of the Eisenhower Expressway running due west from the Loop, in a real sense a realization of some aspects of the Burnham Plan. The second was initiation of the Chicago Area Transportation Study (CATS, 1955-1961), financed jointly with local, state, and Federal funds, which set the prototype model for the comprehensive planning studies connected in the 1960s in every U.S. metropolitan area.

These cities of course were exceptional in terms of on-the-ground accomplishment in instituting transit service. Many more cities, however, gave the transit issue serious study during the 1950s. Atlanta began transportation-policy studies through its Metropolitan Planning Commission in 1955, leading to creation of the Metropolitan Atlanta Rapid Transit Authority (MARTA) and construction of a regional rapid transit system (1971-present).

Similarly in Washington, D.C., transportation planning studies initiated in the 1950s led to Congressional appropriations for METRO construction in the 1970s. The appropriations were delayed for years by efforts to tie them to construction of several freeways opposed by many District residents.

The Bay Area, Philadelphia, Cleveland, and Chicago accomplishments are noteworthy not only because they pioneered the resurgence of rapid transit but also because initial action was taken without the incentive and help of Federal or state monies. The first Federal aid for transit was contained in the Housing Act of 1961, which provided for low-interest loans to states, localities, and other authorities for land acquisition, facilities, and equipment for mass transportation. This was very limited funding for very limited purposes, especially compared with the billions for highway transportation, but it was a start.

Appropriations large enough to aid urban transit on a significant scale date from the Urban Mass Transportation Act of 1964, with appropriations totaling $375 million. Improvements such as the 1965-1968 extension of rapid transit to the

Cleveland Airport were funded under this Act. The Federal share of a project could be as much as two-thirds, provided there was a comprehensive transportation plan. Otherwise the Federal share would be 50 percent. This expanded program of grants, loans, and demonstration projects was conducted by the Urban Mass Transportation Administration of the Housing and Home Finance Agency. Most important, the Act marked a turning point in the generally pessimistic outlook for transit by creating a hospitable climate for planning and action.

The 1966 amendments to the 1964 Act required (1) more rigorous technical studies as a prerequisite to grants, (2) transit management training grants, (3) university research and training grants, and (4) the Secretaries of Housing and Urban Development and Commerce (this was just prior to creation of DOT) to work together in research and development of new transportation modes and systems [13.5].

Urban transit aid started in the early 1960s, but it is appropriate here to note the major Congressional policy for highway financing through the 1960s and 1970s. Appropriations continued high during the 1960s, reflecting the growth in motor-fuel revenues. In 1970 Congress established the Federal-Aid Urban System (FAUS) and increased the Federal share of non-Interstate projects to 70 percent. The 1978 Surface Transportation Assistance Act set 1986 as the date by which all remaining Interstate portions had to be under contract or be de-designated. The Act also made Highway Trust Funds monies available for Interstate System resurfacing starting with fiscal year (FY) 1980. This was a response to the fiscal plight of state agencies caught between lower gas-tax revenues due to more fuel-efficient cars and continuing high inflation. Transportation financing is discussed further in Sec. 14.1.

The 3-C (continuing, comprehensive, coopeartive) metropolitan transportation studies initiated by the 1962 Highway Act (Sec. 13.2.3.4) provided local and Federal officials with persuasive technical support for Federal transit aid. Comprehensive transit plans were prepared in many cities, particularly where private transit was folding. Much more to the point: the impressive computer-modeled travel demand forecasts (Chap. 7) highlighted the social, economic, and political infeasibility in larger cities of meeting the demand primarily with urban freeways.

The Urban Mass Transportation Act of 1970 represented an even greater Federal commitment to transit. It authorized $10 billion over a 12-year period. In addition, a program of 18-year loans was established for purchase of real property or equipment for transit purposes. The 1970 Federal-Aid Highway Act authorized expenditure of highway funds for bus transit projects. The 1973 Act allowed use of Federal-Aid Urban System (FAUS) funds for mass-transportation projects and allowed the substitution of transit projects for withdrawn Interstate segments.

The National Urban Mass Transportation Assistance Act (1974) provided operating assistance for urban transit systems and funded capital equipment for rural communities (under 50,000 population). Rural operating aid (Section 18 funds) came in 1978. In the 14-year period of 1964–1978, urban transit thus was transferred from hundreds of dead or dying private operators to hundreds of subsidized public systems.

13.2.3.2 Elderly and handicapped The widespread movement to provide better accessibility to the elderly and handicapped was reflected in transportation. For wheelchair users, ramps were provided into buildings and curbs ramped at pedestrian crossings of streets. The 1970 Urban Mass Transportation Assistance Act required that transit systems consider the special needs of the elderly and handicapped. The Rehabilitation Act of 1973, Section 504, resulted in a series of regulations to assure that special efforts were made to provide accessibility for the handicapped on all transportation modes [13.4]. This rule, which required lifts or "kneeling" devices on all Federally financed buses to aid access by elderly and handicapped persons, was rescinded in 1981 in favor of a policy giving officials in each locality the authority to decide how to meet the transportation needs of the handicapped. The 1981 DOT Appropriation Act, Section 324, reinforced this by providing that DOT may not require purchase of lifts if the local grant recipient has submitted an acceptable alternative program for providing accessible transit service.

13.2.3.3 Environmental concerns The environmental movement also affected transportation legislation. The 1968 Federal Aid Highway Act, Section 4(f), required that the Secretary of Transportation not approve a program or project requiring use of a publicly owned park, recreation area, or wildlife or waterfowl refuge unless (1) there is no prudent and feasible alternative, and (2) all possible planning had been done to minimize harm.

The cornerstone environmental legislation was the National Environment Policy Action (NEPA) of 1969. Section 203 of the Act requires that, prior to a final decision, the Federal agency responsible for any Federal-aid project prepare and circulate a draft environmental impact statement (EIS) on the project to all affected local, state, and Federal agencies, private concerns, and public groups. The EIS must reflect a systematic interdisciplinary approach to planning and decision-making, including full investigation of alternatives in respect to their social, economic, and (natural and physical) environmental (SEE) impacts. The Act was enforced in nearly 200 court decisions during its first 2 years.

Because Federal-aid highway projects are actually planned by the states, NEPA posed a problem for FHWA. In 1972 this was resolved with Policy and Procedure Memorandum 90-4, "Process Guidelines," which required each state to develop an environmental action plan. The action plan must describe the state's organization, and technical and participatory, procedures for planning and implementing highway projects. Topics to be covered include SEE effects, alternative courses of action, involvement of other agencies and the public, decision-making process, responsibility for implementation, and fiscal and other resources.

In 1970, amendments to the Clean Air Act created the Environmental Projection Agency and authorized it to set ambient-air quality standards. This was significant for transportation, since 60 to 80 percent of the air pollution in metropolitan areas was traceable to motor vehicles. The Act set deadlines, and requires State Implementation Plans (SIPs).

13.2.3.4 Planning and decision-making process The complex problems of technologically, socially, and economically advanced society—pollution, inflation, integration, crime, resource limitations, etc.—required more coordinated approaches to Federal–state–local planning and decision-making. Local and state initiatives were matched by Federal aid and direction. Many of these efforts in the field of transportation planning utilized either "701" or "HPR $1\frac{1}{2}$ percent" funds. The former were initiated in Section 701 of the Housing and Urban Redevelopment Act of 1954, which provided Urban Planning Assistance to communities of under 50,000 population and to metropolitan planning. Highway Planning and Research (HPR) Funds were first authorized by the Federal Highway Act of 1934. Both acts required matching state or local funds according to prescribed formulas.

As Federal aid in many fields increased, there developed numerous incidents of local and state governments and agencies in a metropolitan area using Federally aided projects in uncoordinated and even conflicting ways. Congress responded by requiring comprehensive plans as a prerequisite in most grant programs. For example, the transit grants and loans provided in the Housing Act of 1961 could be obtained only where a plan for a coordinated mass-transportation system, as an integral part of a metropolitan comprehensive plan, had been or was being developed. The Federal-Aid Highway Act of 1962 contained a similar requirement, which, because of the magnitude and prestige of the highway program, had a greater impact. Section 302 of the Act reads in its entirety as follows:

> Highway Act of 1962, Public Law 87-866. It is declared to be in the national interest to encourage and promote the development of transportation systems, embracing various modes of transport in a manner that will serve the States and local communities efficiently and effectively. To accomplish this objective the Secretary shall cooperate with the States, as authorized in this title, in the development of long-range highway plans and programs which are properly coordinated with plans for improvements in other affected forms of transportation and which are formulated with due consideration to probable effect on the future development of urban areas of more than fifty thousand population. After July 1, 1965, the Secretary shall not approve under section 105 of this title any program for projects in any urban area of more than fifty thousand population unless he finds that such projects are based on a continuing comprehensive transportation planning process carried on cooperatively by States and local communities in conformance with the objectives stated in this section.

The 1962 Act provided the foundation for the 3-C (continuing, comprehensive, cooperative) transportation planning process. The Federal Highway Administration (FHWA) in 1971 established annual certification of 3-C processes. Congress continued to refine these comprehensive planning requirements. In the Demonstration Cities and Metropolitan Development Act of 1966, Title II required that Federal funds for any project in 34 program categories, including highways, transit, and airports, be dependent upon (a) existence of a metropolitan body composed of at least 50 percent of local elected officials of general government and (b) the body having at least 60 days in which to recommend approval, disapproval, or otherwise comment on the project. The 1966 Act was further strengthened by the 1968 Intergovernmental Cooperation Act, which required that national, state, regional, and local viewpoints be taken into account, to the extent possible, in planning of Federally assisted programs and projects. Implementing regulations were codified in 1969 and since updated in the Office of Management and Budget (OMB) Circular Letter A-95. The state-wide and area-wide reviews of Federally aided projects have become known as "A-95 reviews."

NEPA, mentioned above, influenced planning and decision-making processes within the Federal system far beyond the environmental areas. It has been highly significant in forcing (1) interagency and intergovernmental planning, (2) attention to the entire range of impacts on social, economic, natural, and physical aspects of the projects' environment and (3) a more open public decision process.

As a result of the A-95 and Environmental Impact Statement requirements, most states have established an interagency, intergovernmental Project Notification and Review System (PNRS) to assure all affected agencies and groups (1) information about any forthcoming decision on a state or Federal aid project and (2) the opportunity to comment.

In 1975 FHWA and UMTA finalized a joint regulation that shifted more transportation decision-making power to the local government in a metropolitan area acting collectively. The regulation required that the shared powers of the governments be exercised through a metropolitan planning organization (MPO) designated by the governor for each urban area (see Sec. 13.4).

Citizen participation, given a legal basis by NEPA and other acts, continues to be a Federal concern. In 1980, DOT set "Final Policy and Guidelines on Citizen

Participation in Local Transportation Planning." The net effect of the Congressional and state actions during this decade had been to force a public decision process that is more comprehensive, in that it takes into account the full range of relevant factors, and more open in that there is a more complete disclosure of the basis for any decision and a better opportunity for participation.

13.3 LOCAL/STATE LEGAL POWERS

While the legislation described above has led to most of the laws and regulations used for national guidance, they certainly have not eliminated the need for state and local actions. The legal tools available at these levels for implementing urban transportation plans are many and varied and include control of land use as it affects transportation. The basis for plan implementation is, of course, governmental authority, which is exercised through a number of important, long-established powers possessed by governments, whether they engage in conscious planning or not. Cities and counties are creatures of the states—that is, they have only that authority conferred on them by the states. Therefore, not all localities will have all these powers, and when they do, the authority may be limited in various ways.

Einsweiler et al. [13.1] have identified over 57 techniques or systems elements for growth guidance at the local level. All could touch, in some way, on transportation planning, but only the most relevant will be discussed here. Particular emphasis will be given to eminent domain, zoning, and official maps.

13.3.1 Fiscal Powers

Most governments (with the exception of some regional planning agencies) have the power to raise revenue, incur debt, borrow, make expenditures, collect and distribute funds, and make appropriations. They can levy taxes and special assessments.

Under the financial power there are three important considerations for planning: (1) the types and amounts of taxes can influence patterns of urban development and redevelopment; (2) similarly, the power to expend money for capital improvements (see Chap. 14) such as transportation, utilities, schools, and parks can shape land development; and (3) the power to spend money to construct and operate public transportation facilities implies the wise exercise of this authority.

Two examples of the use of fiscal powers are special assessments and creation of development districts. In the former, the cost of a specified facility, such as a local road improvement, is assigned either fully or in part of the owners of the adjacent, benefitting property. Development districts might be created around, say, a transit station. The allowed development in the station area then could be subjected to a special (usually additional) tax.

13.3.2 The Power of Eminent Domain[1]

Eminent domain is the power used by governments to acquire land or easements. The latter involves the right of utilization of land (e.g., for underground utilities) without actually owning it. The process of application of eminent domain is called condemnation. In the early history of highway building of America, this process was rarely used since land generally was available through voluntary sale by the owner or in many instances by owner dedication of private land to the public

---

[1] Adapted in part from [13.15].

use. In recent years, however, these two forms of acquisition have not been able to fulfill increasing transportation land needs. Thus the exercise of eminent domain has become a measure of frequent use.

Through the power of eminent domain a unit of government, within its jurisdiction, may take private property for public use without the consent of the owner, subject to payment of just compensation as prescribed by law. States are granted this power directly by the Constitution, but political subdivisions of states—counties, cities, towns, and villages—must receive such power by delegation of the state legislature.[2]

Condemnation law can be divided into two very broad categories: administrative and judicial. The administrative method involves the condemner filing official papers showing the plot of land with proposed improvement and a financial offer to the landowner or deposit in court of an award of compensation. Should the landowner wish to contest either the right to take the property or the amount given in compensation, then court proceedings may begin by his or her action with the burden of proof being the responsibility of the landowner.

The judicial approach requires that condemnation be started by an agency petitioning the court for a judgment on the right to take and the amount of compensation to be awarded. The court must assure itself that if the property owner is not satisfied, at least a full hearing is given. Once the court issues an order of condemnation and the landowner receives payment, the title of the property becomes the possession of the condemner, unless appealed within a specified time.

Statutory provisions in state law may fall under the general framework of either the administrative or the judicial law procedure. However, a great deal of variation exists among the different state constitutions and in enabling legislation for their respective political subdivisions. In general, four questions of condemnation can be identified as key problems in statutory provisions as they apply to states, counties, cities, towns, and special authorities:

1. Who has the authority to condemn?
2. Which property may be taken?
3. Which type(s) of legal estate may be acquired?
4. What procedure is to be followed?

Elements of some of these will be discussed next.

At the state level the two categories of property adopted most commonly are "right-of-way" and that containing "necessary road-building materials." While all states have provisions for acquiring road rights-of-way through condemnation of the actual real estate and any interest therein, only 42 have statutes authorizing the power to condemn land containing such necessary road building materials as gravel and rock (that is, the right to purchase such at a "reasonable" cost).

A third category of more recent origin is condemnation by highway or transportation departments of marginal land. In terms of road building, land is considered marginal if it is not a part of the actual right-of-way. Before the emergence of widespread urbanization as well as the increased use of the automobile, the condemnation of such land was an infrequent practice, but in recent years the benefits of such acquisition have become much more obvious. The acquisition of marginal land may serve the following benefits:

---

[2] In all highway projects, except the National System of Interstate and Defense Highways, condemnation is a power vested only in states and their political subdivisions.

1. To affect the economy, by acquiring an entire tract of land when the necessary portion plus the severance damages to the remainder would involve an equal or greater expenditure than if the entire tract were acquired with the right to salvage the remainder later.
2. To prevent the creation of small, uneconomical remnants of land.
3. To remove unsightly buildings and obnoxious uses and to assist landscaping.
4. To control the use of adjoining property for aesthetic, safety, or future highway development objectives by incorporating restrictive easements on a later sale of property.
5. To diminish right-of-way costs through the sale of acquired marginal land.

About two-thirds of the states have either constitutional provisions or statutes delegating in a specific manner the authority to condemn marginal land.

There are two general ways by which land can be acquired by state highway departments through the process of condemnation. These are the fee simple approach, in which the state has title (full ownership by deed) to the land condemned, and the easement approach, wherein the state has a limited right (i.e., for highway right-of-way only) to use land held in title by the original owner. While all states have provisions giving the state highway or transportation department these options, this is not always the case in cities, towns, and villages. A fee simple acquisition, of course, requires more initial investment, but it is preferred increasingly by state highway departments since easements almost always have reverter clauses requiring the return of the use of property (and therefore loss by the state of any financial investment incurred) to the landowner should the highway in question not be constructed within a specified period of time.

In those states granting counties the power of eminent domain, the authority granted always is less than that retained by the state transportation department. The power given may take numerous forms, but in a very simplified fashion three types seem prevalent:

1. Establish, lay out, open, alter, change, widen, and improve public roads and ferries in the county
2. Condemn property for county roads
3. Acquire property for the state

Cities, towns, and villages, like counties, must be granted condemnation powers by the state. Legislation for these entities is more fragmented than that directed toward counties. In a very general sense, however, the most common form of authority when granted is the power to "lay out, open, extend, widen, and improve public streets and alleys," and in furtherance of such power to acquire property through eminent domain.

A more recent expression of legislative intent in the transportation field can be found in the establishment of functionally separate authorities whose existence concerns a single purpose such as turnpike, bridge, or tunnel, and some more recent mass-transit projects. In general, the power of eminent domain is made explicit in a broad manner.

An interesting extension of the acquisition process is that of land banking. This is purchase in anticipation of urban expansion. The land may be made available for private development or retained in public ownership and leased to private parties or sold with deed restrictions.

13.3.3 The Power of Proprietorship

Governments are engaged in many enterprises and are owners of considerable property. For example, a city owns streets that, as noted, it acquires by dedication, purchase, or condemnation; it owns land and buildings; and it owns or operates utility systems. All three of these broad areas are of importance to planning because of the tremendous bargaining power they allow the city, particularly in dealing with land developers. It could well be that the bargaining position the city possesses by virtue of its power of proprietorship may be more significant in terms of development control than the police power (Sec. 13.3.4).

Two uses of the proprietorship principle are in location of facilities to influence growth and in provision (or not) of access to existing facilities. In the former, roads, sewers, and water systems are placed in certain areas to influence the further location of land developments there. In the latter, access can be obtained only by permit, and such is controlled, both in number and timing, so as not to overtax the developmental and financial capacity of an area. A particular example of this is the limitation of curb-cut access to highways (usually arterials) to reduce the side-traffic "friction" and thus congestion and accidents.

13.3.4 The Police Power

The general police power is the most comprehensive and pervasive of all powers of government. It is a regulatory power that restricts the use of private property in the public interest and can prevent a property use that is harmful to public welfare. In its original and broadest sense, the police power denoted the inherent authority of government to control people and things, but it is not quite so broadly defined today. It is "limited" to the power to establish the social order, protect life and health, secure people's existence and comfort, and safeguard them in the enjoyment of private and social life and the beneficial use of their property.

In some countries, and in some localities in the United States, the plan itself is a legal document emanating from the police power. This means that it has been developed and is supposed to be implemented exactly as indicated. Deviations from the plan are illegal, since they assumedly would not bring about the desired level of public welfare.

Somewhat less stringent (although often still strong) regulations coming from the police power are various types of environmental controls. Three are especially relevant to transportation planning:

1. Controls for flood plains, stream valleys, wetlands, shorelands, slopes, and mountainous areas. These have evolved to protect natural processes, such as flooding, stormwater runoff, and groundwater seepage, or to prevent development in sensitive resource areas. A good illustration is the prohibition of any building (including roads) on the mountain tops and slopes in Belconnen (Fig. 12.6) in Canberra, Australia.
2. Identification of critical areas. If certain areas are found to be particularly susceptible to problems created by new developments, such can be set aside as "critical." An example may be a refuge for an endangered animal species, which could be destroyed by, say, a new road.
3. Pollution controls. These might include air- and water-pollution standards and limits. For instance, the Environmental Protection Agency's "complex source" or indirect source standards dealing with air pollution caused by

traffic at major generators (e.g., large shopping centers or sports stadia) may be employed as a restraint on the location and size of such generators.

The police power is important because no other authority available to a city is so far-reaching with respect to relationships between government and the individual's personal and property interest.

13.3.5 Land-Use Controls

This major category consists of exercise of the police power in zoning ordinances, subdivision regulations, transferrable development rights, and mapped-streets ordinances. Each of these is designed for a specific purpose, and the most effective plan implementation is achieved if they are used to supplement each other. Each has some relevance to transportation planning. These have traditionally been local powers, but recently they have been used increasingly at the state level.

13.3.5.1 Zoning Zoning is one of the major tools of plan implementation. While its original purposes relate to the regulation of land use, it is frequently used to serve a number of transportation related functions—requirement of adequate off-street parking and limitation of intensity of trip generation, among other functions. Zoning is basically [13.16]:

> the governmental regulation of the uses of land and buildings according to districts or zones. It is a means of insuring that land uses within the community are properly situated ·in relation to one another, that adequate space is available for various types of developments, and that the density of development in each area is held at a level which can be properly served by governmental facilities and will permit light, air, and privacy for persons living and working within the community.

To accomplish its purpose, a city can zone with regard to land usage, lot area, population density, size of all yards and open spaces, building setbacks, parking, signs, and billboards. It can prohibit some new uses and eliminate some existing ones. In addition to controlling industrial and commercial noises, fumes, smoke, and particle emissions, zoning can even control erection of structures in the air-space approaches to airports. Zoning has become an infinitely sophisticated tool, with new approaches and techniques being developed all the time to meet new needs emerging in a complex society. Interestingly, planning and zoning often are misunderstood, and many times the words are used interchangeably.

Zoning has been known as a "preventive" device, intended to deter community blight and deterioration by prescribing standards for uses in separate areas and by assisting in the control of new buildings. It requires similar uses in given areas and thereby helps keep out blighting factors and keep up property values in all areas. It must be pointed out here that this concept is changing somewhat, although not rapidly. Some courts are now more prone than ever before to allow zoning "for the future" rather than using it as a preventive tool.

The theory of zoning is to generate land-use improvements by confining specific classes of buildings and uses to certain localities without causing owners undue hardship. Prior to general public acceptance of zoning in this country, haphazard location and use of buildings existed in all our municipalities. Zoning attempted to remedy this situation by imposing regulations that would exclude new uses and structures prejudicial to the preservation of the true character of a neighborhood.

Zoning that regulates the use of land, irrespective of its ownership, aims not primarily at protecting the value of the property of particular individuals, but instead at promoting the welfare of the whole community. Zoning does protect property values, of course, but it is intended to do more: to ensure the availability of adequate light, air, and accessibility to all property. It protects the public health and minimizes the number of fire hazards.

The essential considerations of zoning require the municipality to adhere to basic purposes (including security from fire, panic and other dangers; promotion of health, morals, and general welfare; adequate light and air; prevention of overcrowding and of undue concentration of population; the character of the district and its peculiar suitability for particular purposes) with a view to conserving the value of property and encouraging the most appropriate use of land throughout such municipality.[3]

In Washington, D.C., Paris, and some other cities, zoning ordinances contain limitations on building heights. While these were created in part for aesthetic reasons (e.g., in the case of Washington, D.C., to allow views of the Capitol and other monuments), they also have important traffic-generation impacts. Obviously if building heights are restricted, the number of people working, living, shopping, etc. at any one time in those buildings also is restricted. This may have the effect of increasing trip length, but it helps lower localized congestion.

Two other zoning techniques that may have great relevance to transportation planning, particularly at the neighborhood scale, are planned unit developments (PUDs) and performance standards. PUDs combine some of the attributes of zoning and subdivision regulation (see Sec. 13.3.5.2). A developer is required to submit a plan for the area under question. A mixture of uses and residential buildings (single-family, apartments, townhouses, etc.) is possible, instead of the rather strict delineation between those inherent in zoning. Schemes are worked out in an interactive process between developers and public officials. This is particularly valuable for obtaining better control of road design and use.

The performance standards approach involves a set of standards relating to acceptable levels of nuisance or side effects. These have been employed, for example, in industrial zoning to control noise, glare, and emissions as well as demands for traffic services, sewerage, and the like. Performance standards are a means of specifying acceptable levels of external effect and letting the market determine use, as opposed to specifying precise acceptable uses ahead of time.

13.3.5.2 Subdivision control   Subdivision regulation is crucial, because once large tracts of land are broken into individual parcels, the pattern of development is irretrievably set. Thus a subdivider is taking action that is of great importance to the community—to the homeowner, to the governing body, and to the general public, the taxpayers. It is through subdivision regulations that the community interest is expressed—and protected.

Subdivision control plays a fundamental role in the development of a community because, although a city is something more than a total of its land subdivisions, much of the form and character of the city will be determined by the quality of those subdivisions and the standards built into them.

The purpose of subdivision control is to prevent congestion of population and to provide land development in accord with established design standards, to create sound neighborhood patterns, and to integrate the area involved—sooner or later—into the community as a whole. For this purpose, the regulations usually establish standards to be met in construction of public improvements and often require the developer to provide basic improvements before sale of any lots.

Early attempts at subdivision control are likely to be clumsy and inefficient, as

---

[3] *Thornton v. Village of Redwood* IIIA. 2d 899 (N.J., 1955).

is the case with many new devices, but continued use will prove its value. Maximum effectiveness can be obtained only if there is a land-use plan and a mapped-streets plan (see Sec. 13.3.5.4), because there are prerequisites to good subdivision control. All these tools of planning implementation must be coordinated for best results.

Subdivision control, like zoning, could be carried out without long-range planning, but would be limited to the prevention of obvious mistakes—such as excessive street grades, awkward intersections, and substandard improvements—in an individual plan. This would be subdivision control at its weakest. While this alone would be of considerable value [13.9],

> true measures of success are the creation of sound neighborhood patterns; integration of residential development with other land uses; acquisition of sites for public parks, schools, and other facilities; and the continuation of the transportation network, among others.

Subdivision control, therefore, is an integral part of the planning process, an important tool of plan implementation. If the municipality has not developed a comprehensive plan, intelligent subdivision control is not possible. It is in this broad operation, based on a comprehensive plan, that subdivision control's greatest contribution to orderly community growth can be made.

The statutes of all states, save one, make some sort of provision for subdivision regulation, but an examination of these statutes indicates a wide variation in their provisions, and the extent of enforcement also varies greatly from state to state.

13.3.5.3 Transferable development rights (TDR)   A TDR program provides all landowners with development rights. Then, based on a plan, areas are designated where development is restrained and other areas where such is permitted. Owners in restricted parts can sell their rights through a market mechanism to owners in permitted areas, who may have fewer rights than they need in order to carry out their developments [13.1].

This technique was created in response to the increasingly restrictive use of the police power, which has resulted in "windfalls" and "wipeouts" to individual owners and resulting great pressures on them to be on the former side of the ledger. From the transportation planner's point of view, however, the main concern is for traffic generation and its distribution. A TDR plan should help to identify the potential location of major generators, although there will be some questions of which permitted areas will be developed first (if at all).

13.3.5.4 Official map   Enabling legislation available in most states is a prerequisite for local use of official map powers. The exact provisions vary from state to state. The following procedure is typical:

1. A comprehensive plan or major street plan must be prepared and officially adopted, as evidence that the proposed street or highway is part of a well-considered plan for the community's future.
2. The proposed "taking line" for the ROW must be surveyed or otherwise indicated on property maps of appropriate scale.
3. The local legislative body may adopt the map and prohibit the issuance of building permits or the subdivision of land within the bed of the proposed ROW until the elapse of a waiting period (say 60 days) to permit public acquisition of the needed parcel.
4. When building permits or land subdivisions are requested, they are then checked against the map. If there is no conflict, they are processed as usual. If, however, the proposed ROW is threatened, approval is withheld for not

more than the permitted length of time, and the chief executive and legislative body is notified so that acquisition can proceed.

5. Negotiation with the builder or subidivider for acquisition of the tract at a mutually satisfactory price then takes place. Sometimes the subdivision design can be modified so as to accommodate the ROW, or even dedicate it. Similarly, sometimes the proposed building for which the permit is requested can be relocated so as not to conflict with the ROW. The permit can then be issued.

6. If negotiation fails to produce a satisfactory solution, the local jurisdiction must initiate condemnation proceedings to acquire the ROW.

7. If the condemnation proceedings are not initiated, or fail, the permit or subdivision must be processed at once.

When systematically administered and adequately supported, the official map process has been useful in insuring space for new transportation facilities.

13.4 GOVERNMENTAL ORGANIZATION AND ROLES

The evolutionary development of urban transportation legislation just described has resulted in a complex, intergovernmental organizational structure within the Federal system. In no two metropolitan areas or states is the organization exactly the same. However, organization generally exists at four levels, each with functions that mesh with those of other levels. This concept is summarized in Table 13.2, in which major functions are displayed across the top row. For each function at each level an indication is given of most likely responsibilities (lead role, participant roles, financial support, and supporting role). This table helps to set the framework for the discussion to follow, as well as to demonstrate the complex interconnections within and between governmental levels.

13.4.1 Federal Role

The U.S. Department of Transportation (DOT) administers the financial and regulatory actions set by Congress. It is also responsible for the Interstate system. Other Federal agencies play roles in human services transportation, in urban development, and in assessing the impacts of urban transportation (e.g., through the Project Notification and Review System).

Table 13.2 shows the Federal role as solely financial. This could be somewhat misleading. On its strong financial role the Federal government has piggybacked strong policy leadership in emphasizing national goals and implementing regulations. These goals have included the Interstate System, clean air, safety, transportation for the elderly, handicapped and rural, and a metropolitan-level, shared-power system of urban transportation decision making. The 1980s will probably see some reduction of the Federal role in regulation, but continued requirements for shared metropolitan decisionmaking, systems planning, and short-range urban programming.[4] Federal agencies are expected also to continue a strong role in technology development and dissemination, and in professional training [13.19, 13.20, 13.21, 13.22].

In addition to the Federal transportation agencies, represented in Table 13.1, a

---

[4] Programming, as a process of budget allocation (through a transportation improvement program, of TIP), is discussed in detail in Chap. 14).

Table 13.2 Roles of Governments: Likely Responsibilities for Components of the Urban Transportation Planning Process*

Agency	Strategic & policy planning	Long-range system planning	Subarea & corridor planning	Short-range & project planning	Programming		
					Priority coordination	Budgeting	Project development
Federal	F	F	F	F	F	F	F
State	P	P	L/P	L/P	P	L	L
Region	L	L	P/L	P/L	L	P	S
Local							
County	P	P	P/L	L/P	P	L	L
City	P	P	P/L	L/P	P	L	L
Operating agency	P	P	L/P	L/P	P	L	L

*L = lead role; P = participant role; F = financial support; S = supporting role.
Source: [13.19, p. 61].

number of other Federal agencies have roles legislated into the urban transportation planning and programming process in recent years—Environmental Protection Agency (clean air), Corps of Engineers (waterway permits), Equal Opportunity Office (affirmative action), Health, Education and Welfare (social services), Department of Energy (fuel conservation), Housing and Urban Development (land use) and the Office of Management and Budget (regional budget reviews), among others. While the names of the agencies, and the content of their regulations, will undoubtedly change during the upcoming years, it is now well established that transportation planning, programming, and decision making service a range of national, nontransportation goals.

13.4.2 State Roles

Under the Federal constitution, the states possess the power for planning, design, construction, and operation of Federal-aid and state highways, and administration of Federal and state transit aid.

In most states the internal organization of transportation agencies is either by function or by mode, or most often by some combination of these. Organization by function provides divisions for systems planning, programming, location, design, structures, ROW acquisition, construction/contract management, operations, and maintenance. Organization by mode is most often for highways, transit, and railroads. State transportation agencies also have regional and district offices and frequently contract with local governments for work of a local nature.

13.4.3 Metropolitan/Regional Roles

Metropolitan planning organizations—MPOs—are the newest partners in the Federal transportation organization. MPOs are not a level of government, but rather a forum through which the governments responsible for, or affected by, transportation plans, programs, or projects coordinate their actions. This coordination meets the Federal 3-C requirements (see Sec. 13.2). It is in effect a sharing of the legal power entrusted in a responsible implementing agency, state or local, with other affected governments.

MPOs take many forms depending on local conditions and desires. It may be part of the gubernatorially designated A-95 agency, generally a Council of Governments (COG) or Metropolitan Planning Commission, or it may be an independent transportation planning agency. Typically there are two constituent bodies: a policy committee or board, and a technical committee. The policy committee is composed of elected officials of local general governments and may or may not include the Governor, Commissioner of Transportation, or other state official as a voting member. The Technical Committee consists of the heads of the planning, implementation, and operating agencies in the region. There always is also a citizens advisory committee or other provision for citizen participation. Table 13.2 shows the MPO as the lead agency in several functions. This regional orientation is important because many (but not all) of the components of the urban transportation system have regional significance, impacts, and costs. However, other agencies should be actively involved in the process, and should hold differing responsibilities as a function of the issues and projects considered. In particular, the leadership role in planning should be closely related to responsibilities for implementation.

The Transportation Improvement Program (TIP) continues to be a viable mechanism for achieving intergovernmental agreement on project priorities in support of Federal funding requests. While the setting of priorities is generally a regional function, budgeting for specific local projects should originate with local jurisdictions. This should reflect a "bottom-up" approach to priority setting [13.19]. Often, there is a conflict between Federal requirements and local motivations. Recent sympathies have been that national goals and policies insofar as possible be recommendations, supported by procedural guidelines (i.e., nonmandatory). Planning and programming tasks "become significant *only* if they are strongly linked to locally-defined goals" [13.19] that can provide the basis for local decisions to pursue them. Examples of such issue areas are air quality and the transportation needs of the elderly and handicapped. To relate to local concerns as effectively as possible, urban transportation planning and programming must become more "strategic" [13.19, 13.20]. Strategic planning focuses on how to respond to a given problem situation, on defining the desired goals and objectives, and on selecting the best policies for accomplishing the objectives.

The regional MPO is also the lead agency for urban systems planning. Systems planning will become more closely tied to the information needs of local decision-makers. In particular, it would test major actions for their impacts and system effectiveness as candidates for the transportation improvement program (TIP). The MPO also has the lead role in setting priorities and coordination of the TIP.

In two areas, the regional agency might be designated lead agency under particular circumstances. In subarea/corridor planning situations where land-use considerations are dominant, a local government might have the leading interest. However, where the main concern was alternative route locations or modes, the state is more likely to be the appropriate lead agency. Short-range (TSM) and project-planning activities are more directly in the implementation "pipeline" and would tend to have the ultimate implementing agency leading. In these two functions, the MPO might lead only under limited circumstances, such as when many jurisdictions are involved and/or when it possesses particular expertise. An example might be a car-pooling project.

13.4.4 Local Roles

Local governments—cities and counties—while legal creatures of the state, have long built and maintained roads. Traffic engineering has also been a forte of urban governments. The authority and interest is still there; only the funding is more shy. Table 13.2 shows cities and counties as lead agencies for projects that are locally funded, of local scope, or where the state has delegated the lead. The 1962 3-C act and the implementation of the 1973 law to require MPOs and TIPs have also given local governments a strong new role in initiating and programming projects. Operating agencies, like transit or parking authorities, whether public or private, have roles similar to those of local governments [13.19–13.21].

Although not represented in Table 13.2, it is not unusual for local nontransportation agencies to be involved in planning or programming a particular transportation project that affects their interest. These actors may include agencies responsible for comprehensive planning, land use, parks, education, recreation, social services, environment, or economic development. Transit operations may be organized as a department of local government, contracted for by a local government, or vested in an independent, usually multijurisdictional, authority.

13.5 IMPLEMENTATION PROCESS

Some earlier sections (see 11.1 and 11.4) may have left the impression that transportation planning and implementation are simply two sequential steps in the development of a transportation project or system. Actually, development of a transportation project or system is a complex, multisequenced process, each stage of which may come in parallel or in series with others. In all states the process is conceptualized and published for public information in the transportation agency's "environmental action plan."

Typically, the process will have the elements shown in Fig. 13.1. In addition there usually are two major subprocesses central to implementation: (1) the intergovernmental "approval process" and (2) the involvement of public groups and other agencies at key points in the process. In the discussion to follow, the principal FHA approval points are indicated with (F) and community involvement points with (C). After a brief outline of the major stages in development of a transportation project or system, we will look at the approval and involvement subprocesses.

1 Strategic planning (direction-setting) phase Strategic planning is the determination of goals and broad policies for achieving them (Chap. 12). It usually involves legislative and interest groups. It is needed whenever an agency's mission requires redefining, when new technical, political, social, or economic problems affect transportation (particularly when the tradeoffs between goals must be rethought), and when financial resources are scarce. Strategic planning has become a continuous necessity for many agencies. Several states (e.g., Pennsylvania and

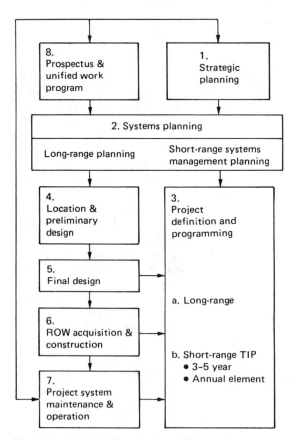

Fig. 13.1 Transportation development process.

Florida) have established units for this function. As described in Sec. 13.4, these activities are conducted by, or with, the locality directly affected, usually a local government or MPO. In the process of strategic planning, (a) problem analysis (C) is often the seedbed of project ideas, and (b) goals and objectives (C) become linked to the project.

2 *Systems planning stage* Systems planning involves four major components:

a. Prospectus and intergovernmental agreement (F) for MPO activity
b. Systems planning study (F,C), which ascertains that the project would be an integral part of the future transportation network, and that the network is a desirable, economical plan for accommodating future travel demand
c. Long-range program of projects (C)—15 to 25 years
d. Short-range systems management planning (F,C), a parallel step to the preceding three, which focuses on operational improvements in the present transportation system

3 *Project definition and programming* This is done by local and state transportation officials through the MPO and has four primary sections:

a. Preliminary short-range program of projects (C), based on local and state priorities for 3 to 10 years; updated annually with active local government participation.
b. Preliminary level of project significance, in respect to social, economic, and environmental (SEE) impacts; is publicized widely to secure involvement (F,C) for budgetary purposes.
c. Final level of project significance, and cost estimates for location, preliminary design, EIS, involvement, final design, ROW purchase and construction.
d. Final short-range program of projects (F), scheduling funds by years and source. Based on estimated project cost and difficulty, and with a firm annual element for the next year.

4 *Location and preliminary design* After a project has been identified, the following have to be prepared:

a. Descriptions of alternative locations, and sketch designs
b. Social, economic, and environmental (SEE) impact evaluations (C,F) of alternative locations
c. Draft EIS and location public hearing (C,F)
d. Final EIS and location preliminary design report (F)
e. Local government approval
f. Area-wide, interagency (A-95) review (C)

5 *Final design* The most detailed step then involves:

a. Final design: Constructing drawings and specifications
b. Evaluation and certification that the design is compatible with SEE requirements (C,F)
c. Design public hearing (C), final design report (F); local government approval and arrangements for street closing, variations, household relocation, detours, etc.
d. Supplemental project specification and engineering inspection (F)

6 *ROW acquisition and construction* The last step requires five actions:

a. Right-of-way acquisition; household and business relocation.
b. Contract preparation for letting (F)
c. Advertisement and awards of contracts (F)
d. Project construction (C)
e. Inspection and project acceptance (F)

7 Operation and maintenance (C).

13.5.1 The Approval Process

Pushing a planned program toward implementation requires a major intergovernmental coordination. The progress generally is measured by the securing of approvals at various stages of project development. While the preceding descriptions focused on FHWA (F) approval, more generally officials of all affected governments participate to assure that the project meets planning, engineering, environmental, and administrative requirements. As was noted earlier, efforts were intensified in the early 1980s to streamline regulations governing the approval process.

13.5.2 Community Involvement of Local Implementors and Interest Groups

Three other local sets of actors, in addition to the planning staff, can be vital to prompt, economical implementation and should be involved in planning well before the time for implementation. One set is the agencies who will be involved in implementation. Examples might be housing agencies for relocation of families, education officials who may have to relocate or modify schools and service districts, transit operators who may need to adjust mass-transportation services, utility companies with affected facilities, and environmental protection and redevelopment authorities whose plans are affected by the transportation project. Elected officials are also involved, since their approval of transportation plans in general is a condition for implementation. The final set of actors who can make or break implementation is the affected public groups. General support from these groups is the basis for approval by elected officials. Even a single-, well-organized group may trigger legislative disapproval or time- and money-wasting litigation.

The concepts and techniques of involving these actors in the transportation planning process were discussed in Chap. 11. In this section we want only to point up those problems that particularly affect implementation and outline a general strategy for laying the essential groundwork for implementation during the planning effort. Some of the reasons for implementation failures most cited by transportation officials are [13.3]: disinterest by officials because the transportation plan fails to address the needs and goals of the community and/or of the decision-makers; lack of adequate, understandable information on the plan; turnover of officials during planning or before implementation; their short-range political orientation—the need for immediate results "before the next election"; lack of public understanding; and lack of a protransportation constituency.

All of these problems are addressed by a general strategy, variously called "constituency building," "consensus building," or "policy chain" [13.1]. In a broad sense the strategy is a community-wide version of the widely-used management by objectives (MBO), with the following eleven guidelines applying:

Process design

1. Identify and maintain involvement of key persons—power holders, elected officials, opinion leaders, planners, implementers, affected groups, etc.

2. Create and utilize channels of interaction between the key persons.
3. Provide technical support. Clarify roles and responsibilities of participants, including the ultimate decision-makers in subsequent steps.

Advance planning

4. Identify motivating needs and values, and problems constraining their realization.
5. Devise and adopt explicit goals, objectives, and strategies, based on the anticipated consequences of various alternatives.
6. Detail the objectives and strategies into an explicit plan of what is to be achieved (e.g., a comprehensive development plan, as in Sec. 13.7).

Management planning

7. Develop and approve a 3- to 10-year, year-by-year improvement program of the action steps to be taken in relation to the goals, objectives, and strategies; specify cost, financing sources, responsibilities, timing, etc.
8. Prepare detailed final plans for the improvement objectives—projects, programs, organizations, annual budgets, etc.

Operations

9. Execute the projects, programs, budgets, and other objectives.
10. Monitor not only the performance of the operations, but also exogenous factors and the general condition of goal achievement.
11. Involve all key persons in reassessment of needs and values problems, goals, objectives, and strategies in light of the surveillance results.

Comparison of the foregoing steps with the approval process for development of a transportation project reveals the ease with which they can be combined. This kind of continuing involvement can build a procommunity interest, profuture constituency. The key is directly relating local needs and values to project and program implementation, thus making the development and realization of plans over the years easy and worthwhile to follow. As pointed out in Chap. 11, the technical and nontechnical aspects of the process can be meshed for this purpose. A policy chain thus has been forged [13.1] that can facilitate the realization of transportation plans.

In summary, the cast of actors in implementation is much the same as that in planning. The roles are different. Without an appreciation of the diversity of these actors and their often conflicting interests, the process will appear to be only a complex, sophisticated technical activity. Actually the pressures from the participants in the process can be frequently conflicting and centrifugal. A good implementation process is one that generates out of the conflict and centrifugality a course of effective, cohesive action. To do this it gives an appropriate role to each of the actors and yet facilitates timely, goal-oriented transportation decision making. Timeliness is critical; tardiness in implementation decisions can result in loss of funds, unmet transportation needs, loss of political support, and loss of agency credibility. Goal orientation is also critical, since individual decisions must add up to quality transportation service. Occasionally, transportation programmers become paranoid about particular group pressures, but mostly the latter are recognized as an inevitable part of democratic decision-making, a part to be blended, sometimes offset, with sound technical work and leadership.

13.6 LOCAL COORDINATION

Perhaps one of the most difficult tasks in organizing for implementing plans is that of coordination. In fact, this can be so complicated in many cases that a coordination plan often is a necessity in itself. The extent of the problem is described in a report by the General Accounting Office in which 114 Federal programs that provide financial assistance to the transportation of people were identified [13.13]. Note that these are only Federal programs, and also do not include efforts on that level through income transfer, tax expenditure, and credit programs.[5] The difficulties of coordination thus can be quite sizeable.

Coordination often must take place between agencies at a given level of government, between different levels, and between public, quasi-public, and private organizations. Each type of coordination varies somewhat from the others, with some concerned primarily with funding, some with organization, others with legislation and regulation, and still others with technical assistance, training, and the like. Federal regulations, for example, may spell out for a particular program [13.14, p. 6]:

1. Target group eligibility
2. Program administration responsibilities
3. Types of funding arrangements and amounts
4. Provider eligibility
5. Payment methods
6. Service accountability

While these regulations may be reasonable for each individual program, they may be inconsistent with regulations for others, with the result that coordination is substantially hindered. A look at Table 14.3 shows, for instance, that some programs only can serve certain clients, some provide capital grants only to nonprofit organizations, and so on.

While coordination concerns are high for just about every transportation endeavor, they are particularly problematic in human-services transportation, which will provide the main mode for illustration here.

There are many ways in which coordination can take place. One classification of "services integration linkages" is presented in Table 13.3. Notice that this scheme includes joint planning (dimension 1) and transportation (dimension 25) efforts as linkages in themselves. This highlights the idea that transportation can be viewed as only one element in a coordination plan involving many different services.

At the state level there seem to be four major groups of linkages for broad coordination of specialized transportation services [13.6]:

1. Legislation or executive orders
2. Interagency committee or task force
3. Interagency agreements
4. Technical assistance: state to local

An investigation done by the Social Services Research Institute [13.6] found that 33 of 50 states had interagency committees, 11 of which were required by law or executive order. Furthermore, 15 states provided wide-ranging technical assistance to many providers, and 9 used state funding as a coordinating mechanism.

---

[5] For definitions and discussions, see Chap. 14.

Table 13.3 Dimensions of Services Integration Linkages

1. Joint planning: The joint determination of total service delivery system needs and priorities through a structured planning process.
2. Joint development of operating policies: A structured process in which the policies, procedures, regulations, and guidelines governing the administration of a project are jointly established.
3. Joint programming: The joint development of programmatic solutions to defined problems in relation to existing resources.
4. Information sharing: An exchange of information regarding resources, procedures, and legal requirements (but not individual clients) between the project integrator and various service providers.
5. Joint evaluation: The joint determination of effectiveness of service in meeting client needs.
6. Coordinated budgeting/planning: The integrator sits with all service providers together or individually to develop their budgets, but without any authority to ensure that the budgets are adhered to; or the traditional service agencies develop their budgets together.
7. Centralized budgeting: A centralized authority develops the budgets for the traditional service agencies with the authority to ensure that they are adhered to; may or may not include central-point funding.
8. Joint funding: Two or more service providers give funds to support service; most often in a broad programmatic fashion.
9. Purchase of service: Formal agreements that may or may not involve a written contract between the integrated system and some other party or among agencies to obtain or provide service; generally a fee-for-service arrangement.
10. Transfer of budget authority: Funds are shifted from one agency within the integrated system to another agency in that same system.
11. Consolidated personnel administration: The centralized provision of some or all of the following: hiring, firing, promoting, placing, classifying, training.
12. Joint use of staff: Two different agencies deliver service by using the same staff; both agencies have line authority over staff.
13. Seconding, cross-agency assignment: One or more employees are on the payroll of one agency but under the administrative control of another.
14. Organizational change across the agencies: Service agencies in the integrated system or newly created agencies receive staff or units from another agency in the system and/or an umbrella organization is created.
15. Organizational change within the agency: Reorganization of agency staff or organizational units involving changes internal to each organization only (may be similar changes in each agency).
16. Colocation of central offices: Central administrative offices for two more agencies at the locale are relocated at a single site.
17. Colocation of branch functions: Several agencies colocate personnel performing branch as opposed to centralized administrative functions at a single site.
18. Outstationing: Placement of a service provider in the facility of another service agency; no transfer of line authority or payroll responsibility takes place.
19. Joint record keeping: The gathering, storing, and disseminating of information about clients.
20. Joint grants management: The servicing of grants.
21. Central support services: The consolidated or centralized provision of services such as auditing, purchasing, exchange of material and equipment, and consultative services.
22. Satellite services: Service provided wherever personnel from one service agency are restationed so as to increase the number of site agencies in the integrated network.
23. Joint outreach: The systematic recruitment of clients.
24. Joint intake: The process resulting in the admission (including determination of eligibility) of a client to the provision of direct service.
25. Joint transportation: Provision of transportation to clients on a joint basis.

Table 13.3 Dimensions of Services Integration Linkages (*Continued*)

---

26. Referral: The process by which a client is directed or sent for services to another provider, by a system that is in some way centralized.
27. Diagnosis: The assessment of overall service needs of individual clients.
28. Follow-up: The process used to determine whether clients receive the services to which they have been referred and to shepherd the client through the service system.
29. Case conference: A meeting between the staff of two or more agencies who provide service to a given family for the purpose of discussing that family either generally or in terms of a specific problem, possibly determining a course of action and assigning responsibility among the agencies for implementing the solution.
30. Case consultation: A meeting of staff members of agencies who provide service to a given family for the same purposes as specified in "case conference" above.
31. Case coordinator: The designated staff member having prime responsibility to assure the provision of service to a given client by multiple autonomous providers.
32. Case team: The arrangement in which a number of staff members, either representing different disciplines or working with different members of a given family, work together to relate a range of services of autonomous providers to a given client. The primary difference between case conferences and case teams is that the former may be ad hoc whereas the latter involve continuous and systematic interaction among the members of the team.
33. Joint data system: A multiagency machine or computerized record keeping system containing as a minimum information regarding clients contacted and treated.

Source: [13.8,pp. 48-51].

One interesting example of a service coordination effort, including transportation, was that undertaken by the Community Life Association (CLA) in Hartford, Connecticut [13.8, pp. 16-18]. The main fiscal linkage was a pooled fund, made up of public and private monies contributed to the CLA (e.g., from the City of Hartford and the United Way). The CLA requested these funds on the basis of detailed program plans. The money then was employed by the CLA case managers to issue service purchase orders against the pool to obtain most necessary services on behalf of a client, within the context of the CLA program objectives.

While the service purchasing power was not directly in the hands of the client, all decisions to initiate and continue purchase of service from an agency were controlled by the client, relying to a greater or lesser extent on information and advice provided by the CLA case managers [13.8, p. 16].

Two major advantages of the pooled fund concept were that (1) monies could be employed for services not provided by public agencies, or on an interim basis for those which could be obtained only after significant delay (e.g., during long application/processing periods); and (2) there was close association between service provision and client demand and satisfaction.

On the latter, when it was determined that the service was at fault, the purchasing power provided by the pool allowed the client's dissatisfaction to affect change in service delivery. The client, with the case manager, could attempt to resolve the problem with the original agency, could shift to another agency for service, or if dissatisfaction was more broadly with a type of service, could elect an alternative method of service. The case managers recorded these decisions, documenting client preference and demands for certain agencies and certain services [13.8, p. 16]. This process allowed managers to identify service gaps and inadequacies and aid in the development of appropriate new or expanded services.

The major disadvantages of the CLA concepts were (1) the very complex and expensive accounting procedure; and (2) tight controls on certain types of funding. An illustration of the second was transportation in the WIN (Work Incentive) program. This could not be provided until a registrant referred to the CLA became a certified WIN participant.

In summary, the CLA concept involved a package of 12 of the 33 linkages described in Table 13.3:

1. Joint planning	20. Joint grants management
3. Joint programming	24. Joint intake
6. Coordinated budgeting/planning	25. Joint transportation
7. Centralized budgeting	26. Referral
8. Joint funding	28. Follow-up
9. Purchase of service	31. Case coordinator

Joint funding (dimension 8), of course, was the main linkage.

Another coordination mechanism that has had much exposure in human service transportation is the brokerage concept.[6] Usually the agency in charge of such a system neither owns nor directly operates any of the vehicles. Instead it contracts with existing community transportation providers, both profit and nonprofit, to give services in the vehicles already owned by them (and possibly new ones purchased for the brokerage activity itself).

A good example of brokerage was that undertaken in Pittsburgh. The Port Authority of Allegheny County (PAT), in charge of regional transit, contracted with a private firm, ACCESS, to act as a third-party broker. The firm in turn contracted with individual providers, largely using per vehicle·hour charges.

All elderly and handicapped citizens in the ACCESS service area were eligible for the service; however, only those citizens so disabled that they would not use regular PAT transit service were eligible for the PAT subsidized fare. Individuals who wished to be certified for eligibility for this subsidy made an appointment with the Easter Seal Society, which was under contract to ACCESS to screen applicants. The association used a mock-up of the front end of a regular transit coach; if an individual could not mount the first step, he or she was certified as eligible for the subsidized fare. Approximately two-thirds of those so certified were in wheelchairs; the other third used walkers or other devices and were semi-ambulatory (note that certification patterns are not equivalent to ridership patterns). By October of 1980, ACCESS had certified over 1800 persons for the fare subsidy.

Those riders eligible for the PAT subsidy (a directed user-side subsidy) purchased a book of ride tickets or scrip for 25% of the face value. They used this scrip when they purchased a ride with ACCESS. ACCESS carriers and drivers were not permitted to take money from clients.

Various agencies could purchase service from ACCESS for their own clients. ACCESS tried to be very flexible and responsible to the needs of agencies. An agency could have a formal contract for service or simply an oral or written understanding that sets up a monthly charge account. Agency-sponsored trips generally were based on the same fare schedule used to compute all other trips—that schedule being designed to reflect shared-ride service characteristics. However, many trips were not shared-ride, simply because demand patterns did not allow such grouping. Several agencies noted this phenomenon and asked for, and received, discounts when more than one of the clients rode together.

---

[6] Much of the discussion here has been adapted from [13.10].

ACCESS discounts allow a certain percent savings over the computed fare for each agency client riding with other agency clients. That percentage discount only continues to the point where that figure equals the vehicle·hour charge that ACCESS is paying to its contractors; at that point the agency is simply assessed the vehicle·hour charge as the fare for all clients. The procedure is designed to prove that ACCESS's policy is that everyone, including agencies, should pay the full cost of transporting their clients.

The Pittsburgh brokerage concept can be seen to involve to some extent about 14 of the 33 linkage types in Table 13.3. Rosenbloom and Warren [13.10] conclude that this approach, as opposed to what existed before, results in better, and in some cases cheaper, services. Most agencies find it easier to deal with the broker than with local transportation providers or to own vans and provide their own services. Most also incur lower costs. While it was not clear that specialized services are advantageous versus fixed-route transit (with, say, wheelchair lifts), since no direct comparisons were possible, it was true that the specialized services were meeting the needs of a large number of citizens. Also, there has been relatively little demand for increased fixed-route transit.

13.7 COMPREHENSIVE URBAN DEVELOPMENT PLANNING

Simultaneously with development of transportation systems, in each community a more comprehensive development process is also conducted. The two processes must be closely coordinated. This is not always easy since different interest groups with different goals stand behind each process.

Many of the obstacles to the implementation of transportation plans listed at the beginning of this chapter can be traced directly to the isolation of the transportation planners from community life. Until the automobile age, urban transportation planning was inseparable from general urban development planning, popularly called "city planning." This close relationship was evidenced in the plans of ancient and medieval cities, and in the early American settlements on the eastern seaboard. It was not until the rise of separate professional bureaucracies during the middle of the twentieth century that lack of coordination between the two fields became a problem.

As mentioned earlier, a major trend and transportation policy objective, especially since the early 1960s, has been closer coordination of urban transportation planning with other aspects of urban development. Starting with the 1962 3-C requirement, Federal law through MPOs has given substantial powers to local elected officials for transportation planning and decision-making. In the mid-1970s these officials were also charged with preparing the annual urban-area Short-Range Transportation Improvement Program (TIP), which determines priorities for both capital and operating improvements. These roles for local officials have led to transportation planning again becoming a more integral part of comprehensive urban development planning.[7]

Urban development planning was "master" planning in the early 1900s, and today still answers to a variety of names such as comprehensive planning, general

---

[7]The assistance provided in preparation of this section by two recent books [13.17, 13.18] is acknowledged. Particular thanks are due F. Stuart Chapin, Edward J. Kaiser, Frank Beal, and Elizabeth Hollender.

planning, policy planning, or, in the 1970s, growth management. Development planning is the principal legal device used by urban governments to guide physical development, redevelopment, and community change in desired, coordinated, economic manner. The development plan is a guide to decision by the city's or county's planning commission, legislative body, chief executive, and staff. It also guides private decision-making either directly or indirectly through zoning or other land use controls. Like the transportation-planning process, the urban-development planning process at both the local government and MPO levels serves a number of purposes, particularly those of:

1. Explicit policy. The process functions to formulate community goals and objectives and strategies for attaining them.
2. Decision guidance. The process provides a development plan and/or other guidance for the many public and private decisions that continually shape the future city.
3. Fulfill legal requirements. In many states localities are required to have a development plan. In addition, a plan is a prerequisite for many types of Federal aid.

In most areas the urban development process, centered in local planning agencies and coordinated through the MPO (or other metropolitan agency), will afford the transportation planner a number of opportunities to further transportation objectives. Preceding sections of this chapter discussed two such opportunities of a major nature—transportation–land-use balance, and ROW protection and acquisition. In this section we will identify six more opportunities that can be derived from working with local planning officials: planning information, goals and objectives, impact analysis, advanced planning, plan flexibility, and public support.

13.7.1 Planning Information

Local and regional planning agencies often provide land-use data for transportation planning. They may also be a good source of other types of information useful in transportation planning:

- Possible future land-use development
- Neighborhood and subarea plans
- Traffic and transportation problems
- Community values, goals, and objectives
- Population, employment, and land-use forecasts
- Land holding-capacity estimates
- Official plans and programs
- Local leadership and power structure

13.7.2 Goals and Objectives

Two frequent complaints about transportation planning is that there is a lack of sensitivity to local needs, values, and objectives, and a lack of creativeness toward transportation solutions to nontransportation problems. Local planning officials, technical and elected, and citizens can be viewed as a resource for identifying needs and values, formulating goals and objectives, and developing innovative ways of achieving them. The planning process serves as a forum in which transportation planners and the many other professionals and interest groups can join to set collective goals, objectives, and strategies for future development. The planning process at this direction-setting stage is one of discussion, debate, and conflict resolution. This lays the basis for coordinated community actions that mobilize financial, regulatory, and other resources for achievement of the goals. This kind of consensus building facilitates the integration of transportation plans into the fabric of the community.

Thus, increasingly, the development planning process is a guidance system for the strategic management of an urban community's future [13.17, pp. 60, 486–487]. The term "strategic" denotes that there has been a deliberate determination of realistic, achievable goals, describing the desired future state of the community, and an appropriate set of policies for achieving those goals.

Strategic growth management, compared with earlier forms of urban planning, is characterized by more explicit goals regarding the amount and kind of growth, appropriate plans and powers for achieving the goals, informed and active public interest groups, and consistent decision-making. This kind of rationality is more akin to that traditionally found in transportation planning. The transportation planner is likely to find the planning climate in growth-management communities particularly conducive to his or her role in transportation system development and management.

13.7.3 Impact Analysis

Another problem in implementation is that there is sometimes a lack of awareness of the social, economic, and environmental impacts of transportation plans on the part of the planners, and even more so on the part of affected groups. Sometimes the reverse complaint is made—that the effect on transportation plans of a new subdivision, or apartment, business or industrial complex was not analyzed or communicated. Comprehensive urban planning provides a framework for identifying and analyzing impacts in both directions and in communicating them.

13.7.4 Advance Planning

The importance attached to particular impacts will depend on a familiarity with community values and goals, needs, and objectives based on prior groundwork. Minimizing adverse impacts and developing beneficial ones will also require lead time for the necessary planning. For example, schools can be located away from major streets so as to minimize school-crossing disruption of traffic. Through streets need not have the hazards of residential driveways; they can be planned so dwellings have access to side streets. Relocation can be minimized, and right-of-way for a future street protected through land-development controls. Similarly, right-of-way for separate bicycle paths can be secured through land-development planning and controls. In contrast to older concepts of rigid master plans for transportation and urban development, modern planning provides opportunities for on-going, advanced joint efforts to obtain more optimal and implementable solutions.

13.7.5 Plan Flexibility

Officials concerned with implementation of transportation plans experience problems of changed conditions concerning land development, financing, energy, urban lifestyles, etc., and the difficulty of adapting plans to these changes. A number of approaches have been developed for coping with this set of implementation obstacles. These approaches include forecasting of alternative futures, contingency planning for these futures, and surveillance of actual developments to determine which future is actually occurring. These efforts, of course, require monitoring of local conditions, as well as an advanced level of staffing and institutional sophistication, which can sometimes be provided by the local comprehensive planning team.

13.7.6 Public Support

Local comprehensive planning can also aid in building a "policy chain" of public support for transportation. Many local groups have a stake in the planning process and the results. For example, the Chamber of Commerce may find the plan a useful tool for assuring prospective firms that utilities, roads, and living amenities are or will be in place. Central business district interests can show that the transportation improvements they need are a legitimate part of the regional network. Neighborhood associations see that road improvements will divert traffic from their areas, rather than disrupting them. Generally speaking, the urban development process provides a forum for the broad range of local interest groups to collaboratively formulate common interest goals and plans that they have a stake in achieving. Transportation planners have an obvious interest in this kind of joint planning.

13.8 SUMMARY

The main concern of this chapter has been the legal and organizational aspects of transportation plan implementation. The first section described Federal legislation of importance. A trend was seen toward such greater Federal overview of planning, particularly through such processes as environmental impact statement (EIS) requirements. Local and state legal mechanisms then were discussed, with emphasis on relevant growth-management tools.

Organizational aspects can be quite complicated, particularly where coordination is needed. Thirty-three different service integration linkages were described and illustrated as means of coordination. A formal means for integration is the approval process, which covers the stages of:

1. Direction setting
2. Systems planning
3. Project definition and programming
4. Location and preliminary design
5. Final design
6. Right-of-way acquisition, and construction

This process actually is part of a larger effort developed to bring in greater preimplementation involvement of local implementors and interest groups. In the end, these both should fit into, and gain from, the more comprehensive urban development planning process, described in the last section.

BIBLIOGRAPHY

13.1 Einsweiler, R. C., et al: "Comparative Description of Selected Municipal Growth Guidance Systems," in R. W. Scott (ed.) *Management of Control of Growth*, vol. II, The Urban Land Institute, Washington, D.C., 1975.

13.2 Johnson, E. R., G. G. Huebner, and L. G. Wilson: *Transportation: Economic Principles and Practice*, Appleton-Century-Crofts, New York, 1940.

13.3 Schnediman, F., J. S. Silverman, and R. C. Young Jr.: *Management and Control of Growth: Techniques in Application*, vol. IV, The Urban Land Institute, Washington, D.C., 1978.

13.4 Smerk, G. M.: *Urban Transportation: The Federal Role*, Indiana University Press, Bloomington, Ind., 1965.

13.5 Smerk, G. M.: *Readings in Urban Transportation*, Indiana University Press, Bloomington, Ind., 1963.

13.6 Office of Human Development Service, Department of Health and Human Services: *Coordination of Specialized Transportation: An Inventory of State Activities*, Washington, D.C., June 1981.

13.7 U.S. Department of Health, Education, and Welfare and U.S. Department of Transportation (Region IV): *Planning and Coordination Manual*, vol. 1, Atlanta, Jan. 1979.

13.8 Agranoff, R.: *Dimensions of Services Integration*, Human Services Monograph Series no. 13, Project SHARE, Rockville, Md., April 1979

13.9 John, D.: *Managing the Human Services "System": What Have We Learned from Services Integration?*, Human Services Monograph Series No. 4, Project SHARE, Rockville, Md., Aug. 1977.

13.10 Rosenbloom, S., and D. Warren: "A Comparison of Two Brokerages: The Lessons to be Learned from Pittsburgh and Houston," Transportation Research Record (forthcoming).

13.11 U.S. Department of Health, Education, and Welfare (and other agencies): *Planning Guidelines for Coordinated Agency Transportation Services*, Washington, D.C., April 1980.

13.12 Department of Transportation: *Coordinating Transportation Services for the Elderly and Handicapped*, vol. 1, Government Printing Office, Washington, D.C., May 1979.

13.13 U.S. General Accounting Office: *Hindrances to Coordinating Transportation of People Participating in Federally Funded Grant Programs*, vol. 1, Government Printing Office, Washington, D.C., Oct. 17, 1977.

13.14 U.S. Department of Health, Education, and Welfare (Atlanta): *Transportation Authorities in Federal Human Services Programs*, Atlanta, January, 1976.

13.15 Feldman, H. G.: *Condemnation of Property for Highway Purposes*, Highway Research Board Special Report 32, Washington, D.C., 1958.

13.16 League of Kansas Municipalities: *Planning Tools: Theory, Law, Practice*, vol. 15, Topeka, Kans., 1962.

13.17 Chapin, F. S., and E. J. Kaiser: *Urban Land Use Planning* (3rd ed.), University of Illinois Press, Urbana, Ill., 1979.

13.18 Beal, F., and E. Hollender: "City Development Plans," in F. S. So, I. Stollman, F. Beal, and D. S. Arnold (eds.): *The Practice of Local Government Planning*, The International City Managers Association, Washington, D.C., 1979.

13.19 Schofer, J. L., "Future of the Urban Transportation Planning Process," *Proceedings, TRB Conference on Transportation Planning in the Eighties*, Workshop Report (preliminary), Washington, D.C., November, 1981.

13.20 Shunk, G.: "Preliminary Workshop Report Long-Range Regional Transportation Planning," *TRB Conference on Transportation Planning in the Eighties*, November, 1981, p. 5.

13.21 Wachs, M.: "Highlights of TRB Conference" (preliminary), *Urban Transportation Planning in the 1980's*, 1982.

13.22 Manheim, M. L.: "Toward More Programmatic Planning," in G. E. Gray and L. A. Hoel (eds.), *Public Transportation: Planning, Operations and Management*, Prentice-Hall, Englewood Cliffs, N.J., 1978.

EXERCISES

13.1 Identify either in the literature or in a real-world case in your community a human-services transportation project or program. For this, describe the types of services integration linkages being employed (Table 13.3) and indicate which other ones might be productive (and why).

13.2 Pick a particular local/state legal power (Sec. 13.3). Describe it in about 250 words and then discuss the conditions under which it may be advantageous or disadvantageous to transportation-plan implementation (another 500 words).

13.3 For each of the following types of projects list (a) the implementing powers that most certainly will be needed and (b) those that might be required:

1. Freeway construction that penetrates the central core of a large city.
2. A new rail rapid transit system in the same large city.
3. Use of express lanes on surface streets and freeways to connect two urban centers.
4. An outer-loop bypass through rural land uses.
5. An intersection redesign requiring no new ROW.

13.4 Suggest a "cure" for each of the problems of implementation noted in Sec. 13.1.

14 Implementation II: Finance and Budgeting (Programming)

Of all the powers to implement transportation plans, those of government to tax and/or spend are certainly the most important. This chapter deals with the finance of urban transportation—types and amounts of revenues and expenditures, as well as techniques for estimating, assessing, and programming (budgeting) the two.[1] As will become apparent, transportation revenues and expenditures come from and are directed to a wide variety of programs, so that a comprehensive listing and assessment of all is a difficult task.

14.1 REVENUES

Income to finance urban transportation comes from seven major sources:

User charges, in the case of automotive travel, include gasoline taxes, vehicle registration fees, drivers' license fees, tolls, vehicle sales taxes, etc. For transit, these would be the fares or fees charged.

General funds, which are the common central treasuries existing at all levels of government, from which discretionary budgets are fashioned by the executive and adopted by the legislature. Into the general fund may go property, sales, and income taxes, as well as a wide variety of other revenues.

Borrowings, including bond issues and short-term notes as well as loans from one government to another, or to individuals.

Investment income, interest income or profit on the purchase and sale of securities, particularly while waiting for other funds to be needed.

---

[1] Techniques for expenditure evaluation are presented in detail in Chaps. 10 and 11.

Grants (or grants-in-aid), intergovernmental transfer payments that usually flow from Federal to state and state to local governments, but also may flow to individuals or private firms.

Government purchase agreements, whereby a governmental agency agrees to buy a given amount of goods or services from a given contractor. That agency or firm then may use that agreement as collateral in borrowing money.

Tax expenditures, the tax-revenue losses attributable to special deductions or exclusions, or exemptions from gross income or a special credit, preferential rate of tax, or deferral of a tax liability [14.2].

The distinctions among these sources become blurred in some cases. For instance, a firm might be able to get a loan to start a taxi service from, say, the Small Business Administration at below-market interest rates. If so, the interest payments saved really represent a "grant" from that agency.

The relationship between the different sources can be very complex, and also can change rapidly. As an example of the former, consider again the small firm trying to start a taxi service. If new taxis are purchased, the firm is eligible, under current law, for an investment tax credit (a tax expenditure) of up to 10%. In other words, if a new taxi costs $10,000, then $1,000 could be deducted directly from the income taxes of the firm. Once in business, the taxi may haul Medicaid (low-income, medically needy) recipients, in which case the cost of the trip will be reimbursed by Federal and state governments (a grant-in-aid). Naturally the taxi company will have standard fares (user charges), which in some cases may be subsidized temporarily from the general fund of a locality (in order, say, to keep the company in business to provide transportation to the elderly and handicapped). Here is an example, then, of a case where five sources of revenue are tapped. The complexity comes when the laws and regulations governing these different sources increase in number and change, which they invariably do.

Suppliers and/or recipients of governmental transportation finance may be private firms and individuals as well as other governmental agencies. The taxi firm in the above illustration was a recipient. It might also be a supplier, at least in an indirect sense, if it provided, say, rides to the elderly at a discount, for which it was not reimbursed.

To make the picture even more confusing, consider also the billions of dollars in income support transfer payments that government provides to individuals. Such programs as Social Security and Supplemental Social Security, Aid to Families with Dependent Children, and General Assistance all provide funds directly to individuals to spend for the most part as they deem appropriate. These are no small sums. In fiscal year 1977 the Disability, Retirement, and Survivors Insurance parts of Social Security provided about $82 billion to individuals [14.7, p. 210]. Assuming about 20% of this was spent on transportation (automobiles, fuel, transit fares, etc.) and 75% of that in urban areas, the result would be $12 billion for urban transportation. This appears to exceed the amount of funds coming from direct government expenditures for highways and transit.

A comparison of revenues among the different sources is not possible, given the particular difficulty of determining how much of certain tax expenditures, loans, and the like are spent on urban transportation. Even some direct expenditures are difficult to categorize, especially outside of transportation departments. Many social-service programs, for example, fund transportation but usually do not keep track of expenditures under that heading.

Given these difficulties, we will try to present figures, very rough in some cases, on urban transportation revenue sources and amounts at each level of government for highways, transit, human services, and private individuals and firms. The communality among all of these is that government plays some role in the financial picture.

14.1.1 Highway Finance

Table 14.1 shows receipts and disbursements for all highways by level of government in 1979. The total is about $37.5 billion. After the Highway Act of 1956, Federal highway user fees were placed in a Highway Trust Fund (instead of the General Fund of the U.S. Treasury). These monies then were dedicated to use only for highways (and more recently in part for transit). As can be seen in the second column of Table 14.1, the fund generated over $8 billion in 1979. This includes close to $1 billion in returns from investment of the funds while waiting for approved expenditures. Most of the $15.1 billion generated by state agencies also comes under the rubric of trust funds.

14.1.1.1 Federal level The taxes and rates required to generate the Federal Highway Trust Fund revenues are indicated in Table 14.2. The major source is the motor fuels (gasoline and diesel fuel) tax, which generated $4.8 billion of the 1979 trust funds [14.12, p. 50]. These revenues have been leveling off in recent years, however, as travel has declined slightly and vehicle efficiency has increased substantially. In 1977, for instance, the figure was close to $4.7 billion.

"Apportionment" is the term given to the geographic or jurisdictional allocation (division) of appropriated funds. Highway Trust Fund monies are apportioned to the states and thence to urban areas for all Federal systems.

Forty-five percent of the Interstate expenditures were to be in urban areas, and apportionments have been running about 10 percent below that figure. However, upon completion, the urban portion of the Interstate System will approximate 15 percent of the system and account for roughly 45 percent of the overall expenditures. Urban roads are also financed by the so-called ABCD program [Federal aid primary (FAP), secondary (FAS), and urban highways].

As of 1970, the present apportionment scheme for the ABCD program from the Highway Trust Fund to each state conforms to the following four formulas:

1. Federal-aid primary system (FAP)—one-third area; one-third rural post miles, with a minimum apportionment of 0.5 percent per state.
2. Federal-aid secondary system (FAS)—one-third area; one-third rural population; and one-third post miles, with a minimum apportionment of 0.5 percent per state.
3. Urban extentions of FAP–FAS systems based on urban population in all urban places above 5,000 people, and with no minimum apportionment per state.
4. New (1970) Federal-aid urban system based on urbanized area, population 50,000 or more [14.12, pp. 30–31].

14.1.1.2 State level Pie charts showing state receipts and disbursements for highways in 1979 are displayed in Fig. 14.1. As Table 14.1 indicates, states actually generate almost as much revenues for highways as all other governmental levels combined ($18.46 billion versus $18.99 billion in 1979). To these are added Federal Trust Funds monies, since the Federal Highway Administration does very little construction on its own.

Table 14.1 Total Receipts and Disbursements for Highways, All Units of Government[1] (in millions of dollars)

Item	Federal government				State agencies and D.C.	Counties and townships	Municipalities	Total
	Federal highway administration		Other Federal agencies	Total Federal				
	Highway trust fund	Other funds						
	Receipts by collecting agencies							
Imposts on highway users[2]								
Motor-fuel and vehicle taxes	7,054	–	–	7,054	13,867	88	138	21,147
Tolls	–	–	–	–	1,255	39	225	1,519
Parking fees	–	–	–	–	–	3	110	113
Subtotal	7,054	–	–	7,054	15,122	130	473	22,779
Other taxes and fees:								
Property taxes and assessments	–	–	–	–	–	1,140	1,020	2,160
General fund appropriations	–	672	1,632	2,304	1,016	1,300	2,600	7,220
Other taxes and fees	–	–	19	19	413	48	170	650
Subtotal	–	672	1,651	2,323	1,429	2,488	3,790	10,030
Investment income and other receipts	962	6	200	1,168	966	265	500	2,899
Total current income	8,016	678	1,851	10,545	17,517	2,883	4,763	35,708
Bond issue proceeds (par value)[3]	–	–	–	–	943	200	600	1,743
Grand total receipts	8,016	678	1,851	10,545	18,460	3,083	5,363	37,451
Intergovernmental payments:								
Federal government:								
Highway trust fund	–7,444	–	–	–7,444	7,311	8	125	–
All other funds	–	–502	–1,264	–1,766	670	580	516	–
State agencies:								
Highway-user imposts	–	–	–	–	–3,538	2,172	1,366	–
All other funds	–	–	–	–	–400	240	160	–
Counties and townships	–	–	–	–	94	–154	60	–
Municipalities	–	–	–	–	146	8	–154	–
Subtotal	–7,444	–502	–1,264	–9,210	4,283	2,854	2,073	–

					Disbursements by expending agencies			
Funds drawn from or placed in reserves[4]	−250	−109	−3	−362	316	−154	284	84
Total funds available	322	67	584	973	23,059	5,783	7,720	37,535
Capital outlay:								
On rural state-administered highways	—	—	—	—	8,377	6	—	8,383
On municipal extensions of state highways	—	—	—	—	3,607	—	20	3,627
On local rural roads	—	—	—	—	706	1,530	—	2,236
On local municipal roads and streets	—	—	—	—	490	25	2,150	2,665
Not classified by system	68	60	445	573[5]	—	—	—	573
Subtotal	68	60	445	573	13,180	1,561	2,170	17,484
Maintenance and traffic services:								
On rural state-administered highways	—	—	—	—	3,638	20	—	3,658
On municipal extensions of state highways	—	—	—	—	820	—	70	890
On local rural roads	—	—	—	—	44	3,200	—	3,244
On local municipal roads and streets	—	—	—	—	26	12	2,920	2,958
Not classified by system	—	—	131	131	—	—	—	131
Subtotal	—	—	131	131	4,528	3,232	2,990	10,881
Administration and research	254	7	8	269	1,577[6]	460	435	2,741
Highway law enforcement and safety	—	—	—	—	1,778	230	1,350	3,358
Interest on debt	—	—	—	—	1,056	115	295	1,466
Total current disbursements	322	67	584	973	22,119	5,598	7,240	35,930
Debt retirement (par value)[3]	—	—	—	—	940	185	480	1,605
Grand total disbursements	322	67	584	973	23,059	5,783	7,720	37,535

[1] This table summarizes and consolidates data reported in greater detail in the FA, SF, LF, UF, LB, and UB table series. Data for Federal and state agencies are final; those for counties and municipalities are estimates subject to revision when data for all local units are available. Tables HF-1 and HF-2 for 1978 contain final data for all units of government.

[2] Excludes amounts allocated for nonhighway purposes. Motor-fuel and vehicle taxes are also net after refunds and collection expenses. Parking fees are amounts in excess of parking costs and considered available for highways.

[3] Issue and redemption of short-term notes or refunding bonds are excluded. Interest is excluded. Premiums and discounts on sale of bonds are included with "Investment income and other receipts." Redemption premiums and discounts are included with "Interest of debt."

[4] Minus signs indicate that funds were placed in reserves.

[5] Includes $20.5 million paid to territories.

[6] Includes $117.2 million of Federal-aid highway funds for research planning.

Source: [14.12, p. 39].

Table 14.2 Federal Tax Rates on Motor Vehicles and Related Products

	As of Jan. 1, 1981	Schedules change Oct. 1, 1984
Gasoline (cents/gal)	4	1.5
Diesel fuel (cents/gal)	4	1.5
Lubricating oil (cents/gal)	6	6
Buses (% manufacturers' sales price)	10	5
Trucks (% manufacturers' sales price)	10	5
Trailers (% manufacturers' sales price)	10	5
Parts and accessories (% manufacturers' sales price)	8.5	5
Tires (cents/lb)	10	5
Tubes (cents/lb)	10	9
Tread rubber (cents/lb)	5	0
Federal use (dollars/1000 lb, vehicles over 26,000 lb)	3	0

Source: [14.12, p. 47].

The main sources, and approximate 1979 amounts (in $ billion), of state user charges/revenues are:

Gasoline and special fuels taxes   9.8
Motor vehicle registration   4.7
Driver license fees   0.3
Certificate of title fees   0.2

As can be seen, gasoline and special fuels taxes are largest in amount. These come from a tax at the pump of anywhere from $0.05 to $0.11 per gallon in 1979 (Texas had the lowest; several states had the highest). This source of revenues, like that at the Federal level, has leveled out, and in many states has started to decline in quantity. In response, some states have given thought to having a gasoline tax as a percentage of the price per gallon instead of a fixed number of cents. In this way, assuming fuel costs continue to rise, so also will automotive fuel tax revenues.

Metropolitan areas receive state highway assistance in several ways—through state construction and maintenance of local roads; through direct grants-in-aid to localities (particularly for maintenance); and through payments for certain toll facility bonds. States also have to use funds to match Federal monies going for urban highways. For example, ABCD projects generally require 75 percent Federal and 25 percent state funds. Capital costs for the Interstate program are 90-10, respectively. In 1978 over $2.5 billion in Federal and state funds went to the construction and maintenance of roads in metropolitan areas [14.48, Table LF-14].

14.1.1.3 Local highway finance According to Table 14.1, 1979 saw local governments receive $8.4 billion in revenues for highways and related purposes. Some of the $3.1 billion part of this going to counties and townships went to rural entities, so we might estimate that a total of approximately $6.5 billion was received and spent in metropolitan areas.

Localities receive their incomes from several sources. In addition to transfers from state funds, these sources include:

Real-estate property taxes
Personal-property taxes
Special assessments

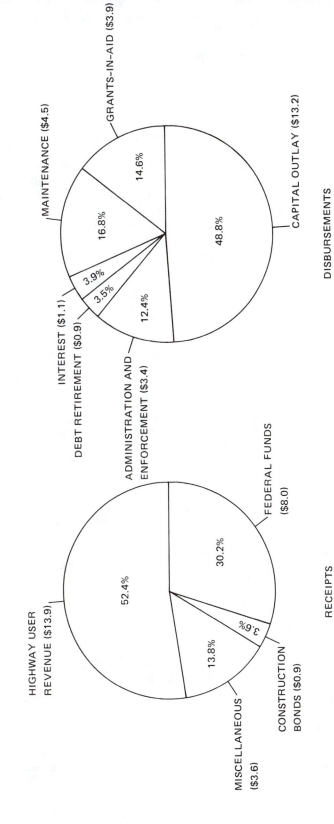

Fig. 14.1 State receipts and disbursements for highways, 1979 ($billions). [14.12, p. 66].

General-fund appropriations
Local tags
Road and crossing tolls
Traffic fines
Bond sales

For example, most cities and counties in Virginia have a personal-property tax, which typically might amount annually to 1 percent of the fair market value of automobiles and trucks a family owns. Most cities in that state also charge a "tag" (sticker) fee, which runs from $10 to $20 for an average automobile.

Few localities dedicate funds received from these listed sources specifically for highway purposes. Instead, they are placed in the overall purse (general fund). It therefore is difficult to establish a direct relationship between those funds received and those expended. Nevertheless, it is estimated that in 1978, SMSAs spent almost $2.3 billion for capital outlays and $3.0 billion for maintenance of streets [14.48]. They also spent $1.7 billion for administration and "other," as well as $1.2 billion for debt service.

14.1.2 Transit Finance

The future of many transit systems looks bleak if further sources of funding are not found. Revenues from the farebox continue to decline as a proportion of expenses. Total passenger revenue in 1970 was $1.64 billion, while expenses were $2.0 billion—bringing a revenue-to-cost ratio of 0.82. But in 1978 passenger revenues increased only to $2.27 billion while total expenses jumped to $4.71 billion [14.1, pp. 20–21], for a ratio of 0.48. The latter figure, by the way, includes only a very small proportion of new capital costs. The 1978 operating deficits were made up by local, state, and Federal operating assistance of $977.7, $564.3, and $689.5 million, respectively.

14.1.2.1 Federal transit finance[2]   Since 1961 the Federal government has provided grants of various types to aid cities in their mass transportation programs. The major Federal programs by type of aid are described below. The section numbers refer to parts of the Urban Mass Transportation Act of 1964, as amended, which is the major act guiding the expenditure of Federal transit funds. Almost all of the money used by the Federal government comes from the General Fund.

Section 3 (capital improvements)   The Federal government will pay 80% of the net project cost for capital improvements. Only legally authorized public bodies (states, municipalities, political subdivisions, public commissions, and so forth) in urbanized areas (places of more than 50,000 population) are eligible to receive capital grants or loans. Private transportation companies may participate in Federally aided capital projects, but they must do so through a public agency.

The kinds of project eligible for capital grants and loans involve acquisition, construction, and reconstruction or improvement of facilities and equipment for use in mass-transportation services. These include purchase of buses or other transit vehicles, construction of garages and other maintenance and storage facilities, offices, shelters for passengers, parking lots, terminal structures, and equipment for the facilities mentioned previously. Capital grant funds also may be used for the acquisition of the assets of a private operator. The proceeds of the capital grant

---

[2] Much of the discussion in this and the next two sections has been adapted from [14.18, pp. 65–67].

may be used for the purchase of land or to rent or lease facilities or personal property needed for an efficient and coordinated transportation system.

Low-interest-rate loans can be obtained under the capital program, but they have been used relatively infrequently because a loan is obviously less attractive than an outright grant. Besides, the requirements are the same for a loan as for a grant, so there has been no real incentive to apply for a loan.

Section 5 (operating and capital aid) Operating aid is made available on a 50% Federal, 50% local basis to urban areas only. The money for operating aid is provided through a formula grant process. The formula money may be used at local option to either cover operating deficits or pay for capital improvements. If used for capital improvements, the Federal share is 80 percent and the local share 20 percent. Nationally, transit agencies have chosen to use formula funds primarily for operating aid.

Section 6 (research, development, and demonstration) The proportion of this aid varies and may be as high as 100 percent Federally funded. For local areas, Section 6 provides an opportunity to try out ideas through practical, real-life demonstrations, by means of the Service and Methods Demonstration Program. Most of the funding for the overall program, however, is based on initiatives developed by the Federal government and contracted out to consultants.

Section 8 (planning aid) Under this program, which is called technical assistance, the Federal government will pick up 80 percent of the cost of transit planning; local and state agencies must pay the remaining 20 percent. While this is a discretionary program, a sizable proportion of the funds are allocated to urbanized areas and states by a nonstatutory formula.

Highway trust fund Highway trust fund money from the Federal Aid for Urban Systems (FAUS) category may be used for transit capital improvement purposes. It also is possible to use money set aside for the Interstate system for mass-transportation capital purposes. The use of Interstate system money in this fashion has been commonly undertaken only where it is difficult or impossible to build a road because of opposition to the highway construction. Rather than forego the funds, a public transportation improvement is made. This use of Interstate switchover funds has been made most notably in larger cities, such as Boston and Philadelphia.

In 1978, Federal aid for capital grants amounted to about $2.0 billion, most of which came from Section 3 ($1.4 billion) and Interstate transfers ($0.6 billion). Operating grant approvals totaled $0.7 billion [14.1, pp. 41, 42].

14.1.2.2 State transit finance[3] Many states have special financial provisions relating to conventional transit. According to a 1973 study, there were 27 states with provisions for tax relief pertaining to property, income, or bonds of the local operating agency. In addition, 16 states had laws governing motor-fuel tax exemptions or refunds for transit operations. Six states also provided reduced-fare support for school children or elderly persons consisting of direct reimbursement of the difference between the regular and the reduced fare [14.20].

Many states, like the Federal government, offer assistance for capital purchases. In a 1975 survey, 18 states indicated that they provided a portion of the local share to help local jurisdictions obtain Federal capital grants. Cost-sharing ratios varied among states. Four states were found to provide all local matching funds, five

---

[3]Much of the discussion in this and the following section is adapted from [14.9, pp. 501-504].

provided over half the local share, and four more provided half. Some states varied their contributions toward the local share among local recipients [14.21, pp. 14–19].

The same survey indicated that 14 states provided operating assistance to eligible local bodies. Since operating deficits are a pervasive problem for conventional transit in the United States, the states often apply allocation procedures for disbursing such subsidy funds. Allocation has been based on the amount of the deficit, on demographic criteria, and on measures of the candidates system's performance. In addition, the percentage of subsidy has varied; for example, Connecticut and Rhode Island have paid all operating losses, Connecticut requiring revenues to equal or exceed 60% of operating costs. The intent in New Jersey is to pay 75 percent of operating losses on buses, in addition to all the (avoidable) losses on commuter rail. Massachusetts has paid half of total costs providing that revenues equal at least one-third of this. Maryland has assumed all the operating losses of the Baltimore system, but assumed half elsewhere in the state.

At the time of the survey, Michigan and New York State employed performance criteria for allocation. New York, in addition to the commuter rail subsidy previously reported, has provided 2 cents per passenger, 7 cents per bus·mile, and 8 cents per rail rapid transit car·mile. Michigan has provided up to 33 percent of operating costs, the funds being allocated by a formula: half distributed according to percentage of urban population of the state's total urban population, and the other half prorated by a similarly based share of annual transit vehicle·miles.

California developed a program for allocating transit funds to areas on the basis of sales-tax revenues by county of origin [14.22]. California's 1971 Transit Development Act (TDA) was quite complex, and we discuss it in somewhat more detail than other state programs for it illustrates a number of options. The TDA applied the California sales tax to the sale of gasoline and made 0.25 cents/gal of the 5 cents/gal sales tax (5% of total such revenues) available for transit development and operation. In general, 85 percent of TDA revenues could be used for operating purposes provided that local government supplied equal matching funds for operating assistance. TDA funds collected by the state are returned to the county of origin. In certain urban counties, the funds must be used for transit purposes. In other counties they may, at local discretion, be used for street and road purposes, after finding that there are no transit needs in the area that can be reasonably met.

The method for allocating TDA funds within the urban regions has varied. For example, the Metropolitan Transportation Commission (MTC) has had considerable authority for determining allocations in the San Francisco Bay area. With several operators having jurisdictions that cover parts of more than one county and some having overlapped jurisdictions, these allocations can be both sensitive and difficult. MTC has allocated these funds on the basis of services provided. In practice, service levels have been determined by examination of specific areas. If an area is served by a route no more than 0.25 miles (.04 km) away with headways of 15 to 30 min during the peak, if there is relatively frequent service for 10 hr per day, 5 days a week, and if there is some night and weekend service, then the area qualifies for all its allocation. Allocations among operations within the same county have proved to be complex. The Bay Area Rapid Transit District serves three of the MTC counties and has received about 25 percent of the funds for providing regional services for each of these. Much of the remainder has gone to the two operators—Alameda-

Contra Costa Transit District and the San Francisco Municipal Railway—that provide local transit services in their jurisdictions.

Subsidy programs at the state level are constantly changing as new legislation is enacted. The trend is for more state programs, primarily to help local jurisdictions qualify for Federal funds and secondarily to provide funding or relief in areas less well covered by Federal programs.

14.1.2.3 Local transit finance[4]   The number of local bodies providing assistance to conventional transit has continued to expand rapidly. Local governments have responded to state and Federal programs by creating bodies appropriate to receive grants or to activate powers delegated by states. In 1973, 23 states reportedly permitted cities or other public bodies to tax within their jurisdictions to obtain funds for transit subsidies [14.20]. Some states permitted the formation of special transit districts or other similar bodies for the purpose of owning, operating, and maintaining transit systems. Frequently, however, the formation of local bodies or levies of special taxes must be approved by a referendum.

A variety of mechanisms are used by local agencies. Most frequently, the funds necessary to cover the deficit are appropriated through special or general property taxation or other local taxes. In a few cities, transit has been operated by utility companies, which provide indirect transit subsidies from utility revenues. Frequently, profitable charter or bus operations are a source of indirect transit subsidy [14.23]. In a few places, facilities or equipment are owned by local government agencies and are leased to private operators at below-market rates. Various contract operations have become moderately popular.

Many cities have sought new forms of levies to obtain revenues. The levy could be of almost any form applied to almost any service, sale, transaction, or activity. For example, in Massachusetts a special statewide sales tax is levied against cigarettes especially to support the Massachusetts Bay Transportation Authority (MBTA).

Some localities have a portion of the sales tax devoted to transit. In states that allow local option taxes, a so-called "wheel tax" often is levied on each automobile belonging to residents. This tax may be a nominal sum of a dollar or two, or may range as high as $35 or $50. These funds often are earmarked for special purposes, such as highway improvements. Household taxes for transit are levied in some places, with a fixed fee per household collected on the utility bill each month. A tax on hotel and motel rooms is another type of levy [14.18, pp. 59, 60].

The mechanisms used in some of the larger cities in Korea are interesting here. In most there is a very heavy tax on the purchase of automobiles. Part of the revenues is used to pay for subway construction and bus subsidies. The tax helps to keep down auto usage and at the same time support a viable alternative.

New York City in 1975 offered a scenario with an interesting variety of organizational arrangements and subsidies [14.24]. Transit was provided by the Metropolitan Transportation Authority (MTA), the Port Authority Trans-Hudson Corporation (PATH) under the Port Authority of New York and New Jersey, and by several privately owned bus companies operating in Manhattan and Queens. PATH provides rapid-transit commuter service between Manhattan and New Jersey and is organizationally distinct from MTA. MTA operates rail rapid transit and bus-transit authorities, commuter rail lines, airports, and a bridge and tunnel authority. MTA

---

[4] Some of the discussion here is taken from [14.18, p. 59].

finances operational deficits in some of these components from surpluses in others. Additional funds come from the New York State and Federal Programs that have already been described. Capital funds for MTA have also been provided from state bond issues. The city reimburses some operators for reduced fares for elderly and for free transit for police and firemen. The city and state reimburse operators for reduced and free fares for school children. The city pays the full cost of the New York City Transit Police, owns the property of the two transit authorities, and pays the principal and interest on their debt. These arrangements serve to suggest the complexity of financial and institutional relationships in large urban areas.

In the Seattle, Washington, area, King County residents voted to form a metropolitan municipal corporation, a Washington State option. The resulting Municipality of Metropolitan Seattle (Metro) was established to provide both water-pollution control and public transit within its service area. It has provided some additional transit services outside its area on a contractual basis and some special scheduled services for major sports events in the surrounding area. State law also permitted local transit agencies to levy a 1 percent motor vehicle excise tax, which is actually a transfer of funds from an already existing statewide 2 percent tax. Transit bodies, in order to quality for the 1 percent motor vehicle use tax revenues within their boundaries, were required to raise at least as much revenue by any of a list of possible "clearly local" taxes. Metro chose a sales tax of 0.3 percent as its local funding device. Metro took over the bus system from the city, which then provided funds to pay for some services, most notably a fare-free program in the downtown area, in addition to the local share of certain park-and-ride lots and bus access lanes on state and Federal-aid highways.

14.1.3 Bond Financing

Finance of both highways and transit systems depends to some extent on borrowing money, usually through bonds issued by public agency. Table 14.1 shows, for instance, that in 1979 all units of government received about $1.7 billion for highway purposes from bond issue proceeds. Transit systems, particularly rail, have benefitted from borrowing. In most cases interest on government-supported bonds can be lower than in the private sphere because the interest payments are tax-free to the private lender. This is particularly helpful to individuals and firms in higher tax brackets.

There are four major types of bonds:

Revenue bonds are used when the transit operation is revenue-producing and may be expected to pay off its debt through the fare box.

The issuance of revenue bonds requires that repayment money be placed in a sinking fund. Revenues, therefore, must be sufficient to cover operating costs as well as the sinking fund allotments. In a time of greatly increasing cost, it may be difficult to hold fares or tolls within reasonable levels. Because the payment of revenue bonds is based on the revenues earned by the undertaking, they can be somewhat more risky. Therefore, the rate of interest required for revenue bonds usually is higher than that required for general obligation bonds.

General obligation bonds are backed with the full faith and credit of the taxing power of the issuer. In other words, a state or locality will pay off the debt if no other way is found. This, of course, is particularly useful for transit.

General obligation bonds in many instances must receive voter approval through a referendum. If the sum involved is large, gaining approval may be difficult.

Special assessment bonds have the debt paid off by a special assessment on property owners who stand to benefit in a more-than-ordinary manner by the service provided through the funds raised from the bonds. In the case of the small-scale transit operation, special assessment bonds might be used to finance the construction of a downtown terminal building if merchants and businesspeople there were expected to gain special benefits from the presence of such a terminal. Nearby properties would then be assessed at a higher rate than those elsewhere.

Municipal assistance bonds can be utilized when a city can procure a lower interest rate than a public transit operation only empowered to issue revenue bonds. The public transit enterprise would then reimburse the city, which in turn would pay off the bonds. Even if a transit agency could issue general obligation bonds, the municipal route usually would be followed if it were able to obtain a lower rate of interest.

14.1.4 Human-Services Transportation Finance

Much support for human-services transportation starts at the Federal level. National assistance in this area has become highly complicated. This can be expected since transportation is an integral part of such services—those being served need some way of getting to the point of reception (or the providers need to get to the recipients). In fact, one study done in 1974 of rural human-services transportation found over 70 Federal programs under which financial support could be obtained [14.11]. Presumably this number would be even higher for urban areas.

Table 14.3 gives a brief summary of some of the larger programs from which Federal assistance can be obtained. These are shown by department and authorizing statute. A comparison shows rather quickly that there are many inconsistencies that make coordinated transportation provision difficult [14.11, pp. D-3 to D-4]:

1. Incongruence in service delivery boundaries
2. Cost-allocation requirements
3. Policy and compliance issues
4. Matching-rate variation
5. Statewideness requirements
6. Single-state agency requirements
7. Need for regulatory waivers
8. Variations in legal, administrative, contractual, and auditing agreements.
9. Need for new, existing, redirected, or discretionary monies

As a small illustration, Section 16(b)(2) funds are discretionary. This means that they are sought through competitive proposals, so that no agency can be assured of funding. Moreover, the monies are only for capital expenditure (e.g., vans). On the other hand, Title III funds are allocated to each agency on a formula basis and can be employed for some operating expenses.

14.1.5 Grants and Loans to Individuals and Firms

As the interdependence among governmental programs grows and techniques for analyzing these linkages get stronger, it will become clearer that government grants or loans to private individuals and firms play a major role in urban transportation, especially on the demand or user side. As noted before, cash income transfer programs are substantial in magnitude. To these must be added the special benefits of loan programs. Naturally most of the programs in these two categories have restricting rules and regulations, some of which may prevent direct expenditures on

Table 14.3 Sample Federal Funding and Requirements in Human-Services Transportation

Department statute	Description	Federal match	Nonfederal match
	Department of Transportation		
Urban mass transportation act of 1964, as amended 1. Section 16(b)(2)	Discretionary capital grants for elderly and handicapped	80%	20% cash
	Department of Health, Education, and Welfare		
Older Americans Act of 1965, as amended 1. Title III Area planning and social services program	Formula grants for state and community programs on aging & area planning and social services	State planning and administration—75% Administration of area plans—75% Social services under area plans—90% Social services not under area plans—75%	25% cash or in kind, must be met by state sources 25% cash or in kind 25% cash or in kind
2. Title III, Part C*	Formula grants, nutrition services for the elderly	90%	10% cash or in kind
Rehabilitation Act of 1973, as amended 1. Title 1 Vocational rehabilitation	Formula grants for vocational rehabilitation services	80%	20%
The developmental disabilities services and facilities construction act of 1970, as amended 1. Formula grant program	Services for the developmentally disabled Formula grants	75%; 90% for projects located in urban or rural poverty areas	25% cash or 10% in kind total non-federal share may be provided in kind
Social Security Act of 1935, as amended 1. XIX Medicaid-Medical assistance programs	Medical assistance programs Formula grants	Federal medical assistance 50–83% Administration of state plan—50%	Varies, related to state's medical assistance percentage—cash 50% cash
2. Title XX Social services for individuals and families	Formula grants for social service programs for individuals and families	90% family planning services 75% other services†	10% cash 25% cash

(*See table footnotes on pages 506 and 507*).

Funding flow	Joint funding	Maintenance of effort
	Department of Transportation	
Federal to state to local	0	0
	Department of Health, Education, and Welfare	
Federal to state to local, formula grant	Provision in statute, only for discretionary R&D activities. Provision in law exempts all programs under the Older Americans Act from any authority under the Joint Funding & Simplification Act of 1974	Prohibits reduction of expenditures, including nonfederal share, for any activity funded under Title III, from one year to the next
Federal to state, formula grant	Provision in act and regulations (only utilized for discretionary R&D projects) same as Title III of Older Americans Act	Amounts payable to States (allotments) for VRS for any fiscal year shall be reduced by the amount (if any) by which expenditures from nonfederal sources are less than the nonfederal expenditures for this program for FY 1972
Formula grants to state for (1) basic support and (2) advocacy systems	No provision	Federal funds must be used to supplement, or increase purposes for which Federal funds made available; not to supplant Federal funds. Aggregate level of state, local, and nonprofit funds for activities under the state plan shall be at least no lower for any fiscal year than for the immediately preceding fiscal year
Formula grants: states—quarterly allowance based on state's federal assistance percentage	No provision	No provision
Formula grants, states	0	The aggregate level of service expenditures made directly from state and local funds cannot be less than the aggregate expenditures for the provision of services appropriated during FY73 or FY74, whichever is less

(*See table footnotes on pages 506 and 507*).

Table 14.3 Sample Federal Funding and Requirements in Human-Services Transportation
(*Continued*)

Department statute	Description	Federal match	Nonfederal match
	Community Services Administration		
Community Services Act of 1974, as amended 1. Title II Community Action Program	Discretionary grants for community action programs	Declining match—80%‡ FY76, 60% FY77; Exception for CAP agencies receiving annual allotments of $300,000 or less, 75% FY76, 70% FY77	20% (1976) 40% (1977) 25% (1976) 30% (1977) in cash or in kind
2. Title V Head Start (Administered by HEW)	Discretionary grants for Head Start programs	80%	20% cash or in kind

*PL95-748 made Title VII Part C of Title III (with separate authorizations).
†State expenditures for social services can be Federally funded at a 75/25 match only up to the State's share of the national total of $2.5 billion a year.

transportation. On the average, however, they permit fairly flexible spending (especially the cash transfer programs).

14.1.5.1 Cash income transfers   Table 14.4 highlights eight major Federal programs offering cash to individuals. Also shown is a small sample of associated state programs. Localities participate in some of these efforts. The total shown for just eight programs at the Federal level is $108.3 billion. The Institute for Socioeconomics Studies estimates that all Federal cash transfer programs totaled, in Fiscal Year 1977, $153.8 billion [14.7, p. 13]. The corresponding figure for states was $34.5 billion [14.8, p. 13]. If, as was assumed earlier, 15 percent of these funds went for urban expenditures for transit and taxi fares, purchase and operation of automobiles, and the like, the resultant sum would be $28.3 billion.

Some of these programs are not intended to subsidize urban transportation. The purpose of one type of housing program, for example, is to provide cash to reduce monthly home rental payments. But, if such housing payments are less than they were before, then some additional money is available for other items—like transportation. This highlights the concept of *fungibility*, or the displacement of funds to be employed for other purposes. Thus, even if the objective of the program is not direct support of transportation, some of the funding could end up doing this anyway.[5]

[5] The same holds for subsidies to transportation. A reduced transit fare, for instance, allows the resultant savings to be spent for other items.

Funding flow	Joint funding	Maintenance of effort
	Community Services Administration	
Federal to local, discretionary grants	Provision in law, same as Older Americans Act, Rehabilitation Act	Under the law, services provided by CAP agencies must be in addition to and not in substitution for services previously provided without Federal assistance. Funds or other resources under Community Action Programs must not be diminished in order to provide the nonfederal share
Federal to local, discretionary grant	No provision	Head Start programs must increase and supplement existing levels of local support. These projects may not replace projects previously funded by nonfederal sources. Resources formerly directed to existing Head Start-related efforts must not be diverted to the Head Start grantee in order to meet the nonfederal share requirements

‡Pending legislation amending the Community Services Act may eliminate declining match—Federal match would be 80%.
Source: [14.14, pp. 7-9].

Some cash transfer programs have rules that appear to affect urban mobility directly. As an illustration, many state programs for SSI, AFDC, and Food Stamps have "asset" limits. A potential recipient cannot own a car worth more than, say, $3500. Or, alternately, his or her total assets (including car, house, furniture, etc.) cannot exceed a certain amount. On the other side, some cash transfer programs allow "deductions" for work travel expenses to help keep the recipient below the eligible income limit. To our knowledge the interrelated impact of these programs on urban transportation and mobility has not been studied in any depth.

14.1.5.2 Loans The second type of transfer to individuals and private firms is through loans. A sample of such loans at the Federal level is presented in Table 14.5. Many have short program lives, and may no longer be in existence at the time of this reading. Others may have taken their place, however, since there has been a tendency to increase governmental loan programs as an alternate to grants, for which there is no return of money to the treasury. Congress and many state legislatures are just now beginning to identify, collate, and analyze the impacts of various loan programs they have approved over the years.

Loan programs can involve one or more of four basic elements:

1. Direct government loans: The loan is made through a governmental agency, which has a specific fund set aside for that purpose.
2. Insurance: Either government or the individuals or firms pay an insurance fee, which is put into a fund to cover potential defaults.

Table 14.4   Estimated Expenditures for Selected 1977 Federal/State Cash Transfer Programs Possibly Related to Urban Transportation

Description	Amount ($ million)
Federal programs	
1. Supplemental security income (SSI)	5,299
2. Special benefits for persons age 72 and over	236
3. Social security—retirement insurance	52,364
4. Social security—disability insurance	11,625
5. Social security—survivors insurance	18,888
6. Aid to families with dependent children (AFDC)	5,718
7. Federal–State unemployment insurance	13,490
8. Special unemployment assistance	691
Total	108,311
State programs	
1. Optional state supplementation for SSI (Alabama)	13
2. General assistance (California)	62
3. Aid to families with dependent children (Connecticut)	75

Source: [14.7].

Table 14.5   Sample of Federal Loan Programs Relevant to Urban Transactions

Agency*	Description of program	Amount† ($ million)
SBA	Handicapped assistance loans	13
Commerce	Business development loan guarantees (guarantees)	1
HUD	Property improvement loan insurance for improving all existing structures and building of new nonresidential structures (insurance)	1,225
DOT	National Capital Transportation Act revenue bond guarantee program (guarantee)	1,200
DOT	Railroad rehabilitation and improvement financing (guarantee)	400
ERDA	Electric and hybrid vehicle research, development, demonstration, and production (guarantee)	60
SBA	Economic opportunity loans (guarantee)	50
SBA	Regular business loan (guarantee)	2,000
SBA	Surety bond guarantee (guarantee)	25
SBA	Small business investment company program (guarantee)	377
USRA	Acquisition and modernization loans (guarantee)	—
USRA	Loans for railroads in reorganization (guarantee)	51
USRA	Loans to state, local, or regional transportation authorities (guarantee)	33

*SBA = Small business administration; Commerce = U.S. Department of Commerce; HUD = U.S. Department of Housing and Urban Development; DOT = U.S. Department of Transportation; USRA = U.S. Railway Association.

†Depending on the program, this will be the amount of funds for loans themselves, or the amount guaranteed and/or insured. The figures are either for 1977 or 1978.

Source: [14.10].

3. Guarantees: Government assures that it will pay off the loan of a bank or other loaning agency should there be a default.
4. Interest subsidies: Loans are provided at less than market interest rate, with government making up the difference in interest payments.

The first element is becoming more common as agencies such as the Federal Finance Bank gain skill and experience in "buying" and making loans. Insurance premiums are common in housing mortgage (loan) programs where there is a small down-payment and banks want assurance that they will be repaid in case anything happens to the payer. The insurance premiums go into a fund to cover these exigencies. The HUD insurance program for nonresidential structures, listed in Table 14.5, is of this type. Presumably a taxi or private transit company, as somewhat risky endeavors, might be eligible to get insurance, and thus loans, under that program.

Guarantees are valuable in that they lower the risk to lenders, and at most times reduce the interest rate accordingly. To see this, note that the equation for yearly payments (Y) of principal (P) plus interest on that principal over N years at an annual percentage (fraction) interest rate i, is:

$$Y = Pi \left[1 - \frac{1}{(1 + i)^N} \right] \tag{14.1}$$

Suppose a taxi company wanted to borrow $1,000,000 but could not get a guarantee and thus had to pay an annual percentage interest rate of 15 percent ($i = 0.15$). This brings a yearly payment of:

$$Y = 1,000,000(0.15) \left[1 - \frac{1}{(1 + 0.15)^{25}} \right] = \$154,669.40$$

or $3,867,485 over the life of the loan.

Now if a guarantee is available, the bank is willing to drop the interest rate to, say, 12 percent. This gives an annual payment of $127,499.97, or $3,187,499 over 25 years. The difference is approximately $680,000, or about two-thirds the amount of the principal itself. This guarantee thus is equivalent to a grant of about $680,000.

Interest subsidies act in the same way. In the preceding example, if a government agency were empowered to make up the difference between the market rate and a 10 percent annual percentage rate, the savings to the firm would be $1,133,203, or more than the principal of the loan.

As can be seen in Table 14.5, many loans are made with small businesses in mind. Particular emphasis has gone to businesses run by minorities and the handicapped, who have in some cases thus been helped to obtain personnel and equipment to bid on construction work for highway and transit (especially rail) projects.

Some loans have benefitted transportation agencies and authorities directly. This is the case, for instance, for the National Capital Transportation Act Revenue Bond Guarantee Program, which presumedly resulted in lower interest costs and thus savings to local taxpayers in the Washington Metropolitan area.

14.1.5.3 Tax expenditures As defined earlier, tax expenditures are [14.2]:

the tax revenue losses attributable to special deductions, exemptions, or exclusions from gross income or a special credit, preferential rate of tax, or deferral of tax liability.

These apply to taxes of most types at most levels of government. Tax expenditures accrue to both corporations and individuals. To our knowledge these never have been analyzed in terms of their impact on urban transportation, but the potential for a significant effect is great. As in cash transfers, the fungibility of tax savings indirectly allows more individual and corporate discretion in the ways the savings are spent.

Table 14.6 provides a selection of tax expenditure types and estimated amounts at the Federal level in 1979. Some relate directly to transportation. To illustrate, if firms or individuals make certain improvements to reduce transportation barriers to the handicapped, they can "expense" these. This means that instead of depreciating, say, a wheelchair lift over 10 years, it can be done all in 1 year. Another example involves railroad retirement benefits. Some of these are paid directly from the Federal treasury without any previous contributions by recipients. Since the recipients do not have to pay taxes on their benefits, they end up getting tax-free income.

Other tax expenditures are particularly valuable in helping to purchase vehicles and/or defray business travel expenses in urban areas (as elsewhere). Investment tax credits, for example, can be obtained when a new car is purchased for business purposes. In 1979, if the vehicle were assumed to have a service life of 7 years or more, a tax credit of 10 percent of the price could be sought. On a $7,000 vehicle, this would mean $700 would be subtracted directly from any Federal tax payments. This type of financial incentive obviously leads to greater investment in autos and trucks, which means more travel.

Table 14.6   Estimated 1979 Federal Tax Expenditures for Selected Programs Related to Urban Transportation

Program description	Estimated tax expenditure ($ million)	
	Corporations	Individuals
1. Expensing of R&D expenditures	1,520	30
2. Deductivity of interest or consumer credit	–	2,350
3. Capital gains	575	7,990
4. Investment tax credit	12,320	2,725
5. Credit for employment of AFDC recipients under work-incentive (WIN) programs	20	0
6. Expensing of architectural and transportation barriers to the handicapped	10	10
7. Exclusion of railroad retirement system benefits	0	280
8. Exclusion of interest on general-purpose state and local debt	3,865	2,150
9. Deductibility of nonbusiness state and local taxes	0	9,440
10. Deductibility of nonbusiness state gasoline taxes	0	840
Totals	18,310	25,814
Grand total		44,125

Source: [14.26].

Individual transportation also gains from deductibility of interest or consumer credit (borrowing), much of which goes for automobile purchase loans. The same holds for the deduction of sales taxes on vehicle purchase and of gasoline taxes.

Other tax expenditures simply add directly to disposable income, which then can be employed in part for transportation-related purposes. The reduced tax on capital gains (e.g., sales of stock, housing, and equipment)—only 40 percent of such gains is taxed—means that more funds are available since fewer taxes are paid.

As mentioned, at this time no one has done an analysis of the impact of tax expenditures on urban transportation. On the Federal level, however, the Congressional Budget Office estimates that in Fiscal Year 1981 Federal tax expenditures will amount to $206 billion [14.27]. Assuming again that 15 percent of these end up being used for urban transportation, this amounts to $31 billion. This rather large sum does not include state- or local-level tax expenditures.

14.1.6 Combinations

A wide variety of combinations of approaches are possible. Some of these were discussed before under the transit finance section. One more illustration here should help to highlight the innovativeness, and complexity, involved. This approach to help finance urban transit—especially purchase of new vehicles—involves investment tax credits for private leasing firms. In a scheme fostered by new Federal tax laws in 1981, a firm (or, preferably, a limited partnership because of its tax status) could borrow funds to purchase buses. The firm then would lease them to a city-government-owned operation. The firm thus would be eligible for tax deductions on loan interest payments as well as accelerated depreciation. Importantly, it might also be able to obtain an investment tax credit of 10 percent of the purchase price of the vehicles. The city could gain, in part, because the firm could offer the vehicles at a lower price, given its tax savings.[6]

An alternate to this, apparently being suggested in New York, would involve public sale of industrial development bonds (the interest payments on which are tax-free to purchasers). The proceeds are loaned to a private firm, which buys the vehicles. It then leases them back to the city, and receives the tax benefits mentioned above.

14.1.7 Estimating Revenues

Given the large number and variety of revenue sources, it often becomes a difficult matter to forecast the revenues to be derived from them. Some, of course, represent one-time grants, which may never be repeated. Others, of more concern here, depend on the response to changes in user charges or fares. The basic estimation procedure is simple in concept, but more difficult in practice. Specifically,

$$\text{Revenue} = (\text{fare or charge}) \times (\text{ridership}) \qquad (14.2)$$

The difficulty is in obtaining the ridership figure, since it may depend, in part, on the fare or charge. In other words, ridership may go down as fares increase, and vice versa. The travel models described in Chap. 7 would help in forecasting the likely response.

---

[6] This approach was outlined by Jack Mason, finance director of Tri-Met, Portland, Oregon. It has not been undertaken before, and possibly might have some legal hitches.

As an illustration of revenue-forecasting procedures, we have simplified an example and process described by Fujita et al. for Honolulu [14.6]. There, in 1978, a proposal was made to raise adult cash fares from $0.25 to $0.50 and student cash fares from $0.10 to $0.25 a ride. This was to be accompanied, however, by initiation of a monthly prepaid bus pass—$15.00 for adults and $7.50 for students. Transfers were to continue to be free.

People familiar with transit in Honolulu were asked to give their best estimates of the likely change in the system under this new fare system. We have simplified their responses to be:

Base 1978 annual adult passenger trips (millions):	32.5
Base 1978 annual student passenger trips (millions):	15.0
Fraction of adult passenger trips involving passes:	0.47
Fraction of student passenger trips involving passes:	0.51
Number of monthly adult trips per pass:	40
Number of monthly student trips per pass:	40
Adult single cash-fare ridership shrinkage:	15%
Student single cash-fare ridership shrinkage:	25%

For instance, the last item was estimated to be relatively high because school bus fares in some rural sections still would stay at $0.10 per trip.

Four types of ridership had to be estimated: adult pass and cash fare, and student pass and cash fare. These then would be multiplied by the respective fares, as in Eq. (14.2), and summed to get the total revenue. For the first, the number of adults expecting to take the pass option was

$$(0.47)(32.5) = 15.28 \text{ million per year}$$

With 40 trips per pass, this becomes

$$15.28/40 = 0.38 \text{ million passes per year}$$

which, at $15.00 per pass, gives

$$15.00(0.38) = \$5.73 \text{ million per year}$$

The corresponding calculations for student passes result in

$$0.50(15.0)(7.50)/40 = \$1.43 \text{ million annually}$$

For the cash fares, the ridership must be decreased by the appropriate shrinkage factor. For adults, the number of cash fare passengers is

$$(1 - 0.47)(32.5) = 17.23 \text{ million passengers per year}$$

times the shrinkage,

$$17.23(1 - 0.15) = 14.64 \text{ million passengers per year}$$

times the fare per passenger,

14.64(0.50) = $7.32 million

For students, the corresponding figures are

(1 − .051)(15.0)(1 − 0.25)(0.25) = $1.38 million

This brings the annual total from all four sources to:

Total revenue = 5.73 + 1.43 + 7.32 + 1.38 = $15.86 million

One of the advantages of these procedures is that sensitivity analyses can be done. That is, difference values in (usually uncertain) input figures can be assumed, and the impact of these values on revenues predicted. Thus, if in the example the fraction of adults selecting passes were to rise to, say, 0.65, total revenues could be found via the above types of calculations to be $15.56 million. Since this sum is slightly less than that when the fraction is 0.47, we might conclude that, at least from a revenue viewpoint, it would be better if the prepaid pass program were not too well utilized.

Actual figures for the Honolulu system showed monthly revenues in 1980 coming in at the yearly rate of about $17.50 million. The overage appears to be due in great measure to an overall increase in passenger ridership from 47.5 million to about 54.0 million.

14.1.8 Assessing Revenue Sources

The impacts of taxes, fares, and other revenue development mechanisms obviously are broader than simple generation of funds. Sometimes it is necessary to assess these kind of impacts as much as, say, the direct effects of transportation projects. In doing so there are at least five factors that might be considered:

1. Revenue generated
2. Efficiency
3. Equity (among modes, among income, age, racial, etc. groups, and among links in the network)
4. Demand and/or consumption (use)
5. Nonuser impacts

"Efficiency" here means that the revenue-collection process should be easy to administer and low in cost relative to the amount of revenues. Fuels taxes fit these conditions. In most states they actually are collected at wholesale locations (e.g., at large oil-storage depots), which may be very few in number. Bills then are sent to the oil companies. In Virginia the administrative costs are less the 1 to 2 percent of revenues.[7] This approach can be contrasted to toll roads, where collection costs can be 10 to 15 percent of revenues.

Equity generally is evaluated in terms of the relative burden placed on different groups. As noted above, these groups can be identified by income, age, sex, race, etc., or by mode or by link in the network. Fuels taxes, for example, are notoriously regressive insofar as their distribution by income class is concerned. This is demonstrated in Table 14.7, which contains data from urban areas nationwide.

[7] Annual report of the Virginia Division of Motor Vehicles, Richmond, Virginia (every year).

Table 14.7 Gasoline Tax Burden by Income Group

1970 median household income class ($1,000)	Average income within class ($)	Average annual auto miles per household*	Annual gasoline tax payments per household† ($)	Gasoline tax as percentage of income
0–3	1,800	3,100	32	1.8
3–4	3,500	5,775	60	1.7
4–5	4,500	7,820	81	1.8
5–6	5,500	10,416	108	2.0
6–7.5	6,750	11,526	120	1.8
7.5–10	8,750	14,396	150	1.7
10–12	11,000	16,104	167	1.5
12–15	13,500	18,300	190	1.4
15–20	17,500	24,080	250	1.4
20–25	22,500	30,020	312	1.4
25+	28,000	32,000	333	1.2

*Based on interpolations from data found in U.S. Department of Transportation, Federal Highway Administration, *Nationwide Personal Transportation Study* (11 vols.), Washington, D.C., 1974 and 1975.

†Assuming that autos driven by those in each income class average 12.5 mpg and the gasoline taxes are $0.09/gal state and $0.04/gal Federal.

Gasoline payments were calculated by assuming an average fuel consumption of 12.5 mpg and a total tax of $0.13 per gallon. For instance, households with $7,500 to $10,000 of 1970 income drove 14,396 miles on average. This leads to a gasoline tax expenditure of:

$$\frac{14{,}396 \text{ miles/yr}}{12.5 \text{ miles/gal}} \times \$0.13/\text{gal} = \$150/\text{yr}$$

which, as a percentage of income, is:

$$\frac{\$150}{\$8750} = 0.017 = 1.7\%$$

According to Table 14.7, the gasoline tax is a relatively constant percentage of 1970 income for those households with less than $10,000 annual income, but then drops by about a third to 1.2 percent for households with yearly incomes over $25,000. Although the percentage of income is not substantial in either case, a larger burden on the rich would seem to be more appropriate in terms of their ability to pay.

Lentze and Ugolik have done equity analyses for transit fares using 1975 data from the Albany, New York, transit system [14.5]. They found that the young pay the highest fare per mile, while the oldest pay the least (Table 14.8). More generally, the short-trip, non-peak hour, non-work, inner-city urban rider who generally is less well off in society and is dependent on the bus paid more per mile than others.

Taxation or setting of user charges for one mode also can offset the viability of other modes. Free transit, for instance, has been employed in part to reduce travel by auto. Similarly, it has been suggested that urban highways, particularly in CBD areas, are "undertaxed" in that each vehicle is not charged enough to pay the additional cost of the highways required to support its use.[8] It then is argued that transit fares similarly should be "subsidized" to keep the competition equal.

Taxes and charges also can affect the actual use of the transport system. The travel models described in Chap. 7 help in estimating these effects. As a quick illustration Dyert, Holec, and Hill found in a survey of 24 cities that the median loss in transit ridership was about 0.35 percent for each 1 percent increase in fares [14.4, p. III-15]. The loss depends, of course, on the availability of alternate modes, demographic characteristics of the population, type of fare structure, and so on.

Nonuser impacts can be as broad as any from direct expenditures. Loss of transit ridership, for example, might lead to slightly increased air pollution. Or, in another vein, attempts to impose large fuel taxes in one locality may lead to decreased purchases from service stations within that locality and resultant economic hardships. An exhaustive analysis might take into account most of the impact factors discussed in Chap. 8.

14.2 EXPENSE ESTIMATION

On the other side of the ledger from revenues are the expenses needed to construct/purchase, operate, and maintain transport systems. In some cases it is not necessary for planning purposes to do detailed estimation of expenses. Most state transportation departments, for instance, have a long list of highway construction projects suitable for funding. It thus is a matter of doing the highest priority projects first each year until funds run out. The planner only needs to know that the projects being proposed for a given metropolitan area do not create the need for expenditures far beyond any likely funding levels. In other words, many highway construction (and also many transit capital expenditure) projects are discretionary. They can be delayed or eliminated if sufficient funds are not available.

Although the distinctions are not always clear, it can be said that most maintenance (highway and transit) as well as transit-operating functions represent uncontrollable (relatively speaking) expenses. They must be sustained at or about

---

[8] For some of the arguments, see [14.4, pp. IV-31 to IV-35].

Table 14.8 Variations in Average Transit Fare per Mile by Age Group

Age group (yr)	Portion of sample (%)	Average fare/ mile (cents)
0–17	14.9	22.2
18–24	24.9	15.9
25–44	26.7	17.1
45–64	23.8	19.1
65+	6.2	12.8
No response	3.5	21.3

Source: [14.5].

certain levels each and every year. They cannot be zero one year, and substantially higher the next. These uncontrollable expenditures thus must be estimated to make sure they do not rise substantially beyond annual revenues.

Committed expenses can be predicted in several ways [14.15]:

1. Direct factor method
2. Regression techniques
3. Causal-factor method

As a simple example of the first, suppose that the Honolulu bus system illustrated in the previous section had a total-operating-expense-per-total-revenue-passenger ratio of 41.9 cents in 1975, for 38.61 million annual revenue passengers. It is desired to find total operating expenses in 1979. First it is necessary to estimate or assume growth rates for both the ratio and revenue passengers. Past experience might indicate that a good guess for these would be, say, 9.4 and 6.2 percent, respectively. These would lead to 1979 levels of

$$41.9(1.094)^4 = 60.1 \text{ cents operating expense/revenue passenger}$$

$$38.61(1.062)^4 = 49.1 \text{ million revenue passengers/year}$$

The exponent 4 in this case represents a compounding of the growth rate for the four years between 1975 and 1979. The estimate of total operating expenses in 1979 then would be calculated as

$$(49.1)(0.601) = \$29.51 \text{ million}$$

The second method, involving regression,[9] represents a more statistically sophisticated approach to determining the relationship between, say, operating expenses and factors like vehicle·hours operated, top operators' wage rate, average daily ridership, and the like. Since several of these factors can be taken into account, this approach represents a more realistic assessment of the multitude of variables affecting costs. The regression approach also is advantageous in that data from several systems can be employed, thereby expanding the experiential base for expenditure prediction.

A regression relationship based on information from 33 conventional bus transit systems has been found to be [14.15, p. V.126];

$$\$OE = 2.66 \text{ (VHT)(W)} - 92.99 \tag{14.3}$$

Operating expenses (OE in 1,000s of current dollars) are shown as a function of annual vehicle·hours operated (VHT)[10] and the top operator's wage rate, (W, in current dollars per hour). The value of r is quite high at 0.9963, which verifies the fact that most of the data points fall very close to the regression line.

Suppose that the top operator on the Honolulu bus system received an hourly wage of $6.00 in 1974. Further assume that this wage rate was expected to go up 10 percent per year from 1974 to 1979. This brings it in the latter year to

---

[9] See App. A for a description of correlation and regression.
[10] Assumed to be revenue hours.

$$\$6.00(1.10)^5 = 6.00(1.61) = \$9.66/hr$$

The transit system anticipates running 1,100 (thousand) revenue vehicle hours in 1979. With these figures and Eq. (14.3), it then can be estimated that operating expenses in that year will be

$$\$OE = (2.66)(1,100)(9.66) - 92.99 = \$28,172\text{thousand} = \$278.17\text{million}$$

Note that one of the benefits of this approach is that analyses can be made of small variations in service levels to determine the impact of expenditures. Thus, even though the bus system in our example may be committed to about 1,100 (thousand) revenue vehicle hours of service (and thus the corresponding expenditures), this commitment is not absolute. It possibly could be, say, 1,000 (thousand) hours, which would require expected expenditures of only

$$\$OE = (2.66)(1,000)(9.66) - 92.99 = \$25,602\text{thousand} = \$25.60\text{million per year}$$

The third method—the so-called "causal-factor" approach—is a simplified combination of the previous two. Expenses first are divided into categories, as exemplified in Table 14.9. Ratios then are established for each category based on

Table 14.9   Allocation of Transit Operating Expenses to Four Causal Factors

		Causative factor: $ allocated			
Expense category	Annual expense ($)	Vehicle· miles operated ($)	Vehicle· hours operated ($)	Vehicles in service ($)	Vehicle operators ($)
Fuel for revenue vehicles	1,000	1,000			
Tires for revenue vehicles	100	100			
Repair of revenue vehicles	6,000	6,000			
Servicing of revenue vehicles	2,000			2,000	
Operator's wages	10,000		10,000		
Operator's fringe benefits	2,000				2,000
Scheduling*	400			200	200
Total	21,500	7,100	10,000	2,200	2,200

*Scheduling costs are split between vehicles and operators in this example on the assumption that half the time of the scheduling personnel is spent in scheduling the vehicle and half in scheduling the vehicle operator. Annual operational data:

Vehicle·miles operated:	21,500
Vehicle·hours operated:	2,150
Vehicles required:	6
Vehicle operators required:	8
Average system speed:	10 mph
Mileage based costs:	$ 0.33 per vehicle·mile
Hourly based costs:	$ 4.65 per vehicle·hour
Vehicle based costs:	$366.67 per vehicle
Operator based costs:	$275.00 per vehicle operator

Source: [14.15, p. V.128].

one or more "causative factors." To illustrate, in Table 14.9 fuel for revenue vehicles obviously depends most closely on vehicle·miles operated. Operator's wages would not be a function of this factor, but of vehicle·hours operated. For some categories the relationship might be with two or more causative factors, in which case regression could be employed to find the connection.

14.3 COMPARING REVENUES TO EXPENSES

After determining both revenues and expenses, the next obvious step is to compare them. If the former exceeds the latter, a not-too-common situation, various options are available. These include reducing fares, adding services, constructing facilities, buying buses, and so on. If expenses exceed revenues, however, difficult decisions must be made about generating new revenues, reducing service, or a combination.

The models in the preceding two sections can help provide an example. Assume that there also can be capital expenditures for buses, which in 1979 cost $97,819 apiece. The bus system wishes to purchase 100 of these.

Revenues are estimated to be $15.86m; operating expenses to be $28.18m; and the 100 buses will cost $9.78m. The difference between revenues and the expenditures consequently is

$$\$15.86m - \$28.18 - \$9.78m = -\$22.10m$$

A substantial subsidy thus will be required.

Some recalculations show, however, that if (1) base adult trips rise to 40m/year and base student trips to 18m/year, (2) fares are raised to $20 per adult per month and $10 per student pass, (3) only 10 new buses are purchased, and (4) revenue vehicle·hours are reduced to 1,000 (thousand), then the subsidy need only be about $4.23m, a significantly lower figure. Techniques like those presented above allow the planner and analyst to make these kinds of important comparisons.

14.4 PROGRAMMING OF IMPROVEMENTS

Programming is the matching of proposed projects with available funds to accomplish long- and short-range goals [14.28]. It sets the work to be performed and the objectives, for a fiscal quarter or year, with regard to most effective use of anticipated monies so as to achieve regional and agency goals. The objectives to be accomplished are the advancement of construction and transportation system management improvement projects, through the successive stages of implementation—planning, design, ROW purchases, construction, etc. [14.29].

Unfortunately, in many agencies planning and programming have been divorced [14.28, 14.34]. "Programming is definitely a planning function...the pivotal implementation step in the planning process" [14.28]. Actually, planning and programming each can, and should, be visualized as an aspect of each other. Both perspectives are right and neither is complete without the other [14.8]:

> Planning is associated with such terms and concepts as long-range, goals and objectives, idealistic, uninhibited, unconstrained, and policy-oriented. Programming is more associated with such terms and concepts as short-range, fiscal constraints, priorities and project-oriented.

The traditional loose relationship between planning and programming is represented in Fig. 13.1. Programming simply picked up where planning left off. As the transportation climate changed during the 1970s, both activities became more closely interrelated so as to provide an integrated information basis for transportation decisions. The working relationships between planning and programming will be discussed throughout the chapter sections that follow. Later in the chapter we will explicitly note how the newer decision-making model (Chaps. 11 and 13) has changed the planning–programming relationship from that in Fig. 13.1.

14.4.1 Dimensions and Definitions

Programming of transportation improvements in and for a metropolitan area is a complex process. Simplistically, one may conceive of programming as the meshing of three flows—projects, funds, and goals—through a multidimensional matrix. Figure 14.2 represents this concept in three main dimensions. One input flow is projects, proposed improvements identified by system and project planning. The second flow is funding, estimated through financial planning, and finally determined by legislative appropriations and socioeconomic conditions. A third flow, in line with the definition of programming at the start of this section, is goals and policies [14.28]. These guide the prioritization of projects and their matching with funds. Policies include Federal, state, and local laws and regulations, and agency goals, objectives, and policies. The application of goals and policies to programming will be discussed at several points in the next two sections.

The three dimensions of forces that meld the three input flows in Fig. 14.2 are structure/involvement, program process, and program categories. Structure/involvement is the set of roles assumed by the various participants in the programming process. It involves the internal organization of the metropolitan agency and its working relationships with local governments, transit agencies, private transportation suppliers, and the state transportation agency and other agencies. In addition, there are state legislatures, transportation users, special-interest groups—highway contractors, land developers, trucking firms—and "public interest" groups such as environmentalists, neighborhood associations, and civic groups. This cast of participants and their roles in programming were discussed in Chap. 13.

Programming process covers the activities of the programming participants to blend the three input flows to produce an effective program of transportation improvements. Program categories are a policy device to help match funds with projects in such a way as to achieve legislative and agency goals. The range in types of transportation improvement projects—highways, transit, paratransit, operations, maintenance, reconstruction, etc.—does not always neatly fit the official program funding categories, such as Interstate, Primary, Secondary, Urban, Resurfacing, Maintenance, Bridge, Safety, Transit Capital, Transit Operations, etc. Most states have six to eight highway categories, and many have more; Tennessee has 25 [14.45].

The reader should be aware that urban transportation programming processes and techniques, despite Federal regulations, are far from uniform or routine. The discussion that follows outlines the main themes, but there are as many variations as there are metropolitan areas. Policies guiding programming in each urban area are complicated by the peculiarities of the particular state or states in which it is located. State constitutions, laws, and regulations are varied, reflecting historical origins, the current number, size, and diversity of metropolitan areas, the rural-urban mix, snowbelt–sunbelt, and other socioeconomic factors. Similar factors

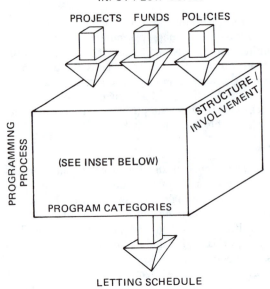

INPUT FLOW-GOALS

PROJECTS   FUNDS   POLICIES

STRUCTURE / INVOLVEMENT

PROGRAMMING PROCESS

(SEE INSET BELOW)

PROGRAM CATEGORIES

LETTING SCHEDULE

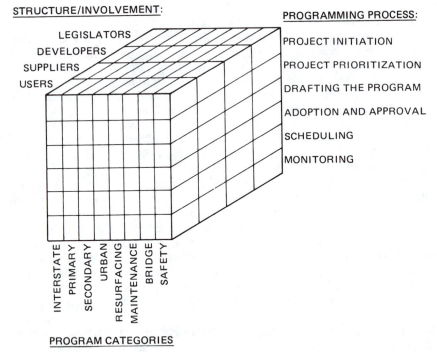

STRUCTURE/INVOLVEMENT:

LEGISLATORS
DEVELOPERS
SUPPLIERS
USERS

PROGRAMMING PROCESS:

PROJECT INITIATION

PROJECT PRIORITIZATION

DRAFTING THE PROGRAM

ADOPTION AND APPROVAL

SCHEDULING

MONITORING

INTERSTATE PRIMARY SECONDARY URBAN RESURFACING MAINTENANCE BRIDGE SAFETY

PROGRAM CATEGORIES

Fig. 14.2 Three dimensional conceptual model of programming. Adapted from [14.45, p. 7].

within each MPO will affect its organization, cast of actors, programming process, goals, and other policies. Funding sources and program categories also vary, being determined by state and local, as well as Federal, policies. In particular, state laws differ concerning the split in responsibilities between state and local levels of government. The foregoing factors also affect transportation improvement programming in each metropolitan area with respect to the structure and involvement of participants, and the process and technique.

14.4.2 The TIP Regulations

FHWA–UMTA joint regulations governing transportation programming were finalized in the *Federal Register* in November 1974 and include the following required elements [14.46]:

1. A "transportation plan consisting of a transportation systems management element and a long-range element" (TSME and LRE).
   a. The LRE provides "for the long-range transportation needs of the urbanized area," identifies "new transportation policies and transportation facilities or major changes in existing facilities" and "shall be consistent with the area's comprehensive long-range land use plan, urban development objectives, and the area's overall social, economic, environmental, system performance and energy conservation goals and objectives."
   b. The TSME provides "for the short-range transportation needs of the urbanized area by making efficient use of existing transportation resources and providing for the movement of people in an efficient manner" and identifies "traffic engineering, public transportation, regulatory, pricing, management, operational and other improvements to the existing urban transportation system not including new transportation facilities or major changes in existing facilities."
2. Transportation improvement program (TIP) and annual element (AE):
   a. The TIP is "a stage multilayer program of transportation improvement" that is "recommended from the transportation systems management and long-range elements of the transportation plan . . . ." The 3–5-year TIP

      shall (a) Identify transportation improvements recommended for advancement during the program period; (b) Indicate the area's priorities; (c) Group improvements of similar urgency and anticipated staging into appropriate staging periods; (d) Include realistic estimates of total costs and revenues for the program period; and (e) Include a discussion of how improvements recommended from the long-range element and the transportation systems management element . . . were merged into the program.

   b. The AE of the TIP is "a list of transportation improvement projects proposed for implementation during the first program year," including all Federally funded and TSM projects. "Federally funded projects shall be initiated for inclusion in the annual element at all stages in the development of the transportation improvement . . . ."
   c. The types of actions to be included in the TIP include, but are not necessarily limited to

      engineering related to the acquisition or construction of transportation facilities; acquisition of rights-of-way, construction, and reconstruction of highways, busways, fixed guideways; fringe parking facilities; major street improvements; transit rolling stock acquisitions; TOPICS projects; bicycle and pedestrian facilities; major revisions in levels of transit service and transit route structures; initiation of exclusive and preferential bus and carpool lanes; staggered work hours; measures to encourage carpooling; regulation of parking supply and costs; and projects to meet the special needs of the elderly and handicapped.

3. A planning work program, including a prospectus and a unified planning work program (UPWP):
   a. The prospectus establishes an ongoing cooperative organization structure, "a multiyear framework within which the unified planning work program is accomplished," which includes "A summary of the planning program including discussion of the important transportation issues facing the area" and provides "A general description of the status and anticipated accomplishments" of the various technical and planning activities, and "A description of the functional responsibilities of each participating agency" including the governor–designated metropolitan planning organization (MPO).
   b. The unified planning work program shall "Annually describe all urban transportation and transportation-related planning activities anticipated within the area during the next 1- or 2-year period regardless of funding sources . . . ."

4. In addition to these planning and programming requirements, present procedures require project-specific documents:
   a. Project applications—for Federal funding with certifications of compliance with numerous legislatively mandated requirements (e.g., civil rights, public hearings, environmental impact statements).
   b. Technical documents—such as technical studies of project alternatives and draft and final environmental impact statements.
   c. Alternatives analysis (AA)—for major mass transportation investment projects that involve "new construction or extension of a fixed guideway system (rapid rail, light rail, commuter rail, automated guideway transit) or a busway . . . ." (An analysis of alternatives is required for all projects as a consequence of the National Environmental Policy Act of 1969).

14.4.3 Programming Process

Figure 14.3 introduces the main elements and linkages in the Programming Process. The five blocks represent functional units, or sets of actors, in state and/or MPO agencies dealing with the particular element, although the legislative body also would come under block II in its appropriation activities.

Arrows in the diagram represent critical working linkages between the functional units. While only four are shown, in actuality there are relatively strong connections between all five blocks. Problems in the relationships indicated by the linkages sometimes develop because the participating staff are usually in different organizational units. Probably the most frequent communications breakdown of this kind occurs because two of the functions, systems planning and urban-transportation programming, are typically organizational units of the MPO while the others are typically in the state transportation agency in the capital, usually geographically separate. The policy-making function is also divided between MPO, state agency, and legislature.

The ideal programmer in this process is a catalyst and a referee. He or she draws on a wide range of participants' views, technical information, and professional judgment to creatively implement the goals and policies of the agency, of the

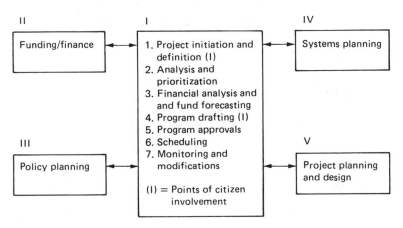

Fig. 14.3   Program process and linkages.

legislature, and, within limits, of other agencies. While this must be done with the skills of an accountant, the task is far more than a mechanical matching of funds and projects.

Short-range transportation programming is cyclical. TIPS are basically updated annually with a firm budget for the next fiscal year (FY), and a tentative designation of projects for funding in the subsequent several years. Adjustment in priorities and schedules may be made more often (quarterly in some agencies). The activities to complete or operationalize a project of any magnitude must be scheduled over several FYs. The activities (or stages) for a construction project include corridor or location studies; Federal, state, and local approvals; hearings; design and design approvals; ROW plans; appraisals and taking; relocation, contract advertising, and contract award; and construction. The programmer, and the responsible decision officials, must look 5 years ahead to schedule properly for two. On major projects, such as urban freeways or rapid rail extensions, the programmer schedules for 10 years to have a reasonable schedule for 5.

Obviously, despite the most careful scheduling, any number of events may occur to change priorities or force adjustments in the program. Hence, the progress of a project through its various stages is closely monitored, and revisions in the program are made as indicated. The other key points to keep in mind concerning programming are [14.28]:

1. The program is rarely new; it usually contains commitments from previous years and to other agencies or groups.
2. The projects are in all stages of development, from basic planning studies to final design. At any point and for any number of reasons, a project may be stopped temporarily and thrown off schedule.
3. The funds available may be restricted to certain categories of use, although there may be some flexibility with regard to transferring funds between categories or reassigning projects to different categories.
4. Priorities may be constantly changing because of changing philosophies, transportation needs, economic conditions, energy availability, political conditions, and other factors affecting individual or collective priorities.

Scheduling of project development activities as just described is done only in a short-range (SR) TIP. Most MPOs maintain some form of mid-range and long-range program or plan to provide a link between the SRTIP, and longer-range systems planning, need studies, and lower priority projects [14.33]. The programming process outlined in Table 14.10 is basically that for the SRTIP with linkages to longer range programming (or "staging") noted.

14.4.3.1 Project initiation and definition initiation Project ideas are generated primarily by the technical activities of the staffs of the MPO and member jurisdictions. The three major sources are systems planning, special studies, and trained observations. Systems planning broadly includes needs studies, short-range TSM studies, long-range systems studies, and other continuous technical-planning efforts. Special studies respond to unique events or situations such as a proposed new shopping center, a parking survey, or a subarea circulation study. Trained observations are the regular, sometimes periodic, surveillance of facilities and operations by regular, trained staff. This is usually done for maintenance and safety purposes.

While all of these technical activities may precipitate many candidate projects, systems planning generally has several features that lead directly to project initiation. Needs studies are intended to identify present or future deficient segments of an existing transportation network and very rough estimates of the cost of improvements needed to bring this network to a desired level of service. Specific proposals are not necessarily included [14.28]. Long-range systems plans (20 years, more or less) generally include a staged list of projects [14.28, 14.33]. Shorter

Table 14.10 Outline of the Programming Process

1. Project initiation and definitions
    1.1 Initiation
        1.1.1 Technical sources: planning studies, special studies, and trained observations
        1.1.2 Nontechnical sources
    1.2 Initial listing
    1.3 Preliminary analysis
    1.4 Planning report (project description)
    1.5 Combined listing (new projects)
2. Project analysis and prioritization
    2.1 Technical prioritizing
    2.2 Nontechnical prioritizing
    2.3 Feedback from project planning and development
    2.4 Integrated listing (old and new projects)
3. Financial analysis and fund forecasting
    3.1 Categorical grants
    3.2 Geographical distribution
    3.3 Fiscal-year projections (fund forecasting)
    3.4 Manpower analysis
    3.5 Financial modifications
4. Drafting the TIP
    4.1 Short-range, 3–5-years TIP
    4.2 Middle- and long-range program or list
5. Program approvals
6. Scheduling
7. Monitoring and modifying

Source: Adapted from [14.28].

range TSM plans also generate specific project proposals. Increasingly, system plans suggest alternatives suited to differing contingencies, resulting in identification of alternative project possibilities [14.33].

Nontechnical sources produce "political" or "citizen" projects, which seem to be relatively few in number. "No matter who requests a project or how ridiculous the project may appear, most jurisdictions provide a courteous discussion with the initiator" [14.28]. The discussion may lead to identification of a possible project, recognition that a project in the program will meet the need, or acceptance of a modification. If the initiator persists, however, the project is recorded for future screening.

These various technical and nontechnical sources produce initial listings of candidate projects. Initial listings may be maintained by many transportation professionals in the various metropolitan jurisdictions—city and county engineers, traffic engineers, transportation planners and land-use planners, transit agencies, the state transportation district or regional offices, as well as the MPO central staff.

Preliminary analysis: The initiating agency, the potential implementing agency, or the MPO staff generally assembles available or easily obtained information on initiated projects, and prepares a planning report. The available information may include regularly recorded and computed data such as need studies, sufficiency ratings, bridge ratings, and accident data. The planning report is a brief description, usually to a standard form, of the project. It "defines" the project. The report contains fundamental data on location, termini, length, accidents, ballpark cost estimate, etc. The form will usually call for some estimate of the project's importance to the system, its potential impacts, and controversialism. Generally, project reports as completed are computerized by the MPO in what is called the project information system.

Combined listing: On some regular basis, the MPO staff formally consolidates the lists form the various jurisdictions and agencies. Usually there is a call for candidate projects to initiate the annual cycle of updating the TIP. The combined list generally does not include projects that are on the previous year's short-range TIP (this is done in the integrated listing, step 2.4). The list does have projects initiated in previous years, but not yet on the TIP; a computerized project information system facilitates this. There are usually few entirely new projects in any given year. However, initiating agencies may change the priority with which they regard a project. Such indications will help determine whether, and how much, further analysis is given a candidate project.

14.4.3.2 Analysis and prioritization Prioritizing is the process of producing a rank order of projects and project sections based on analyses of technical and nontechnical, and quantifiable and nonquantifiable, factors [14.28]. The basic process has two phases: first, technical prioritization, according to quantifiable and sometimes qualitative factors and resulting in an initial priority ranking, and second, modification of this order by nontechnical factors. The latter modifications may occasionally be political but are most often nonquantitative professional judgments that could not be suitably incorporated in the technical analyses. Since the product of the two steps is an integrated, prioritized listing of all projects, new as well as those on the previous TIP, feedback from project planning and development activities is an essential input to this step.

Step 2 does not necessarily begin in a uniform, orderly manner after step 1. The contrasting time frame of the two steps is one reason. Project initiation is sporadic, although it reaches a culmination as the annual program undate deadline approaches. Analysis for prioritization is more continuous. An internal feedback loop is a second reason. Given time constraints, those projects with the highest apparent priority are given the more intensive critical analysis. For example, a project may be studied, considered inferior to an alternative project, and the study terminated. But when the alternative project proves more costly or more impacting than originally assessed, analysis of the original project may be resumed [14.28].

In the project analysis step, the programmer must work closely with other organizational units and agencies and secure their participation (Fig. 14.3). Within tight time constraints, he or she must use all available information. Systems planners will often have data or insights on projects. Policy officials can clarify goals and goal tradeoffs. Project, systems, and environmental planners and local officials will have insights into likely levels of impacts. Planners and designers, when projects have advanced to their stages, may have vital new information on alternatives, costs, impacts, public support or opposition, and other critical prioritization factors.

Technical prioritizing The first stage of technical prioritizing is the development and review of relatively uncomplicated project ratings such as sufficiency ratios, accident totals or rates, and traffic volumes. These ratings provide a first, straightforward screening of projects; more advanced techniques can then be applied to the surviving projects. For the first screening, sufficiency ratings have been in use in many states for over 25 years [14.28]. They are often employed on a biennial basis. The earliest sufficiency ratings considered only structure survey and safety. They were essentially "deficiency ratings," and in the 1980s have a new importance in those states where the objective is to "maintain the existing system to prevent further service deterioration" [14.28, 14.35].

Table 14.11 outlines the evolution of priority ratings, based on the 1973 FHWA report "Objective Priority Programming Procedures" [14.36]. The second generation of techniques—priority ratings in the table—provides a better basis for prioritizing projects by adding additional data, such as accident totals and rates,

Table 14.11 Evolution of Transportation-Project Prioritizing Techniques

1. Sufficiency (deficiency) ratings
2. Priority ratings
   a. Addition of safety factors (accident totals, rates, specific locations)
   b. Addition of capacity factors
   c. Addition of economic factors (e.g., benefit/cost, cost-effectiveness, displacement of businesses, jobs during construction, direct routings, use of air rights)
   d. Addition of quantifiable social and environmental factors (e.g., displacement of families, air pollution)
   e. Addition of nonquantifiable SEE factors
      i. Social (e.g., disruption, proximity, neighborhood cohesion, minority-elderly-handicapped impacts)
      ii. Economic (e.g., economic base, mobility, accessibility, employment after construction, land-use impacts)
      iii. Environmental (e.g., aesthetics, effect on natural resources, water pollution, vibration, noise)
3. Option-evaluation techniques
   a. Comprehensive
   b. Sketch planning

Source: [14.36, p. 28].

volume/capacity (V/C) ratios, and social, economic, and environmental (SEE) impact ratings.

All of these items are relevant to prioritizing, but there are complications. First, how should one provide the priority listing? Should the items be combined, or should deficiency scores of V/C ratios or some other scale be the basis for ranking (with other data and rankings only for information)? If items are to be combined, should they be weighted the same? If not, who should weigh them? Since weighing is judgmental, and would probably vary from one staff member to another, does this defeat the purpose of having a basic objective technical valuation?

Another complication is in the addition of preliminary assessments of social-economic–environmental (SEE) factors, which are very pertinent since they affect project cost and feasibility. The problem here is that, prior to the project-planning stage that produces technical, quantitative assessments of SEE impacts, any quantitative values will be judgmental and vary among individuals. Most states now tackle this problem with a panel of local and state planners and engineers familiar with each project, assigning it a "level of project significance" or "SEE impact level." Among the states this varies from three to eight categories. Reports by FHWA [14.36] and the Transportation Research Board [14.37] review priority rating techniques.

The option-evaluation techniques in Table 14.11 are a third generation of more sophisticated prioritizing methods. Comprehensive techniques attempt to provide optimal project selection based on maximum return on investment [14.38, 14.39, 14.40]. Sketch planning attempts "to test and analyze policies and programs at a broad scale without delving into the specifics of project development and implementation" [14.37, 14.41].

Table 14.12 lists factors now generally considered in technical prioritizing [14.28]. It also affords an overview of the evaluation of prioritization techniques. Those factors at the top of the list were incorporated into the early sufficiency ratings of the 1950s. Consideration of the quantifiable, and then the unquantifiable, SEE factors dates from about 1970.

Although technical prioritization of projects by third-level staff may be overridden by the nontechnical factors in the minds of higher level decision-makers, its importance should not be underestimated. The majority of programming decisions are made firmly and finally based on technical analyses. Furthermore, many unpromising projects are screened out by technical evaluation. Finally, technical evaluation provides an initial ranking, which can then be adjusted judgmentally, intuitively if need be, for nontechnical factors.

Nontechnical prioritization Once the basic technical prioritization is established, the order can be adjusted for nonquantifiable factors. While some of the following factors have technical aspects, most considerations under this broad heading generally required little or no analysis to weigh their relative importance. The weighing is done in the judgment of programmers or higher level decision-makers. A survey by Campbell [14.28] identified the following factors in this category:

1. Political commitments
2. Legislative mandate (line-item budget for specific projects)
3. Emergency
4. Special emphasis
5. Commitments to other agencies

Table 14.12 Factors Involved in Priority Ratings

Quantifiable factors

Physical condition (deterioration): road surface, pavement structure, foundation, shoulders, drainage
Geometrics: pavement width, shoulder width
Alignment: horizontal, vertical
Bridges: condition rating, operating rating
Safety rating: accident totals and/or rates
Benefit/cost rating
Cost-effectiveness index
Recreational use
Social families displaced
Economic: businesses displaced, direct routings, jobs during construction, use of air rights
Environmental: air–noise–water pollution

Nonquantifiable factors

Social: neighborhood cohesion, minority–elderly–handicapped impacts, disruption, proximity
Economic: build vs. no-build, economic base, mobility, accessibility, employment after construction
Environmental: Effect on natural resources, aesthetics, water pollution, vibration, noise
Land-use impacts: future development, community standards
Transportation need
Uncertainty: public support, court cases

Interrelationships

Impacts on connecting facilities
Impacts on competing facilities
State construction
System continuity
Agreements and commitments (other agency plans)

Source: [14.28].

6. Project interdependencies (system continuity–connectivity) [14.42]
7. Position in pipeline (project readiness)

The Campbell survey indicated most jurisdictions seemed satisfied that positive political input far outweighs politics in the worst sense. There seems to be relatively little of the latter, and even in these cases the project had sufficient technical merit that it would probably have been built later anyway. Further, positive political input provides a users' view of the transportation system [14.28]:

> It is a valuable check on technical prioritizing. In many cases it illuminates items the technical system has missed. In those myriad programming decisions where a multifaceted value judgement is called for ... political input is the only source provided by a democratic society.

Position in the pipeline is definitely a factor. Projects in advanced states of planning, design ROW acquisition, etc., have a built-in priority not only because of the years of public investment in them, but also because of the lead time necessary to develop an alternative.

Feedback from project planning and development Communications and teamwork between the programmer and the project planner needs to be very close and open. The term "project planning" describes the development of an urban project between the planning report and final design. Other terms used in some states include "project development," "project analysis," and "location and preliminary design." Usually, project planning will not be a lead responsibility of the MPO; this role is normally taken by the ultimate implementing agency. There are at least the following five overlapping major activities in project planning:

1. Development of alternatives, including joint development with land use, environment, or other considerations
2. Environmental analysis of SEE factors, and at some point an environmental impact statement (EIS)
3. Community and technical interaction (citizen participation)
4. Interaction with other agencies
5. Preliminary cost estimates

These data from project planning are then fed back to programming and systems planning for reevaluation of a project's cost-effectiveness for the system as a whole and for its individual benefit/cost ratio. This information can indicate the need for project delays, tabling, or expansion or reduction. Many minor projects, especially in rural areas, do not require as much in-depth development.

Feedback from project planning and design may come at two points in the programming process. First, traditionally, the projects in project planning are already in the program. Feedback is thus used to establish their priority in relation to the previously unprogrammed projects. Second, more and more, in all forms of transportation, in a period of limited funding, projects are receiving serious analysis so that their potential can be better ascertained before inclusion in the published program.

This is related to the practice of "over programming" or, more precisely, "overproducing." To be ready for the inevitable contingency of project delays or tabling, the project development bureau may have twice as many projects under study as construction funds would permit. So that there are no credibility gaps with the public over these "program alternatives," an alternative project may not be listed in the official program [14.28].

Integrated project listing The product of the prioritization analysis in step 2 is a completed listing of all projects. Previous lists and published programs will have been reviewed and new projects included. In addition to program and project alternatives, the unpublished listing may contain "system alternatives" to protect future decision flexibility under changing conditions.

14.4.3.3 Financial analysis and fund forecasting This third step generally is conducted in parallel with step 2, with the results of the two flows of activities—projects and funds—merged into a preliminary program in step 4 (Figs. 14.2 and 14.3). Two fundamental issues are addressed in step 3: how much money will be available, and how much will be required? For the current year these questions have firm answers, subject only to revenue shortfalls and project events. Forecasting of fund availability for the next 5–10 years is a very difficult, complex, hazardous task, yet essential to the realistic commitment of projects to expensive stages of project development and to coordination of transportation with other objectives of regional development. Since the bulk of financial analysis and forecasting for metropolitan transportation is done in state agencies, and in any case must be integrated with the state procedures, the following discussion will cover these procedures.

Categories, apportionment, and allocations Fund forecasting starts with anticipating Congressional and state legislative appropriations, as well as the less important nonlegislative sources (Sec. 14.1). All sources are and will be complicated by certain program categories (for example, see Sec. 14.1.1.1). Broad categories are explicitly subdivided into further categories. Appropriations also often contain necessary geographic apportionment formulas. The states criticize the Federal government for the rigidity imposed by the categories and formulas, and the cities and MPOs criticize the same practice by the state legislatures. Although some long-standing categories are giving way to block grants, the fact is that legislators have to assure that goals of their constituents will be met. Program categories and apportionment formulas are a complicating fact of life for programming. Thus [14.28],

> Fund forecasting requires estimating the amount, timing, and restrictions of federal funding; estimating state matching amounts and their timing; estimating manpower requirements and various related expenditures; relating commitments to cash flow; and constantly monitoring and modifying perhaps dozens of accounts, project timing, and staging.

Some idea of the complexity of fund forecasting is offered by Fig. 14.4. The critical role of fund forecasting in a time of inflating costs and declining revenues is suggested in Fig. 14.5. The product of fund forecasting is a set of allocations—monies estimated to be available for financing projects within program and geographic categories for the next 5-10 fiscal years (FY).

Financial requirements and modifications On the other side of the ledger are

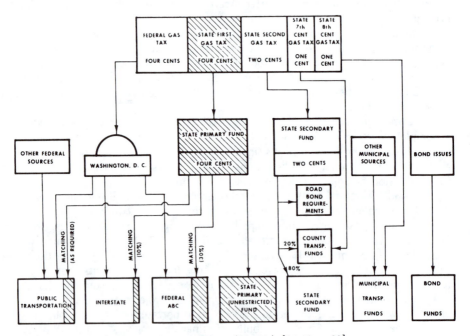

Fig. 14.4 Fund distribution of motor fuel tax (Florida). [14.43, p. 28].

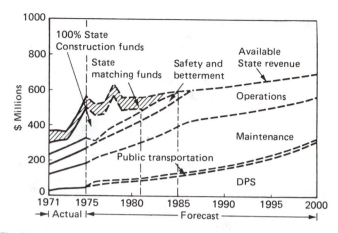

Fig. 14.5 Application of available state revenues (Texas). [14.44, p. 28].

the monies needed. Current estimates of project costs come from programming and project planning and design activities. However, additional manpower analysis is needed since some costs are administrative and not project-related, and others are a function of timing. For example, manpower analysis determines the need for outside consultants, or intra-agency personnel shifts, due to anticipated overloads from concurrent projects. This aspect of financial analysis and forecasting blends into the financial modifications that must be made to develop and maintain a viable program. These modifications are shown as feedback lines in Fig. 14.3 and will be discussed in connection with steps 4 and 7.

14.4.3.4 Drafting the program In step 4, the prioritized list of projects (step 2) and the fund forecast by fiscal years (step 3) are melded into a first-draft program. In so doing the previous list of project divides into two parts: a 3–5-year short-range, relatively firm TIP, and the remaining list of possible future projects. The TIP naturally gets the most attention. It relates project advancement—planning, preliminary engineering, ROW acquisition, design, and construction—with funding by year, and with all necessary cross indexing—fund source, geographic distribution, etc. Each category is overprogrammed (whether published or not) to meet contingencies and prevent the lapsing of funds. The TIP is not a new program as much as it is a reorientation of, and an addition to, last year's program. After the programmer prepares the first-draft program, a series of staff meetings is held to revise and improve the program so that the participating jurisdictions will (hopefully) make few, if any, changes. Alternatively, in some small regions, the state agency or a dominant local jurisdiction will take the lead in collaboration with the staff and other jurisdictions.

There are separate feedback loops, shown on Fig. 14.3, for each of the two parts of the original list. A satisfactory TIP generally requires many iterations back from step 4 to short-range goals and policies to determine their relative weights influencing project selection, and to step 3 to make financial modifications. Available funds are matched with available desired projects, and typically on first iteration the desired projects substantially exceed the funds (overprogramming). *NCHRP Synthesis Report 48* [14.28] describes the hard questions to be answered:

On the project side: not all short-range goals can be accomplished. Should projects be partially built, if possible, or further sectionalized? Should the level of system development be reduced for a less important system? On the financial side: are more funds forecast in a certain category in future years, thus enabling these projects to be put off? Are new federal or state programs developing? Are there other sources of funding for some projects? Will the crunch encourage the legislative body to consider a new bond issue?

Financial modifications are possible. For example, projects not intended originally as Federal-air ones may be moved to that category, and the 100 percent state monies thus made available to balance geographic distribution.

The second feedback loop is the remaining projects from the integrated listing (Sec. 14.4.3.2) that were not included in the TIP. These are cycled back to Systems and Project Planning for further study, which may in turn result in revision of the long-range plan and program [14.33]. In this way there is a regular, project-oriented feedback from the current decision policy climate to long-range and TSM planning.

14.4.3.5 Program approvals The process of review and approval of the TIP, and the structure of the formal and informal bodies participating, will vary by region. In any case, since the MPO is not a government, but a shared-power forum (Chap. 13), the approval process is not the same as within state or local government. Generally there are two levels of approvals—technical and policy. The informal dynamics, as well as the participation, of nongovernmental interests can be very important. Some of the participatory strategies and techniques discussed in Chap. 11 will be appropriate in this and preceding steps.

14.4.3.6 Scheduling The TIP usually contains a schedule of major phases of the project linked to FY funding. These phases, such as planning, design, ROW acquisition, and construction, when measured in percent completion will also give some basis for monitoring. Some jurisdictions with responsibilities for implementing projects develop much more specific schedules that include, for example, steps in planning, feasibility and environmental studies, dates for public hearings, Federal and state approvals, dates for advertising and awarding contracts, and construction progress. In short, all project implementation mileposts are scheduled.

Scheduling is thus "the process of developing a timetable of operations to carry out the short-range program" [14.28]. If a project's advancement falls behind schedule, the programmer must be ready to substitute. The schedule thus is a management tool for implementing and monitoring the program. The more complete and detailed the schedule, and the more closely it is monitored, the easier it will be for the programmer and the MPO officials to make timely program modifications.

14.4.3.7 Monitoring and modifying Monitoring is "the process of checking the actual progress and comparing it with the scheduled progress. A program is a living document" [14.28]. A TIP, for all the care in its preparation, is probably out of date the day it is printed.

When monitoring indicates serious slippage in a project's advancement through its various stages, modification of the program will be necessary. Modification may also be indicated by a wide range of other circumstances, such as new high-priority projects; earthquakes, landslides, or other emergencies; public controversy; new prioritizing techniques (and therefore new priorities); new, excessive, project-cost estimates; inflation; changes in anticipated revenues or allocations; a new administration with new philosophy; and new jurisdictional or MPO goals and policies. Program modification is represented by feedbacks to financial analysis and forecasting (step 3), and to goals and policies.

14.5 SUMMARY

Programming is definitely a step in the planning process, necessary to assure that the goals of the early steps are realized. Conversely, planning can be seen as the handmaiden of programming, an aid to better immediate decisions between program, project, and system alternatives.

In this chapter we have attempted to show how programming can link planning closely to funding and implementation decisions so as to achieve transportation goals. There are three major planning flows into programming—projects, funds, and goals. Goals and policies govern the matching of projects and funds. Previous chapters discussed how policy planning develops agency goals and policies, and how systems planning, both long-range and short-range, capital and operational, identifies needed projects. The projects are further developed in project planning and design.

In this chapter we saw (a) how financial planning starts by identifying and forecasting the wide range of transportation revenues from Federal, state, local, and private sources, (b) how fund forecasting is complicated by the necessary categorical and geographical constraints placed on appropriations by Congress and state legislatures, (c) how capital, maintenance, and operating costs are estimated, and (d) how revenues and costs are compared and balanced in the programming process. The programming process, like the planning process, consists of identifiable steps, which are both technical and participatory.

Only by closely linking planning and programming can decision-makers provide effective transportation systems. Major decisions are, of necessity, political decisions, but like all implementation can, and must be, the culmination of an orderly planning process.

BIBLIOGRAPHY

14.1   American Public Transit Association: *Transit Fact Book (1978–1979)*, Washington, D.C., Dec. 1979.

14.2   Committee on the Budget, United States Senate: *Tax Expenditures*, Government Printing Office, Washington, D.C., 1978.

14.3   Congressional Budget Office, Congress of the United States: *Urban Mass Transportation: Options for Federal Assistance*, Government Printing Office, Washington, D.C., Feb. 1977.

14.4   U.S. Department of Transportation, Office of the Secretary: *Public Transportation Fare Policy*. National Technical Information Service, Springfield, Va., 1977.

14.5   Lentze, C. B., and W. R. Ugolik: "Who Pays the Highest and Lowest Transit Fares?," *Preliminary Research Report No. 136*, N.Y. State Department of Transportation, Planning Research Unit, Albany, N.Y., Feb. 1978.

14.6   Fujita, A., et al.: "Fare Charges and Prepaid Pass Programs: Honolulu's Experience," paper presented at the Transportation Research Board Meeting, Washington, D.C., Jan. 1981.

14.7   The Institute for Socioeconomic Studies: *An Inventory of Federal Income Transfer Programs*, White Plains, N.Y., 1978.

14.8   The Institute for Socioeconomic Studies: *An Inventory of State and Local Income Transfer Programs*, White Plains, N.Y., 1980.

14.9   McGillivray, R. G., and R. E. Kirby: "Financing Public Transportation," in G. E. Gray and L. A. Hoel (eds.), *Public Transportation: Planning, Operations and Management*, Prentice-Hall, Englewood Cliffs, N.J., 1978.

14.10 U.S. Congress, House of Representatives, Subcommittee on Economic Stabilization of the Committee on Banking, Finance and Urban Affairs: *Catalog of Federal Loan Guarantee Programs*, Government Printing Office, Washington, D.C., 1977.

14.11 U.S. Department of Health, Education and Welfare, Office of the Regional Director (Atlanta): *Transportation Authorities in Federal Human Services Programs*, Atlanta, Jan. 1976.

14.12 U.S. Department of Transportation, Federal Highway Administration: *Highway Statistics, 1979*, Government Printing Office, Washington, D.C., 1980.

14.13 U.S. Department of Transportation, Federal Highway Administration: *Highway Needs Report 1972*, Government Printing Office, Washington, D.C., 1972.

14.14 U.S. Department of Transportation, Office of the Secretary: *Coordinating Transportation Services for the Elderly and Handicapped*, vol. 1, Government Printing Office, Washington, D.C., 1979.

14.15 U.S. Department of Transportation, Urban Mass Transportation Administration: *Analyzing Transit Options for Small Urban Communities: vol. 2, Analysis Methods*, Government Printing Office, Washington, D.C., 1979.

14.16 U.S. Department of Transportation, Urban Mass Transportation Administration: *Financing Transit: Alternatives for Local Government* (Summary Volume), Government Printing Office, Washington, D.C., 1980.

14.17 U.S. Department of Transportation, Urban Mass Transportation Administration: *Innovative Transit Financing*, Government Printing Office, Washington, D.C., 1979.

14.18 U.S. Department of Transportation, Urban Mass Transportation Administration: *Mass Transit Management: A Handbook for Small Cities*, Part 1, Government Printing Office, Washington, D.C., 1980.

14.19 U.S. Department of Transportation, Urban Mass Transportation Administration: *Small City Transit: Summary of State Aid Programs*, Government Printing Office, Washington, D.C., 1976.

14.20 Hart, W. D.: *Public Financial Support for Transit*, Technical Study Memorandum No. 7, Highway Users Federation for Safety and Mobility, Washington, D.C., Sept. 1973.

14.21 Carstens, R. L., C. R. Mercier, and E. J. Kannel: "Current Status of State Level Support for Transit," *Transportation Research Record 589*, 1976.

14.22 Jones, D. W., J. Mollenkopf, and H. Rowen: *Transit Operating Assistance: Options for a Second Generation Program of State Aid*, Stanford University Transportation Program and Center for Interdisciplinary Research, Stanford, Calif., Feb. 1976.

14.23 American Public Transit Association: *Transit Financial Assistance, Reported for Calendar/Fiscal Year 1974 and Calendar/Fiscal Year 1973*, Washington, D.C., May 1976.

14.24 U.S. Department of Transportation, Urban Mass Transportation Administration: *Feasibility of Federal Assistance for Federal Urban Mass Transportation Operating Costs*, Washington, D.C., Nov. 1971.

14.25 Kirby, R. F., and G. K. Miller: "Some Promising Innovations in Taxicab Operations," *Transportation*, vol. 4, no. 4, pp. 369–386, Dec. 1975.

14.26 U.S. Office of Management and Budget: *Special Analyses, Budget of the United States Government*, Government Printing Office, Washington, D.C., 1980.

14.27 Congress of the United States, Congressional Budget Office: *Tax Expenditures: Current Issues and Five Year Budget Regulations 1981–1985*, Government Printing Office, Washington, D.C., 1980.

14.28 National Cooperative Highway Research Program: *Priority Programming and Project Selection:* Synthesis of Highway Practice No. 48, 1978.

14.29 Stearns, P. N., and D. A. Hodgens: "Programming and Scheduling Highway Improvements." *Highway Planning Tech. Rep.*, no. 4, Bureau of Public Roads, Washington, D.C., Feb. 1969.

14.30 Schofer, J. L.: "Future of the Urban Transportation Planning Process," *Proceedings, TRB Conference on Transportation Planning in the Eighties*, Workshop Report (preliminary), Washington, D.C., 1981.

14.31 Shunk, G.: "Preliminary Workshop Report Long-Range Regional Transportation Planning," *TRB Conference on Transportation Planning in the Eighties*, p. 5, Transportation Research Board, Washington, D.C., 1981.

14.32 Wachs, M.: "Highlights of TRB Conference" (preliminary), *Urban Transportation Planning in the 1980s*, Transportation Research Board, Washington, D.C., 1982.

14.33 Manheim, M. L.: "Toward More Programmatic Planning," in G. E. Gray and L. A. Hoel (eds.), *Public Transportation: Planning, Operations and Management*, Prentice-Hall, Englewood Cliffs, N.J., 1978.

14.34 Transportation Research Board: "Issues in Statewide Transportation Planning," *Report of a Conference in Williamsburg, Va.,* TRB Special Report 146, 1974.

14.35 Knox, R. R., et al.: "Programming Highway Improvements in the New Funding Environment." *Transportation Research Record* 599, 1976.

14.36 U.S. Department of Transportation, Federal Highway Administration: *Objective Priority Programming Procedures,* Report No. DOT-FJ-11-7882, Washington, D.C., March 1973.

14.37 Bellomo, S. J., et al.: "Evaluating Options in Statewide Transportation Planning/ Programming," *NCHRP Report* 179, 1977.

14.38 Melinyshn, W., et al.: "Transportation Planning Improvement Priorities: Development of a Methodology," *Highway Research Record* 458, 1973.

14.39 Shortreed, J. H., and R. F. Crowther: "Programming Transport Investment: A Priority-Planning Procedure," *Transportation Research Record* 574, 1976.

14.40 Gruver, J. E., et al.: "Highway Investment Analysis Package," *Transportation Research Record* 599, 1976.

14.41 Rihani, F. A., et al.: "Statewide Transportation Planning: The North Carolina Experience (1974/75)," paper presented at 55th annual meeting, Transportation Research Board, Jan. 1976.

14.42 Juster, R. D., and W. M. Pecknold: "Improving the Process of Programming Transportation Investments." *Transportation Research Record* 599, 1976.

14.43 Florida Department of Transportation: "Fund Distribution of Motor Fuel Tax, Transportation Financing and Programming," Tallahassee, Fla., Sept. 1975.

14.44 Yancy, M. L.: "LASM Forecasting and Allocation," paper presented at AASHTO Subcommittee on Computer Technology, Birmingham, Ala., Nov. 1976.

14.45 U.S. Department of Transportation, Federal Highway Administration: *Seven Approaches to Highway Programming,* Washington, D.C., May 1981.

14.46 Federal Highway Administration and Urban Mass Transportation Administration: "Transportation Improvement Program," *Federal Register,* vol. 39, no. 217, November 8, 1974. Can be found in Title 23, *Code of Federal Regulations* (CFR) 450.

14.47 U.S. Department of Transportation, Federal Highway Administration, *Highway Expenditure within Standard Metropolitan Statistical Areas,* Washington, D.C., Jan. 1981.

EXERCISES

14.1 In the revenue estimation example given in Sec. 14.1.7, supposed that the adult single cash fare were $0.75 and the student $0.40. In line with these, adult passes would be $22.50 and student $12.00. Associated shrinkages are assumed to be 0.20 and 0.30. What would be the estimated total revenue and that from each fare type?

14.2 A private firm is anticipating setting up a small taxi/transit company. The market interest rate for a 20-year, $500,000 loan is 14%, but a loan guarantee from the Small Business Administration will allow the bank to lower that rate to 11%. How much will the guarantee save the firm in interest payments over the life of the loan? (Assume repayment once a year.)

14.3 For the revenue/expenditure example given in Sec. 14.3, assume the following 1979 conditions and estimate the ratio of expense to revenues and the subsidy required in 1979:

Base adult trips (m/yr)	40
Base student trips (m/yr)	18
Revenue vehicle hours (1000)	1000
New buses purchased	10

All other conditions are as in examples in Secs. 14.1.7, 14.2, and 14.3.

14.4 Looking at Table 14.7, suppose that it were possible to implement a city "tag" tax according to the schedule below that varied with the amount of household auto mileage driven. Assuming no reduction in travel because of this tax, what would be the burden (tax as a percent of income) for each household income category of the tag tax plus the gasoline tax?

Annual auto mileage, in thousands	0-6	6-12	12-18	18-24	24+
"Tag" tax ($/yr)	50	100	200	400	800

14.5 A proposed freeway project would complete a circumferential belt and would have high benefit–cost value to the system as a whole. In the early phases of project planning it becomes clear that the freeway will also have serious SEE impacts. With reference to Fig. 14.4, what would, or might, be the consequences for the programming of the project? In other words, how might the programmer handle a high-impact project?

14.6 The TIP for the next budget year schedules implementation of the following projects:

1. ROW acquisition for a circumferential freeway segment.
2. Acquisition of much needed replacement buses for the metropolitan transit authority.
3. Four street-widening or intersection improvement projects.
4. Car-pooling and van-pooling development efforts at major employment centers.

Now comes information that Federal aid will be lower than anticipated and that a shortfall in state gas-tax revenues is anticipated due to more fuel-efficient cars. These events may actually presage a period of reduced transportation financing. Without being specific about dollar amounts, speculate about (a) what might be appropriate short-range goals for a period of austerity and (b) what effect the new short-range goals might have on each of the four projects.

15 Transportation System Operations and Maintenance

The planning of transportation facilities requires basic familiarity with the operation and maintenance of such facilities. Without proper response to these rudimentary requirements, planning activity can be severely hampered. Besides, with funding likely to be more limited in the future, operations and maintenance will become of greater concern in planning itself.

The following discussions focus on various aspects of the control, operation, and maintenance of urban traffic and transit systems. The requirements for such systems can be described in terms of the following taxonomy, which, as noted in Chap. 12, can be used to describe generically just about any urban transport system:

1. The *network*—the physical traveled way (if any) that comprises the nodes (switching points) and links over which trips are made.
2. The *vehicle*— as described by the characteristics (e.g., size) of the passenger and cargo space, suspension system, etc.
3. The *terminal*— with its loading, unloading, storage, and maintenance areas.
4. The *control system*—which may consist of physical elements, but also legal, organizational, and fiscal elements.

In the discussions that follow, these components are used as a basis for comparison and contrast.

15.1 URBAN TRAFFIC CONTROL AND OPERATIONS

The operation of urban street systems is based on a number of control techniques that essentially serve the same purpose—that of eliminating conflicts between opposing flows of traffic that, in effect, compete for use of the same space. In the broadest sense, these traffic-control techniques can be said to be directed toward the achievement of the following goals [15.1]:

1. To maximize efficiency—in terms of minimum delay, maximum capacity, and minimum stops—relative to the movement of people and goods over the street system
2. To maximize safety
3. To maximize reliability
4. To provide road users with necessary and sufficient information to aid them in the efficient utilization of the traffic facilities
5. To minimize maintenance costs.

Techniques include the utilization of traffic-control devices, such as stop and yield signs, traffic signals, traffic signs, and pavement markings. Properly installed and adjusted, traffic-control devices can minimize delays and accidents to motorists, as well as improving the overall capacity and performance of the urban street network.

Stop and yield signs are used to control traffic at important intersections, thereby giving definite priority to one of the approaches. The major flow suffers almost no delay, and the minor flow is held up waiting for acceptable gaps or openings in the main flow. As the main flow increases, the number of acceptable gaps decreases, and hence minor-flow vehicles are delayed longer. Or, if the minor flow increases, queues (lines) will develop, and minor-flow vehicles will be delayed both in crossing the main flow and in queuing.

By contrast, the use of signal control will reduce the delays to side-street vehicles waiting for acceptable gaps and will often improve overall safety at the locations. It may also increase the overall throughput at the intersection. Delays to the main-flow vehicles may, however, increase.

15.1.1 Intersection Control

The simplest form of traffic signal is the pretimed signal. It permits traffic to proceed, and directs it to stop, in accordance with a single, predetermined time schedule (cycle), or a series of such schedules. Traffic flows are assigned the right-of-way by a fixed sequence of phases. The advantages of pretimed signals include simplicity of equipment; provision for easy adjustment, servicing, and maintenance; and the capability of being coordinated to provide continuous flow of traffic at a definite speed along a certain route. The pretimed signal does not, however, recognize short-term fluctuations in traffic demand and, as a result, often causes excessive delay to vehicles and pedestrians during off-peak periods.

A traffic-actuated signal has the capability of varying operation in accordance with the demands of traffic. It does this through detection of vehicles or pedestrians on one or more approaches. The advantages of traffic-actuated operation of signals include reduction of total vehicle delay and increase in capacity (when properly timed), adaptability to short-term fluctuations in traffic demand, and continuous, traffic-responsive operation under low traffic demand conditions. The primary disadvantage of these signals relates to cost, since installation expenses are from two to five times those of pretimed signals. Moreover, maintenance and inspection costs of actuated controllers and detectors are higher because of the complexity of the equipment.

Traffic-actuated signalization can involve semi-actuated, fully actuated, or volume–density control. The first requires vehicle detectors on one or more, but not all, approaches to an intersection. Detectors are usually placed on the minor street approaches with low demand but some short sporadic peaks. Operating characteristics include:

1. Major-street traffic receives a minimum green time.
2. Major-street green time extends indefinitely after the minimum time unless interrupted by a vehicle arrival on the minor street.[1]
3. Minor-street traffic receives green time after the actuation, providing the major-street traffic has obtained its minimum green time.
4. Minor-street traffic has a minimum initial green period.
5. Minor-street green is extended by additional traffic actuations (if any) until a preset maximum is reached.
6. On the minor street, additional actuations after the maximum green time has been exceeded will be remembered and the green will be returned after major street traffic has received its minimum green time.
7. Amber (yellow) intervals are preset for both phases.

Fully actuated traffic-control equipment is used at isolated intersections that serve approximately equal traffic volumes but where the distribution of traffic between approaches fluctuates. This system thus has the capability of reacting to the traffic demand on all approaches. Operating characteristics include:

1. Detectors on all approaches.
2. Each green-time phase has a preset initial interval that allows for starting of standing vehicles.
3. Green time is extended by a preset vehicle interval for each actuation after the expiration of the initial interval.
4. Green extension is limited by a preset maximum.
5. Amber intervals are preset for each phase.

Volume–density type traffic control provides the maximum demand responsiveness for signalization of isolated intersections. Green time is allocated on the basis of relative volumes on approach legs. Not only does the volume-density controller respond in a predetermined fashion to an actuation, but it records and retains information on volumes, queue lengths, and delay times. It also records the time gap between vehicle arrivals on an approach and measures it against a maximum standard, thus providing the "density" control function. The operating characteristics include:

1. Detectors on all approaches.
2. Each phase has a certain assured green time as set by three dials on the controller:
   a. Minimum green interval
   b. Number of actuations before minimum green starts to increase
   c. Increase of minimum green for each added actuation
3. Passage time (the extended green time created by each additional actuation after the assured green time has elapsed) is set as the time required to travel from the detector to the stop line. This interval can be reduced when a predetermined lower limit is reached, when:

---

[1] Otherwise known as an "actuation" of the detector on that street.

a. The red-phase vehicles have waited a preset time.
b. The number of vehicles waiting on the red phase exceeds a preset value.
c. The number of green-phase vehicles per 10 sec is less than a preset value.
d. "Platoon carry-over percentage" is established. This is a predetermined proportion of the previous green-period traffic. It is applied to the number of vehicles waiting on the red phase, thus insuring a more prompt return to the green phase when the next platoon of vehicles hits the detector.
e. Amber intervals are set for each direction.
f. Each phase has a recall switch that operates in the same manner as described above for the fully actuated controller.

15.1.2 Arterial Street Control

Individual signalized intersections can be interconnected to form systems. Urban arterial streets often are controlled to provide progressive traffic flow along the arterial. The basic approach is to assure that vehicles are released in platoons from a signal and then travel in these platoons to the next signal. Thus, by the establishment of a time relationship between the beginning of the green phase at an intersection and that at each subsequent one (taking account of the speed of the platoon), continuous flow along an arterial street can be achieved, along with a reduction in delay.

If attention is given only to traffic movement in a single direction, one-way progressive movement is easily accomplished. The problem of providing progression in both directions is much more difficult. Good two-way operation is dependent upon intersection spacing that, ideally, must be fairly constant; such conditions rarely exist. Techniques have been created for developing signal-timing plans with the assistance of a digital computer on the basis of inputs defining the existing traffic parameters [15.1].

15.1.3 Network Traffic Control

The most sophisticated signal systems are those in which entire networks are computer-controlled. A truly traffic-responsive signal system should employ decisions that result from continual sampling of traffic volumes and densities. This requires gathering and processing of pertinent data, determination of settings, and actuation of the controls in real time. A high-speed computer thus is essential. It would work through specially prepared program equations, based on some form of delay-minimization criterion, to compute the optimum signal display at a given time for the entire network. A number of cities have experimented successfully with this form of signal control and have shown that it can appreciably reduce delays in ways not possible for existing controllers.

The Federal Highway Administration has established a major research project in Washington, D.C., that provides for computer control of 200 signalized intersections. The system contains 512 vehicle detectors and extensive data-processing, communications, and display equipment. Experiments have resulted in the development of three generations of network traffic-control strategies [15.1].

15.1.4 Freeway-Related Traffic Control

Since freeways are designed as controlled-access roadways and are intended to provide a high level of a service, the freeway user does not expect traffic-control restrictions as on city streets and arterial highways. Nonetheless, conditions can and do develop that result in a decrease in the level of service and an increase in congestion. Consequently it becomes necessary to provide some degree of control so that freeway operations do not deteriorate completely and, in parallel, so that local streets and arterials do not become completely saturated as well.

The factors that lead to freeway congestion can be categorized as geometric design, operational, and random [15.2]. Geometric design factors refer to physical features of the roadway that can contribute to capacity reductions and thereby congestion. These can include reduction in the number of freeway lanes, changes in horizontal and vertical alignment (i.e., sharp curves and steep grades), and reduced lateral clearance. Operational factors relate to travel demands that result in more traffic than the roadway can accommodate. This is analogous to a stream overflowing its banks as water inflow exceeds the stream bed capacity. The highway system cannot overflow, but it can become so congested that traffic is delayed and stopped.

The specific traffic conditions that result in congestion can include unrestrained ramp access, exit-ramp queues, and heavy weaving movements. It should be recognized that traffic conditions often can also be related to specific geometric design characteristics. For example, inadequate ramp spacing can result in excessive weaving movements.

The final category (random factors) relates to events that are unpredictable in terms of time and/or location (such as accidents or changes in vehicular speeds) and thus result in congestion. Environmental factors such as rain, sunrise or sunset, glare, or the like may also result in a flow disturbance that, under congested conditions, could result in a random major disruption.

Several concepts of freeway traffic control associated primarily with operational factors are presented next [15.2].

Restricted entry refers to a variety of actions that can be taken to limit or meter the number of vehicles that can pass through freeway entrance ramps. The most severe limitation involves physically closing the ramp to traffic. Metering can be accomplished through the use of pretimed or traffic responsive signals. Pretimed control simply fixes the rate at which vehicles are permitted to enter the freeway. Traffic-responsive signal control must be based on a procedure that accounts for a number of factors, including freeway volume, density, and capacity, as well as ramp volumes and lines.

A strategy that involves a series of ramps along a freeway is called integrated ramp control. This level of control is highly complex both in terms of the traffic variables that must be taken into account and the process for performing the control function.

Mainline control refers to the use of variable-speed mechanisms and informational signing to direct drivers in response to changes in traffic characteristics. Variable-speed and lane controls are the major ones presently in widespread use. The former is used to limit the speed of traffic on a freeway to an appropriate level based upon theoretical volume–speed–density relationships.[2] Lane controls are used primarily to direct drivers to or away from a particular lane in advance of a disturbance or incident there. Controls of this type require motorist adherence to be effective. Studies [15.2] have shown that many motorists do not consider variable-speed limits as regulatory (as opposed to advisory). As a result they do not decrease their speed in accordance with posted limits unless there is an obvious reason to do so.

[2] See Chap. 5

A further level of freeway traffic control is that which involves a complete corridor, that is, the freeway and adjacent and parallel surface streets or other freeways [15.4]. The principle of a two-route corridor system is shown in Fig. 15.1 (one direction only). The essential activities include [15.4]:

1. Diversion of traffic before entering the corridor (advisory or mandatory)
2. Diversion of traffic using the corridor (advisory or mandatory)
3. Protection against oversaturation on freeways by means of ramp metering or ramp closure
4. Prevention of accidents by means of warning systems, speed control, and lane control

The controls illustrated in Fig. 15.1 generally require the following facilities:

1. A detection system, which provides the primary data and allows certain predictions to be made
2. A control center, which transforms the detector information into control measures
3. A communication system, which presents the decisions of the control center to the driver
4. A data-transmission system linking the detection equipment to the center and the center to the communications system

Regulation of traffic within a corridor is accomplished with three major methods that are not mutually exclusive—network, ramp, and linear control. The first involves the utilization of an areawide traffic-control system to prevent overloading of the freeway through the diversion of certain traffic flows or parts of them to adjacent routes with excess capacity. Ramp controls have been discussed previously; they act to limit the number of vehicles entering the freeway. Linear control, involving the use of variable-message signs, is aimed at commanding lane usage and traffic speeds to optimize corridor throughput and safety.

15.2 RAIL AND RAIL TRANSIT CONTROL SYSTEMS

Although the primary objective of railway signaling originally has been safety, wider development and use of signaling and control devices and techniques have resulted in greater economy, efficiency, and flexibility within the system. This has been achieved through increases in speed, greater line capacity, consolidation of interlockings, remote control and optimum use of trackage (sometimes affecting a reduction), automation of marshalling yards, and automatic train control.

15.2.1 Rail Signaling and Control Principles

Railway signaling and control, having evolved from the original use of policemen to modern, sophisticated, and complex systems, is a highly specialized technique and as such does not lend itself to other transport systems [15.5]. Obviously, as long as a single train is in possession of a stretch of track, no signaling or control is needed; but as soon as more than one is involved, prevention from collision becomes necessary. The original method of handling traffic was by a time interval, whereby successive trains were dispatched after a specified length of time. This method, however, was neither speedy nor efficient. As a result, it was not long before an interval of space became established as the major regulatory element, such

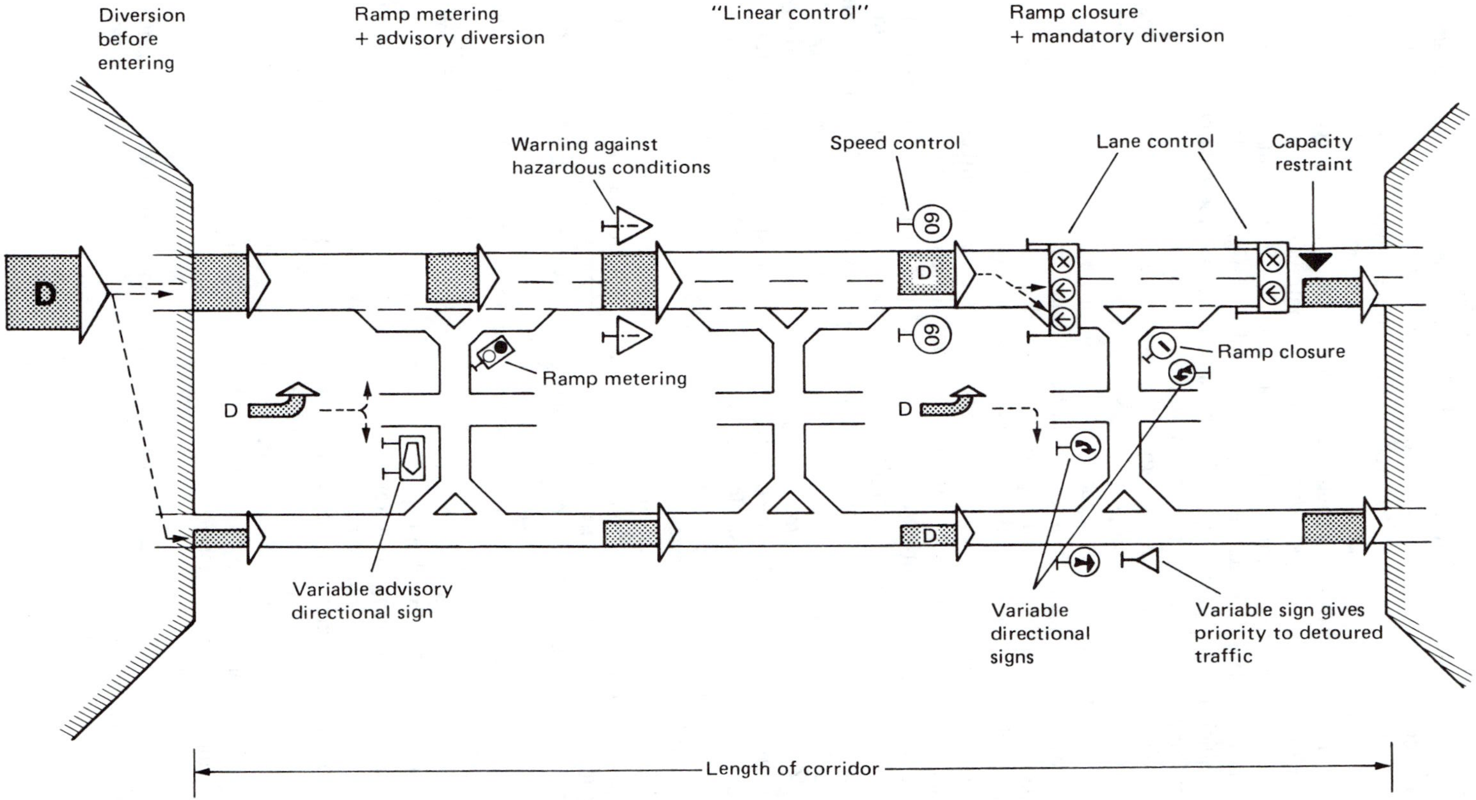
Diversion
before
entering
Ramp metering
+ advisory diversion
"Linear control"
Ramp closure
+ mandatory diversion
Warning against
hazardous conditions
Speed control
Lane control
Capacity
restraint
60
60
D
D
D
D
Ramp metering
Ramp closure
Variable advisory
directional sign
Variable
directional
signs
Variable sign gives
priority to detoured
traffic
Length of corridor

Fig. 15.1 Schematic of corridor controls (one direction only). [15.4, p. 15].

that the trackage was divided into convenient sections, each one under the control of an individual with some form of signal. Fundamentally, only one train at a time could occupy one section. If it were blocked by a train (or part of a train), a second was not permitted to proceed. The space interval became known as block section. It still exists today, both on lines commanded by the older mechanical methods and on those with high speeds and the most modern signaling.

With improved track (due largely to continuously welded rail) and longer switches, train speeds are rising. Rail traffic control thus has to be geared accordingly. This involves not only mechanical design, but also more appropriate and quicker train-spacing control. The aim is to provide complete route selection and immediate clearing of signals under safe conditions. Brake control is also becoming more important, so that the link between the train and the control system assumes larger proportions. Speed and its reciprocal, brake power, are important factors in line capacity. As traffic increases, utilization of the line also has to increase. Signaling is one method (but not the only one) of increasing line capacity. Power signaling installations are used not only to concentrate control in the hands of a central authority, thereby obtaining higher efficiency and greater safety, but also to economize on personnel and other resources. The greater use of electronic devices for supervision and transmission of commands and data has enabled areas under control to be extended considerably. The ultimate objective is one central control for a complete railway system. It must be accepted that the need for technical maintenance personnel is thereby increased, both in number and skills, but the overall result is usually well worthwhile [15.5].

Centralized traffic control (CTC) is one means by which a single operator can directly regulate an almost unlimited number of stations and length of track. Usually control is over a single line, with the operator having a complete display of all conditions on a diagram before him or her. CTC systems have been widely adopted in the United States.

Railways, with movements in a single dimension, appear to lend themselves nicely to automatic operation. But practice is complicated by the large number of separate linear maneuvers involved. Train movements for main-line, suburban, urban, etc., services must be accommodated along with a variety of ancillary processes such as loading, unloading, revenue collection and checking, booking of traffic, and marshalling. Safety cannot be sacrificed; on the contrary, an automatic railway must be safer than a conventional one.

The desire in recent years to improve the quality of service of rapid transit, combined with the availability of new components and techniques, has accelerated the pace of technological innovation in rapid transit. A part of that innovation has been in the field of *automatic train control* (ATC). This is the term applied to a broad range of functions that deal with the control of motion and the assurance of certain aspects of safety of vehicles and switches in a rail transit system. ATC generally provides the four functions of train detection, train operation, train supervision, and system communications. These may be performed by people, machines, or both. Each of these functions and its individual components are outlined in Table 15.1.

Automation offers real advantage in rail transit operations, bringing improvements and financial savings to justify its implementation. The main advantages of rail automation are:

Table 15.1 Automatic Train Control Functions

ATC function	Subfunction	Description
ATP (Automatic Train Protection) — That portion of the system that prevents collisions and derailments. Traditionally, its functions/requirements override all other ATC functions/requirements either through design or procedure.	Train and track surveillance	Usually performed by personnel on the train. Involves monitoring conditions on the track ahead and in the vicinity of the track as well as monitoring certain safety-related conditions on-board the train.
	CO)LLIS	Functions which prevent trains from colliding with each other.
	Train separation and interlocking	Prevent situations from arising in which trains on crossing or merging tracks could collide. Also concerned with safe operation of trains through switches.
	Overspeed protection	Assures that train will not exceed command speed. Guards against derailments or other accidents due to excess speed. Does not control train speed; simply prevent command speed from being exceeded.
ATO (Automatic Train Operation) — Controls movement of the train as directed by ATS, but always within constraints of ATP.	Train velocity regulation	Involves control of train speed within constraints of ATP. For efficient transit system operation, it is desirable that a variety of speeds be available and that trains operate at or near intended speed.
	Programmed stopping (station stopping)	Brings train to a stop within specified area in a station. Tolerance on stop-point ranges from tens of feet in some open platform systems to a few inches in a system with both platform doors and car doors.
	Door control and starting	Involves opening/closing doors and starting trains once doors are closed. Typically ATC includes command/monitoring of these functions but not their execution.
ATS (Automatic Train Supervision) — Includes both management and housekeeping functions as described in the right-hand columns.	Schedule design/implementation	Concerned with operation of entire transit system. Functions deal with providing needed services and efficient utilization of equipment and manpower. Directs establishment of routes and movement of trains on the basis of information received through the communications system, and within safety constraints of ATP.

Table 15.1 Automatic Train Control Functions (*Continued*)

ATC Function	Subfunction	Description
ATS (Automatic Train Supervision) (*Continued*)	Yard train control	Involves assembly and organization of trains in yards, to meet service requirements on the lines.
	ATC system maintenance	That part of system maintenance related to ATC function. ATS may play a role in identifying the need for maintenance (scheduled or unscheduled) throughout the ATC system.
SC (System Communications) Includes communications functions necessary for operation of ATS, ATO, and ATP systems, as well as other voice/video communications requirements.	Passenger–attendant communications	Communications within trains; may be simple, one-way public address facilities or two-way communication system.
	Passenger–central control communications	Virtual necessity for unmanned systems and possibly desirable for some manned systems.
	Vehicle–central control communications	A more general case of passenger–central control communications. May include private facilities for motormen and/or data communications.
	Platform surveillance	Not strictly an ATC function, but possibly important to overall system security. May include fare-collection surveillance.
	Maintenance force communications	Provides for rapid response to abnormal events. May require radio link and/or external telephone lines.
	Routine business communications	May utilize private lines or ordinary public telephone systems.

Source: [15.43].

546

1. Safety is increased since human error is eliminated; automatic systems replace the human operator as the controlling element either partially or completely—at most, the operator's task is reduced to a supervisory function.
2. Economy is achieved through optimum train operation, requiring less energy, and protecting equipment against undue wear.
3. Line capacity is maximized by increasing traffic density and reducing or eliminating delays.
4. Punctuality is aided since the schedule is kept under constant supervision and adjusted as necessary.
5. Traveling comfort is maintained at the same high level for all trains, as acceleration and especially braking are no longer dependent on individual driver skill.
6. Saving of personnel is affected by their release from routine work, thus reducing recurrent costs.

15.2.2 Urban Rail System Control

Older rail systems generally have the lowest levels of automation, while newer ones have the newest equipment and the highest levels. Historically, the conversion from manual to automatic train control in rail rapid transit has been incremental and has followed a more-or-less common course for all systems. The major technological steps along the road to automation are summarized in Table 15.2.

At the essentially manual level, train protection, operation, and supervision are carried out by operators and towermen or central supervisors with little or no aid from automatic equipment. This is done either by rules or procedures alone or with the aid of advisory wayside signals. There are no automatic stop-enforcing mechanisms either on the wayside or on board the train. Dispatching is carried out by personnel at terminals or control towers along the routes, using either a written schedule or timing devices that act as prompters to signal train departures. Route assignment and interlocking control are accomplished by manually activated equipment that may have some automatic safety features but that are entirely controlled by human operators. Communications are by means of visual signals (lights, hand signals, posted civil speeds, etc.) or by telephone from stations and towers to central control.

For wayside train protection, signals with trip stops form the basis for automatic train protection (ATP). These assure separation of following trains and prevent conflicting moves at interlockings. Incorporation of timing devices with the trip stops also provides equipment-enforced overspeed prevention. While train protection thus becomes automatic, train operation is still completely manual. Train supervision also remains an essentially manual activity, although track circuits and signals used primarily for train protection do permit some automation of route interlocking and dispatching—usually in the form of semi-automatic devices (i.e., manually activated but automatically operated).

Carborne train protection, consisting of (1) cab signaling through coded track circuits and (2) automatic stopping and speed-limit enforcement, represents the same level of ATP as wayside signals with trip stops. Cab signaling generally is considered a higher level of automation, however, since it also provides some automatic aids to train operation—principally continuous display of speed information to assist the operator in running the train and stopping at stations. Other aspects of train

Table 15.2 Levels of Automation

Level	Characteristics	Example
Essentially manual	Train protection by rules and procedures Train operation manual (with or without the aid of advisory wayside signals) Train supervision to towermen and/or central dispatchers	Chicago Transit Authority (CTA) (Ravenswood and Evanston Lines)
Wayside train protection	Wayside block signals with trip stops for train separation and overspeed protection Train operation manual Supervision manual with some automation of dispatching and route interlocking	New York City Authority (NYCTA)
Carborne train protection	Cab signals and equipment-enforced train protection Train operation manual Supervision as above	CTA
Automatic train operation (ATO)	Automatic train protection (ATP) as above Train operation either completely automatic or with manual door operation and train starting Train supervision as above	Port Authority Transit Corporation (PATCO)
Automatic train supervision (ATS)	ATP and ATO as above Train supervision automatic (or mostly so) under central computer control	Bay Area Rapid Transit (BART)
Unmanned operation	ATP, ATO, ATS as above No on-board operator System manned only by small number of central control personnel	AIRTRANS
Full automation	ATP, ATO, ATS as above, with automatic, not manual backups for each Skeleton force at central control Yard operation automated	None

operation are still essentially manual. Cab signaling does not necessarily lead to any increase in the automation of supervisory functions, nor is it accompanied by any change in the communications systems.

The major advantage of cab over wayside signaling is that it permits evolution to automatic train operation (ATO). All of the information needed to operate the train automatically is either inherent in the cab signal system or readily available through modular additions. At this level, the human is removed from the speed control, station stopping, door control, or starting loops—or any combination of them. The human no longer functions as an operator but as an overseer of carborne control systems.

The newest transit systems in this country—the Bay Area Rapid Transit, the Port Authority Transit Corporation, and the Washington Metro—all have ATO. The new systems under development in Atlanta and Baltimore will also have it.

Along with ATO, there is often (but not necessarily) an increase in the level of *automatic train supervision* (ATS) functions. These include automatic dispatching, route assignment, and performance level modification. ATS requires a rather complex and sophisticated communication network, not only for voice messages but also for the interchange of large quantities of data among automatic system elements on a real-time basis. The distinguishing feature of ATS, however, is the use of a central computer (or computers) to process and handle data, make decisions, and formulate instructions.

The Bay Area Rapid Transit system was the first rail rapid transit system to make extensive use of ATS. The new Washington Metro system also has highly automated train supervision based on computer control.

At all the levels of automation described previously there is at least one operator on board each train and some supervisory personnel in central control. While these people are not part of the normal control loop, they do exercise important functions as overseers of automatic equipment and back-ups in case of failure or emergency. A more advanced form is unmanned operation, with all ATP or ATO functions performed by automatic devices. The few remaining human operators in the system are at central control, but even these personnel may be reduced in number as more supervisory tasks are allocated to machines.

No rail rapid transit system in the United States, or anywhere in the world, is now operating at this level or automation. The technology to do this, however, is available and has been applied in various people-mover systems, such as the Morgantown Personnel Rapid Transit (PRT) and several airport transportation systems.[3] A notable example of an unmanned airport system is AIRTRANS at the Dallas–Fort Worth Airport, where small unmanned transit vehicles circulate on fixed guideways over a complex of interconnecting routes. The entire system is operated and supervised from a central location by a few persons aided by a train-control computer [15.6]. AIRTRANS is discussed further in a later section.

Full automation, or complete removal of people from control of transit system operations (even from central control), probably is neither technically feasible nor desirable. For safety and continuity of operation, it will always be necessary to have someone to monitor the system and intervene to restore operations or assist passengers in an emergency. The number of such supervisors could be only a handful, however, and it is doubtful that they could never conduct normal operations manually as a backup to automatic systems.

15.2.3 Automated Guideway System Control

As mentioned, an example of a modern, operating guideway transit system is the AIRTRANS facility at the Dallas–Fort Worth Regional Airport [15.6-15.8]. It is designed to carry passengers and employees (in separate cars), transport all interline baggage and mail, remove all trash from the terminals to a common incinerator, and deliver commissary supplies from a common warehouse to the terminals. AIRTRANS has 13 miles of guideway, 53 stations, an operational maintenance facility, 68 vehicles plus 13 gasoline-powered tugs, and a central control point to provide surveillance and emergency override (when the action is safe) over the automatic operation.

The automatic control system is divided into three subsystems: (1) automatic vehicle protection AVP); (2) automatic vehicle operation (AVO); and (3) central control (CC) [15.7]. The conventional block system is used with vital relays located both on-board and along the guideway. Computer failure will allow vehicles to

---

[3] For descriptions of some, see Chap. 12.

bunch but not collide, because spacing control is entirely separate from the computer and software.

Vehicle route information is stored in an on-board control logic assembly. This device responds to an interrogation from the wayside every 0.2 sec and sends back route as well as malfunction information. The wayside controls decode route information and set the switches to the proper position. The correct speed command for each vehicle, depending on its location and other traffic, is transmitted to the vehicle by the fixed-block control system from the wayside control units. The latter are made up of standard fail-safe relays that bring the system to a safe halt in the event of an emergency condition. The system is designed for a nominal operating speed of 25 ft/sec (27.4 kph).

Vehicle operating safety relies on a nominal 5-block control system. The guideway is divided into 708 blocks by insulators spaced at intervals along the signal rail. The average block is 90 ft long, but ranges from 45 to 240. In the terminal sidings, 45-ft blocks are used to allow closer vehicle spacing, permissible at the lower siding speeds. The vehicle maximum stopping distance under emergency conditions is 165 ft. In a 5-block system, 1 block is allowed for emergency stopping. During any operation at least 1 full block must separate the vehicles. At a high-speed cruise, 5 blocks separate the vehicles. A "proceed-at-full-speed" signal is sent to the vehicle from the wayside whenever its separation is 5 full blocks or more and it is cruising at high speed. When the separation becomes less than 4 blocks, a signal is sent to the vehicle to slow to medium speed (14 ft/sec); for separations less than 2 blocks, the command is to stop. This ensures at least 1 clear block between queuing vehicles. In a high-speed case, the vehicles have a minimum separation of 450 ft (137.16 m). At 25 ft/sec, the minimum headway is 18 sec.

The central control console, from which the system is supervised, is located in the central heating and air conditioning building. The console shows the status of the system and permits the operator to override the automatic operation of the system if necessary and interference is safe.

The display route map shows the location and status of each vehicle. TV screens permit viewing of all passenger terminals. Two-way voice communication is possible with any or all vehicles. Malfunction information is also displayed. Another route map shows the status of the power distribution system.

The supervisor does not operate the system but may add or subtract trains-in-operation, change the routes, and dispatch services crews. A printed copy of all operator actions and indicated malfunctions of vehicles or stations is available via the central-control line printer.

15.2.4 System Safety and Reliability

Two major requirements of rail and guideway transit are safety and reliability. The potential for future technological advances is closely related to these requirements. Another area, maintenance activity, can be appropriately included in a general analysis. Table 15.3 summarizes the major safety critical functions identified for each of five elements of rail and guideway systems. The complex nature of safety and reliability starts to become apparent. For example, the command and control equipment is commonly distributed throughout the entire system: at the central control facility, along the wayside (including the stations), and on-board the vechiles. Also, it includes all the antennas, inductive coils, and other noncontact signal-transmission devices necessary for transferring and transmitting control and/or monitoring data/information between various portions of the overall command and

**Table 15.3 Major AGT Subsystems
and Their Safety Critical Factors**

A. Command and control
  1. Train presence and detection
  2. Route and switch interlocking
  3. Overspeed protection
  4. Rollback protection
  5. Zero-speed detection
  6. Vehicle door interlocks
  7. Station-alignment detection and interlocking
  8. Collision avoidance
  9. Lateral guidance control
B. Vehicle subsystem
  1. Guidance/steering and switching
  2. Friction braking
  3. Electric power collection and distribution
  4. Safety provisions
c. Guideway subsystem
  1. Support of loads
  2. Guidance and switching
  3. Weather-effects mitigation
  4. Emergency provisions
D. Station subsystem
  1. Facilitation of internal passenger movement
  2. Emergency provisions
E. Power distribution subsystem
  1. Power distribution and control
  2. Safety provisions

Source: Adapted from E. S. Cheaney, *Safety in Urban Mass Transportation: Research Report,* UMTA Report RI-06-0005-75-1, Washington, D.C., Jan. 1975.

control system. In some systems, the running rails also serve as the conductors for the transmission of signals between wayside and the vehicles and for the train-presence detection circuitry.

Of the major functions performed by the command and control subsystem, only those associated with automatic train protection (ATP) are considered to be safety-critical.

15.2.5 New Safety Technology

Technological change impacts safety by spawning new hazards that must be identified and controlled before accidents occur. For example, BART's employment of aircraft-type hydraulically actuated disc brakes rather than traditional air brakes necessitated a close examination of the possible hazards in a new environment.

The absence of prior experience with novel equipment or applications forces the safety activities and tasks into an active, forward-thinking style. That is, safety problems must be anticipated through testing and abstract analysis methodologies. This is in contrast to the general tenor of the safety programs associated with systems in the operational stage of the life cycle. There, the safety problems are mainly reacted to as they arise.

15.3 BUS TRANSIT OPERATIONS

According to the American Public Transit Association (APTA), there were 1,003 transit systems in the United States in 1978. As a result, over 75 percent of the U.S. population had access to fixed-route service [15.9]. Operating nearly 69,000 revenue vehicles, these systems transported nearly 8 billion passengers in that year. In comparison, intercity buses carried 355 million passengers, intercity railroads 79 million, and domestic aircraft 257 million.

The transit industry is owned primarily by the American public through their city, county, regional, and state governments. Nearly half the systems in the United States, carrying 91% of all transit passengers, are owned by public agencies.

Urban bus systems vary in size and purpose. One attempt at characterizing urban bus fleets divided them into three size categories [15.11]. Table 15.4 indicates that the vast majority of bus systems have a fleet of less than 100 buses. A majority are part of the 20 large city systems. Table 15.4 also presents a breakdown of each typical fleet by type of route, which reflects the following service definitions:

1. Arterial route. Operating through developed areas into the central business district, allowing for at least five stops along every route mile.
2. Express route. Traversing a limited-access highway or operating "closed-door" for a portion of its distance.
3. Suburban route. Connecting an outlying area and the central business district, allowing for less than five stops per mile along a portion of its route mileage, but not for operating express.
4. Circulator route. Operating within a central business district or other major activity center (university, hospital complex, airport, etc.).
5. Crosstown route. Traversing developed areas, crossing arterial routes, and linking outlying activity centers, but not the central business district.
6. Feeder route. Operating to a commuter rail, rapid transit, or express bus stop.
7. Tripper route (school/employment). An extra bus operating over a portion of a regular route for the purpose of carrying students or employees, nonexclusively, directly to and from their school/work destination.

Table 15.4 Fleet Usage for Average Bus Systems

	Large system	Medium system	Small system
Definition	500 buses or more	100–500 buses	less than 100 buses
Number of systems	20	60	about 860
Number of buses, nationally	25,000	12,000	12,000
Typical fleet usage (%):			
Arterial service	55	63	70
Express suburban	15	13	10
Circulator service	2	2	0
Crosstown/feeder service	10	5	0
Tripper service/spares	18	17	20

Source: American Public Transit Association, *1974–1975 Statistics,* used to update 1972 Simpson & Curtin National Bus Fleet Inventory.

Table 15.5   Performance Characteristics of Typical Bus Systems

Operating parameters	Large system (500 buses or more)	Medium system (100–500 buses)	Small system (less than 100 buses)
Total number of buses	25,000	12,000	12,000
Schedule speed norms (mph)*			
Arterial	11	12	13
Express/suburban	18	20	22
Circulator	8	9	–
Crosstown/feeder	12	13	–
Schedule adherence norms (%)			
Early	10	10	10
On-time (0–5 min late)	60	70	80
Late	30	20	10

*Bus revenue·miles divided by bus revenue·hours, excluding recovery time.
Source: [15.47].

8. Spares. Extra buses required either to cover maintenance and/or repair needs or being held for contingencies.

In Table 15.4, these eight service definitions are condensed into five types. Thus, for a large system of 500 or more transit buses, it can be expected that the majority (55%) would be committed to arterial services, while significant fleet commitments would be made to express/suburban (15%), crosstown/feeder (10%), and tripper/spares (18%). A relatively small group (2%) of buses can be expected to provide circulation service within major activity centers. For a medium system of 100 to 500 transit buses, arterial service typically predominates, employing about two-thirds of the total bus fleet. Small bus systems (under 100 transit buses) typically have no circulator or crosstown service and only limited suburban coverage, with the vast majority of buses committed to arterial service.

Table 15.5 presents performance characteristics by size of fleet. Although schedule speeds and adherence vary widely among systems, approximate norms for these key operating parameters were hypothesized by Simpson and Curtin [15.47]. As indicated in the table, arterial speeds are considerably lower than express/ suburban, with a slight upward trend in all speeds for operation in smaller, less congested urban areas. In general, schedule adherence also improves in the smaller metropolitan areas because of more predictable traffic conditions.

Many other factors are important to the operation of a local bus system. The complexity can be deduced from the following list:

Organizational structure of the firm (system)
Establishment of goals and objectives for operation of the firm and each of its components
Personnel selected by the firm, how recruited and trained, and methods used in supervision and negotiation with organized labor
Selection of equipment
Maintenance program for equipment and facilities
Data and information gathered and how used for management purposes, both record keeping and decision making

Establishment of routes and schedules

Methods used for communication between and among operating and staff personnel

Transit marketing program

Advertising program

Public information program

Community relations program

A suggested organizational structure based on functional activities for a transit operation is illustrated in Fig. 15.2.

A bus-transit system operates within an environment with a set of "basic characteristics which prescribe the (transit) management task" [15.12]. These characteristics help to better define transit operations for the urban planner and to provide a better understanding of bus-transit capabilities and constraints. Essentially, these characteristics indicate that transit is:

1. A retail enterprise, and therefore competitive and marketable
2. Labor-intensive
3. A cash business
4. Regulated

Each of these characteristics is discussed in more detail in the following paragraphs, drawing on the presentation included in [15.12].

Because bus transit is a retail enterprise, there is a product and a consumer. Bus transit operations must be aimed at the provision of service to the various groups of consumers, whether they are commuters, shoppers, students, or others. An understanding of the travel needs and patterns within the community is essential to the provision of an adequate and appropriate transit route structure. In addition, consideration must be given to the local development activities, including road and street projects and the construction of major new traffic generators. Current emphasis on transportation system management (see Sec. 12.2) at the Federal and local level has helped to provide a focus for these activities.

The transit system operator also must work closely with Federal and state transit administrations to take advantage of available assistance to make capital purchases and obtain operating assistance. Service and the facilities offered to the consumer to carry out the service go hand-in-hand in achieving a final product. The marketing of the product to reach existing and potential system users must also be carried out.

Because more than 70 percent of all transit operating expenses are associated with labor, operations must be designed to utilize that input as efficiently as possible. The introduction of new and specialized transit services, coupled with the constraints of local work rules, schedules, and other conditions, make the ultimate goal of 100 percent correlation between hours in revenue service and payroll hours a unique challenge.

The fact that a transit system is a cash business engenders two major comments. The first concerns the need to account for each passenger payment (fare) and relate this to the cost of service being provided. In most cases local transit authorities require operational subsidies to remain solvent. Nevertheless, farebox revenues must be collected and accounted for in all cases. The second comment concerns the need for appropriate safeguards to ensure that farebox revenues are eventually deposited into the transit authority's bank account.

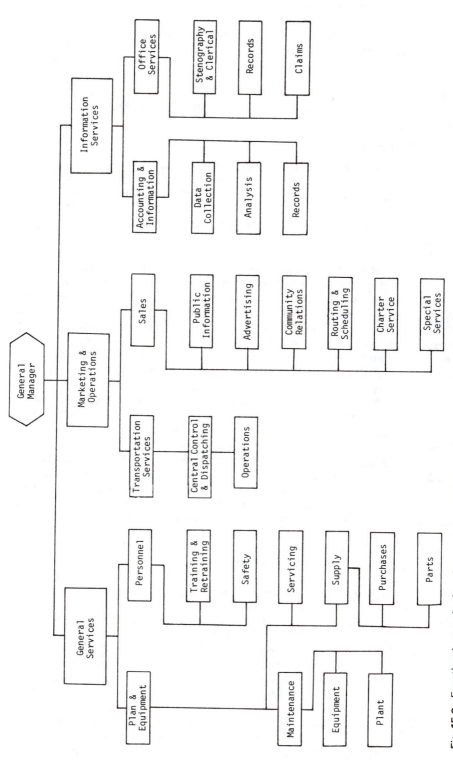

Fig. 15.2 Functional organization for mass transportation. [*An adaptation from George M. Smerk, "Mass Transit Management," Business Horizons 14, no. 6, December 1971, p. 13.*]

The transit industry is a regulated industry. This is more in the form of requirements from funding sources, but the effect is much the same. Federal, state, and local transit and transportation agencies have created a myriad of regulations to be met in the application for capital and operating assistance. Moreover, the consumer and the voter can be expected to exert some control, ultimately, through the farebox and at the polls, especially in relation to requests for increased operating tax levies.

15.3.1 Bus-Route Design and Scheduling

The many activities and aspects of bus-transit systems are closely interrelated. Bus-route design and scheduling is just one aspect of bus operations that can be examined in detail. It is considered here because it is an area that holds significant promise for future improvement, and also because it has been estimated that the bus schedule determines over 80 percent of the transit property's total operating budget.

The planning activities that take place as part of the local transportation planning process (described in more detail in previous chapters) and as part of the transit property's efforts should provide information on the nature and extent of the demand for transit service. This, combined with knowledge of the street network, will provide the transit planner with the basic data set for bus-route planning. It has been pointed out that experience plays the major role in this activity, which essentially is assignment of "vehicles and operators to routes in an economical fashion, in response to existing and latent ridership patterns, and in accordance with written and unwritten labor rules and practices" [15.13].

In an early examination of bus-route planning, it was pointed out that there were no "off-the-shelf" planning procedures to assist in the detailed work involved [15.14]. Consequently, a review of necessary planning considerations, summarized in Table 15.6, was developed. Two very serious problems associated with current scheduling processes have been identified recently [15.13]. First, the amount of paperwork associated with the scheduling process is substantial and increasing steadily each year. In addition to preparing schedules and appropriate route information and instructions for drivers, the scheduling department is being called on to prepare statistical reports on these activities. Second, the number of qualified and experienced schedule-makers is not increasing at the same rate as the need for them. As demands on their time increases, their availability to train new personnel is diminished. The ultimate danger is that appropriately trained individuals will not be available to replace current schedule-makers as they retire.

One partial solution to these related problems is the computer package known as RUCUS. This is a new tool and, as can be expected in an industry not highly computer-oriented, it has yet to be utilized extensively. This is evidenced by the fact that fewer than 20 transit properties are making significant use of Rucus in their operations [15.13]. Moreover, problems involved with converting the massive amounts of scheduling information into computer data bases greatly hampers its utilization by the larger transit properties.

RUCUS has been used primarily "to cut major quantities of runs in preparation for line ups." Translation of this transit-scheduling parlance reveals that the program determines appropriate assignment of operators to vehicles in view of the daily assignment of vehicles to the bus routes. This is done in preparation of an event, held periodically, during which vehicle operators have the opportunity to select their daily runs, the order of operator selection being a function of time in service.

Table 15.6   Planning Considerations for Bus-Route Design and Scheduling

Route location	• Routes should be designed to provide access to major traffic generators • Avoid one-way street operation where possible to provide for better safety and route identity • Avoid areas of local traffic congestion • Attempt to improve the route aesthetics for bus users
Bus-stop and transfer locations	• Should be coordinated through the local traffic engineer • Must reflect location of major traffic generators • Generally there should be no more than seven bus stops per mile • The allocation of the bus stop shelters should be based on passenger demand
Driver allocations	• Drivers need to report for work in time to obtain change, transfers, and all necessary paperwork • Time to reach start of route must be accounted for to meet schedule starts • Standardize procedures for advising drivers of the bus and route allocation • A visual vehicle check by the driver prior to starting is advised • Standby drivers and vehicles must be available • Established hiring procedures require adequate time for educating and training new drivers
Route time schedules	• Actual route travel time should be determined empirically • Driver allocations and availability must be considered in determining service start and stop times • Route schedule time must take account of coordination of buses at transfer locations • Detailed cost estimates of actual route costs must be generated • New routes necessitate driver training and information sessions • New route or schedule changes must be coordinated among drivers, equipment, and the public-information specialists
Bus allocations	• Time must be allowed for maintenance, start-up procedures, driving time to start location, etc. • Generally, the newest and best equipment should have the highest utilization; other equipment can be assigned to peak service and as standbys

Source: [15.14, p. 112].

Reports of up to 3 percent savings of system operating costs and 5 percent in the number of drivers needed have been encouraging. Moreover, during labor negotiations, it has been possible to use RUCUS to simulate the effects of proposed changes in work rules. It is possible, as examples, to obtain quick and accurate estimates of the impacts of changes in a particular policy or rule on operators' working conditions, total pay hours, and the total number of operators and vehicles required.

15.3.2 Paratransit Operations

As discussed in Chap. 12, paratransit services are aimed at personalized, flexible, client-oriented transport by a variety of means that bridge the gap between fixed-route bus systems and the personal automobile. Such services have been categorized in a number of ways; Table 15.7 presents one such classification scheme. Figure 15.3 further identifies the position of paratransit relative to other transit services, indicating the service frequency and acres/coverage regions in which paratransit is most likely to be effective.

Demand-responsive services, such as shared-ride taxi or dial-a-ride, generally require fewer vehicles to provide coverage to relatively low density areas and are particularly effective when origins and destinations are scattered throughout an area. These services can also be used as feeders to more conventional fixed-route service(s).

Ride-sharing via car-pools or van-pools can be an attractive alternative to driving alone for commuters, particularly those not well served by conventional transit. Such commuters include those that live relatively long distances from their work place and/or live in low-density areas. Subscription bus service is another possibility for any group of people travelling to the same destination on a regular basis. Subscription bus, of course, requires more commuters making approximately the same trip in order to be successful.

Many studies of paratransit operations have been undertaken. As an example, the Urban Mass Transportation Administration has sponsored an investigation of computer dispatching for demand responsive transit (DRT). The effectiveness of such dispatching in improving service levels has not yet been completely determined. It would appear, however, that as demand for service increases, computer dispatching will be more efficient than manual dispatching and more cost-effective as well.

While *ride-sharing* commonly includes all modes of passenger transportation other than driving alone (bus, commuter, rail, carpool, etc.), private ride-sharing in the form of car-pooling and van-pooling has regained considerable public interest in the past half decade, particularly for trips that cannot be efficiently served by conventional systems. Theoretically, car-pooling and van-pooling could serve any

Table 15.7 Classification of Paratransit Operations

A. Pre-arranged ride-sharing
  - Car-pools
  - Van-pools
  - Charter and subscription buses
  - Employer-provided transport services
  - Transit clubs

B. Demand-responsive transit
  - Collective/shared taxis
  - Demand-responsive buses
  - Organized hitchhiking

C. Ownership/use schemes
  - Self-drive taxis
  - Specialized car rentals
  - Joint ownership arrangements

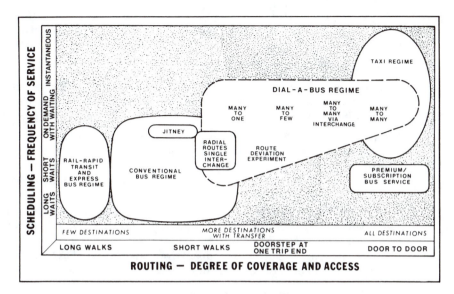

Fig. 15.3 Conceptual framework for Dial-a-Bus services. [15.45, p. 21].

travel purpose. In practice, however, they are generally limited to commuter travel because of the regularity and frequency with which commuter trips are made.

Major problems inherent in these program efforts include vehicle purchase/lease arrangements and the question of insurance. Most typically, van-pooling involves the purchasing or leasing of the vans by the promoter. Thus far, despite the differences in purchasing or leasing vans, after accounting for age and luxury variations among the vans, the break-even distance-based fares charged by various projects are about the same.

Considerable progress has been made in resolving the debate over regulations and state laws that have imposed barriers to ride-sharing in general, and van-pooling in particular. In addition, the Insurance Service Office (ISO), which collects and analyzes statistics and publishes classification and rate guides for the insurance industry, recently announced a policy on such for ride-sharing. The ISO guidelines suggest that the rates for privately owned van-pools be identical to those of private automobiles of equal value and with the same coverage. Van-pools not privately owned (employer and others) should be rated under a new category of "van-pools" that qualifies for the lowest commercial rates times a factor which varies between 1.00 and 1.25 depending on vehicle capacity and van-pool type. This reclassification has permitted considerable savings on insurance costs.

Employer involvement appears to be an important element in a successful ride-sharing program. Some employers have expressed concern that pool participants might be less willing to work overtime when required. Other employers have actively promoted ride-sharing and even offered incentives to those who car-pool or van-pool. Companies are probably more apt to promote ride-sharing when there is an obvious need for it as, for example, to relieve local congestion or a parking shortage. Employees are undoubtedly more willing to ride-share when the organization encourages prompt departures to meet car-pools.

Another barrier to ridesharing is the inconvenience an individual can expect to experience because of the constraints imposed by the group arrangements. Typically disbenefits include: circuitry of travel, required intermediate stops, and schedule adherence.

15.3.3 Bus Priority Measures

Bus priority measures have been instituted in many countries throughout the world, with some distinctive variations and varying terminology as well. We have utilized these definitions:

1. Priority lane: a specially delineated portion of a street reserved for the use of buses or other high-occupancy vehicles (HOVs).
2. Bus street: an urban street entirely reserved for bus use.
3. Busway: a specially delineated portion of highway or freeway that is reserved for buses and HOVs.
4. Bus-lane and busway system: a network of bus lanes and busways coordinated with a system of bus routes to provide a high level of efficiency and performance for buses.
5. Bus priority signal systems: those that give priority to bus movements through detectors or actuators that react to bus presence.

Bus priority lanes can be characterized using the following criteria [15.18]:

1. The direction of traffic on the lane in relation to the direction of the general traffic flow, e.g., with-flow, contra-flow, or reversible.
2. The position of the lane in relation to the direction of the general traffic flow, i.e., to the right of traffic (curb lane), or to the left of traffic.
3. The manner in which the lane is delineated from the other lanes, i.e., pavement markings, temporary barriers, or permanent barriers.

Applications in the United States have shown that the reduction in travel time for buses depends on the length of lane, the number of bus stops along the restricted portion of the route, and the level of enforcement. Express buses operating on contra-flow arterial lanes realize greater time savings than local buses operating on with-flow curb lanes within the CBD, for example [15.19].

Bus streets, or transit malls, have proven successful in a number of applications, most notably Nicollet Mall in Minneapolis. The application will differ, depending on whether bus or pedestrian traffic is the most prevalent (i.e., it may be necessary to separate pedestrian and bus traffic). Two very important problems that may arise in pedestrian areas (whether or not they are served by buses) involve access for delivery vehicles and parking for residents' vehicles.

Bus priority signal systems designed to reduce delays to buses fall into three general categories [15.20]. The first deals with those adjustments that can be made to the signal cycle and/or phasing to account for the presence of buses in the traffic stream, particularly for fixed-time signals. Studies have shown that short signal-cycle lengths (to some limit) are beneficial to buses by reducing delay and minimizing bunching of buses. In addition, where a computerized system of interconnected traffic signals is in place, the basic strategy for signal optimization can be developed on the basis of providing priority to bus movements.

In the second category, individual buses are detected as they approach the signal and the phasing of the traffic lights is affected directly, if necessary and possible, to minimize delay to the bus. Measures of this type often are referred to as signal preemption measures. They may be either active (an optical or radio emitter on the bus and a detector upstream of, or at, the intersection; a transponder on the bus and a loop detector embedded on the roadway) or passive (a detector embedded in the roadway that recognizes the passage of a bus; a sensing device on a trolley wire).

Finally, signals may be used to control the flow of nonpriority traffic (in much the same way as for ramp metering) while buses avoid the restriction by making use of either a bus street or a lane with a separate set of signals timed to favor bus flow [15.21].

15.3.4 Advanced Operational Concepts

A number of new investigations have resulted in advanced operational equipment and/or techniques that could possibly have widespread impact and applications over the next 10 years. The Transbus and Advanced Design Bus (ADB) programs (Chap. 12) have resulted in a marked improvement in the equipment available to transit properties and users. In conjunction, general bus procurement specifications have been developed that include standardized terms and conditions as well as warranty and quality assurance requirements, in addition to the technical specifications.

The automatic vehicle monitoring (AVM) program provides a good illustration of advanced concepts, in this case specialized bus-related equipment expected to improve the operational characteristics of bus systems by providing more reliable service to passengers and reduced operating costs [15.21]. In an AVM system, buses operating on city streets are electronically monitored through equipment that (1) identifies vehicle location, (2) transmits that location and passenger-count information from the vehicle to a control center, and (3) processes the information to determine schedule adherence and recommends the optimum strategy to maintain scheduled service. Figure 15.4 displays a schematic of how AVM systems work.

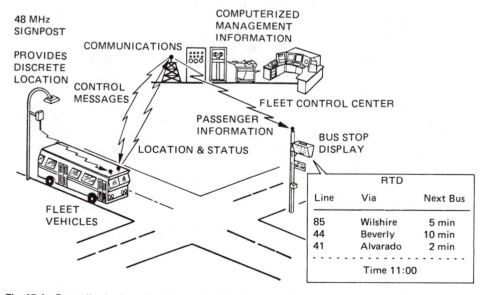

Fig. 15.4 Generalized schematic of operation of an automatic vehicle monitoring system (AVM). [15.21].

These systems can be used to generate a substantial amount of data and information to determine the productivity of each bus in service [15.21]. Federal requirements on transit-system operating data have resulted in a process of transit-performance evaluation that, if coordinated with a daily AVM-based data-gathering effort, could yield significant benefits in terms of the scheduling, operations, planning, and maintenance activities of a transit property. Operational and organizational decisions should be enhanced by the availability of up-to-date information on actual system performance.

15.4 TERMINAL OPERATIONS

Since many intraurban as well as interurban trips involve a change of mode, the importance of the operation of terminal facilities cannot be minimized. In fact, multimodal terminals that provide an interface between combinations of a number of modes have become prominent in transportation plans. As a result, an introduction into the operational characteristics of various terminal facilities is essential to those engaged in the transportation planning process.

15.4.1 Automobile Terminals

As the population of persons and automobiles grows, so too does the need for adequate parking facilities. These are needed off-street simply because increasing travel demand in urban areas requires the fullest utilization of existing streets and highways and because of the high number of urban accidents associated with curb parking (typically 20%). Furthermore, unless urban travel characteristics change drastically in a short period of time, parking requirements in urban areas will remain for some time.[4]

The supply of parking spaces has been shown to increase with city size, although at a decreasing rate. Furthermore, the nature of parking supply alters with city size, as illustrated in Fig. 15.5. Any specific planning analysis must also take into account certain parking characteristics such as parker trip purpose, accumulation of parkers at specific locations and within the general urban area, parking duration, and the parking turnover rates for the type and size of location under study. Furthermore, parking demand for a certain area is influenced by a number of factors including population characteristics, land and building use, alternative transportation modes, traffic access, parking facility congestion, supply shortages, location, and cost.

Information is available to guide transportation planners in establishing sensible parking requirements. Suggested standards such as shown in Table 15.8 are useful for zoning purposes, but it is essential that local analyses be prepared to establish proper values for specific conditions. Moreover, other considerations, such as the desire to reduce the adverse effects of traffic, may affect the kinds of standards set. For example, one conference of experts investigating the possiblities and effects of policies for limiting traffic in urban areas concluded that "it is possible to limit traffic levels by reducing the number of parking spaces, by imposing controls on the hours of which parking spaces are available, by limiting parking duration, by restricting access to certain groups of users, or by charging for parking" [15.44]. The resolution of these two positions, one suggesting minimum standards for parking

---

[4]This presentation relies heavily on the information gathered for the paper *Parking Principles* [15.22], and another entitled *Parking in the City Center* [15.23].

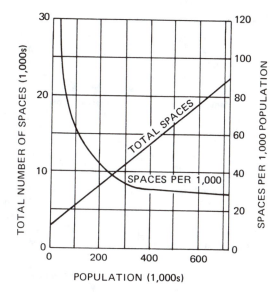

Fig. 15.5 Number of parking spaces in the CBD as related to population of the urbanized area. [15.22].

spaces to be provided and another restricting the total number of spaces, is one element of transportation systems management, as discussed in Sec. 12.2.

A general-purpose public parking facility has as its goals the enhancement of local economic values, increase in economic production, and reduction of needs for streets, in various combinations. As a result, numerous factors determine the appropriateness of specific facility locations; factors include degrees of parking shortages and types of nearby parking generators, user considerations (walking distances, security, and access convenience), development costs, and street-system elements such as capacity, directional flows, and turn restrictions. The specifics of parking-terminal design are determined by the selection of a self-parking or attendant-parking facility. Self-parking is common in new garages and is increasingly acceptable to all types of parkers. It is also possible to develop facilities that combine the parking function with one or many other functions, including retail establishments, offices, schools, and others.

Beyond the simple off-street parking lot, the types of parking terminals include underground, mechanical, and ramp garages. The design of specific parking terminals is dependent on the size, weight, and turning characteristics of the automobile. Building codes throughout the United States have been established to accommodate the loading characteristics required in final designs for both multideck and roof-level parking garages. Typical design configurations are shown in Fig. 15.6.

Parking-terminal designs must also be evaluated in light of certain other needs within the terminal. Sufficient reservoir space must be allocated where drivers are required to stop, so that cars do not block other automobile or pedestrian movements. Walking within the terminal, be it horizontal or vertical, must be considered in the design stage. Also necessary is an adequate lighting system to promote vehicle and pedestrian safety, maximize operating deficiency, and promote user security. In addition, consideration should be given to the requirements of ventilation, heating, fire protection, and adequate informational and directional signing.

Table 15.8 Zoning Standard Guidelines

Use of building or site	Minimum number of parking spaces required
Residential	
Single family	2.0 per dwelling unit
Multifamily*	
Efficiency	1.0 per dwelling unit
One or two bedrooms	1.5 per dwelling unit
Three or more bedrooms	2.0 per dwelling unit
Commercial	
Offices* and banks	3.3 per 1,000 sq ft GFA[†]
General retail	4.0 per 1,000 sq ft GFA
Shopping centers	5.5 per 1,000 sq ft GFA
Restaurants	0.3 per seat
Hotels, motels*	1.0 per rentable room plus 0.5 per employee
Industrial	0.6 per employee
Auditoriums and theaters*	0.3 per seat
Churches	0.3 per seat
College/university	
Good transit access	0.2 per student
Auto access only	0.5 per student
Senior high school	0.2 per student plus 1.0 per staff member
Elementary and junior high school	1.0 per classroom
Hospitals	1.2 per bed

*Exceptions permitted in CBD if adequate public transportation is available.
[†]GFA = gross floor area.
Source: [15.22, p. 39].

15.4.2 Change-of-Mode Terminals

Closely related to automobile parking facilities are terminals used in connection with express bus or commuter train operations. Provision of change-of-mode parking convenient to quick, reliable transit service provides the commuter with an alternative to CBD traffic and parking congestion.

The operation of a change-of-mode facility depends upon the variety of modal changes involved. Consideration must be given to the interrelationships between train boarding, bus or taxi loading, and private automobile pickup–dropoff. The location of areas for these various activities must minimize pedestrian-vehiclular conflicts. This usually can be accomplished according to the following suggested schedule of priorities:

1. Bus loading–unloading
2. Taxi loading–unloading (may intermix with buses or cars)

3. Passenger-car unloading–loading
4. Short-term parking
5. Long-term parking

A suggested plan for a rapid-transit change-of-mode terminal, shown in Fig. 15.7, illustrates these considerations.

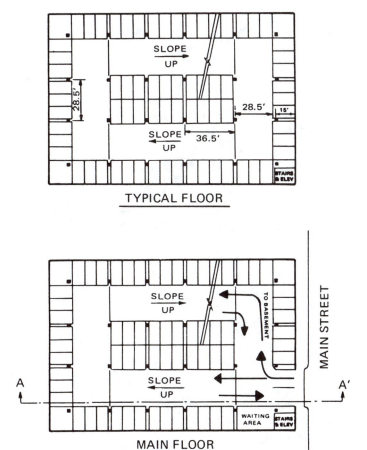

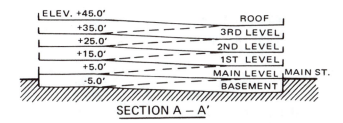

Fig. 15.6 Typical ramp garage design. [15.22, p. xxiv-9].

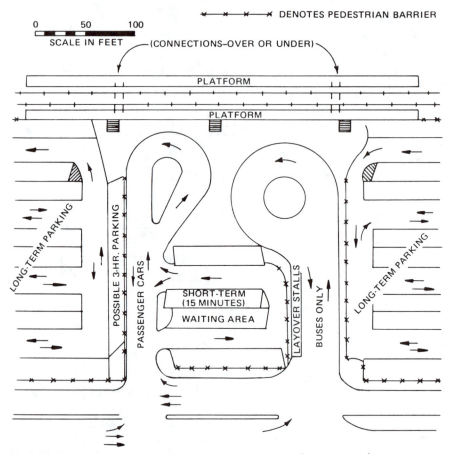

Fig. 15.7 Example of a rapid transit change-of-mode terminal. [15.22, p. 158].

A change-of-mode terminal can be located in outlying residential areas, along an urban freeway, or in the CBD [15.25]. Outlying terminals help intercept motorists and buses in suburban areas and facilitate passenger transfer to express lines. They also provide convenient access for transit patrons. On-freeway terminals can be built on the freeway right-of-way, thereby eliminating the need for express buses to leave the freeway and travel to local streets to suitable location for loading and unloading passengers. CBD terminals provide off-street downtown distribution for radial express bus operations.

15.4.3 Transit Terminals

The design and operation of transit terminal facilities has become an area of increasing concern because of the major investments now being made and contemplated in urban rapid transit, the need to renovate and upgrade many out-dated facilities, and the continued interest in the downtown people-mover (DPM) program [15.25]. The transit terminal plays a major role in the overall effectiveness of transit-system performance, since its major function is to process the flow of passengers between modes. The degree to which it accomplishes this activity smoothly and in a safe and pleasant environment will strongly influence system acceptance. In general, station configurations are based on two primary objectives: to avoid conflicts and to provide adequate capacity. Conflict avoidance requires that the station circulation pattern include only necessary turns and no dead-end corridors, provide for unobstructed walking areas and lines of sight, and include duplicate access routes. Attainment of smooth and continuous pedestrian movement, reduction of conflict-producing situations, and escalators where vertical height exceeds 12 feet also are desirable.

The capacity of a terminal facility will be determined by its "weakest link," or the component with the least capacity. Inadequate components will cause congestion, queues, and general passenger disorientation and discomfort. The principal components of a station that must be taken into account are the stairways, ramps, escalators, platforms, and fare-collection areas. A summary of common guidelines for these components is included in Table 15.9.

In addition to the components just discussed, there are a number of features important to the successful operation of the terminal facility. These, therefore, must be carefully considered in design and operations activities. Typical features are given in Table 15.10.

Recently published guidelines for downtown people-mover (DPM) planning illustrate a number of unique concerns related to DPM terminal design and operation resulting from the nature of such systems [15.29]. Stations located within buildings offer a number of advantages, the most obvious being the removal of the station's bulk and support structures from the public right-of-way. Platforms, fare-collection areas, stairways, escalators, and elevators can be removed from the street and incorporated into the buildings. One can visualize shops and services for the DPM user located in a second-floor lobby of a new office building. Advantages include added revenues for rent of retail space in the building and shorter walking distances for DPM users working in the building. However, problems of construction plus security, institutional barriers involving building codes, and insurance all may lead to difficulties. Obviously such schemes can be better carried out with newly constructed buildings than with older ones adapted for that purpose. In fact, it is

Table 15.9   Common Guidelines for Selected Terminal Facility Components

Component	Characteristic	Standard/guideline
Stairways, ramps, and passageways	Width	54 in.
	Grade	6%
	Stairway riser height	7 in.
	Stairway landing depth	> 6 ft
Escalators	Tread depth	16 in.
	Riser height	8.5 in.
	Maximum inclination angle	30°
	Normal width	32 in. and 48 in.
	Speed	90 fpm to 120 fpm
Platforms	Side platform width	> 12 ft
	Island platform width	12 ft to 17 ft
Fare-collection area	Coin-operated	15 persons/min
	Machine-read tickets	30 persons/min

Source: [15.28, p. 2].

**Table 15.10   Typical Station Features
Associated with Terminal Operation
and Performance**

Passenger processing
  Level-change facilities
  Entrance–exit facilities
  Area provided per person on flow paths
  Travel distances
  Travel paths
  Fare-collection devices
  Vehicle boarding and exiting areas

Passenger orientation
  Directional signs and maps
  Visibility of major destination points
  Information booths

Physical environment
  Air-flow control devices
  Heating and air conditioning
  Lights
  Weather protection

Security
  Police patrols
  Isolated spaces
  Surveillance cameras
  Alarms
  Entry control

Safety
  Number of levels
  Walking distances
  Curbs
  Stairs
  Escalators
  Platform edges
  Lighting

Source: [15.28, p. 1–4].

questionable whether "retrofit" of buildings for DPM stations is practical or even possible. Much may depend on the minimization of the combined guideway and vehicle cross section to allow for threading the DPM through existing structures (see Chap. 12 for an example).

It is possible to examine the relationship between the arrival rate of passengers at a station platform and the rate of acceptance of those passengers by station escalators, stairways, fare-collection systems, or DPM vehicles through the use of simulation models. The Urban Mass Transportation Administration has developed a transit-station simulation computer model that can be used as a design and evaluation tool to assist in developing alternative configurations for intermodal transit stations [15.29]. The computer model is capable of recognizing various station characteristics, including the following details:

1. System technology—vehicles (length of car, number of doors per car, capacity of car, number of cars per train); platform length; and headways (peak and off-peak).
2. Station location—boundaries; elevation; location of guideways; access/egress points; other modes (buses); and existing elements (pedestrian mall).
3. Passenger design volumes—number of inbound/outbound users during the peak/off-peak times by mode.
4. Passenger processing elements—boarding and deboarding procedures: fare-collection system; desired levels of service and performance standards (comfort and safety); waiting areas; handicapped pedestrians; emergency evacuation; and intermodal transfers.

The specification of this data allows the analyst to obtain information regarding schemes to account more realistically for such station components and elements as platform, concourse, and station-area requirements, number of turnstiles, security gates, and fare-card readers needed, time requirements for passenger processing, limitations on passenger processing capacity, and so on.

15.5 TRANSPORTATION SYSTEM MAINTENANCE

System maintenance would be an important concern if only because of the costs involved. For example, 22.7 percent of transit-industry expenses were directed toward maintenance in 1978. The level of these costs indicate that proper functioning of the transportation physical plant depends on adequate facilities and equipment, which must be capable of providing the service levels that the public has come to expect.

The following sections highlight examples of maintenance procedures in highway, bus, and rail systems and provide additional information regarding the requirements for planning and conducting such activities. Where appropriate the discussion describes the requirements in terms of the following generic classes:

1. Inspection: comprises frequent general inspection and periodic inspection at longer intervals and more detailed in nature; both monitor the condition of the maintainable items and establish the necessary maintenance and procedures to keep these items in satisfactory condition.
2. Scheduled preventive maintenance: includes all foreseeable maintenance ideally required to maximize the service life of an item and replacement of components (or the items themselves) when needed. This work is undertaken within scheduled programs.
3. Nonscheduled maintenance and repairs: generally comprise repair occasioned by breakdown or accidental damage.
4. Cleaning or clean-up activities: general removal, as necessary or on a scheduled basis, of debris or litter from the vehicles, travelled way, or stations.

15.5.1 Street and Highway Maintenance

The sheer magnitude of the cost of street and highway maintenance should rank it among the most important activities regarding transportation systems. Approximately one-quarter of all highway funds go to highway maintenance, which is generally performed by the appropriate highway agencies themselves. Moreover, as a highway system expands, the maintenance costs can increase to the point that they exceed the total expenditures previously spent for highways. Once the Interstate Highway System is completed, many states will annually be spending more to maintain their portions of the system than they spent for new highway construction. Since World War II, highway maintenance has emerged as a major industry; it utilizes over 200,000 personnel, 300,000 to 400,000 pieces of equipment, and billions of dollars each year. About half of this total is accounted for by the 50 state highway departments.

The costs of highway maintenance are accelerating at a rapid pace while state tax revenues are decreasing, both at a time when much of the highway system constructed since the mid 1950s is beginning to show a need for revitalization. The information in Table 15.11 reflects the maintenance needs, by major category, of a typical north-central state with approximately 17,500 road miles of Interstate and Federal-aid primary highways. This table generalizes the maintenance costs by program category for fiscal year 1978, but reflects actual conditions.

In examining future maintenance needs, the state made a projection of expenditures for the period from FY 1980 to FY 1985: the projection utilized a modest inflation rate of 6% and specified the elimination of deferred maintenance and the achievement of a satisfactory level of service. The results of this projection (in 1979 dollars), shown in Table 15.11, underscore the significant future requirements of road maintenance.

15.5.1.1 Purpose and organization The purpose of highway maintenance is the preservation of each type of roadway, roadside, structure, and facility as nearly as possible in its original condition as constructed or as subsequently improved, and the operation of highway facilities and services to provide satisfactory and safe transportation. Highway maintenance programs are developed to offset the effects of weather, organic growth, deterioration, traffic wear, damage, and vandalism. Deterioration would include effects of aging, material failures, and design and construction faults. Elements of the highway system that come under maintenance include the roadway surfaces, shoulders, roadsides, drainage facilities, bridges, tunnels, signs, markings, lighting, fixtures, and traffic signals, as well as such services as snow and ice removal, and the operation of roadside rest areas.

Each maintenance organizational structure is probably unique because the maintenance manager's responsibility and authority are influenced by the nature of the organization of which he or she is a subordinate part. In some cases, maintenance and construction are part of the same organization. Moreover, many statewide maintenance organizations are divided along geographic lines, primarily

Table 15.11 Summary of Projected Maintenance Need ($1979 Million)

Fiscal year	Inflation adjusted budget	Level of service adjustment	Equipment requirements	Other requirements	Total maintenance needs
1980	138.0	40.7	19.6	4.7	203.0
1981	146.3	30.1	20.7	6.4	203.5
1982	154.5	32.7	22.0	5.4	214.6
1983	164.2	36.4	13.6	7.4	221.6
1984	173.9	40.3	14.5	6.2	234.9
1985	184.9	44.2	15.3	8.6	253.0

because of the distances involved for travel to work sites. Maintenance crews encompass a highly versatile range of talents and equipment. Functions that require expensive or sophisticated equipment or that are required on a limited basis can be handled by special crews who are centrally located.

The fact that maintenance is performed for the most part by the various highway agencies themselves is in direct contrast to construction, where 95 percent of the work, exclusive of engineering, is done by contract. The principal explanation is that maintenance work is so diverse, so subject to variation from the expected, and on occasion so hurried, that it does not lend itself to competitive bidding. Sometimes maintenance forces also perform "betterment" work that might also be done by contract. Common betterment projects include grading and paving for small line changes, resurfacing, and mulching, painting, or other erosion-control work.

There is a close relationship between design and construction practices and maintenance costs. For example, insufficient pavement or base thickness or improper construction of these elements soon results in expensive patching or surface repair. Shoulder care becomes a serious problem where narrow lanes force heavy vehicles to travel with one set of wheels off the pavement. Improperly designed drainage facilities mean erosion or deposition of materials and costly cleaning operations or other corrective measures. Sharp ditches and steep slopes require hand maintenance, while flatter ditches and slopes permit machines to do the work more cheaply. In snow country, improper location, extremely low fills, and narrow cuts that leave no room for snow storage can create extremely difficult snow-removal problems.

15.5.1.2 Maintenance management[5] Highway maintenance operations have long felt the constraints of tight money, inflation, and the energy crises, because the maintenance activity is constantly responding to an increasing road network and increases in the demands upon its resources. Consequently the planning, budgeting, and scheduling of effective maintenance is receiving closer attention, particularly by the state legislators and executives who are ultimately responsible to the tax-payers. In addition, standards and priorities must be established to assure that appropriate work assignments can be made. The complexity of these various activities taken in conjunction with increasing budget requirements and material and personnel costs have led many state and local agencies to develop maintenance management and productivity units in an attempt to improve upon the maintenance performance.

A good example of one type of maintenance management comes under the heading of pavement management systems (PMS). There does not appear to be any widely accepted definition for a PMS. It is safe to say, however, that a PMS is an ordered and objective process whereby the most serviceable pavements possible are provided at the lowest possible cost to the users.

Several specific benefits of a PMS to transportation administrators have been identified. These are:

1. Improved performance monitoring and forecasting
2. Objective support for legislative funding requests
3. Identifiable consequences of various funding levels
4. Improved administrative credibility
5. A basis for cost allocation to highway users
6. Improved engineering input for policy decisions

---

[5] Much of this discussion is adapted from [15.39].

These might be accomplished for at least two categories of decisions: those involving projects and their priorities for maintenance, and those involving total highway networks and the funds needed to maintain them. The former might be left to the discretion of local engineers and planners, who are most familiar with the pavements under their jurisdictions. At the same time, it is evident that network-wide decisions, such as determining needed revisions in funding levels and the consequence of those levels, must be centralized responsibilities.

The step from individual project to network analysis involves the development and implementation of a valid sampling plan where pavement conditions as a function of time or traffic can be identified. Then, projections of funding needs can be based on estimates of maintenance needs. Since pavement design parameters materially influence the shape of the condition–time curve, it is possible to predict mean ratings as a function of traffic loadings; pavement strength, and time. Such subsequently can be used to examine the consequences of various funding levels.

Elements of a PMS   The AASHO road test conducted in the late 1950s provided the foundation for effective long-range planning of pavement expenditures. During that test, a system of pavement ratings on a scale from 0 to 5 was developed with the following designations:

0 to 1   very poor
1 to 2   poor
2 to 3   fair
3 to 4   good
4 to 5   very good

The system was developed from a series of subjective ratings of various pavements by a panel of road users and was transformed into an objective present serviceability index (PSI), where physical measurements such as roughness, rutting, cracking, and patching are the principal determinants. Further road-test studies showed that a pavement performs in the manner indicated in Fig. 15.8, where the vertical scale is PSI and the horizontal scale may be either time or accumulated traffic loads.

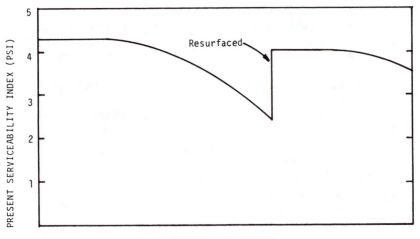

Fig. 15.8 Typical pavement performance curve.

Typically, a pavement loses serviceability (deteriorates) very slowly for several years, then enters a period of rather rapid decline toward total failure. This period of rapid decline is marked by the presence of cracking and deformation and by a decrease in ridability. As indicated in Fig. 15.8, an overlay at some time after the period of rapid deterioration begins can restore the pavement to where a new cycle begins.

Virginia pavements presently are designed to provide a PSI of no less than 2.5 over a 20-year design life. Typically, an overlay is required in 6 to 10 years to avoid an excessive loss of PSI. A sampling of Virginia pavements reviewed for resurfacings during 1980 showed a range in PSI values from approximately 2.1 for primary highways with low traffic volumes to approximately 3.8 for an Interstate.

The serviceability rating system discussed above reflects the user's perception of pavement serviceability. Another approach, preferred by some engineers, is to base pavement ratings on engineering characteristics of pavement. Such ratings, tempered by some measure of the user's perception of ridability, have a time or traffic relationship similar to that shown in Fig. 15.8 for PSI ratings. This approach, known in Virginia as the pavement maintenance rating (MR), is discussed below.

There are three elements of pavement condition to be measured. The first, pavement distress, often is combined with the third, ride quality and serviceability.

Pavement distress may be measured by several methods, most of which use subjective evaluations expressed in numerical terms to determine the number of points to deduct for different frequencies and severity levels of various types of distress. The basic system requires an engineer to ride slowly through the roadway section under consideration and assign subjective rating scores to the whole section. Approximately 5 min are required to rate each 1-mile pavement section. Then a process of mental averaging of the overall pavement condition is followed. Trials of the system in several districts have shown that people will rank the same pavements in the same order of priority but with different rating scores.

The *structural integrity* of a pavement is usually measured with one of several types of dynamic testing units. The unit used in Virginia for about 15 years is the dynaflect, a trailer-type device propelled by a van or pick-up and capable of relatively rapid tests if these are not too frequent. Speed in the testing mode is 3 to 4 mph, with about a 1-min stop at each test site. Between jobs, the device is moved at highway speeds.

There are two major ways to measure *ride quality and serviceability*. One is the called response-type road roughness measurement (RTRRM), and the other is direct profile measurement (DPM). The former is undertaken with a passenger car instrumented to obtain information on the vehicle's response to pavement roughness. It is assumed in this approach that the passenger's assessment of ride quality is proportional to that measured by the instrument. Also, readings on such instruments are highly dependent on the vehicle in which they are mounted, so that meticulous and frequent calibrating is required if measurements are to be meaningful over long periods of time and if differences between vehicles are to be accommodated.

The most popular and versatile means of direct profile measurement is provided by the surface dynamics profilometer (SDP) developed by General Motors in the early 1960s. This device incorporates means for direct measurements in each wheel path of a travel lane. Instrumentation, including an on-board computer, is fully enclosed in a van normally operated by a two-person crew. The SDP is internally calibrated.

Among the uses of the latest version of the SDP are:

1. Measurements of true pavement profile
2. Calibration of RTRRM devices

3. Estimates of overlay thicknesses required to restore serviceability
4. Acceptance-testing of ride quality versus pavement construction of maintenance

Offsetting these multiple uses and the improved reliability of data are the relatively high cost (about $250,000 in 1981) and slow speed (about 20 mph) in traffic of the SDP.

Sampling plan Because the magnitude of urban highway systems virtually precludes total sampling of all pavements, a statistically valid sampling plan is necessary. Such a plan must provide information permitting conclusions on the condition of total highway networks (Interstate, primary, and secondary), yet must involve a reasonable and manageable amount of testing and evaluative field work. The plan also must provide the capability to detect changes in pavement condition with time and accumulated traffic volume.

Studies performed in Texas [15.42] show that a very small, properly stratified random sample of a highway network can yield the information necessary for projections of needed funding levels and consequences of reduced funding. Such samples will not however, permit the prioritizing of individual projects.

15.5.2 Bus Transit Maintenance

The basic principle of maintenance performance, be it with regard to a large transit system or one's own automobile, is the provision of an optimum schedule that achieves a balance between the cost of scheduled maintenance and that resulting from unscheduled repairs. Bus transit system maintenance is guided by this principle and is generally carried out under a system of periodic inspection and maintenance [15.38].

Periodic maintenance is based on observing fixed periods of mileage between inspections and maintenance for the different parts of the bus. Bus manufacturer recommendations may be used as a starting point, with the time periods being subject to review and adjustment according to the accumulated experience with road failures and maintenance costs. Variations from manufacturer recommendations could be attributable to a number of factors, including the types of routes covered by the buses, the average distance between stops, the number of passengers carried, road conditions, traffic density, and of course the design and construction of the bus itself. It is conceivable that this method of trial and error and empirical testing will eventually converge to the most economical schedule.

The advantages of the periodic maintenance system stems from its simplicity. Since the maintenance procedures are repeated in each case, the system has a high degree of certainty, thereby lending itself to standardization and concomitant reductions in labor cost and greater reliability for the maintained systems. As an additional advantage, spare-part inventories can be supported with greater certainty. The Bi-State Transit Company in St. Louis has developed periodic procedures that for many years yielded the company the top award for bus maintenance efficiency. The scheduling of preventive maintenance is based upon the following principal elements [15.35]:

1. Daily servicing, consisting of refueling, cleaning, and checking drivers' reports
2. Mileage inspections at 1,500, 3,000, 9,000, and 27,000 miles
3. Overhauling of units on a condition inspection basis
4. Body overhaul and painting at 4- to 5-year intervals

Increases in bus maintenance activities might require major new or altered facilities and equipment so that operations can be performed efficiently and effectively. For instance, many maintenance facilities have been converted from trolley car to bus operations, and their site and shop layouts do not necessarily provide for efficient maintenance operations [15.38]. Facilities requirements are largely dependent on fleet size and maintenance practice. The latter varies even among properties of similar size. Nevertheless, certain functions are essential and are provided in nearly all properties in one or a number of facilities.

Daily servicing for transit buses is carried out in an inspection or division garage. The preventive maintenance program (periodic inspections and repairs) is the basis for inspection-garage activities. Most involve brake adjustments and engine degreasing in conjunction with inspections. Some properties include tire work, minor body repairs, minor painting, and engine dynamometer testing. Unit rebuild, engine overhaul, significant body repairs, and other major repairs may be performed at a main maintenance facility. A composite inspection garage, not directly representing any existing facility but exemplifying the major component elements required of such a facility, is shown in Fig. 15.9. Separate areas are provided for each of the activities noted above, together with staff and support space and facilities.

The daily servicing of a bus is a routine operation in which vehicles are prepared for revenue operations as part of the activities of an operation garage. Bus servicing includes refueling, interior cleaning, exterior washing, and some minor maintenance checks. Servicing requires almost 50 percent of the labor hours used by the maintenance department. Service facilities therefore deserve careful attention for operational efficiency.

Transit properties large enough to support or require a number of facilities complement the inspection garage(s) with a main maintenance facility which may be located at a unique site or integrated with an inspection garage. Infrequent and major repair functions carried out at these facilities include heavy repairs, engine overhauls, unit rebuilds, major body repairs, painting, upholstery, route and bus-stop sign preparation, brake relining, brake-drum turning, and radiator repair. Main maintenance facilities have greater space and equipment requirements than inspection garages, with total space being distributed according to the individual needs and preferences of the transit property. Figure 15.10 is an example layout for such a facility.

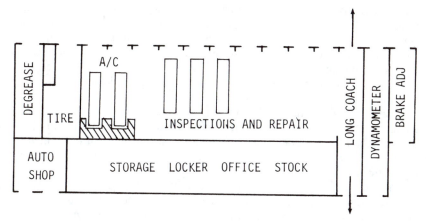

Fig. 15.9 Composite inspection garage.

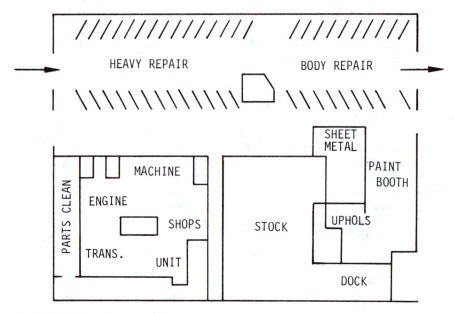

Fig. 15.10 Main maintenance: floor plan.

15.5.3 Rail Transit Maintenance

As new rail systems such as the Washington Metro are implemented and new transportation technologies such as the AIRTRANS people-mover system are put into place, the requirements of system maintenance become strikingly evident. Also, as automation becomes more desirable as a means to reduce operating costs, maintenance and reliability become a component part of system operation as well. Maintenance of structures, stations, and track, in addition to mechanical and electrical equipment, all must be considered.

In a recent study [15.37], a set of guidelines for the optimization of capital expenditures against maintenance costs for subway transit was formulated. These guidelines could be considered to be exemplary for many rail or guideway-type transit systems. The following paragraphs summarize certain essential recommendations and illustrate the breadth of concern as well as the potential for tradeoffs among design considerations, construction costs, and maintenance requirements.

Rail and guideway transit systems are virtually self-contained entities. The trackway consists of a train or vehicles, the rails or guideway, electrical power conduction equipment, a miscellany of sensitive electronic and electrical equipment, and other support services. The station structures may house escalators, elevators, stairways, corridors, fare collection and other supportive mechanical and electrical equipment, and a variety of architectural finishes. All of these items (and possibly others) require considerable maintenance, the scheduling and performance of which are often severely restricted by the presence of trains or vehicles along the trackway and patrons in the stations.

Yen et al. [15.48] enumerated (see Table 15.12) some of the major guideway transit subsystems and the factors (time, use, etc.) most likely to impact subsystem degradation. These can be used to plan and program maintenance activities. Mileage accumulation is likely to be the most important degradation variable for most vehicle-related subsystems. Time can be either calendar or operational, whichever is appropriate for the subsystem under consideration. Frequency of operation and use differ in that the former relates to the number of times a subsystem is operated (doors, warning systems, etc.) and the latter relates to the number of times a component (seat, floor surface, etc.) is utilized by patrons.

Fixed-guideway transit systems exist in an aggressive atmosphere. Chemically charged ground water, shifting ground loads, and the dynamic loading of vehicles and trains take their toll on system structures. Leakage and moisture affect structural and other items, and vandalizing patrons can adversely affect vehicle equipment and architectural finishes. In the formulation of design alternatives, all these factors and their interactions must be assessed. (Structural elements are sometimes subjected to both factors; command and control systems are affected by operational time.)

Environmental conditions are a major cause of deterioration of structures for rail and guideway systems. In particular, underground structures, and much of the

Table 15.12 Factors Expected to Control Maintenance Cycles for Major Subsystems and Components

Subsystem and components	Factors controlling maintenance
Vehicles	
Command and control	Time
Propulsion and control	Mileage, frequency of stops
Running gear and guidance	Mileage
HVAC*.	Time, use (degree-days)
Switching	Frequency of switching events
Doors	Frequency of stops, use
Interior furnishings	Use
Guideway	
Surface	Frequency of vehicle passing, time
Structure	Time (seasonal cycles)
Power rails	Use
Guideway heating	Cycles
Switching	Use
Stations	
Fare collection equipment	Use
Platform screens	Frequency of operation, use
HVAC*	Time, use (degree-days)
Cleaning	Use, time
Escalators	Time
Station graphics	Time
Lighting	Time
Substations and switchgear	Time

*Heating, ventilation, and air conditioning.
Source: [15.48].

fixed equipment within them, are constantly subjected to water or dampness, usually in combination with chemical action and stray currents. With proper design and construction, many structures can be made substantially watertight, thereby minimizing extensive future corrective maintenance and, possibly, replacement costs.

Mechanical and electrical equipment, such as escalators, elevators, pumps, fans, and lighting systems, must be designed in such a way as to perform reliably and dependably. Much of this equipment may be operating for more than 18 hours a day, 7 days a week, with heavy loading during peak hours. The limited time available for maintenance underscores the need for proper design aimed at reliable service.

Architectural layout and the selection and installation of architectural finishes must also be examined with a view toward optimizing installation costs against maintenance costs. Layout planning, in addition to being important to efficient station operation and passenger flow, should have three main objectives: prevention of vandalism, enhancement of safety, and optimization of the cleaning process. Public areas generally should be open and spacious, with sufficient crush space at flow intersection and grade change initiation points. Equipment and surfaces which can be subjected to defacing and vandalism must be kept out of reach of the public.

The establishment of a maintenance strategy should involve a number of steps. An initial one would involve the inspection and maintenance of equipment critical to service. This includes both hardware having profound impacts on operations, and easily failed systems such as doors and systems for command and control, power collection, switching, and braking. Service periods for critical items should be scheduled on a daily, weekly, monthly, etc. basis, as appropriate. Certain less critical items may be left for corrective maintenance. The decision to defer maintenance until a failure occurs must depend on (1) the failure modes experienced, such as the degree to which the failure has caused system degradation or stoppage, or on (2) the ability to monitor the item on a regular basis.

15.6 SUMMARY

The purpose of this chapter has been to familiarize the student of transportation planning with the operation and maintenance of transportation systems and facilities. Although of primary interest to the transportation engineer, operations and maintenance are of importance to the planner because of their impact on the feasibility and implementation of any system plan. Thorough examination of system operation and maintenance requirements and limitations must be carried out in order to ensure that the basic criteria and standards for vehicular and passenger movements will be consistent with the capabilities of the desired equipment and facilities.

The operation of any transportation system requires some form of traffic control of the individual units to ensure safe and efficient movement. Such operation control results in constraints upon the system, together with the need for coordination among systems, particularly where they share a common operating environment. Improved types of control provide for more economic movement and the minimization of delay; automated control for spatially separated systems provides for safe and efficient operation coupled with reduced costs for personnel.

Maintenance of transportation system equipment and facilities is of concern to the planner because any facility subject to usage is bound to deteriorate, and in doing so, the cost of the system as well as the level of service is impacted. The halting of deterioration or the replacement of deteriorated facilities can be costly in economic terms and also in terms of efficient operation of the systems. Excessive delays or stoppages owing to maintenance requirements must be avoided if operation is to be efficient.

BIBLIOGRAPHY

15.1 U.S. Department of Transportation, Federal Highway Administration: *Traffic Control Systems Handbook*, Washington, D.C., June 1976.

15.2 U.S. Department of Transportation, Federal Highway Administration: *Urban Freeway Surveillance and Control—The State of the Art*, Washington, D.C., Nov. 1972.

15.3 Drew, D.: *Traffic Flow Theory and Control*, McGraw-Hill, New York, 1968.

15.4 Organization for Economic Cooperation and Development, Road Research Group: *Research on Traffic Corridor Control*, Paris, Nov. 1975.

15.5 Cunliffe, J. P.: "A Survey of Railway Signalling and Control," *Proceedings of the IEEE*, vol. 56, no. 4, April 1968.

15.6 Corbin, A.: "AIRTRANS: Intra-Airport Transportation System," SAE Paper 730384, Air Transportation Meeting, Miami, Fla., April 1973.

15.7 U.S. Congress, Office of Technology Assessment: *Automated Guideway Transit*, Washington, D.C., May 1975.

15.8 U.S. Department of Transportation, Urban Mass Transportation Administration: *Innovation in Public Transportation*, Washington, D.C., March 1980.

15.9 American Public Transit Association: *Transit Fact Book, 1978–1979*, Washington, D.C., 1979.

15.10 McManus, R.: "Relationships between UMTA's Policies, Programs, and R. D. and D.," in *Proceedings of the Third UMTA R. and D. Priorities Conference*, Cambridge, Nov. 1978.

15.11 Booz Allen Applied Research and Simpson and Curtin: *Transbus Operational, Passenger, and Cost Impacts*, Transbus Document TR 75-002, Bethesda, Md., July 1976.

15.12 Smerk, G.: "The Management of Public Transit," in G. Gray and L. A. Hoel (eds.), *Public Transportation: Planning, Operation and Maintenance*, Prentice-Hall, Englewood Cliffs, N.J., 1979.

15.13 Hinds, D.: "RUCUS: A Comprehensive Status Report and Assessment," *Transit Journal*, Winter 1979.

15.14 Shortreed, J. H., et al.: "Detailed Planning for Transit Routes," in J. H. Shortreed (ed.), *Urban Bus Transit—A Planning Guide*, University of Waterloo, Ontario, Canada, 1974.

15.15 U.S. Department of Transportation, Urban Mass Transportation Administration: *Service and Methods Demonstration Program—Annual Report*, Washington, D.C., Aug. 1979.

15.16 Organization for Economic Cooperation and Development, Road Research Group: *Bus Lanes and Busway Systems*, Paris, 1976.

15.17 NATO Committee on the Challenges of Modern Society: *Bus Priority Systems*, CCMS Report no. 45, Transport and Road Research Laboratory, Crowthorne, England, 1976.

15.18 Robertson, D. I., and R. A. Vincent: "Bus Priority in a Network of Fixed-Time Signals," *Transport and Road Research Laboratory Report* LR 666, Crowthorne, England, 1975.

15.19 Fisher, R., et al.: "Priority Treatment for High Occupancy Vehicles in the United States: A Review of Recent and Forthcoming Report," *ITE Journal*, Jan. 1980.

15.20 Levinson, H. S., W. F. Howey, D. B. Sanders, and F. H. Wynn: *Bus Use of Highways, State of the Art*, NCHRP Report 143, Washington, D.C., 1973.

15.21 General Motors Laboratory, Transportation Systems Division: *Urban Transportation Laboratory—A Progress Report*, Report no. EP 78028, Warren, Mich., May 1978.

15.22 Highway Research Board: *Parking Principles*, Special Report 125, 1971.

15.23 Smith, W., and Assoc.: *Parking in the City Center*, New Haven, Conn., 1965.

15.24 The Eno Foundation for Highway Traffic Control: *Parking Garage Operation*, Saugatuck, Conn., 1961.

15.25 Taylor-Harris, A.: "Transit Centers—A Manual of Improving Transit Services," *ITE Journal*, July 1980.

15.26 Demetsky, M. J., L. A. Hoel, and M. R. Virkler: *Methodology for the Design of Urban Transportation Interface Facilities*, U.S. Department of Transportation, Office of University Research, Washington, D.C., 1977.

15.27 Hoel, L. A., M. J. Demetsky, and M. R. Virkler: *Criteria for Evaluating Alternative Transit Station Designs*, U.S. Department of Transportation, Office of University Research, Washington, D.C., Dec. 1976.

15.28 Hoel, L. A., and E. S. Roszner, "Planning and Design of Intermodal Transit Facilities," *Transportation Research Record* 614.

15.29 U.S. Department of Transportation, Urban Mass Transportation Administration: *DPM: Planning for Downtown People Movers*, vols. I and II, Washington, D.C., April 1979.

15.30 U.S. Department of Transportation, Federal Highway Administration: *Urban Mass Transportation Planning for Goods and Services*, Washington, D.C., June 1979.

15.31 Marconi, W., et al.: "Traffic Engineering and Design to Facilitate Urban Goods Movement, Report of Probe Group I," in G. Fisher (ed.): *Proceedings of the Engineering Foundation Conference on Goods Transportation in Urban Areas*, U.S. Department of Transportation, Washington, D.C., June 1978.

15.32 U.S. Department of Transportation, Urban Mass Transportation Administration: *Urban Goods Movement Program Design*, Washington, D.C., June 1972.

15.33 Way, A., and J. Eisenberg: *Verification of Performance Prediction Models and Development of Data Base, Phase II*, Arizona Department of Highways, report no. FHWA/A2-80/169, Phoenix, Sept. 1980.

15.34 Maas, R., and W. R. Hudson: *Pavement Management Systems*, McGraw-Hill, New York, 1978.

15.35 Bi-State Transit System: *Preventive Maintenance*, St. Louis, Mo., 1968.

15.36 U.S. Department of Transportation, Urban Mass Transportation Administration: *Guidelines for the Maintenance of Downtown People Mover (DPM) Systems*, Washington, D.C., July 1979.

15.37 U.S. Department of Transportation, Urban Mass Transportation Administration: *Rapid Transit Subways—Guidelines for Engineering New Installations for Reduced Maintenance*, Washington, D.C., Jan. 1978.

15.38 U.S. Department of Transportation, Urban Mass Transportation Administration: *Bus Maintenance Facilities—A Transit Management Handbook*, Washington, D.C., Nov. 1975.

15.39 McGhee, K. H.: *An Approach to Pavement Management in Virginia*, report VHTRC 82-R9, Virginia Highway and Transportation Research Council, Charlottesville, Va., July 1981.

15.40 May, A. D.: *The Role of Operational Planning Models in Transportation Systems Management*, Research Report UCB-ITS-RR-78-11, Institute for Transportation Studies, University of California, Berkeley, Dec. 1978.

15.41 Utah Department of Transportation: "Good Roads Cost Less," Salt Lake City, Oct. 1977.

15.42 Mahoney, J. P.: "Measuring Pavement Performance by Using Statistical Sampling Techniques," *Transportation Research Record*, 715, 1979.

15.43 U.S. Congress, Office of Technology Assessment: *Automatic Train Control in Rail Rapid Transit*, Washington, D.C., May 1976.

15.44 Environmental Directorate, Organization for Economic Cooperation and Development: *Better Towns with Less Traffic*, Proceedings of a Conference, Paris, April 1975.

15.45 U.S. Department of Transportation, Urban Mass Transportation Administration: *Demand-Responsive Transit: State of the Art Overview*, Washington, D.C., Aug. 1974.

15.46 Kennedy, N., J. M. Kell, and W. Hombruger: *Fundamentals of Traffic Engineering*, 7th ed., The Institute of Transportation and Traffic Engineering, University of California, Berkeley, Calif., 1969.

15.47 Simpson & Curtin, Inc.: *Transbus Public Testing and Evaluation Program*, Final Report, UMTA-IT-06-0025-76-1, January, 1976.

15.48 Yen, et al.: *Guidelines for the Maintenance of Downtown People Mover (DPM) Systems*, UMTA Report VA-06-0055-81, Washington, D.C., July, 1979.

EXERCISES

15.1 Consider the case of a county road administrator: develop a framework for a maintenance management priority system for programming work elements.

15.2 Road-maintenance management systems require an inventory mechanism to account for responsibilities. List the elements that should be inventoried and discuss alternative methods of making an inventory.

15.3 Your supervisor in the city engineer's office has been asked to attend a review panel meeting regarding a proposed design for an intermodal transfer facility that will include auto parking areas, a bus terminal, and a light rail station. He asks you to assess the need for signing and information aids for such a facility. Prepare a brief for him regarding the signing requirements which he can use as a basis for discussion at the meeting.

15.4 Transit-system scheduling is an important part of system operations. Describe the input requirements and the steps involved in establishing a schedule for a small urban area. Discuss the impact of driver work rules on crew and fleet requirements.

Appendix A
Mathematical Tools

Planners and engineers involved in the study of transportation often use mathematical techniques or tools that differ from those utilized in many other endeavors. For example, calculus traditionally has played an important role in engineering work in such efforts as the analysis of structures, fluids, and soils. But other techniques, such as regression, have greater application to the unique kinds of problems which evolve from the analysis of transportation systems. It is to these latter methods that this appendix is directed. However, the presentation made here will dwell on each one of the several topics only lightly and should not be construed as an all-encompassing work. References to more complete discussions are given at the end of the appendix.

A.1 MEANS AND STANDARD DEVIATIONS

There are several types of averages that can be calculated; the mean, the mode, and the median are three. We will deal only with the mean in this discussion. The mean can be obtained through the equation

$$\bar{x} = \sum_{i=1}^{n} \frac{x_i}{n} \tag{A.1}$$

where $\bar{x}$ = arithmetic mean of n observed values of variable X
  x_i = value associated with observation i of variable X

The simplest way to measure variability would be to take the average of the deviations of each observation from the mean. This procedure is essentially what is done in computing the standard deviation. However, we want to make sure that positive and negative deviations do not cancel themselves out. The usual procedure thus is first to square the deviations, thereby making all positive numbers, and then to sum the squared deviations, divide by the number of observations (minus one, to account for lost degrees of freedom), and finally take the square root of the resulting number (in order to compensate for the squaring done initially). Symbolically, this procedure reduces to

$$s = \pm \sqrt{\sum_{i=1}^{n} \frac{(x_i - \bar{x})^2}{n - 1}} \qquad \text{(A.2)}$$

where s = standard deviation
x_i = value associated with observation i of variable X
$\bar{x}$ = arithmetic mean of n observed values of variable X

A.2 REGRESSION AND CORRELATION

The preceding discussion dealt with only one variable. The main goal to be pursued in this section is that of estimating the value of one variable, *given the value of another*. This situation is common in all endeavors in science and engineering where, for example, we might wish to predict the strain on a steel bar, knowing the stress, or to predict the pressure in a cylinder, knowing the temperature. To establish a relationship between two (or more) variables, two questions must be answered—first, what is the nature of the relationship (linear, nonlinear, direct, inverse, and so forth), and second, how reliable is it for purposes of estimation? Regression and correlation are methods often employed to answer these questions.

In Figs. A.1 and A.2 are presented two scatter diagrams, which are plots of pairs of observations collected in the field. If we were to attempt to fit a straight line among the points of Fig. A.1, that is, to form a *linear relationship* between average number of resident vehicular trips per dwelling unit (the dependent variable) and average income (the independent variable), we would find that the line more than likely would have a positive slope *and* that most points would not lie very close to it. We then would conclude that the general nature of the relationship, as indicated by the slope, is direct, but that due to the probable dispersion of points about the line, the relationship could not be considered to be very reliable for making any estimates of the dependent variable. In Fig. A.2 similar conclusions would be reached except that the relationship would be inverse. What is lacking in the above conclusions, however, is a more precise way to describe the phrases "direct," "inverse," and "not very reliable." The development of such descriptions is the main objective of this section.

A.2.1 Linear Regression

The first question posed above was that of determining or describing the nature of a relationship between two variables, say X and Y, with the purpose in mind of employing the relationship for making future estimates of Y given X. If the

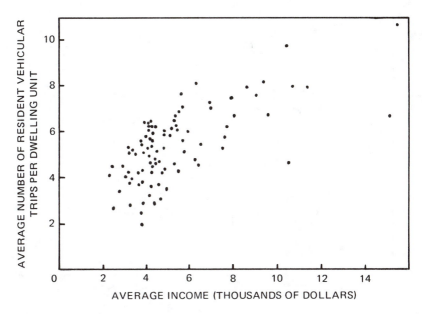

Fig. A.1 Relationship between average income per household and average trip production per household in Washington, D.C. [A.10, p. 10].

relationship appears to be linear, then the procedure to be discussed below can be utilized to give the slope and intercept for the line which "fits" the data best. This procedure is known as *linear regression*, and the resulting line is called a *regression line*.

Referring to Fig. A.3, where several data points are given, we assume that a linear relationship between X and Y is reasonable and that our objective is to find a line of the form

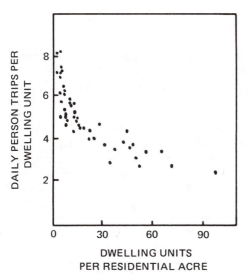

Fig. A.2 Effect of residential density on trip productions by districts in Washington, D.C. [A.10, p. 8].

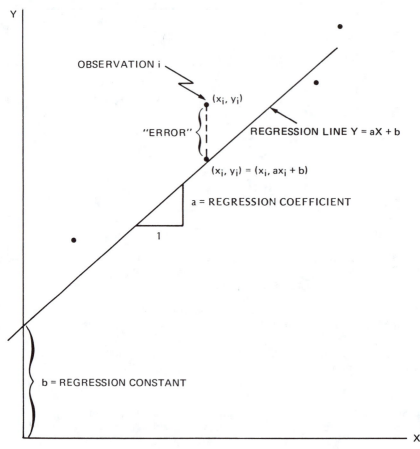

Fig. A.3 Diagram indicating terms associated with regression.

$$Y = aX + b \tag{A.3}$$

which lies closest to all of the points. Thus, a and b are the unknowns to be determined. If any observation point, indexed by i, is denoted by (x_i, y_i), and the corresponding value of Y falling on the regression line directly below (x_i, y_i) is denoted by $\hat{y}_i$, then $(y_i - \hat{y}_i)$ represents the "error" that would be made in estimating Y when $X = x_i$. In other words, knowing x_i, we would estimate that $Y = ax_i + b = \hat{y}_i$, but from our observations we know that $Y = y_i$ when $X = x_i$, so that our estimate is in "error" by the amount $(y_i - \hat{y}_i)$. Adding all of the errors from each observation point then would give the *total error* in estimating Y.

One drawback in utilizing the term $(y_i - \hat{y}_i)$ as a measure of "error" is that there may be both positive and negative values for the term, and these may cancel each other out when the addition is made to find the total error. A similar situation arose in regard to the standard deviation discussed in Sec. A.1, where the tack taken was that of squaring the deviation term in order to obtain all positive numbers. Applying the same strategy here we obtain the quantities $(y_i - \hat{y}_i)^2$ which, when summed, give the error sum of squares (ESS)

$$ESS = \sum_{i=1}^{n} (y_i - \hat{y}_i)^2 \qquad (A.4)$$

Since ESS is a measure of the total error, it is desirable to set the linear regression line (that is, determine a and b) so that ESS is a minimum. Thus, from Eq. (A.4),

$$\hat{y}_i = ax_i + b \qquad (A.5)$$

and then

$$ESS = \sum_{i=1}^{n} (y_i - ax_i - b)^2 \qquad (A.6)$$

This is the quantity to be minimized, using calculus procedures that will not be discussed here. The resulting value for a can be computed from

$$a = \frac{n \sum_{i=1}^{n} x_i y_i - \left(\sum_{i=1}^{n} x_i \right)\left(\sum_{i=1}^{n} y_i \right)}{n \sum_{i=1}^{n} x_i^2 - \left(\sum_{i=1}^{n} x_i \right)^2} \qquad (A.7)$$

Knowing the value of a, we then can find b as

$$b = \frac{\sum_{i=1}^{n} y_i - a \sum_{i=1}^{n} x_i}{n} \qquad (A.8)$$

As a result of all these manipulations, we find that for a given set of observations, $[(x_1, y_1), (x_2, y_2), \ldots, (x_i, y_i), \ldots, (x_n, y_n)]$, we can determine the slope and intercept of the line that has the best[1] fit to the observations from Eq. (A.7) and Eq. (A.8). An example of calculations involving these two equations is presented in Table A.1, where the number of average daily auto trips per dwelling unit (Y) is found to be related to the average number of autos owned per dwelling unit (X) by

$$Y = 5.489X - 2.996 \qquad (A.9)$$

The minimum value of ESS equals 0.1334.

While the actual determination of this equation is an important finding, the greatest benefit is yet to come since it is through the *utilization* of Eq. (A.9) that many useful predictions can be made. If, for instance, it were known that in a certain section of a city the average number of autos owned per dwelling unit in 1995 would be 0.76,[2] then by entering this value for X in Eq. (A.9) the average daily auto trips per dwelling unit could be calculated as

---

[1] "Best" is defined in terms of the criterion of Eq. (A.6).

[2] The question of whether the regression coefficient and constant will be the same in 1995 as in the present is not raised at this point. See Chap. 7 for discussion of this dilemma.

$$Y = 5.489(0.76) - 2.996 = 1.17$$

Thus to repeat what was stated before, by having available a relationship like Eq. (A.10) it is possible to make *predictions*, thereby relieving us from the chore of collecting further data to obtain the desired estimate[3] and also relieving us from the

[3] Notice that the 0.76 value for X is not found in Table A.1. Strictly speaking, this situation should dictate that more data should be collected to determine the value of Y with X set exactly at 0.76.

Table A. 1   (Hypothetical) Sample Data on Auto Trips and Ownership Used to Estimate a Linear Regression Equation

Observation number (i)	Average daily auto trips per dwelling unit (y_i)	Average autos owned per dwelling unit (x_i)
1	1.24	0.75
2	1.31	0.78
3	1.04	0.76
4	1.56	0.84
5	0.53	0.68
6	1.47	0.80
7	0.49	0.65
8	0.86	0.71
9	1.60	0.82
10	1.15	0.75
11	1.47	0.80
12	0.39	0.60
13	0.80	0.70
14	1.72	0.84
$n = 15$	1.80	0.89

$$\sum_{i=1}^{n} y_i = 16.54 \qquad \sum_{i=1}^{n} x_i = 11.20$$

$$n \sum_{i=1}^{n} x_i y_i - \left(\sum_{i=1}^{n} x_i\right)\left(\sum_{i=1}^{n} y_i\right) = 193.4820 - 185.2479 = 8.2341$$

$$n \sum_{i=1}^{n} x_i^2 - \left(\sum_{i=1}^{n} x_i\right)^2 = 126.939 - 125.439 = 1.500$$

$$a = \frac{8.2341}{1.500} = 5.489$$

$$b = \frac{16.54 - (5.489)(11.20)}{15} = -2.996$$

min   ESS $= 0.1334$

situation in which we would have *no number to use* since the event about which we seek information is yet to occur (as is the case in the above example where the auto trips will not be made until 1995). These then are two of the major reasons for developing relationships through regression.

A.2.2 Multiple (Multivariate) Linear Regression

Extensions of the linear regression technique can be made in several directions, two of which will be discussed briefly in the following subsections. The first involves. the expansion of the general model of Eq. (A.3) to the situation where there is more than one dependent variable, as is exemplified by the equation [A.14, p. 220]

$$Y = 3.79X_1 - 0.0033X_2 + 3.80 \qquad \text{(A.10)}$$

where Y = daily person trips per family
$\quad X_1$ = autos per dwelling unit
$\quad X_2$ = dwelling places per residential acre
An equation like Eq. (A.10) can be obtained in an operation similar to that performed in going from Eq. (A.6) to Eq. (A.7) where, as before,

$$\text{ESS} = \sum_{i=1}^{n} (y_i - \hat{y}_i)^2 \qquad \text{(A.11)}$$

but where $\hat{y}_i$ now is the ordinate on a *plane* of the form

$$Y = cX_1 + dX_2 + e \qquad \text{(A.12)}$$

Therefore

$$\text{ESS} = \sum_{i=1}^{n} (y_i - cX_{1i} - dX_{2i} - e)^2 \qquad \text{(A.13)}$$

where X_{1i} = value of variable 1 for observation i
$\quad X_{2i}$ = value of variable 2 for observation i
A solution now could be made for c, d, and e, which would give the equation of the plane that best fits the three-dimensional array of observations. Corresponding calculations can be performed for three or more independent variables.

A.2.3 Nonlinear Regression Equations

Up to this point only linear or supposedly linear functions have been explored. However, many relationships cannot be considered to be linear. The one underlying the data in Fig. A.2 is a good exa: .ple of such a case.

Two nonlinear forms commonly found in relationships relevant to transportation are

$$Y = aX_1^{b_1} X_2^{b_2} \cdots X_n^{b_n} \qquad \text{(A.14)}$$

and

$$Y = b_1 X_1 + b_2 X_1^2 + \ldots + b_n X_1^n + a \qquad \text{(A.15)}$$

The parameters (a and b_1 through b_n) for the first equation can be established using logarithmic transformations of both sides of the equation, whereas those for the second equation are found in a manner similar to that utilized in the preceding section. An explanation for a two-variable situation will be given for the first case and an example of a specific equation for the second.

The general form of Eq. (A.14) when there are two independent variables is

$$Y = aX_1^{b_1} X_2^{b_2}$$ (A.16)

Taking the logarithms of both sides of Eq. (A.16) yields

$$\log Y = \log a + b_1 \log X_1 + b_2 \log X_2$$ (A.17)

Letting $Z = \log Y$; $g = \log a$; $W_1 = \log X_1$; and $W_2 = \log X_2$, the preceding equality becomes

$$Z = g + b_1 W_1 + b_2 W_2$$ (A.18)

which is exactly similar to Eq. (A.12) in Sec. A.2.2. As a consequence, the slopes and intercept of the plane, Eq. (A.18) are equivalent, respectively, to the exponents and coefficient in Eq. (A.14). Subsequently, if a regression plane is fit to the *logarithms* of the values of each multivariate observation, values for $a_1, b_1, and b_2$ can be determined and utilized directly in Eq. (A.14). This procedure thus provides us with a direct technique for establishing the specific form of Eq. (A.14).

In the case of Eq. (A.15), the general relationship Eq. (A.11) becomes

$$\text{ESS} = \sum_{i=1}^{n} (y_i - b_1 x_1 - b_2 x_1^2 - \ldots - b_n x_1^n - a)^2$$ (A.19)

This can be solved for the needed values of the parameters. In one example found in the literature [A.14, p. 221], an empirical relation

$$Y = 38 + 2.53 X_1 - 0.0111 X_2^2$$ (A.20)

where Y = percentage of urban trips via transit
    X_1 = net residential density per acre
was determined using this technique.

A variety of regression techniques not discussed in this subsection can be found in most statistics books. Reference should be made to such often-used techniques as stepwise and simultaneous regression and to hypothesis testing relative to these techniques.

A.2.4 Correlation

The previous sections on regression have shown techniques useful for answering the question posed earlier concerning the *nature* of a relationship between a dependent variable and one or more independent variables—the nature was specified by means of the parameters (coefficients, exponents, and so forth) in the equation. The other question, that of the reliability or accuracy of the relationships, has yet to be answered. The correlation coefficient is a measure or index commonly used for this purpose.

If it were desired to estimate the value of a variable for some time in the future and nothing were known about any inherent relationships, the only reliable prediction method would involve the use of the *mean* from previous measurements of the variable. For example, if it were known from past experience that shopping centers generated an average of 500 trips per acre per day, then we would have no reasonable alternative other than to use the same figure in estimating the number of trips to a shopping center about to be constructed. Nevertheless, previous measurements might have shown considerable *variations* in trip generation rates, say from 200 to 700 trips per acre per day, so that our estimate might not be very accurate. Consequently, it is this *variation* we would like to reduce or eliminate by means of some relationship in order to obtain reliable and accurate forecasts.

The expression

$$TSS = \sum_{i=1}^{n} (y_i - \bar{y})^2 \qquad (A.21)$$

where TSS represents the "total sum of the squared deviations from the mean," or, more concisely, the "total sum of squares," is a good indicator of the *total* variation to be found in the variable[4] Y and thus, from the preceding discussion, also indicates the quantity we would like to reduce as much as possible. Returning to Sec. A.2.1 on linear regression, we find that ESS, the "error sum of squares," is a measure of the variation that remains *after* a linear regression has been performed, so that the quantity $(TSS - ESS)$ would represent the variation that has been *eliminated* by the establishment of the linear relationship between Y and X. Therefore

$$r^2 = \frac{TSS - ESS}{TSS} \qquad (A.22)$$

shows the *part* (or, if multiplied by 100, the *percentage*) of the total sum of squares eliminated by means of the relationship. Thus, r^2, referred to as the *coefficient of determination*, is a good index of the reliability of the relationship. A regression line that lies extremely close to all the observation points will give a low ESS, which will lead to a high value of r^2, whereas a scatter of points around a regression line will give a high ESS and a low r^2.

If the square root of Eq. (A.22) were taken, we would obtain

$$r = \pm \sqrt{\frac{TSS - ESS}{TSS}} \qquad (A.23)$$

where r is the commonly used *correlation coefficient*.

For the case of *linear* regression, the plus sign is employed when the slope of

---

[4] Note that the *average* variation has been presented in Sec. A.1 as being the *standard deviation* and that the TSS in Eq. (A.21) is included in the expression for the standard deviation, Eq. (A.2).

the line is positive, the minus when the slope is negative. It should be noted that the range of r is between -1 and $+1$, since the *least* the ESS can be is 0, which gives $(TSS - 0)/TSS = 1$. Thus, an r of ± 1 corresponds to a relationship with the *highest possible* reliability whereas an r equal to 0 would imply that ESS = TSS so that $(TSS - TSS)/TSS = 0$, which is the *worst possible* reliability.

Figure A.4 gives some representations of linear relationships having various degrees of correlation. These can be utilized to obtain a rough idea of the magnitude and direction of a correlation coefficient in a given situation. However, to calculate an exact value for r, one must return to Eq. (A.23) and its relevant predecessors or else use the following general equation for a linear relationship, which yields the same results

$$r = \frac{n \sum_{i=1}^{n} x_i y_i - \left(\sum_{i=1}^{n} x_i \right)\left(\sum_{i=1}^{n} y_i \right)}{\sqrt{\left[n \sum_{i=1}^{n} x_i^2 - \left(\sum_{i=1}^{n} x_i \right)^2 \right]\left[n \sum_{i=1}^{n} y_i^2 - \left(\sum_{i=1}^{n} y_i \right)^2 \right]}} \tag{A.24}$$

where the x_i and y_i are pairs of observations on the independent and dependent variables, respectively. The sign of r comes out automatically, and no knowledge of the underlying regression equation is needed.

As an example of an application of the correlation coefficient, consider the four data points portrayed in Fig. A.5. The calculations are given in Table A.2. A high, "+" value of r is expected since most of the points lie close to the regression line, which has a positive slope. This prediction turns out to be correct, with the correlation coefficient found to be +0.993. As a consequence, we can feel fairly certain that the regression equation will yield fairly reliable estimates of the average speed of vehicles moving into the main traffic stream (from a ramp) *knowing* the length of the acceleration lane leading onto the main trafficway.

A.2.5 Standard Error of Regression

While the correlation coefficient is one of the most frequently used measures of the reliability or accuracy of a relationship, it has one major drawback—it is a relative rather than an absolute measure. Looking back to Eq. (A.22) and Eq. (A.23) it can be seen that the correlation coefficient deals essentially with the *percentage* of the total sum of squares, TSS, that is eliminated through the regression relationship. There is no indication that the TSS, a measure of initial variability, is large in absolute terms. Nor is there a similar indication for the ESS. The consequence is that a relationship could produce a low ESS and yet have a low r if the initial TSS also were low. In most cases we would expect a low ESS to indicate a fairly reliable relationship.

What is lacking in the correlation coefficient, which deals with *total* deviation and *total* error, is a measure of the *average* deviation of observation points from the regression line. Thus, in a situation similar to that under which the *standard deviation* (average deviation from the mean) was derived,[5] we could define a *standard error of regression*, s_E, as

---

[5] See Sec. A. 1.

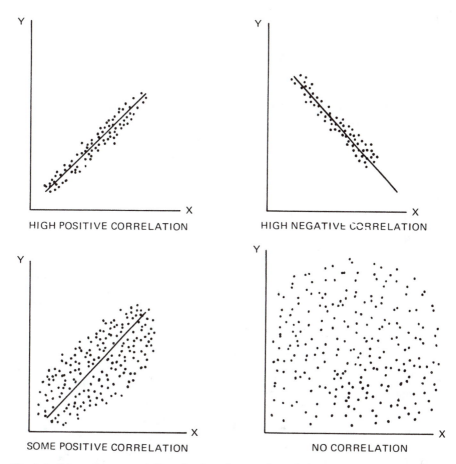

Fig. A.4   Schematic representation of various degrees of correlation.

$$s_E = \pm\sqrt{\frac{\text{ESS}}{n-2}} = \pm\sqrt{\frac{\sum_{i=1}^{n}(y_i - \hat{y}_i)^2}{n-2}} \tag{A.25}$$

where y_i = observation i of variable Y
  $\hat{y}_i = f(x_i)$
  n = number of observations
The error sum of squares, ESS, is divided by the total number of observations (minus 2) to get an "average" error per observation. The square root of this figure then is taken in order to balance the previously performed squaring operation. We subsequently obtain a measure of the average deviations of points from the regression function $Y = f(X)$.

The reason for subtracting 2 from n in the denominator inside the radical is similar to the reason for subtracting 1 when computing the standard deviation. *Assuming a linear regression function with one independent variable,* we would see

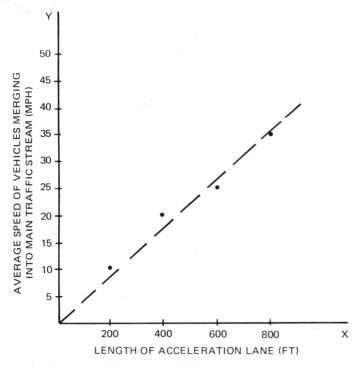

Fig. A.5 (Hypothetical) observed data on average speeds of merging vehicles and lengths of acceleration lanes.

that if only two observations were available, the function would pass through both points. This situation is unsatisfactory since it will happen *no matter what the values corresponding to the two observations.* Therefore, that standard error should reflect this undefined situation by becoming similarly undefined, $\sqrt{0/0}$, when $n = 2$.

An example of the calculations involved in determining a standard error of regression can be constructed for the hypothetical relationship shown in Table A.2. The equation of the linear regression line portrayed there is

Table A.2 Calculation of Standard Error of Regression Using Example Given in Fig. A.5

	x_i	y_i	$\hat{y}_i$*	$y_i - \hat{y}_i$	$(y_i - \hat{y}_i)^2$
	200	10	10.5	−0.5	0.25
	400	20	18.5	1.5	2.25
	600	25	26.5	−1.5	2.25
	800	35	34.5	0.5	0.25
Sum				0	5.00

$$s_E = \pm\sqrt{\frac{5.00}{4-2}} = \pm\sqrt{2.50} = \pm1.59$$

*Calculated from the equation $Y = 0.04X + 2.5$.

$$Y = 0.04X + 2.5 \tag{A.26}$$

The standard error comes out to ±1.59. Standard errors also could be calculated for other than linear or univariate (one-independent-variable) relationships. One unit should be subtracted from n in the denominator of Eq. (A.25) for each parameter (coefficient, exponent, constant, and so forth) in the estimating equation.

A.3 MINIMUM TIME-PATH ALGORITHMS

Minimum time paths [A.13] are needed as inputs to most of the land-use and travel models discussed in Chaps. 6 and 7. The full sequence of steps necessary to determine minimum time paths will be presented here. For an example, we will use the street classification and base map shown in Fig. A.6. After the important thoroughfares have been identified, as in that figure, and zone centroids, corresponding roughly to the geographic centers of the zones, have been located, another map showing the major network links and their measured lengths (top number) and speeds (bottom number) can be prepared (Fig. A.7).[6] The problem then, is to find the minimum travel time path between each and every zone centroid. It is easy to see that without some kind of logical procedure to do this, many tedious and perhaps incorrect calculations would be made before the best path were found.

The most common technique employed for determining minimum travel time paths is that developed by Moore [A.13]. As an example of his approach we will take the overly simplified network in Fig. A.8 representing the highway system in the nine counties of the San Francisco Bay Area. Travel times are marked on the links, and our interest is centered on finding the minimum time route between centroid 1 and centroid 7. We start by determining the minimum path from centroid 1 over all links connecting to it. Thus, min (21, 20, 10) = 10 on the link from 1 to 2. In the second stage, we find the minimum total (from 1) time path from centroids 1 and 2 to all centroids directly connected to 1 or 2. So, min (10 + 15, 10 + 20, 20, 21) = 20, which corresponds to the 1-4 path. Continuing in a similar manner, we take the minimum total time from 1 to 2 to all connecting centroids except 4, from 1 to 4 to all connecting centroids except 2, and from 1 to all connecting centroids except 2 and 4. This gives min (10 + 15, 20 + 10, 20 + 10, 21) = 21, which is the time on the link from 1 to 9. The remaining calculations along with those shown above are presented in detail in Fig. A.10, with the resulting minimum time paths shown together in Fig. A.9. An interesting outcome of Moore's algorithm is that in finding the minimum time path to the farthest centroid, it is also necessary to find the minimum time paths to all other centroids. The output of the algorithm therefore is a set of minimum travel time paths known as a *minimum path tree*, emanating from one centroid. Figure A.10 displays the tree for the first centroid in the network in Fig. A.7. It should be noted that in both the example problem and in Fig. A.10, the tree is developed for only one centroid. A complete analysis would require the building of trees from all centroids.

A.4 COMPUTER APPLICATIONS

Many of the techniques discussed in this appendix and especially in Chaps. 6 to 8 probably owe their existence to the computer, for without it most of the

---

[6] Notice in Fig. A.7 that some centroids might not lie on a major thoroughfare so that some reasonable, yet arbitrary speeds must be given to the travel distance to the nearest actual street nodes.

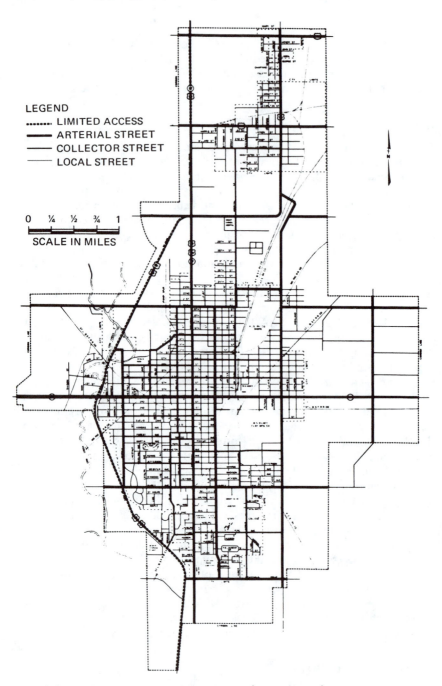

Fig. A.6   Sample street classification and base map. [A.11, p. iv–10].

LEGEND
.... LIMITED ACCESS
━━━ ARTERIAL STREET
─── COLLECTOR STREET
─── LOCAL STREET

0   ¼   ½   ¾   1
SCALE IN MILES

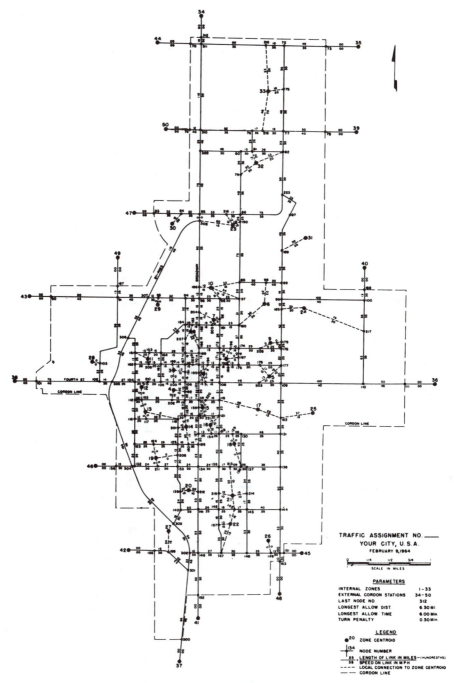

Fig. A.7 Sample network map. [A.11, p. iv-11].

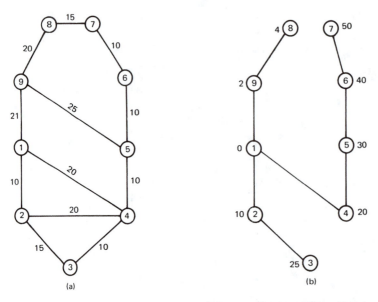

(a)   (b)

Fig. A.8 Simplified representation of the highway network in the San Francisco Bay area.

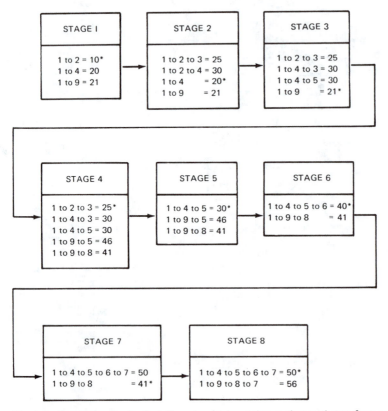

Fig. A.9 Complete set of calculations to obtain minimum time path tree from network in Fig. A.8. (Asterisks indicate minimum time at each stage.)

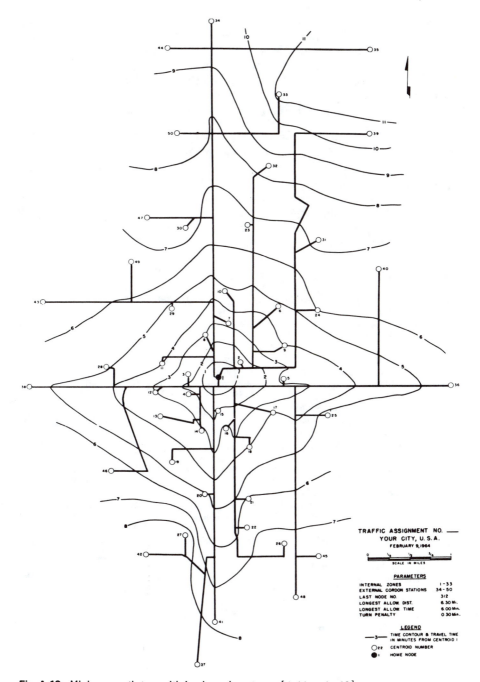

Fig. A.10 Minimum path tree with isochronal contours. [A.11, p. iv-13].

extensive and cumbersome calculations could not be done. In addition, most of the large-scale data handling inherent in the collection and processing procedures presented in Chap. 9 would not be possible. It thus seems appropriate to discuss computer applications briefly in this section.

The types of computers available vary widely, but what is important with respect to the type of analyses discussed here is the "software" (preprogrammed material) provided. It would be extremely time-consuming and expensive for an individual agency or firm to program for itself something like the gravity model or multiple regression or a linear-programming algorithm. In fact, the gravity-model program package supplied as part of the UTPS (Chap. 7) has over 30 subroutines or subprograms, each one of which is somewhat complex. Most agencies therefore make use of the "canned" programs.

The most commonly utilized "canned" programs for correlation and regression are the SAS [A.7] and the SPSS [A.8]. These are readily available and have a wide range of options covering most of the situations that might arise. Minimum time-path algorithms can be found in the UTPS programs.

BIBLIOGRAPHY

A.1 Guilford, J. P.: *Fundamental Statistics in Psychology and Education,* 4th ed., McGraw-Hill, New York, 1965.

A.2 Dickey, J. W., and T. M. Watts: *Analytic Techniques in Urban and Regional Planning,* McGraw-Hill, New York, 1978.

A.3 Krueckeberg, D. A., and A. L. Silvers: *Urban Planning Analysis,* Wiley, New York, 1974.

A.4 Wonnacott, T. H., and R. J. Wonnacott: *Econometrics,* Wiley, New York, 1970.

A.5 Blalock, H. M. Jr.: *Social Statistics,* 2d ed., McGraw-Hill, New York, 1972.

A.6 Ezekiel, M., and K. A. Fox: *Methods of Correlation and Regression Analysis: Linear and Curvilinear,* 3rd ed., Wiley, New York, 1959.

A.7 SAS Institute: *Statistical Analysis System User's Guide,* Cary, N.C. (various years).

A.8 Nie, N. M., et al: *Statistical Package for the Social Sciences,* 2d ed., McGraw-Hill, New York, 1975.

A.9 Baerwald, J. E. (ed.): *Traffic Engineering Handbook* (3d ed.), Institute of Traffic Engineers, Washington, D.C., 1965.

A.10 U.S. Department of Transportation, Federal Highway Administration, Bureau of Public Roads: *Guidelines for Trip Generation Analysis,* U.S. Government Printing Office, Washington, D.C., June, 1967.

A.11 U.S. Department of Commerce, Bureau of Public Roads: *Calibrating and Testing a Gravity Model for Any Size Urban Area,* U.S. Government Printing Office, Washington, D.C., Oct., 1965.

A.12 Institute of Traffic Engineers: *Manual of Traffic Engineering Studies* (3d ed.), Washington, D.C., 1964.

A.13 Moore, E. F.: "The Shortest Path Through a Maze," *Proceedings,* International Symposium on the Theory of Switching, Harvard Univ. Press, Cambridge, Mass., April 2-5, 1957.

A.14 Wohl, M., and B. V. Martin: *Traffic System Analysis for Engineers and Planners,* McGraw-Hill, New York, 1967.

A.15 Hull, C. H., and W. H. Nie: *SPSS Update 7-9,* McGraw-Hill, New York, 1981.

Appendix B
Further Reading

In the main text, not many references were made to journals that cover the wide variety of topics discussed. There are two main reasons for this. First, most of the journals cover state-of-the-art material, which would fit better in an advanced text rather than this one. Second, many of the government publications referenced provide better *summaries* of topics than do the journals. These reasons must be balanced against the generally greater accessibility of journals in libraries. Government reports sometimes are not easy to obtain in a short period of time.

Since many students will be interested in following up on the material discussed in this text, we have listed below many of the main journals of relevance, along with the address of the publisher and the name of the editor. The number of journals has risen rapidly in the last 10 years, so it might be worthwhile when in the library to search even further than those listed below. Often there is a journal dealing in detail with a particular topic (e.g., evaluation) but not specifically aimed at urban transportation. These sometimes can help expand horizons on a particular topic.

American Planning Association Journal. American Planning Association, 1313 East 60th, Chicago, Ill. 60637. Editor: Kenneth Pearlman.

Business Week. McGraw-Hill Publications Co., 1221 Avenue of the Americas, New York, N.Y. 10020. Editor: John L. Cobbs.

Datamation. Technical Publishing Co. (New York), 666 Fifth Avenue, New York, N.Y. 10103. Editor: John L. Kirkley.

Ekistics; reviews on the problems and science of human settlements. Athens

Center of Ekistics, 24 Strat. Syndesmou St., Box 471, Athens 136, Greece. Editor: P. Psomopoulos.

Environment and Planning A. Pion Ltd., 207 Brondesbury Park, London NW2 5JN, England. Editorial Board.

Environment and Planning B. Pion Ltd., 207 Brondesbury Park, London NW2 5JN, England. Editor: Lionel March.

Evaluation Review; a journal of applied social research. Sage Publications Inc., 275 South Beverly Drive, Beverly Hills, Calif. 90212. Editors: Richard A. Berk, Howard E. Freeman.

Futures; the journal of forecasting and planning. IPC Science and Technology Press Ltd., Box 63, Westbury House, Bury Street, Guildford, Surrey GU2 5BH, England. Editor: Ralph Jones.

Futurist; a journal of forecasts, trends, and ideas about the future. World Future Society, 4916 St. Elmo Avenue, Washington, D.C. 20014. Editor: Edmund Cornish.

Growth and Change; a journal of regional development. University of Kentucky, College of Business and Economics, Lexington, Ky. 40506. Editor: Hirofumi Shibata.

ITE (Institute of Transportation Engineers) *Journal.* Institute of Transportation Engineers, 525 School St., S.W., Suite 510, Washington, D.C. 20024. Editor: Nancy D. Angarola.

Journal of Advanced Transportation. Institute for Transportation, Inc., 1410 Duke University Road, Durham, N.C. 27701. Editors: Charles M. Harman, Verne L. Roberts.

Journal of Transport Economics and Policy. London School of Economics and Political Science, Houghton St., Aldwych, London, WC2A 2AE, England.

Journal of Urban Analysis. Gordon and Breach Science Publishers Ltd., 42 William IV St., London WC2, England. Editors: Stanley M. Altman, Edward H. Blum.

Mass Transit. C. Carroll Carter, 555 National Press Building, Washington, D.C. 20045. Editor: C. Carroll Carter.

Passenger Transport. American Public Transit Association, 1225 Connecticut Ave., N.W., Washington, D.C. 20036. Editor: Albert Engelken.

Popular Science; the what's new magazine. Times Mirror Magazines, Inc., 380 Madison Avenue, New York, N.Y. 10017. Editor: Hubert P. Luckett.

Public Finance Quarterly. State Publications Inc., 275 South Beverly Drive, Beverly Hills, Calif. 90212. Editor: Irving Goffman.

Public Policy. (Harvard University, John F. Kennedy School of Government.) John Wiley & Sons, Inc., 605 Third Avenue, New York, N.Y., 10016. Editors: John D. Montgomery, Edith Stokey.

Regional Science Association Papers. Regional Science Association, 3718 Locust Walk, University of Pennsylvania, Philadelphia, Pennsylvania, 19174. Editor: Morgan Thomas.

Regional Studies (Regional Studies Association.) Pergamon Press, Inc., Journals Division, Journals Department, Maxwell House, Fairview Park, Elmsford, N.Y. 10523. Editor: J. B. Goddard.

Specialized Transportation Planning and Practice. Gordon and Breach Science Publishers, Ltd., 42 William IV St., London WC2, England. Editor: William Bell.

Traffic Engineering and Control. Printerhall Ltd., 29 Newman Street, London W1P 3PE, England. Editor: Keith Lumley.

Traffic Quarterly. The Eno Foundation, Box 55, Saugatuck Station, Westport, Conn. 06880. Editor: Wilbur S. Smith.

Transit Journal. American Public Transit Association, 1225 Connecticut Avenue N.W., Washington, D.C. 20036. Editorial Board.

Transportation; an international journal devoted to the improvement of transportation planning and practice. Elsevier Scientific Publishing Co., Box 211, 1000 AE Amsterdam, Netherlands. Editor: Martin G. Richards.

Transportation Engineering Journal. American Society of Civil Engineers, 345 East 47th Street, New York, N.Y. 10017.

Transportation Planning and Technology; reviews and communications. Gordon and Breach Science Publishers, Ltd., 42 William IV Street, London WC2, England, Editor: N. Ashford.

Transportation Research. Part A: General. Pergamon Press, Inc., Journals Division, Maxwell House, Fairview Park, Elmsford, N.Y. 10523. Editor: Frank A. Haight.

Transportation Research. Part B: Methodological. Pergamon Press, Inc., Journals Division, Maxwell House, Fairview Park, Elmsford, N.Y. 10523. Editor: Frank A. Haight.

Transportation Research Records. Transportation Research Special Reports. National Cooperative Highway and Transportation Research Project Reports. Transportation Research News. All from: Transportation Research Board, 2101 Constitution Ave., N.W., Washington, D.C. 20418. Editor: Hugh M. Gillespie.

Index